Field Theory

MONOGRAPHS AND TEXTBOOKS IN PURE AND APPLIED MATHEMATICS

1. *K. Yano,* Integral Formulas in Riemannian Geometry (1970) *(out of print)*
2. *S. Kobayashi,* Hyperbolic Manifolds and Holomorphic Mappings (1970) *(out of print)*
3. *V. S. Vladimirov,* Equations of Mathematical Physics (A. Jeffrey, editor; A. Littlewood, translator) (1970) *(out of print)*
4. *B. N. Pshenichnyi,* Necessary Conditions for an Extremum (L. Neustadt, translation editor; K. Makowski, translator) (1971)
5. *L. Narici, E. Beckenstein, and G. Bachman,* Functional Analysis and Valuation Theory (1971)
6. *D. S. Passman,* Infinite Group Rings (1971)
7. *L. Dornhoff,* Group Representation Theory (in two parts). Part A: Ordinary Representation Theory. Part B: Modular Representation Theory (1971, 1972)
8. *W. Boothby and G. L. Weiss (eds.),* Symmetric Spaces: Short Courses Presented at Washington University (1972)
9. *Y. Matsushima,* Differentiable Manifolds (E. T. Kobayashi, translator) (1972)
10. *L. E. Ward, Jr.,* Topology: An Outline for a First Course (1972) *(out of print)*
11. *A. Babakhanian,* Cohomological Methods in Group Theory (1972)
12. *R. Gilmer,* Multiplicative Ideal Theory (1972)
13. *J. Yeh,* Stochastic Processes and the Wiener Integral (1973) *(out of print)*
14. *J. Barros-Neto,* Introduction to the Theory of Distributions (1973) *(out of print)*
15. *R. Larsen,* Functional Analysis: An Introduction (1973) *(out of print)*
16. *K. Yano and S. Ishihara,* Tangent and Cotangent Bundles: Differential Geometry (1973) *(out of print)*
17. *C. Procesi,* Rings with Polynomial Identities (1973)
18. *R. Hermann,* Geometry, Physics, and Systems (1973)
19. *N. R. Wallach,* Harmonic Analysis on Homogeneous Spaces (1973) *(out of print)*
20. *J. Dieudonné,* Introduction to the Theory of Formal Groups (1973)
21. *I. Vaisman,* Cohomology and Differential Forms (1973)
22. *B. -Y. Chen,* Geometry of Submanifolds (1973)
23. *M. Marcus,* Finite Dimensional Multilinear Algebra (in two parts) (1973, 1975)
24. *R. Larsen,* Banach Algebras: An Introduction (1973)
25. *R. O. Kujala and A. L. Vitter (eds.),* Value Distribution Theory: Part A; Part B: Deficit and Bezout Estimates by Wilhelm Stoll (1973)
26. *K. B. Stolarsky,* Algebraic Numbers and Diophantine Approximation (1974)
27. *A. R. Magid,* The Separable Galois Theory of Commutative Rings (1974)
28. *B. R. McDonald,* Finite Rings with Identity (1974)
29. *J. Satake,* Linear Algebra (S. Koh, T. A. Akiba, and S. Ihara, translators) (1975)

30. *J. S. Golan,* Localization of Noncommutative Rings (1975)
31. *G. Klambauer,* Mathematical Analysis (1975)
32. *M. K. Agoston,* Algebraic Topology: A First Course (1976)
33. *K. R. Goodearl,* Ring Theory: Nonsingular Rings and Modules (1976)
34. *L. E. Mansfield,* Linear Algebra with Geometric Applications: Selected Topics (1976)
35. *N. J. Pullman,* Matrix Theory and Its Applications (1976)
36. *B. R. McDonald,* Geometric Algebra Over Local Rings (1976)
37. *C. W. Groetsch,* Generalized Inverses of Linear Operators: Representation and Approximation (1977)
38. *J. E. Kuczkowski and J. L. Gersting,* Abstract Algebra: A First Look (1977)
39. *C. O. Christenson and W. L. Voxman,* Aspects of Topology (1977)
40. *M. Nagata,* Field Theory (1977)
41. *R. L. Long,* Algebraic Number Theory (1977)
42. *W. F. Pfeffer,* Integrals and Measures (1977)
43. *R. L. Wheeden and A. Zygmund,* Measure and Integral: An Introduction to Real Analysis (1977)
44. *J. H. Curtiss,* Introduction to Functions of a Complex Variable (1978)
45. *K. Hrbacek and T. Jech,* Introduction to Set Theory (1978)
46. *W. S. Massey,* Homology and Cohomology Theory (1978)
47. *M. Marcus,* Introduction to Modern Algebra (1978)
48. *E. C. Young,* Vector and Tensor Analysis (1978)
49. *S. B. Nadler, Jr.,* Hyperspaces of Sets (1978)
50. *S. K. Segal,* Topics in Group Rings (1978)
51. *A. C. M. van Rooij,* Non-Archimedean Functional Analysis (1978)
54. *L. Corwin and R. Szczarba,* Calculus in Vector Spaces (1979)
53. *C. Sadosky,* Interpolation of Operators and Singular Integrals: An Introduction to Harmonic Analysis (1979)
54. *J. Cronin,* Differential Equations: Introduction and Quantitative Theory (1980)
55. *C. W. Groetsch,* Elements of Applicable Functional Analysis (1980)
56. *I. Vaisman,* Foundations of Three-Dimensional Euclidean Geometry (1980)
57. *H. I. Freedman,* Deterministic Mathematical Models in Population Ecology (1980)
58. *S. B. Chae,* Lebesgue Integration (1980)
59. *C. S. Rees, S. M. Shah, and C. V. Stanojević,* Theory and Applications of Fourier Analysis (1981)
60. *L. Nachbin,* Introduction to Functional Analysis: Banach Spaces and Differential Calculus (R. M. Aron, translator) (1981)
61. *G. Orzech and M. Orzech,* Plane Algebraic Curves: An Introduction Via Valuations (1981)
62. *R. Johnsonbaugh and W. E. Pfaffenberger,* Foundations of Mathematical Analysis (1981)
63. *W. L. Voxman and R. H. Goetschel,* Advanced Calculus: An Introduction to Modern Analysis (1981)
64. *L. J. Corwin and R. H. Szcarba,* Multivariable Calculus (1982)
65. *V. I. Istrătescu,* Introduction to Linear Operator Theory (1981)
66. *R. D. Järvinen,* Finite and Infinite Dimensional Linear Spaces: A Comparative Study in Algebraic and Analytic Settings (1981)

67. *J. K. Beem and P. E. Ehrlich,* Global Lorentzian Geometry (1981)
68. *D. L. Armacost,* The Structure of Locally Compact Abelian Groups (1981)
69. *J. W. Brewer and M. K. Smith, eds.,* Emmy Noether: A Tribute to Her Life and Work (1981)
70. *K. H. Kim,* Boolean Matrix Theory and Applications (1982)
71. *T. W. Wieting,* The Mathematical Theory of Chromatic Plane Ornaments (1982)
72. *D. B. Gauld,* Differential Topology: An Introduction (1982)
73. *R. L. Faber,* Foundations of Euclidean and Non-Euclidean Geometry (1983)
74. *M. Carmeli,* Statistical Theory and Random Matrices (1983)
75. *J. H. Carruth, J. A. Hildebrant, and R. J. Koch,* The Theory of Topological Semigroups (1983)
76. *R. L. Faber,* Differential Geometry and Relativity Theory: An Introduction (1983)
77. *S. Barnett,* Polynomials and Linear Control Systems (1983)
78. *G. Karpilovsky,* Commutative Group Algebras (1983)
79. *F. Van Oystaeyen and A. Verschoren,* Relative Invariants of Rings: The Commutative Theory (1983)
80. *I. Vaisman,* A First Course in Differential Geometry (1984)
81. *G. W. Swan,* Applications of Optimal Control Theory in Biomedicine (1984)
82. *T. Petrie and J. D. Randall,* Transformation Groups on Manifolds (1984)
83. *K. Goebel and S. Reich,* Uniform Convexity, Hyperbolic Geometry, and Nonexpansive Mappings (1984)
84. *T. Albu and C. Năstăsescu,* Relative Finiteness in Module Theory (1984)
85. *K. Hrbacek and T. Jech,* Introduction to Set Theory, Second Edition, Revised and Expanded (1984)
86. *F. Van Oystaeyen and A. Verschoren,* Relative Invariants of Rings: The Noncommutative Theory (1984)
87. *B. R. McDonald,* Linear Algebra Over Commutative Rings (1984)
88. *M. Namba,* Geometry of Projective Algebraic Curves (1984)
89. *G. F. Webb,* Theory of Nonlinear Age-Dependent Population Dynamics (1985)
90. *M. R. Bremner, R. V. Moody, and J. Patera,* Tables of Dominant Weight Multiplicities for Representations of Simple Lie Algebras (1985)
91. *A. E. Fekete,* Real Linear Algebra (1985)
92. *S. B. Chae,* Holomorphy and Calculus in Normed Spaces (1985)
93. *A. J. Jerri,* Introduction to Integral Equations with Applications (1985)
94. *G. Karpilovsky,* Projective Representations of Finite Groups (1985)
95. *L. Narici and E. Beckenstein,* Topological Vector Spaces (1985)
96. *J. Weeks,* The Shape of Space: How to Visualize Surfaces and Three-Dimensional Manifolds (1985)
97. *P. R. Gribik and K. O. Kortanek,* Extremal Methods of Operations Research (1985)
98. *J.-A. Chao and W. A. Woyczynski, eds.,* Probability Theory and Harmonic Analysis (1986)
99. *G. D. Crown, M. H. Fenrick, and R. J. Valenza,* Abstract Algebra (1986)
100. *J. H. Carruth, J. A. Hildebrant, and R. J. Koch,* The Theory of Topological Semigroups, Volume 2 (1986)

101. *R. S. Doran and V. A. Belfi,* Characterizations of C*-Algebras: The Gelfand-Naimark Theorems (1986)
102. *M. W. Jeter,* Mathematical Programming: An Introduction to Optimization (1986)
103. *M. Altman,* A Unified Theory of Nonlinear Operator and Evolution Equations with Applications: A New Approach to Nonlinear Partial Differential Equations (1986)
104. *A. Verschoren,* Relative Invariants of Sheaves (1987)
105. *R. A. Usmani,* Applied Linear Algebra (1987)
106. *P. Blass and J. Lang,* Zariski Surfaces and Differential Equations in Characteristic $p > 0$ (1987)
107. *J. A. Reneke, R. E. Fennell, and R. B. Minton.* Structured Hereditary Systems (1987)
108. *H. Busemann and B. B. Phadke,* Spaces with Distinguished Geodesics (1987)
109. *R. Harte,* Invertibility and Singularity for Bounded Linear Operators (1988)
110. *G. S. Ladde, V. Lakshmikantham, and B. G. Zhang,* Oscillation Theory of Differential Equations with Deviating Arguments (1987)
111. *L. Dudkin, I. Rabinovich, and I. Vakhutinsky,* Iterative Aggregation Theory: Mathematical Methods of Coordinating Detailed and Aggregate Problems in Large Control Systems (1987)
112. *T. Okubo,* Differential Geometry (1987)
113. *D. L. Stancl and M. L. Stancl,* Real Analysis with Point-Set Topology (1987)
114. *T. C. Gard,* Introduction to Stochastic Differential Equations (1988)
115. *S. S. Abhyankar,* Enumerative Combinatorics of Young Tableaux (1988)
116. *H. Strade and R. Farnsteiner,* Modular Lie Algebras and Their Representations (1988)
117. *J. A. Huckaba,* Commutative Rings with Zero Divisors (1988)
118. *W. D. Wallis,* Combinatorial Designs (1988)
119. *W. Więsław,* Topological Fields (1988)
120. *G. Karpilovsky,* Field Theory: Classical Foundations and Multiplicative Groups (1988)
121. *S. Caenepeel and F. Van Oystaeyen,* Brauer Groups and the Cohomology of Graded Rings (1988)
122. *W. Kozlowski,* Modular Function Spaces (1988)

Other Volumes in Preparation

Field Theory

Classical Foundations and Multiplicative Groups

Gregory Karpilovsky

University of the Witwatersrand
Johannesburg, South Africa

Marcel Dekker, Inc. New York • Basel

Library of Congress Cataloging-in-Publication Data

Karpilovsky, Gregory
Field theory : classical foundations and multiplicative groups / Gregory Karpilovsky.
p. cm. -- (Monographs and textbooks in pure and applied mathematics ; vol. 120)
Bibliography: p.
Includes index.
ISBN 0-8247-8029-9
1. Class field theory. 2. Groups, Theory of. I. Title II. Series.
QA247.K324 1988
512'.32--dc19 88-18914

MARCEL DEKKER, INC.
270 Madison Avenue, New York, New York 10016

Current printing (last digit):
10 9 8 7 6 5 4 3 2 1

PRINTED IN THE UNITED STATES OF AMERICA

For Helen, Suzanne, Elliott

Preface

The twin aims of this book are to provide a young algebraist with a reasonably comprehensive summary of the material on which research in field theory is based and to show how this material is used in the current research pertaining to multiplicative groups of fields. The treatment is by no means exhaustive - this would have been a Sisyphean task, since the subject has become so extensive and is growing almost from day to day.

The book can be roughly divided into two parts, which will not preclude, however, some strong interrelations between these. The first part (Chapters 1-3) introduces classical foundations of the theory of fields, with an emphasis on refinements and extensions achieved in scope of relatively recent developments. Among these, we exhibit canonical fundamental units of certain classes of pure cubic fields, prove Kneser's theorem on torsion groups of separable field extensions and establish a theorem of Schinzel which provides necessary and sufficient conditions for the Galois group of a binomial $X^n - a$ to be abelian. A separate section is devoted to a result of Isaacs concerning the degrees of sums in a separable field extension. If G is the Galois group of a finite Galois extension E/F, an interpretation of a distinguished subgroup of $H^3(G,E^*)$ is provided by using the methods of Eilenberg and MacLane. We also include some recent results pertaining to radical extensions. Among them, we examine a cogalois correspondence discovered by Greither and Harrison in 1986.

The second part of the book (Chapter 4) is devoted exclusively to the study of multiplicative groups of fields. After proving some preliminary results, we investigate the isomorphism class of F^*, where F is a distinguished field such as local or global. Concentrating on field extensions E/F, we then prove Brandis's theorem which asserts that if F is infinite and $E \neq F$, then E^*/F^* is not finitely generated. The rest of the chapter is based on works of May, who made fundamental contributions to the subject. Numerous examples are given to illustrate that the results obtained are the best possible. Special attention is drawn to the study of fields whose multiplicative groups are free modulo torsion. These results have a number of important applications in the study of unit groups of group algebras and the isomorphism problem. For this reason a complete account of May's contributions to the topic is given.

This monograph is written on the assumption that the reader has had the equivalent of a standard first-year graduate algebra course. Thus we assume a familiarity with basic ring-theoretic and group-theoretic concepts and an understanding of elementary properties of modules, tensor products and fields. Apart from a few specific results and the general knowledge we have presupposed, the book is entirely self-contained. We have included, for the convenience of the reader, a chapter on algebraic preliminaries. This chapter provides a brief survey of topics needed later in the book. We have tried to avoid making the discussion too technical. With this view in mind, maximum generality has not been achieved in those places where this would entail a loss of clarity or a lot of technicalities.

A word about notation. As is customary, Theorem 2.1.2 denotes the second result of Section 1 of Chapter 2; however, for simplicity, all references to this result within Chapter 2 are designated as Theorem 1.2. A systematic description of the material is supplied by the introductions to individual chapters and therefore will not be repeated here.

I would like to express my gratitude to my wife for the encouragement she has given me in the preparation of this book. For answering specific queries on topics contained in the text I am indebted to W. May. Finally, my thanks go to Lucy Rich for her excellent typing.

Gregory Karpilovsky

Contents

Field Theory

1
Preliminaries

The aim of this chapter is twofold. First, to establish various notational conventions for the rest of the book. Second, to provide some preliminary results pertaining to polynomial rings. Because we presuppose a familiarity with various elementary group-theoretic and ring-theoretic terms, only a brief description of them is presented. Many readers may wish to glance briefly at the contents of this chapter, referring back to the relevant sections when they are needed later.

1. NOTATION AND TERMINOLOGY

The aim of this section is to establish various notational conventions that we shall use throughout the book.

Let A and B be arbitrary sets. A *map*

$$f : A \longrightarrow B$$

is a function that associates with each element $a \in A$ a unique element $b \in B$, $f : a \longmapsto b$. This b, the image of a under f, is denoted by $b = f(a)$. Let

$$f : A \longrightarrow B \quad \text{and} \quad g : B \longrightarrow C$$

be maps. The composite map $A \longrightarrow B \longrightarrow C$ is denoted by $g \circ f$. Recall that $g \circ f$ acts according to the rule

$$(g \circ f)(a) = g[f(a)] \qquad \text{for all } a \in A$$

If $f : A \longrightarrow B$ is any map, then $f|A'$ denotes the restriction of f to a subset A' of A. The *inclusion map* $i : A' \longrightarrow A$ is defined by

$$i(a) = a \qquad \text{for all } a \in A'$$

If A,B are sets, we write $A \subseteq B$ if A is a subset of B and $A \subset B$ if A is a proper subset of B. If $A \subseteq B$, $B\text{-}A$ denotes as usual the set of elements

of B not contained in A. The cardinality of the set A is denoted by $|A|$
The diagram

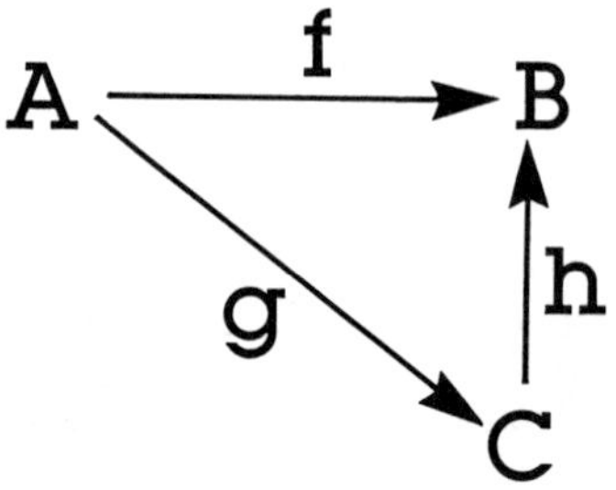

signifies that A, B and C are sets and that $f : A \longrightarrow B, g : A \longrightarrow C$ and $h : C \longrightarrow B$ are maps. The diagram *commutes*, or is said to be *commutative*, in case $f = h \circ g$. An arbitrary diagram is *commutative* if we get the same composite maps whenever we follow directed arrows along different paths from one set to another set in the diagram.

Unless explicitely stated otherwise, all groups are assumed to be multiplicative. We use the same symbol 1 for both the identity element and the identity subgroup of a group G. The multiplicative group of a field F is denoted by F^*. If X is a subset of G, $\langle X \rangle$ will denote the subgroup of G generated by X. A cyclic group of order n is denoted by $\mathbb{Z}_n$. If G is finite, the least common multiple of the orders of the elements of G is called the *exponent* of G. If every element of G is of finite order, G is called a *torsion group*, while G is *torsion-free* if all its elements, except for 1, are of infinite order.

Let H be a subgroup of G. A subset of G containing just one element from each left coset xH is called a *left transversal* for H in G, and right transversals are defined correspondingly.

Let $\{G_i\}_{i \in I}$ be a family of groups and let $\prod_{i \in I} G_i$ be their direct product. The subgroup $\oplus_{i \in I} G_i$ of $\prod_{i \in I} G_i$ consisting of all (g_i) with finitely many g_i distinct from 1, is called the *direct sum* (or *restricted direct product*) of the groups G_i. A sequence of groups and homomorphisms

$$\cdots \longrightarrow G_{i-1} \xrightarrow{f_{i-1}} G_i \xrightarrow{f_i} G_{i+1} \longrightarrow \cdots$$

is said to be exact at G_i if

$\mathrm{Ker} f_i = \mathrm{Im} f_{i-1}$; the sequence is called *exact* if it is exact at every group. In particular, $1 \longrightarrow G \xrightarrow{f} H$ is exact if and only if f is injective, while $G \xrightarrow{f} H \longrightarrow 1$ is exact if and only if f is surjective. If we are given a 3-terms exact sequence

$$1 \longrightarrow G_1 \xrightarrow{f} G_2 \xrightarrow{g} G_3 \longrightarrow 1$$

also called a *short exact sequence*, we see that

$$G_1 \cong f(G_1) \quad \text{and} \quad G_3 \cong G_2/f(G_1)$$

Let N be a normal subgroup of G and let $\pi : G \longrightarrow G/N$ be the natural homomorphism. Then the short exact sequence

$$1 \longrightarrow N \longrightarrow G \xrightarrow{\pi} G/N \longrightarrow 1$$

is also called the *natural exact sequence*.

Suppose now that A is an abelian group. Then the set $t(A)$ of elements of finite order in A is a subgroup of A, called the *maximal torsion subgroup* of A. It will also be called the torsion subgroup of A, if no confusion can arise.

For any integer $n > 0$, the map $A \longrightarrow A$ sending every element a of A to a^n is a homomorphism whose image and kernel are denoted by A^n and $A[n]$, respectively. We shall say that A is *divisible* if $A^n = A$ for all $n \in \mathbb{N}$ and n-divisible if $A^n = A$. Note that a p-group is divisible if and only if it is p-divisible. Typical examples of divisible groups are:

(i) The multiplicative group F^* of an algebraically closed field F

(ii) The group Z_{p^∞} of the p^nth complex roots of 1 with n running over all natural numbers.

(iii) The additive group $\mathbb{Q}/\mathbb{Z}$ which is isomorphic to the group of all complex roots of 1.

Given a prime p, we denote by A_p the *p-component* of A consisting of all elements of A whose order is a non-negative power of p. Note that if A is *torsion*, i.e. if $t(A) = A$, then there is a set π of prime numbers such that $A = \oplus_{p\in\pi} A_p$.

All rings in this book are associative with $1 \neq 0$ and subrings of a ring R

are assumed to have the same identity element as R. Each ring homomorphism will be assumed to respect identity elements.

Let R be a ring. The mapping $\mathbb{Z} \longrightarrow R$ defined by $n \longmapsto n\cdot 1$ is a ring homomorphism whose image is called the *prime subring* of R; its kernel is an ideal $m\mathbb{Z}$ for a unique $m \geqslant 0$, called the *characteristic* of R and denoted by char$R$.

Let R be a ring. An element $x \in R$ is *nilpotent* if $x^n = 0$ for some positive integer n. An ideal J in R is *nil* if every element of J is nilpotent, while J is *nilpotent* if there is a positive integer n such that $J^n = 0$, where J^n is the product of J with itself n times. An element e of R is *idempotent* if $e^2 = e$. Two idempotents u,v are *orthogonal* if $uv = vu = 0$. A nonzero idempotent is *primitive* if it cannot be written as a sum of two nonzero orthogonal idempotents.

An element u of a ring R is said to be a *unit* if $uv = vu = 1$ for some $v \in R$. The set of all units of R, denoted by $U(R)$, constitutes a group called the *unit group* of R.

Given an arbitrary ring R, we denote by $J(R)$ the Jacobson radical of R, i.e. the intersection of all maximal left ideals of R. We say that R is *semisimple* if $J(R) = 0$.

Let $\{R_i\}_{i\in I}$ be a family of rings and let R be the direct product set $\prod_{i\in I} R_i$. We can define addition and multiplication on R by the rules

$$(x_i) + (y_i) = (x_i+y_i) \qquad\qquad (x_i)(y_i) = (x_iy_i)$$

It is straightforward to verify that R is a ring; we shall refer to R as the *direct product of the family* $(R_i), i \in I$. For all $i \in I$, the projection of R onto R_i is a ring homomorphism, the injections $R_i \longrightarrow R$ preserve addition and multiplication but not 1 and so are not ring homomorphisms.

Let R be a commutative ring. Then an *R-algebra* is a ring A which is at the same time an R-module such that

$$r(xy) = (rx)y = x(ry) \qquad \text{for all} \quad r \in R \quad \text{and} \quad x,y \in A$$

It follows directly from the definition that $rx = (r\cdot 1)x$ for $r \in R$, $x \in A$, and that the map sending r to $r\cdot 1$ is a homomorphism of R into the centre $Z(A)$ of A. Conversely, if $f : R \longrightarrow A$ is a homomorphism of R into $Z(A)$, then A can be regarded as an R-algebra by setting

$$rx = f(r)x \qquad \text{for all } r \in R,\ x \in A$$

We close this section by presenting some information pertaining to commutative rings. In what follows R *denotes a commutative ring*. We say that R is *noetherian* if every ascending chain of ideals breaks off, or equivalently, if every ideal is finitely generated. A ring R is called *artinian* if every descending chain of ideals breaks off. The set $N(R)$ of all nilpotent elements of R constitutes an ideal called the *nilradical* of R. We say that R is *reduced* if $N(R) = 0$. Note that $N(R) \subseteq J(R)$ and thus $N(R) = J(R)$ if and only if $J(R)$ is a nil ideal. We say that R is *local* (respectively, *semilocal*) if it has precisely one maximal ideal (respectively, if it has only finitely many maximal ideals). An element x of R is a *zero divisor* if $xy = 0$ for some nonzero $y \in R$; in case $x \neq 0$ and x is a zero divisor, we say that x is a *proper zero divisor*.

An *integral domain* is a commutative ring without proper zero divisors. An ideal P of a commutative ring R is *prime* if R/P is an integral domain, while P is *primary* if every zero divisor in R/P is nilpotent.

Let S be a subset of a commutative ring R. We say that S is *multiplicative* if it contains 1 and if, for any x,y in S, $xy \in S$. The *quotient ring* R_S associated with a multiplicative set S consists of all elements of the form r/s with $r \in R$ and $s \in S$. By definition, $r/s = r'/s'$ if there exists $s_1 \in S$ such that

$$s_1(s'r-sr') = 0$$

Observe that if $0 \in S$, then $0/1$ is the only element of R_S; for this reason we shall assume that $0 \notin S$. Multiplication and addition in R_S are defined by

$$(r/s)(r'/s') = rr'/ss' \ , \quad r/s+r'/s' = (s'r+sr')/ss'$$

The mapping $\psi_S : R \longrightarrow R_S$ defined by $\psi_S(r) = r/1$ is a homomorphism, called *canonical;* ψ_S is injective if and only if S contains no zero divisors.

Suppose now that A is a commutative R-algebra and that S is a multiplicative subset of A. Then the ring A_S may be endowed with an R-algebra structure by setting

$$r(a/s) = ra/s \qquad r \in R,\ a \in A,\ s \in S$$

in which case ψ_S becomes a homomorphism of R-algebras. The R-algebra A_S enjoys the following easily verified universal property.

1.1. PROPOSITION. Let B be a commutative R-algebra. Then, for any R-algebra homomorphism $f : A \longrightarrow B$ such that every element of $f(S)$ is a unit, there exists one and only one R-algebra homomorphism $h : A_S \longrightarrow B$ which renders commutative the following diagram:

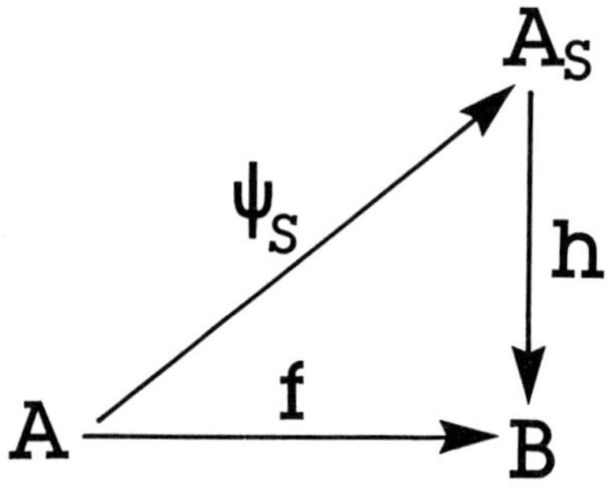

Especially important is the case where S is the complement of a prime ideal P. Here one often writes R_P instead of R_S. With every ideal I in R we associate the expanded ideal I^e of R_S given by

$$I^e = \{i/s \mid i \in I,\ s \in S\}$$

The following property is easily verified.

1.2. PROPOSITION. For any prime ideal P of R the correspondence $I \longmapsto I^e$ is a bijection between the prime ideals of R contained in P and the prime ideals of R_P.

It follows from the above that P^e is a unique maximal ideal of R_P; hence R_P is a local ring. This ring R_P is also called the *local ring* of R at P, and the process of forming R_P is called *localization*.

Finally, assume that R is an integral domain. Then $S = R - \{0\}$ is a multiplicative subset of R and R_S is obviously a field containing an isomorphic copy of R. We shall refer to R_S as the *quotient field* of R. We close by recording the following elementary fact.

1.3. PROPOSITION. Let R be an artinian integral domain. Then R is a field.

Proof. Fix $0 \neq x \in R$ and observe that the descending chain

$$Rx \supseteq Rx^2 \supseteq \ldots$$

must terminate. Hence $Rx^n = Rx^{n+1}$ for some $n \geqslant 1$, so $x^n = rx^{n+1}$ or $x^n(1-rx) = 0$. Since R has no proper zero divisors, $1 - rx = 0$ and thus r is a unit. ■

2. POLYNOMIAL ALGEBRAS

A *monoid* is a set G, with an associative binary operation, and having an identity element 1.

Let R be a commutative ring and let G be a monoid. The *monoid algebra* RG of G over R is the free R-module on the elements of G, with multiplication induced by that in G. More explicitly, RG consists of all formal linear combinations

$$\Sigma x_g \cdot g \qquad (x_g \in R,\ g \in G)$$

with finitely many $x_g \neq 0$ subject to

(i) $\Sigma x_g \cdot g = \Sigma y_g \cdot g$ if and only if $x_g = y_g$ for all $g \in G$

(ii) $\Sigma x_g \cdot g + \Sigma y_g \cdot g = \Sigma(x_g + y_g) \cdot g$

(iii) $(\Sigma x_g \cdot g)(\Sigma y_h \cdot h) = \Sigma z_t \cdot t$ where $z_t = \sum_{gh=t} x_g y_h$

(iv) $r(\Sigma x_g \cdot g) = \Sigma(rx_g) \cdot g$ for all $r \in R$

It is straightforward to verify that these operations define RG as an associative R-algebra with $1 = 1_R \cdot 1_G$, where 1_R and 1_G are identity elements of R and G, respectively. With the aid of the injective homomorphisms

$$\begin{cases} R \longrightarrow RG \\ r \longrightarrow r \cdot 1_G \end{cases} \qquad \begin{cases} G \longrightarrow RG \\ g \longrightarrow 1_R \cdot g \end{cases}$$

we shall in the future identify R and G with their images in RG. With these identifications, the formal sums and products become ordinary sums and products. For this reason, from now on we drop the dot in $x_g \cdot g$.

Let $x = \Sigma x_g g \in RG$. Then the *support* of x, written $\mathrm{Supp}x$, is defined by

$$\mathrm{Supp}x = \{g \in G | x_g \neq 0\}$$

It is plain that $\mathrm{Supp}x$ is a finite subset of G that is empty if and only if $x = 0$.

2.1. PROPOSITION. Let A be an R-algebra, let G be a monoid and let

$$\psi : G \longrightarrow A$$

be any map satisfying

$$\psi(1) = 1 \text{ and } \psi(xy) = \psi(x)\psi(y) \qquad \text{for all } x,y \in G$$

Then the map $\psi^* : RG \longrightarrow A$ defined by

$$\psi^*(\Sigma x_g g) = \Sigma x_g \psi(g)$$

is a homomorphism of R-algebras. In particular, if ψ is injective and A is R-free with $\psi(G)$ as a basis, then $RG \cong A$.

Proof. Because RG is R-free with G as a basis, ψ^* is a homomorphism of R-modules. Let

$$x = \Sigma x_g g \text{ and } y = \Sigma y_g g$$

be two elements of RG. Then

$$\begin{aligned}\psi^*(xy) &= \psi^*(\sum_{a,b\in G} x_a y_b ab) = \Sigma x_a y_b \psi(a)\psi(b)\\ &= \sum_{a\in G} x_a \psi(a) \sum_{b\in G} y_b \psi(b)\\ &= \psi^*(x)\psi^*(y),\end{aligned}$$

as asserted. ∎

Assume that G is a free abelian group freely generated by the set

$$\{X_i | i \in I\}$$

Then each element of G can be uniquely written in the form

$$\prod_{i\in I} X_i^{n_i} \qquad (n_i \in \mathbb{Z}) \tag{1}$$

with only finitely many $n_i \neq 0$. The submonoid of G consisting of all elements (1) in which $n_i \geq 0$ for any $i \in I$ is called a *free commutative monoid* freely generated by $\{X_i | i \in I\}$. The corresponding monoid algebra over a commutative ring R is denoted by $R[(X_i)_{i\in I}]$. We shall refer to $R[(X_i)_{i\in I}]$ as the *polynomial ring* in the indeterminates X_i with coefficients in R. For $I = \{1,2,\ldots,n\}$, we write $R[X_1,X_2,\ldots,X_n]$ instead of $R[(X_i)_{i\in I}]$. In particular, the polynomial ring in the indeterminate X with coefficients in R will be denoted by $R[X]$. The elements $\prod_{i\in I} X_i^{n_i}$ of $R[(X_i)_{i\in I}]$ ($n_i \geq 0$ and $n_i = 0$ for all but finitely many $i \in I$) are called *monomials*. Thus by definition, $R[(X_i)_{i\in I}]$ is R-free freely generated by all monomials. The *degree* of the monomial (1) is defined by

$$\deg(\prod_{i\in I} X_i^{n_i}) = \sum_{i\in I} n_i$$

If f is a nonzero polynomial in $R[(X_i)_{i\in I}]$, then we define the *degree* of f, written $\deg f$, to be the maximum of the degrees of the monomials in the support of f. Thus $\deg f = 0$ if and only if f is a nonzero element of R. If $f = 0$, then we say that its degree is $-\infty$.

Let d be a nonnegative integer and let f be a nonzero polynomial in $R[(X_i)_{i\in I}]$. We shall say that f is *homogeneous* of degree d (or is a *form* of degree d) if all monomials in $\mathrm{Supp} f$ are of degree d.

Let $f(X)$ be a nonzero polynomial in $R[X]$ of degree n. Then

$$f(X) = a_0 + a_1 X + \ldots + a_n X^n$$

with $a_i \in R$ and with $a_n \neq 0$. We call a_n the *leading coefficient* of f and a_0 its *constant term*. We also say that f is *monic* if $a_n = 1$.

Let A be an R-algebra and let B be any subset of A. Then the elements

$$\sum_\nu r_\nu \prod_{b\in B} b^{\nu(b)} \qquad (r_\nu \in R,\ \nu(b) \geq 0)$$

with finitely many r_ν and $\nu(b)$ distinct from zero form a subalgebra of A. This subalgebra, denoted by $R[B]$, (or simply $R[a]$ if $B = \{a\}$) is obviously the smallest subalgebra containing B. For this reason, we refer to $R[B]$ as the subalgebra of A generated by B. If $A = R[B]$, then we say that A is *generated* by B.

Let A be an R-algebra and let B be any subset of A. We say that B is *algebraically independent* over R (or that the elements of B are algebraically independent over R) if the elements

$$\prod_{b \in B} b^{\nu(b)} \qquad (\nu(b) \geqslant 0)$$

are R-linearly independent. In the special case where $B = \{a\}$, we see that a is algebraically independent over R if and only if $\{1, a, a^2, \dots, a^n, \dots\}$ is an R-linearly independent set. If $a \in A$ is algebraically independent over R, then we also say that a is *transcendental* over R. In the case where a is not transcendental over R, we say that a is *algebraic* over R. Thus a is algebraic over R if and only if there exists a nonzero polynomial $f(X) \in R[X]$ such that $f(a) = 0$.

We next provide some elementary observations pertaining to polynomial rings.

2.2. PROPOSITION. Let A be any commutative R-algebra.

(i) Any map $\{X_i | i \in I\} \longrightarrow A$, $X_i \longmapsto a_i$ can be extended to a homomorphism

$$R[(X_i)_{i \in I}] \longrightarrow A$$

of R-algebras. In particular, if A is generated by $\{a_i | i \in I\}$, then A is a homomorphic image of $R[(X_i)_{i \in I}]$.

(ii) The above homomorphism is injective if and only if the elements a_i are algebraically independent over R

(iii) $A \cong R[(X_i)_{i \in I}]$ if and only if there is a generating set $\{a_i | i \in I\}$ for A which is algebraically independent over R. In particular, $A \cong R[X]$ if and only if A is generated by a transcendental element.

Proof. (i) Let G be the monoid generated by the X_i, $i \in I$. Setting

$$\psi\left(\prod_{i \in I} X_i^{n_i}\right) = \prod_{i \in I} a_i^{n_i} \qquad (n_i \geqslant 0) \qquad (2)$$

it follows that $\psi(1) = 1$ and $\psi(xy) = \psi(x)\psi(y)$ for all $x, y \in G$. Now apply Proposition 2.1

(ii) Direct consequence of (2) and the definition of algebraic independence.

(iii) Direct consequence of (i) and (ii). ∎

2.3. PROPOSITION. Let $I = J \cup K$ be a disjoint union and let $S = R[(X_j)_{j\in J}]$. Then

$$R[(X_i)_{i\in I}] \cong S[(X_k)_{k\in K}] \qquad \text{as } R\text{-algebras.}$$

and, in particular,

$$R[X_1,\dots,X_n] \cong R[X_1,\dots,X_{n-1}][X_n] \qquad \text{for all } n \geqslant 2$$

Proof. By Proposition 2.2(i), the map $X_i \longmapsto X_i$, $i \in I$ extends to a homomorphism $R[(X_i)_{i\in I}] \longrightarrow S[(X_k)_{k\in K}]$ of R-algebras. Since its image contains S and each X_k, $k \in K$, the given homomorphism is surjective. Moreover, the elements $X_i, i \in I$ of $S[(X_k)_{k\in K}]$ are obviously algebraically independent over R. Now apply Proposition 2.2(ii). ∎

2.4. PROPOSITION. Let $f(X)$ and $g(X)$ be nonzero polynomials in $R[X]$ and let a_n and b_m be the leading coefficients of $f(X)$ and $g(X)$, respectively. If at least one of a_n, b_m is not a zero divisor in R, then

$$\deg(fg) = \deg f + \deg g$$

and the leading coefficient of fg is $a_n b_m$.

Proof. Write

$$f(X) = a_0 + a_1X + \dots + a_nX^n,\ g(X) = b_0 + b_1X + \dots + b_mX^m$$

Then

$$f(X)g(X) = a_0b_0 + \dots + a_nb_mX^{m+n}$$

and the result follows. ∎

2.5. COROLLARY. Let R be an integral domain. Then, for any set I, $R[(X_i)_{i\in I}]$ is also an integral domain.

Proof. We may harmlessly assume that I is finite. Furthermore, by Proposition 2.3, it suffices to show that $R[X]$ is an integral domain. The latter being a consequence of Proposition 2.4, the result follows. ∎

2.6. PROPOSITION. Let $f(X)$ and $g(X)$ be two nonzero polynomials in $R[X]$ of degrees m and n, respectively. Put $k = \max(m-n+1,0)$ and denote by a the leading coefficient of $g(X)$. Then there exist polynomials $g(X)$ and $r(X)$ such that

$$a^k f(X) = q(X)g(X) + r(X)$$

where either $r(X) = 0$ or $\deg r(X) < n$. Moreover, if a is not a zero divisor in R, then $q(X)$ and $r(X)$ are uniquely determined.

Proof. If $m < n$, then $k = 0$ and we may take $q(X) = 0$, $r(X) = f(X)$. We may therefore assume that $m \geqslant n-1$ in which case $k = m-n+1$. To prove the first part, we argue by induction on m. The case $m = n-1$ being trivial, assume $m \geqslant n$. Then $af(X) - bX^{m-n}g(X)$ has degree at most $m-1$, where b is the leading coefficient of $f(X)$. Applying induction hypothesis, there exist polynomials $q_1(X)$ and $r_1(X)$ such that

$$a^{(m-1)-n+1}(af(X) - bX^{m-n}g(X)) = q_1(X)g(X) + r_1(X)$$

where either $r_1(X) = 0$ or $\deg r_1(X) < n$. This proves the first assertion, by taking

$$q(X) = ba^{m-n}X^{m-n} + q_1(X), \; r(X) = r_1(X)$$

Assume that a is not a zero divisor and that

$$a^k f(X) = q'(X)g(X) + r'(X)$$

where either $r'(X) = 0$ or $\deg r'(X) < n$. Then

$$(q(X) - q'(X))g(X) = r'(X) - r(X)$$

If $q(X) - q'(X) \neq 0$, then the left side has degree at least n, since the leading coefficient of $g(X)$ is not a zero divisor. However, this is impossible since $r'(X)-r(X)$ is either zero or has degree less than n. Hence $q(X) = q'(X)$ and so $r'(X) = r(X)$. ■

2.7. COROLLARY. Let $f(X)$ and $g(X)$ be two nonzero polynomials in $R[X]$ and let the leading coefficient of $g(X)$ be a unit of R. Then there exist unique polynomials $q(X), r(X) \in R[X]$ such that

$$f(X) = g(X)q(X) + r(X)$$

where either $r(X) = 0$ or $\deg r(X) < \deg g(X)$.

Proof. Direct consequence of Proposition 2.6. ∎

2.8. COROLLARY. Let $f(X)$ be a nonzero polynomial in $R[X]$. Then, for any given $r \in R$, $f(r) = 0$ if and only if $X - r$ is a divisor of $f(X)$.

Proof. If $f(X) = (X-r)g(X)$ for some $g(X) \in R[X]$, then obviously $f(r) = 0$. Conversely, assume that $f(r) = 0$ and put $g(X) = X - r$. Since $g(X)$ is a monic polynomial of degree 1, it follows from Proposition 2.6 that $f(X) = q(X)g(X) + r_1$ for some $r_1 \in R$ and some $q(X) \in R[X]$. But then $0 = f(r) = r_1$, as required. ∎

If X is an indeterminate, an element $r \in R$ such that $f(r) = 0$ will be called a *root* of $f(X)$.

2.9. COROLLARY. Let R be an integral domain and let $f(X) \in R[X]$.

(i) If $a_1,\ldots,a_m$ are distinct roots of $f(X)$ in R, then

$$(X-a_1)(X-a_2)\ \ldots\ (X-a_m) \text{ divides } f(X)$$

(ii) If $f(X) \neq 0$, then the number of roots of $f(X)$ in R does not exceed $\deg f(X)$.

Proof. We first note that (ii) is a consequence of (i) from considerations of degree. To prove (i), we argue by induction on m. The case $m = 1$ being a consequence of Corollary 2.8, assume that the statement is true for $m - 1$ roots. Then

$$f(X) = (X-a_1)\ \ldots\ (X-a_{m-1})q(X)$$

and therefore $f(a_m) = (a_m-a_1)\ \ldots\ (a_m-a_{m-1})q(a_m)$. Since R is an integral domain, $q(a_m) = 0$ so that $X - a_m$ divides $q(X)$, by Corollary 2.8. This completes the proof. ∎

Recall that a *principal ideal domain* (*PID*) is an integral domain in which all ideals are principal.

2.10. COROLLARY. Let F be a field. Then $F[X]$ is a principal ideal domain.

Proof. We know, from Corollary 2.5, that $F[X]$ is an integral domain. Let I be a nonzero ideal of $F[X]$ and let g be a nonzero element of I of smallest possible degree. If $f \neq 0$ is an element of I, then by Corollary 2.7, $f = qg + r$ for some $q,r \in F[X]$, where $\deg r < \deg g$ if $r \neq 0$. But $r = f - qg \in I$, hence $r = 0$ as required. ■

Let R be an integral domain. As in $\mathbb{Z}$ we define $a|b$ (for any $a,b \in R$) to mean

$$b = ar \quad \text{for some} \quad r \in R$$

This is equivalent to the requirement $(b) \subseteq (a)$. Observe that an element $u \in R$ is a unit if and only if $u|1$. The units are trivial divisors since they are divisors of every element of R. If $a|b$ and $b|a$ (or, equivalently, if $(a) = (b)$), then we shall say that a and b are *associated*. It is easy to see that a and b are associated if and only if $a = bu$ for some unit u of R.

A nonzero element p of R is said to be a *prime* if (p) is a prime ideal. Expressed otherwise, a nonzero $p \in R$ is a prime if p is a nonunit such that

$$p|ab \quad \text{implies} \quad p|a \quad \text{or} \quad p|b$$

By an *irreducible element*, we understand a nonunit which is not a product of two nonunits.

A prime is always irreducible: if a prime p satisfies $p = ab$, then by definition, $p|a$ or $p|b$, say $p|a$. It follows that $a = pr = abr$ for some $r \in R$, and $a \neq 0$ (as divisor of p). Hence $br = 1$, by cancellation, so b is a unit.

A *greatest common divisor* (*GCD*) of elements $a_1,\dots,a_n \in R$ is an element $r \in R$ such that r divides each a_i and r is divisible by each common divisor of $a_1,\dots,a_n$.

The *GCD* is not unique: if r is a *GCD*, then so is any r' associated to r. Conversely, if r and r' are both *GCD*s of $a_1,\dots,a_n$, then r' is associated to r. We shall ignore the distinction between associates and denote any one of the *GCD*s of $a_1,\dots,a_n$ by $(a_1,\dots,a_n)$.

A *GCD domain* is an integral domain in which each pair of nonzero elements

has a greatest common divisor.

As we have remarked earlier, a prime is always irreducible. The following lemma shows that the converse is also true, provided R is a GCD domain.

2.11. LEMMA. Let R be a GCD domain. Then every irreducible element of R is a prime.

Proof. Suppose p is irreducible and that $p|ab$ for some a,b in R. Since p is irreducible and (p,a) is a divisor of p, either $(p,a) = p$ or $(p,a) = 1$. Similarly, $(p,b) = p$ or $(p,b) = 1$. Now $(p,a) = (p,b) = 1$ would imply $(p,ab) = 1$, which is impossible. Thus either $p|a$ or $p|b$ as asserted. ■

We next exhibit a useful characterization of GCD domains.

2.12. PROPOSITION. Let R be an integral domain. Then R is a GCD domain if and only if the intersection of any two principal ideals of R is a principal ideal.

Proof. To prove sufficiency fix nonzero elements $a,b \in R$. By hypothesis, $(a) \cap (b) = (c)$ for some $c \in R$; hence $ab = cr$ for some $r \in R$. We claim that $r = (a,b)$. Indeed, since $c = ar_1 = br_2$ for some $r_1,r_2 \in R$, we have $ab = ar_1r = br_2r$ and so $r|a, r|b$ by cancellation. If d is another common divisor of a and b, say $a = ds_1$ and $b = ds_2$, put $m = ds_1s_2$. Then $m \in (a) \cap (b)$, hence $m = cs$ say, and so $cr = ab = dm = cds$. It follows, by cancellation, that $r = ds$ and sufficiency is therefore established.

To prove necessity, let $d = (a,b)$ and write $a = dx, b = dy$. Then it is easily verified that $(x,y) = 1$ and $(xz,yz) = z$ for any $z \in R$. The latter implies that if $x|yz$, then $x|z$. If $xz_1 = yz_2 \in (x) \cap (y)$, then $x|yz_2$ and so $x|z_2$. This shows that $z_2 \in (x)$ and $yz_2 \in (xy)$ so that $(x) \cap (y) = (xy)$. Consequently,

$$(a) \cap (b) = d((x) \cap (y)) = (dxy)$$

as asserted. ■

A *unique factorization domain* (UFD) is an integral domain in which every element not zero or a unit can be written as a product of irreducible elements, and given two complete factorizations of the same element.

$$x = y_1 y_2 \dots y_r = z_1 z_2 \dots z_t \qquad (y_i, z_j \text{ irreducible})$$

then $r = t$, and after suitably renumbering the z's, y_i is associated to z_i.

Note that if R is a *UFD*, then R is a *GCD* domain. In particular, by Lemma 2.11, *every irreducible element of* R *is a prime*. Indeed, given two nonzero elements $x, y \in R$, we can write

$$x = up_1^{\alpha_1} p_2^{\alpha_2} \dots p_t^{\alpha_t}, \quad y = vp_1^{\beta_1} p_2^{\beta_2} \dots p_t^{\beta_t} \qquad (\alpha_i \geqslant 0, \beta_i \geqslant 0)$$

where u and v are units and where $p_1, \dots, p_t$ are pairwise nonassociated irreducible elements. It follows that any common divisor of x and y is of the form $wp_1^{\delta_1} p_2^{\delta_2} \dots p_t^{\delta_t}$ where w is a unit and $\delta_i \leqslant \min\{\alpha_i, \beta_i\}$. Setting $\gamma_i = \min\{\alpha_i, \beta_i\}$, we infer that

$$(x, y) = p_1^{\gamma_1} p_2^{\gamma_2} \dots p_t^{\gamma_t}$$

proving that R is a *GCD* domain.

Our next aim is to show that the *UFD*s are precisely those *GCD* domains in which the ascending chain condition for principal ideals (*ACCP*) is satisfied.

2.13. PROPOSITION. Let R be an integral domain. Then the following conditions are equivalent:

(i) Every nonzero nonunit of R is a product of primes

(ii) R is a *UFD*

(iii) R is a *GCD* domain in which *ACCP* is satisfied.

Proof. (i) ⇒ (ii): Since any prime is irreducible, we need only prove uniqueness of factorization. Let

$$c = a_1 a_2 \dots a_r = b_1 b_2 \dots b_s$$

be two factorizations into primes. We must prove that $r = s$ and that after suitably renumbering the b's, b_i is associated to a_i. We use induction on r; for $r = 1$ there is nothing to prove (because s must then also be 1), so let $r > 1$. A simple inductive argument shows that, if a product of more than two vactors is divisible by a prime p, then so is one of the factors. Therefore

$a_1|b_k$ for some k, say $k = 1$ (by renumbering the $b's$) and so $a_1 = b_1u$ for some unit u of R. Dividing $a_1 \dots a_r = b_1 \dots b_s$ by b_1, we obtain $ua_2 \dots a_r = b_2 \dots b_s$. It therefore follows, by induction, that $r - 1 = s - 1$ and that we can renumber the $b's$ so that a_i is associated to $b_i (2 \leq i \leq r)$. But this also holds for $i = 1$, hence the uniqueness of factorization.

(ii) ⇒ (iii): As has been observed earlier, R is a *GCD* domain. Bearing in mind that there are only finitely many of pairwise nonassociated divisors of $x \in R$, we conclude that (x) is contained in only finitely many principal ideals of R. Consequently, *ACCP* is satisfied in R.

(iii) ⇒ (i): By Lemma 2.11, every irreducible element of R is a prime. Therefore, given a nonunit $x \neq 0$, we need only prove that x is a product of irreducible elements. We shall first show that x has an irreducible divisor. If x is irreducible, there is nothing to prove. Otherwise let $x = x_1y_1$ where x_1, y_1 are nonunits. Then either x_1 is irreducible or $x_1 = x_2y_2$ where x_2, y_2 are nonunits. Continuing this process, we obtain the ascending chain

$$(x) \subseteq (x_1) \subseteq (x_2) \subseteq \dots$$

By hypothesis, this chain must break off; if x_n is the last term, x_n is irreducible, and $x_n|x$.

We now let $x_n = p_1$ and write $x = p_1c_1$. If c_1 is irreducible, then the proof is complete. Otherwise we have $c_1 = p_2c_2$ where p_2 is irreducible. Continuing in this way, we obtain the ascending chain

$$(x) \subseteq (c_1) \subseteq (c_2) \subseteq \dots$$

This breaks off with an irreducible element $c_n = p_{n+1}$ Accordingly, there exist irreducible elements $p_1, p_2, \dots, p_{n+1}$ such that

$$x = p_1c_1 = p_1p_2c_2 = \dots = p_1p_2 \dots p_{n+1}$$

and the result follows.

2.14. COROLLARY. Every *PID* is a *UFD*.

Proof. Assume that R is a *PID*. Owing to Propositions 2.12 and 2.13, it suffices to verify that R satisfies *ACCP*. To this end, let

$$(a_1) \subseteq (a_2) \subseteq \cdots$$

be an ascending chain of principal ideals of R. Then the union is an ideal generated by b, say, and if $b \in (a_k)$, then

$$\bigcup_n (a_n) = (b) = (a_k)$$

Thus $(a_k) = (a_{k+1}) = \ldots$, as required. ∎

The following observation is often useful.

2.15. LEMMA. Let R be an integral domain. Then, for any set I, the unit groups of R and $R[(X_i)_{i\in I}]$ are the same.

Proof. We need only verify that if an element u of $R[(X_i)_{i\in I}]$ is a unit, then u is a unit of R. But if $uv = 1$, then $u,v \in R$ by comparing degrees, hence the result. ∎

Let F be a field. Owing to Lemma 2.15, a nonzero $f(X) \in F[X]$ is a nonunit if and only of $\deg f(X) \geqslant 1$. Thus $f(X)$ is irreducible if and only if $\deg f(X) \geqslant 1$ and $f(X)$ is not a product of two polynomials of degree $\geqslant 1$.

2.16. COROLLARY. Let F be a field. Then $F[X]$ is a UFD and every polynomial $f(X) \in F[X]$ of positive degree can be written as

$$f(X) = a\, f_1(X)^{n_1} \cdots f_r(X)^{n_r} \qquad (n_i \geqslant 1, a \in F)$$

where $f_1(X),\ldots,f_r(X)$ are distinct irreducible monic polynomials uniquely determined by $f(X)$.

Proof. Direct consequence of Corollaries 2.14 and 2.10. ∎

Let F be a field and let $f(X) \in F[X]$ be a monic polynomial of positive degree. Then, by Corollary 2.16,

$$f(X) = f_1(X)^{n_1} \cdots f_r(X)^{n_r} \qquad (n_i \geqslant 1) \qquad (3)$$

where $f_1(X),\ldots,f_r(X)$ are distinct irreducible monic polynomials uniquely determined by $f(X)$. We shall refer to (3) as the *canonical decomposition* of $f(X)$.

2.17. PROPOSITION. Let F be a field, let $f(X) \in F[X]$ be of degree $\geqslant 1$, and let

$$f(X) = \prod_{i=1}^{m} (f_i(X))^{e_i}$$

be the decomposition of $f(X)$ into the product of powers of distinct irreducible polynomials. Then for $R = F[X]/(f(X))$ the following properties hold:

(i) $R \cong \prod_{i=1}^{m} F[X]/(f_i(X)^{e_i})$

(ii) $J(R) \cong \prod_{i=1}^{m} (f_i(X))/(f_i(X)^{e_i})$

(iii) $R/J(R) \cong \prod_{i=1}^{m} F[X]/(f_i(X))$

Proof. We first note that (iii) is a consequence of (i) and (ii). Since R is a finite-dimensional F-algebra, we have $N(R) = J(R)$. Therefore (ii) is a consequence of (i). To prove (i), note that both sides are of the same F-dimension. On the other hand, the natural projection $F[X] \longrightarrow \prod_{i=1}^{m} F[X]/(f_i(X)^{e_i})$ has kernel $(f(X))$. So the proposition is true. ∎

The rest of this section is devoted to showing that if R is a *UFD*, then so is $R[(X_i)_{i\in I}]$ for an arbitrary set I.

Let R be a *GCD* domain. Then one immediately verifies that any finite number of elements in R have a *GCD*. Furthermore

$$(a,b,c) = ((a,b),c) = (a,(b,c)) \qquad (a,b,c \in R)$$

$$(a_1,\dots,a_n) = ((a_1,\dots,a_{n-1}),a_n) \qquad (a_i \in R)$$

We shall say that a nonzero polynomial f in $R[X]$ is *primitive* if the *GCD* of its coefficients is 1.

2.18. LEMMA. (Gauss's Lemma). Let R be a *GCD* domain. If f and g are primitive polynomials of $R[X]$, then so is fg.

Proof. Write f and g in the form

$$f = \sum_{i=0}^{m} a_i X^{s_i} \qquad g = \sum_{i=0}^{n} b_i X^{t_i}$$

where $s_0 < s_1 < \dots < s_m$, $t_0 < t_1 < \dots < t_n$ and each a_i, b_j is nonzero. We must show that if $d \neq 0$ is a nonunit of R, then d fails to divide some coefficient of fg.

Suppose first that $(a_0,d) = (b_0,d) = 1$. Then we have $(a_0b_0,d) = 1$ and so d does not divide a_0b_0, a coefficient of fg. We may therefore assume that either $(a_0,d) \neq 1$ or $(b_0,d) \neq 1$; say for definiteness $(a_0,d) \neq 1$. Now consider the sequence $d_0,d_1,\dots,d_m$ defined by

$$d_j = (a_0, a_1, \ldots, a_j, d)$$

Then $d_0 \neq 1$ and because f is primitive, there exists a smallest integer $i \geqslant 1$ such that $d_i = 1$. Since d_{i-1} is a nonunit divisor of d, it suffices to verify that d_{i-1} does not divide some coefficient of fg. Hence we may assume that d divides $a_0, a_1, \ldots, a_{i-1}$ and that $(a_i, d) = 1$. Again, we choose k to be minimal with respect to the property $(b_0, b_1, \ldots, b_k, d) = 1$. Replacing d by $d' = (b_0, b_1, \ldots, b_{k-1}, d)$ (if $k = 0$, then $d = d'$), we can also assume that d divides $b_0, b_1, \ldots, b_{k-1}$ and that $(b_k, d) = 1$.

Let $\bar{\ } : R[X] \longrightarrow [R/(d)][X]$ be the natural homomorphism. Because d divides $a_0, a_1, \ldots, a_{i-1}, b_0, b_1, \ldots, b_{k-1}$, we have

$$\bar{f} = \sum_{j=1}^{m} \bar{a}_j X^{s_j} \qquad \bar{g} = \sum_{j=k}^{n} \bar{b}_j X^{t_j}$$

Bearing in mind that $(a_i, d) = (b_k, d) = 1$, we must also have $(a_i b_k, d) = 1$. Therefore $d \nmid a_i b_k$ so that the coefficient $\bar{a}_i \bar{b}_k$ of $\bar{f}\bar{g}$ is nonzero. It follows that $\overline{fg} = \bar{f}\bar{g} \neq 0$ and hence d fails to divide some coefficient of fg. So the lemma is true. ∎

We now apply the foregoing to provide a large class of GCD domains

2.19. PROPOSITION. Let R be a GCD domain. Then the ring $R[X_1, \ldots, X_n]$ is also a GCD domain.

Proof. In view of Proposition 2.3, it suffices to prove that $R[X]$ is a GCD domain. Denote by F and T the quotient field of R and the set of primitive polynomials in $R[X]$, respectively. We argue first that for $s \in S = R - \{0\}$ and for any $t \in T$,

$$R[X]s \cap R[X]t = R[X]st \tag{4}$$

The inclusion $R[X]st \subseteq R[X]s \cap R[X]t$ always holds, so let

$$h = fs = gt \in R[X]s \cap R[X]t$$

We write $g = s_1 t_1$ where s_1 is the GCD of the coefficients of g and $t_1 \in T$. By Gauss's lemma, $tt_1 \in T$ and so s_1 is the GCD of the coefficients of h and $s | s_1$. This shows that $h = s_1 t_1 t \in R[X]st$ and therefore proves (4).

Regarding $R[X]$ as a subring of $F[X]$ we argue next that

$$F[X]t \cap R[X] = R[X]t \qquad \text{for all} \quad t \in T \tag{5}$$

For let $g \in R[X]$ be such that $g = ft$ for some $f \in F[X]$. Then there exists $s \in S$ and $f_1 \in R[X]$ such that $sg = tf_1$. Note that

$$\text{if } (a) \cap (b) = (ab), \text{ then } a|bc \text{ implies } a|c \qquad (a,b,c \in R) \tag{6}$$

It follows from (4) and (6) that $s|f_1$, so $g \in R[X]t$ and hence $F[X]t \cap R[X] \subseteq R[X]t$. The opposite inclusion being trivial, (5) is established.

Let f and g be nonzero polynomials in $R[X]$. Owing to Proposition 2.12, the result will follow provided we show that the ideal $R[X]f \cap R[X]g$ is principal. To this end, we write $f = s_1t_1$ and $g = s_2t_2$ for some $s_1, s_2 \in S$ and $t_1, t_2 \in T$. Since R is a GCD domain, Proposition 2.12 tells us that $Rs_1 \cap Rs_2 = Rs$ for some $s \in S$. Now the elements of S are units of $F[X]$, so $F[X]f = F[X]t_1$ and $F[X]g = F[X]t_2$. Taking into account that $F[X]$ is a GCD domain (Corollary 2.16 and Proposition 2.13), we conclude that there is a $t \in T$ such that

$$F[X]t_1 \cap F[X]t_2 = F[X]t \tag{7}$$

It will next be shown that $R[X]f \cap R[X]g = R[X]st$ and this will finish the proof. Because $t \in F[X]t_i$ for $i = 1,2$, we deduce from (5) that $t \in R[X]t_1$; hence

$$st \in R[X]st_i \subseteq R[X]s_it_i$$

and therefore $R[X]st \subseteq R[X]f \cap R[X]g$. To prove the reverse inclusion, write $h \in R[X]f \cap R[X]g$ in the form $h = xy$ where $x \in S$ and $y \in T$. Then $s_i|xy$ for $i = 1,2$, and, by (4), $R[X]s_i \cap R[X]y = R[X]s_iy$. It follows from (6) that $s_i|x$ and hence $s|x$. On the other hand, applying (5) and (7), we have

$$\begin{aligned} F[X]h \cap R[X] &= F[X]y \cap R[X] \\ &= R[X]y \subseteq F[X]f \cap F[X]g \cap R[X] \\ &= F[X]t \cap R[X] = R[X]t \end{aligned}$$

so that $t|y$. Accordingly, $st|xy$ and so

$$R[X]f \cap R[X]g \subseteq R[X]st$$

as required. ■

We are now ready to prove the following result.

2.20. THEOREM. Let R be a *UFD*. Then, for any set I, $R[(X_i)_{i\in I}]$ is also a *UFD*.

Proof. We first claim that it suffices to consider the case where I is finite; if sustained, it will follow, from Proposition 2.3, that we need only prove that $R[X]$ is a *UFD*.

Assume that $0 \neq f \in R[(X_i)_{i\in I}]$. Then $f \in R[X_1, X_2, \ldots, X_n]$ for some $n \geq 1$ and if $f = g_1 g_2$ with $g_1, g_2 \in R[(X_i)_{i\in I}]$, then $g_1, g_2 \in R[X_1, \ldots, X_n, X_{n+1}, \ldots, X_m]$ for some $m \geq n$. We may regard $R[X_1, \ldots, X_n, \ldots, X_m]$ as the polynomial ring $R[X_1, \ldots, X_n][X_{n+1}, \ldots, X_m]$, in which case f has degree zero and hence both g_1 and g_2 are of degree zero. This shows that $g_1, g_2 \in R[X_1, \ldots, X_n]$ and hence f is an irreducible element of $R[(X_i)_{i\in I}]$ if and only if f is an irreducible element of $R[X_1, \ldots, X_n]$. This substantiates our claim.

By the foregoing, we are left to verify that $R[X]$ is a *UFD*. By Propositions 2.13 and 2.19, we need only show that any ascending chain

$$(f_1) \subseteq (f_2) \subseteq \cdots \subseteq (f_n) \subseteq \cdots$$

of principal ideals of $R[X]$ breaks off. Clearly, we can assume that all $f_n \neq 0$, in which case $\deg f_1 \geq \deg f_2 \geq \ldots \geq \deg f_n \geq \ldots$. Therefore there is an integer $m \geq 1$ such that $\deg f_n = \deg f_m$ for all $n \geq m$. Consequently, we may assume that $\deg f_1 = \deg f_2 = \ldots$ in which case there is a nonzero r_{n+} in R such that for all $n \geq 1$, $r_{n+1} f_{n+1} = f_n$. In particular if b_n is the leading coefficient of f_n, then $r_{n+1} b_{n+1} = b_n$ $(n = 1, 2, \ldots)$, so we obtain an ascending chain

$$(b_1) \subseteq (b_2) \subseteq \cdots \subseteq (b_n) \subseteq \cdots$$

This breaks off, by Proposition 2.13; hence there is an integer $m \geq 1$ such that $(b_n) = (b_m)$ for all $n \geq m$. Accordingly, $(r_{m+j} b_{m+j}) = (b_{m+j})$ for all $j \geq 1$, and so for all $j \geq 1$, r_{m+j} is a unit. Thus $(f_m) = (f_{m+1}) = \ldots$, as desired. ■

We next provide an application of Gauss's lemma to polynomial rings over unique factorization domains. Let R be a *UFD*, let P be a complete set of pairwise nonassociated primes of R and let F be the quotient field of R. Let $f(X) \in F[X]$ be a nonzero polynomial, written

$$f(X) = a_0X^{s_0} + a_1X^{s_1} + \dots + a_nX^{s_n} \qquad (a_i \neq 0,\ s_i \neq s_j \text{ for } i \neq j)$$

Then there exist $p_1, p_2, \dots, p_m$ in P such that

$$a_i = p_1^{\lambda_{1i}} p_2^{\lambda_{2i}} \dots p_m^{\lambda_{mi}} u_i$$

where $u_i \in U(R)$, $\lambda_{ki} \in \mathbb{Z}$, $0 \leqslant i \leqslant n$, $1 \leqslant k \leqslant m$. We define the *content* of f, written cont(f), to be the product

$$\text{cont}(f) = p_1^{s_1} p_2^{s_2} \dots p_m^{s_m} \qquad (s_k = \min\{\lambda_{ki} \mid 0 \leqslant i \leqslant n\})$$

or any multiple of this product by a unit of R. Thus cont(f) is well-defined up to multiplication by a unit of R. The following properties are immediate consequences of the definition of cont(f):

(a) If $f(X) \in R[X]$, then cont(f) is the greatest common divisor of the coefficients of f.

(b) cont$(f) = 1$ if and only if f is a primitive polynomial of $R[X]$.

(c) $f(X) = cf_1(X)$ where $c = \text{cont}(f)$ and $f_1(X)$ is a primitive polynomial of $R[X]$.

2.21. THEOREM. Let R be a *UFD* and let F be the quotient field of R

(i) If f, g are nonzero polynomials in $F[X]$, then cont(fg) = cont(f)cont(g)

(ii) If $0 \neq f(X) \in R[X]$ has a factorization $f(X) = g(X)h(X)$ in $F[X]$, then

$$f(X) = c_g c_h g_1(X) h_1(X) \qquad (g_1(X), h_1(X) \in R[X])$$

where $c_g = \text{cont}(g), c_h = \text{cont}(h)$, $g = c_g g_1$, $h = c_h h_1$ and $c_g c_h \in R$. In particular, if f with deg ≥ 1 cannot be decomposed in $R[X]$ as a product of two nonconstant factors, then f is irreducible in $F[X]$.

(iii) If $f(X)$ is a monic polynomial in $R[X]$ and $f(X) = g(X)h(X)$, where $g(X)$ and $h(X)$ are monic polynomials in $F[X]$, then $g(X), h(X) \in R[X]$.

Proof. (i) By (c), we may write $f = cf_1$ and $g = dg_1$ where $c = \text{cont}(f)$,

$d = \text{cont}(g)$, and f_1, g_1 are primitive polynomials in $R[X]$. Since $\text{cont}(fg) = cd\,\text{cont}(f_1 g_1)$, the required assertion follows by Lemma 2.18.

(ii) It suffices to notice that, by (i), $c_g c_h$ is the content of $f(X) \in R[X]$, hence $c_g c_h$ is in R, and that, by (c), $g_1(X), h_1(X) \in R[X]$.

(iii) In the notation of (ii), let λ_1 and μ_1 be the leading coefficients of g_1 and h_1, respectively. Since g and h are monic, we have $c_g = \lambda_1^{-1}$ and $c_h = \mu_1^{-1}$ and, by (ii) $\lambda_1^{-1}\mu^{-1} \in R$. Hence

$$c_g = \mu_1(\lambda_1^{-1}\mu_1^{-1}) \in R \quad \text{and} \quad c_h = \lambda_1(\lambda_1^{-1}\mu_1^{-1}) \in R$$

and therefore both $g(X)$ and $h(X)$ are in $R[X]$. ∎

Let I be an ideal of a commutative ring R. Then the natural homomorphism $R \longrightarrow R/I$ induces a surjective ring homomorphism $R[X] \longrightarrow (R/I)[X]$ whose kernel $I[X]$ consists of all polynomials with coefficients in I. We refer to this homomorphism as the natural map.

2.22. THEOREM. Let P be a prime ideal in an integral domain R and let

$$f(X) = r_0 X^n + r_1 X^{n-1} + \dots + r_n \in R[X] \qquad (n \geq 1)$$

be such that

$$r_0 \notin P,\ r_n \notin P^2 \ \text{and} \ r_i \in P\ ,\ 1 \leq i \leq n$$

Then $f(X)$ cannot be decomposed in $R[X]$ as a product of two nonconstant factors.

Proof. Assume by way of contradiction that $f(X) = g(X)h(X)$ with $g(X), h(X) \in R[X]$ and $\deg g(X) \geq 1$, $\deg h(X) \geq 1$. By hypothesis

$$f(X) \equiv r_0 X^n (\text{mod } P[X])$$

Since R/P is an integral domain, by looking at the image of $f(X)$ in $(R/P)[X]$, we deduce that

$$g(X) \equiv \alpha X^k (\text{mod } P[X]) \quad \text{and} \quad h(X) \equiv \beta X^{n-k} (\text{mod } P[X])$$

for some $1 \leq k < n$ and some $\alpha, \beta \in R-P$. This implies that the constant terms of $g(X)$ and $h(X)$ belong P, hence $r_n \in P^2$, a contradiction. ∎

2.23. COROLLARY. (Eisenstein's Criterion). Let R be a UFD, let F be the quotient field of R and let

$$f(X) = r_0X^n + r_1X^{n-1} + \dots + r_n \in R[X] \qquad (n \geq 1)$$

Assume that there is a prime p of R such that

$$r_0 \not\equiv 0(\text{mod } p),\ r_n \not\equiv 0(\text{mod } p^2) \text{ and } r_i \equiv 0(\text{mod } p),\ 1 \leq i \leq n$$

Then $f(X)$ is irreducible in $F[X]$.

Proof. Apply Theorems 2.22 and 2.21(ii). ■

2.24. EXAMPLE. Let k be a nonzero square-free integer $\neq \pm 1$. Then, for any integer $n \geq 1$, the polynomial X^n-k is irreducible over $\mathbb{Q}$. Indeed, take any prime divisor p of k with $p^2 \nmid k$ and apply Corollary 2.23. ■

2.25. EXAMPLE. Let p be a prime number. Then the polynomial

$$f(X) = X^{p-1} + X^{p-2} + \dots + X + 1$$

is irreducible over $\mathbb{Q}$. Indeed, it suffices to prove that $f(X+1)$ is irreducible over $\mathbb{Q}$. To this end, note that the binomial coefficients $\binom{p}{t}$, $1 \leq t \leq p-1$ are divisible by p. Since

$$f(X+1) = \frac{(X+1)^p - 1}{(X+1) - 1} = X^{p-1} + \binom{p}{1}X^{p-2} + \dots + \binom{p}{p-1}$$

the required assertion follows from Corollary 2.23. ■

2.26. EXAMPLE. Let F be a field and let α be an element of some field containing F such that α is transcendental over F. Denote by K the quotient field of $F[\alpha]$. Then, for any integer $n \geq 1$, the polynomial

$$X^n - \alpha \quad \text{is irreducible in } K[X]$$

This is a direct consequence of Corollary 2.23 and the fact that the ring $R = F[\alpha]$ is a UFD such that α is a prime element of R. ■

2.27. PROPOSITION. (Reduction Criterion). Let P be a prime ideal of an integral domain R, let F and $\bar{F}$ be the quotient fields of R and $\bar{R} = R/P$, respectively, and let

$$\bar{\ } : R[X] \longrightarrow \bar{R}[X]$$

be the natural map. If $0 \neq f \in R[X]$ is such that $\deg f = \deg \bar{f}$ and $\bar{f}$ is irreducible in $\bar{F}[X]$, then f cannot be decomposed in $R[X]$ as a product of two nonconstant factors.

Proof. Suppose $f(X) = g(X)h(X)$ with $g(X), h(X) \in R[X]$. Then $\bar{f} = \bar{g}\bar{h}$. Since $\deg\bar{g} \leqslant \deg g$ and $\deg\bar{h} \leqslant \deg h$, our hypothesis implies that we must have equality in these degree relations. Hence from the irreducibility of $\bar{f}$ in $\bar{F}[X]$ we conclude that g or h is an element of R, as required. ■

Let $R \subseteq S$ be commutative rings and let $\alpha_1, \ldots, \alpha_n$ be algebraically independent elements of S over R. Let X be a variable over $R[\alpha_1, \ldots, \alpha_n]$. Then the polynomial

$$f(X) = (X-\alpha_1) \ldots (X-\alpha_n) \quad \text{over} \quad R[\alpha_1, \ldots, \alpha_n]$$

can be written in the form

$$f(X) = X^n - s_1 X^{n-1} + \ldots + (-1)^n s_n$$

where each $s_i = s_i(\alpha_1, \ldots, \alpha_n)$ is a polynomial in $\alpha_1, \ldots, \alpha_n$. In fact

$$s_1 = \alpha_1 + \alpha_2 + \ldots + \alpha_n$$

$$s_2 = \alpha_1\alpha_2 + \alpha_1\alpha_3 + \ldots + \alpha_{n-1}\alpha_n$$

- - - - - - - - - - - - - - - - - -

$$s_k = \sum_{i_1 < i_2 < \ldots < i_k} \alpha_{i_1}\alpha_{i_2} \ldots \alpha_{i_k}$$

- - - - - - - - - - - - - - - - - -

$$s_n = \alpha_1\alpha_2 \ldots \alpha_n$$

The polynomials $s_1, \ldots, s_n$ are called the *elementary symmetric polynomials* $\alpha_1, \ldots, \alpha_n$. It is clear that s_i is homogeneous of degree i in $\alpha_1, \ldots, \alpha_n$. Let S_n be the symmetric group of degree n. If $\sigma \in S_n$ and

$$f(\alpha_1, \ldots, \alpha_n) \in R[\alpha_1, \ldots, \alpha_n],$$

define f^σ by

$$f^\sigma(\alpha_1, \ldots, \alpha_n) = f(\alpha_{\sigma(1)}, \ldots, \alpha_{\sigma(n)})$$

We say that the polynomial f is *symmetric* if $f^\sigma = f$ for all $\sigma \in S_n$. It is obvious that the set of symmetric polynomials is a subring of $R[\alpha_1,\ldots,\alpha_n]$ containing R and the elementary symmetric polynomials $s_1,\ldots,s_n$. The result below shows that it contains nothing else.

Let $X_1,\ldots,X_n$ be variables. We define the *weight* of a monomial

$$X_1^{t_1} \cdots X_n^{t_n}$$

to be $t_1 + 2t_2 + \ldots + nt_n$. The *weight* of an arbitrary polynomial $g(X_1,\ldots,X_n)$ is defined as the maximum of the weights of the monomials occuring in g.

2.28. THEOREM. Let $R \subseteq S$ be commutative rings and let $\alpha_1,\ldots,\alpha_n$ be algebraically independent elements of S over R. If $f(\alpha_1,\ldots,\alpha_n) \in R[\alpha_1,\ldots,\alpha_n]$ is symmetric of degree d, then there exists a polynomial

$$g(X_1,\ldots,X_n) \in R[X_1,\ldots,X_n]$$

of weight $\leqslant d$ such that

$$f(\alpha_1,\ldots,\alpha_n) = g(s_1,\ldots,s_n)$$

Furthermore, the elementary symmetric polynomials $s_1,\ldots,s_n$ of $\alpha_1,\ldots,\alpha_n$ are algebraically independent over R.

Proof. We argue by induction on n. The case $n = 1$ being trivial, assume that the result is true for $n - 1$ variables. Substituting $\alpha_n = 0$ in

$$(X-\alpha_1) \cdots (X-\alpha_n) = X^n - \alpha_1 X^{n-1} + \ldots + (-1)^n s_n$$

we find

$$(X-\alpha_1) \cdots (X-\alpha_{n-1})X = X^n - s_{1,0}X^{n-1} + \ldots + (-1)^{n-1} s_{n-1,0}X$$

where $s_{i,0}$ is the expression obtained by substituting $\alpha_n = 0$ in s_i. Note that $s_{1,0},\ldots,s_{n-1,0}$ are all elementary symmetric polynomials in $\alpha_1,\ldots,\alpha_{n-1}$. To prove the first assertion, we now proceed by induction on d. The case $d = 0$ being trivial, assume $d > 0$ and that our assertion is valid for polynomials of degree $< d$. Given a polynomial $f(\alpha_1,\ldots,\alpha_n)$ of degree d, we may find a polynomial $g_1(X_1,\ldots,X_{n-1})$ of weight $\leqslant d$ such that

$$f(\alpha_1,\ldots,\alpha_{n-1}, 0) = g_1(s_{1,0},\ldots,s_{n-1,0})$$

Observe that $g_1(s_1,\ldots,s_{n-1})$ has degree $\leqslant d$ in $\alpha_1,\ldots,\alpha_n$. Therefore the polynomial

$$f_1(\alpha_1,\ldots,\alpha_n) = f(\alpha_1,\ldots,\alpha_n) - g_1(s_1,\ldots,s_{n-1})$$

has degree $\leqslant d$ (in $\alpha_1,\ldots,\alpha_n$) and is symmetric. Since $f_1(\alpha_1,\ldots,\alpha_{n-1},0)=0$, f_1 contains α_n as a factor. But f_1 is symmetric, so it contains $\alpha_1\ldots\alpha_n$ as a factor. The conclusion is that $f_1 = s_n f_2(\alpha_1,\ldots,\alpha_n)$ for a suitable symmetric polynomial f_2 whose degree is $\leqslant d-n < d$. Applying the induction hypothesis, we may therefore find a polynomial g_2 in n variables and weight $\leqslant d-n$ such that

$$f_2(\alpha_1,\ldots,\alpha_n) = g_2(s_1,\ldots,s_n)$$

It follows that

$$f(\alpha_1,\ldots,\alpha_n) = g_1(s_1,\ldots,s_{n-1}) + s_n g_2(s_1,\ldots,s_n)$$

and each term on the right has weight $\leqslant d$, proving the first assertion.

To prove the second assertion, we argue by contradiction and choose a nonzero polynomial $\lambda(X_1,\ldots,X_n) \in R[X_1,\ldots,X_n]$ of least degree such that

$$\lambda(s_1,\ldots,s_n) = 0$$

We may write λ as a polynomial in X_n with coefficients in $R[X_1,\ldots,X_{n-1}]$,

$$\lambda(X_1,\ldots,X_n) = \lambda_0(X_1,\ldots,X_{n-1}) + \ldots + \lambda_k(X_1,\ldots,X_{n-1})X_n^k \qquad (*)$$

and claim that $\lambda_0 \neq 0$. Indeed, otherwise we can write

$$\lambda(X_1,\ldots,X_n) = X_n\mu(X_1,\ldots,X_n)$$

for a suitable polynomial μ, and thus $s_n\mu(s_1,\ldots,s_n) = 0$. But then $\mu(s_1,\ldots,s_n) = 0$ and μ has degree smaller than the degree of λ, contrary to our choice of λ.

We now substitute s_i for X_i in (*) and obtain

$$0 = \lambda_0(s_1,\ldots,s_{n-1}) + \ldots + \lambda_k(s_1,\ldots,s_{n-1})s_n^k$$

This is a relation in $R[\alpha_1,\ldots,\alpha_n]$ and we substitute 0 for α_n in this relation. In this way we obtain

$$0 = \lambda_0(s_{1,0},\ldots,s_{n-1,0})$$

which is a nontrivial relation between the elementary symmetric polynomials in $\alpha_1,\ldots,\alpha_{n-1}$. This provides the contradiction we have been seeking and so the proof is complete. ■

3. INTEGRAL EXTENSIONS

All rings in this section are assumed to be *commutative*. Assume that R is a subring of S. An element $s \in S$ is said to be *integral* over R if it satisfies an equation of the form

$$s^n + r_1 s^{n-1} + \dots + r_n = 0 \qquad (r_i \in R) \qquad (1)$$

or, equivalently, if s is a root of a monic polynomial in $R[X]$. We say that S is *integral* over R or that $R \subseteq S$ is an *integral extension*, if all elements of S are integral over R. The following observation provides a criterion that is often easier to use.

3.1. LEMMA. Let $R \subseteq S$ and let $s \in S$. Then s is integral over R if and only if $sM \subseteq M$ for some finitely generated R-submodule M of S.

Proof. Assume that s is integral over R. If s satisfies an equation (1) of degree n, take M to be the R-submodule spanned by $1, s, \dots, s^{n-1}$. Conversely, assume that $sM \subseteq M$ for some finitely generated R-submodule M of S. We may write $M = Rs_1 + Rs_2 + \dots + Rs_n$ for some $n \geqslant 1$ and some $s_1, \dots, s_n$ in S. We have equations

$$ss_i = \Sigma\lambda_{ij}s_j \qquad (\lambda_{ij} \in R,\ 1 \leqslant i,j \leqslant n) \qquad (2)$$

Let us bring all the terms in (2) to the left-hand side. The theory of linear equations applies, and if Δ is the determinant

$$\begin{vmatrix} s-\lambda_{11} & -\lambda_{12} & \dots & -\lambda_{1n} \\ -\lambda_{21} & s-\lambda_{22} & \dots & -\lambda_{2n} \\ - - - & - - - & - - - & - - - \\ -\lambda_{n1} & -\lambda_{n2} & \dots & s-\lambda_{nn} \end{vmatrix}$$

we find that $\Delta = 0$. Expanding Δ gives us an equation that shows that s is integral over R. ■

3.2. COROLLARY. If $R \subseteq S$, then the elements of S integral over R form a subring of S.

Proof. Assume that $u, v \in S$ are integral over R. We must show that both

$u+v$ and uv are also integral over R. By Lemma 3.1, $uM \subseteq M$ and $vN \subseteq N$ for some finitely generated R-submodules M,N of S. Then the product MN consisting of all finite sums $\Sigma m_i n_i$, $m_i \in M$, $n_i \in N$ is a finitely generated R-submodule of S. Since $(u+v)MN \subseteq MN$ and $(uv)MN \subseteq MN$, the result follows by virtue of Lemma 3.1. ∎

Let R be a subring of S. Then the subring of S consisting of elements integral over R is called the *integral closure* of R in S. We say that R is *integrally closed* in S if R is the integral closure of R in S.

3.3. PROPOSITION. Let R be a subring of S and let u be a unit of S. Then u^{-1} is integral over R if and only if $u^{-1} \in R[u]$.

Proof. If u^{-1} is integral over R, then

$$u^{-n} + r_1 u^{-(n-1)} + \dots + r_n = 0 \qquad (r_i \in R) \qquad (3)$$

Multiply (3) through by u^n and rearrange to get

$$u(r_1 + r_2 u + \dots + r_n u^{n-1}) = -1$$

whence $u^{-1} \in R[u]$. The argument is reversible. ∎

3.4. COROLLARY. Let R be a subring of S and let $u \in R$ be a unit of S. Then $u^{-1} \in R$ if and only if u^{-1} is integral over R. In particular, $R \cap U(S) = U(R)$ if S is integral over R.

Proof. Since $u \in R$, we have $R[u] = R$. Now apply Proposition 3.7. ∎

3.5. COROLLARY. Let R be an integral domain contained in a field F. If F is integral over R, then R is a field.

Proof. For $u \neq 0$ in R we have that u^{-1} is integral over R. By Proposition 3.3, $u^{-1} \in R[u] = R$, as required. ∎

3.6. PROPOSITION. Let R be a subring of S. Then the following conditions are equivalent:

(i) S is a finitely generated R-module

(ii) S is a finitely generated ring over R and is integral over R

Proof. (i) ⇒ (ii): This is a direct consequence of Lemma 3.1

(ii) ⇒ (i): Assume that S is generated as a ring over R by $u_1,\ldots,u_k$ and suppose that the equation showing u_i to be integral has degree n_i. Then the elements

$$u_1^{r_1}u_2^{r_2}\ldots u_k^{r_k} \qquad (0 \leqslant r_i \leqslant n_i-1)$$

span S over R. ■

3.7. PROPOSITION. Let $R \subseteq S$ be rings and u an element of a ring containing S. Suppose that u is integral over S and that S is integral over R. Then u is integral over R.

Proof. By hypothesis,

$$u^n + a_1u^{n-1} + \ldots + a_n = 0 \qquad (a_i \in S)$$

for some $n \geqslant 1$. Let $R_1 = R[a_1,\ldots,a_n,u]$. Then one immediately verifies that R_1 is a finitely generated R-module. Now apply Lemma 3.1. ■

An integral domain is said to be *integrally closed* if its integrally closed in its quotient field.

3.8. PROPOSITION. Let S be a multiplicative subset of an integral domain R. If R is integrally closed, then so is R_S.

Proof. Suppose that the element u in the quotient field of R_S is integral over R_S; we have to show that $u \in R_S$. By hypothesis,

$$u^n + (a_1/s_1)u^{n-1} + \ldots + (a_n/s_n) = 0 \qquad (a_i \in R, s_i \in S)$$

Put $s = s_1s_2\ldots s_n$ and $t_i = s/s_i$. Then

$$su^n + t_1a_1u^{n-1} + \ldots + t_na_n = 0$$

Multiplying the above equality by s^{n-1} gives an equation asserting that su is integral over R. Hence $su \in R$ and $u \in R_S$, as required. ■

3.9. PROPOSITION. Let R be a *GCD* domain. Then R is integrally closed.

Proof. Let F be the quotient field of R. Suppose $\lambda \in F$ is such that

$$\lambda^n + r_1\lambda^{n-1} + \ldots + r_n = 0 \qquad (r_i \in R) \qquad (4)$$

Write $\lambda = a/b$ with $a,b \in R$. Then there exist $s,t \in R$ such that $a = (a,b)s$,

$b = (a,b)t$, and $(s,t) = 1$. Substituting $\lambda = s/t$ in (4), we obtain

$$s^n + r_1 s^{n-1}t + \dots + r_n t^n = 0$$

whence t divides s^n. Since $(s^n,t) = 1$, it follows that t is a unit of R. Hence $\lambda \in R$ and the result follows. ■

4. TENSOR PRODUCTS

Let R be a commutative ring and U,V,W any R-modules. A map

$$f : U \times V \longrightarrow W \tag{1}$$

is said to be *bilinear* if f is linear in each argument, i.e.

$$f(r_1u_1+r_2u_2,v) = r_1f(u_1,v) + r_2f(u_2,v)$$
$$f(u,r_1v_1+r_2v_2) = r_1f(u,v_1) + r_2f(u,v_2)$$

Our aim is to construct an R-module M and a bilinear map $\lambda : U \times V \longrightarrow M$ which is universal for all bilinear mapping (1), in the sense that to any bilinear map f as in (1) there exists a unique homomorphism $f^* : M \longrightarrow W$ which renders commutative the following diagram:

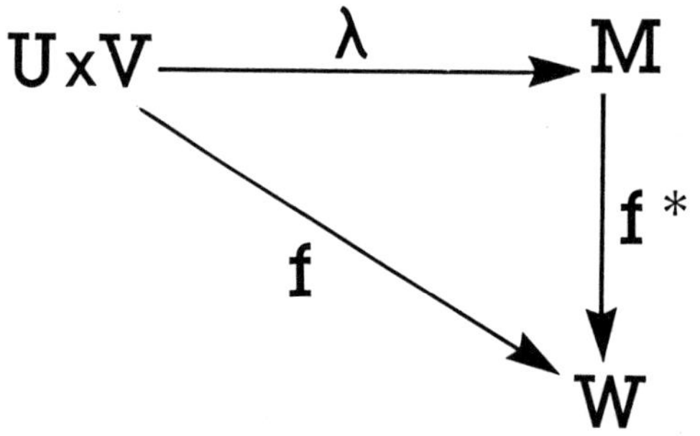

An R-module M with these properties is called a tensor product of U and V and is denoted by $U \otimes_R V$ or simply $U \otimes V$. If it exists it is clearly unique up to isomorphism, and we shall speak of *the* tensor product.

The following standard procedure illustrates the existence of M. Let F be a free R-module with $U \times V$ as a basis and let N be the submodule generated by all elements of the form

$$(r_1u_1+r_2u_2,v) - r_1(u_1,v) - r_2(u_2,v)$$
$$(u,r_1v_1+r_2v_2) - r_1(u,v_1) - r_2(u,v_2)$$

for $u,u_1,u_2 \in U$, $v,v_1,v_2 \in V$ and $r_1,r_2 \in R$. Then $U \underset{R}{\otimes} V$ is defined as the factor module F/N. The image of (u,v) under the natural homomorphism $F \longrightarrow F/N$ is denoted by $u \otimes v$. Thus the R-module $U \underset{R}{\otimes} V$ consists of all finite sums

$$\Sigma u_i \otimes v_i \qquad (u_i \in U, v_i \in V)$$

and the elements $u \otimes v$ satisfy the relations

$$(r_1u_1+r_2u_2) \otimes v = r_1(u_1 \otimes v) + r_2(u_2 \otimes v) \qquad (2)$$
$$u \otimes (r_1v_1+r_2v_2) = r_1(u \otimes v_1) + r_2(u \otimes v_2)$$

The map $\lambda : U \times V \longrightarrow U \underset{R}{\otimes} V$ defined by $\lambda(u,v) = u \otimes v$ is therefore bilinear. Furthermore if $f : U \times V \longrightarrow W$ is any map, then f determines a homomorphism $f' : F \longrightarrow W$ given by

$$f'(\Sigma r_{u,v}(u,v)) = \Sigma r_{u,v} f'(u,v)$$

If we assume that f is bilinear, then $N \subseteq \text{Ker} f'$ and so the map

$$f^* : U \underset{R}{\otimes} V \longrightarrow W$$

defined by

$$f^*(u \otimes v) = f(u,v)$$

is a required homomorphism.

4.1. PROPOSITION. Let V and W be free R-modules with bases $\{v_i\}$ and $\{w_j\}$, respectively. Then $V \underset{R}{\otimes} W$ is a free R-module with basis $\{v_i \otimes w_j\}$. Moreover, every element $x \in V \underset{R}{\otimes} W$ can be written uniquely as the finite sum

$$x = \sum_i v_i \otimes x_i \qquad (v_i \in V,\ x_i \in W)$$

and as

$$x = \sum_j x'_j \otimes w_j \qquad (x'_j \in V,\ w_j \in W)$$

Proof. Given $v \in V$ and $w \in W$, we have $v = \Sigma r_i v_i$ and $w = \Sigma r'_j w_j$ for suitable $r_i, r'_j \in R$, where both sums are, of course, finite. Therefore, by (2),

we have

$$v \otimes w = \sum_{i,j} r_i r'_j (v_i \otimes w_j)$$

Thus the R-module $V \underset{R}{\otimes} W$ is generated by $\{v_i \otimes w_j\}$. Suppose now that

$$\Sigma r_{ij}(v_i \otimes w_j) = 0 \qquad (r_{ij} \in R)$$

and fix some subscripts, say $i = k$, $j = s$. Since $\{v_i\}$ is a basis for V, there exists an R-linear map $\lambda_k : V \longrightarrow R$ such that $\lambda_k(v_i) = 0$ for $i \neq k$ and $\lambda_k(v_k) = 1$. Similarly, there exists $\mu_s : W \longrightarrow R$ with $\mu_s(w_j) = 0$ for $j \neq s$ and $\mu_s(w_s) = 1$. Then the map $f : V \times W \longrightarrow R$ defined by $f(v,w) = \lambda_k(v)\mu_s(w)$ is bilinear. Hence there exists an R-linear map $f^* : V \otimes W \longrightarrow R$ with

$$f^*(v \otimes w) = \lambda_k(v)\mu_s(w)$$

Finally,

$$0 = f^*(\Sigma r_{ij}(v_i \otimes w_j)) = \Sigma r_{ij} f^*(v_i \otimes w_j) = r_{ks}$$

and $\{v_i \otimes w_j\}$ is a basis for $V \otimes W$.

The remaining assertions follow easily by grouping terms. Thus if $x \in V \otimes W$, then x is uniquely given as the finite sum $x = \sum_{i,j} x_{ij}(v_i \otimes w_j)$ and hence by (2),

$$\begin{aligned} x &= \sum_i v_i \otimes (\sum_j x_{ij} w_j) \\ &= \sum_j (\sum_i x_{ij} v_i) \otimes w_j \end{aligned}$$

as required. ■

Let B and C be *commuting subalgebras* of an algebra A, i.e. let $bc = cb$ for all $b \in B$, $c \in C$. Then the set BC consisting of all finite sums $\Sigma b_i c_i$ with $b_i \in B$, $c_i \in C$, is the smallest subalgebra containing B and C. If $A = BC$ then we say that A is *generated* by B and C.

Now assume that V and W are R-algebras. Because they are R-modules, $V \underset{R}{\otimes} W$ exists, and we assert that this R-module is also an R-algebra.

4.2. PROPOSITION. Let V and W be R-algebras. Then $V \underset{R}{\otimes} W$ is also an R-algebra with multiplication given by

$$(v_1 \otimes w_1)(v_2 \otimes w_2) = (v_1 v_2) \otimes (w_1 w_2) \qquad (v_1, v_2 \in V,\ w_1, w_2 \in W)$$

Furthermore, the maps $v \longmapsto v \otimes 1$ and $w \longrightarrow 1 \otimes w$ are R-algebra homomorphisms of V and W, respectively, into $V \otimes W$, which are injective if V and W are R-free, and the images $V \otimes 1$ and $1 \otimes W$ are commuting subalgebras that generate $V \otimes W$.

Proof. Let F be a free R-module with $V \times W$ as a basis. Define multiplication in F distributively using

$$(v_1, w_1)(v_2, w_2) = (v_1 v_2, w_1 w_2)$$

Then F is obviously an R-algebra. Furthermore, the submodule N used in the definition of $V \underset{R}{\otimes} W$ is an ideal of F. Therefore $V \underset{F}{\otimes} W = F/N$ inherits the algebra structure of F. In particular, since (v,w) maps to $v \otimes w$ this says that

$$(v_1 \otimes w_1)(v_2 \otimes w_2) = (v_1 v_2) \otimes (w_1 w_2)$$

Finally, by the foregoing and (2) it is clear that the maps $v \longmapsto v \otimes 1$ and $w \longmapsto 1 \otimes w$ are R-algebra homomorphisms of V and W, respectively, into $V \otimes W$. Moreover, if V and W are R-free then by Proposition 4.1, these homomorphisms are injective. The remaining assertion being obvious, the result follows. ∎

We shall refer to the homomorphisms in Proposition 4.2 as *canonical*. The next result provides a universal characterization of tensor product of algebras.

4.3. PROPOSITION. Let A_1, A_2 and B be R-algebras and let $\psi_i : A_i \longrightarrow B$ be an R-algebra homomorphism, $i = 1,2$, such that $\psi_1(A_1)$ and $\psi_2(A_2)$ are commuting subalgebras of B. Then there exists one and only one R-algebra homomorphism

$$h : A_1 \underset{R}{\otimes} A_2 \longrightarrow B$$

which renders commutative the following diagram:

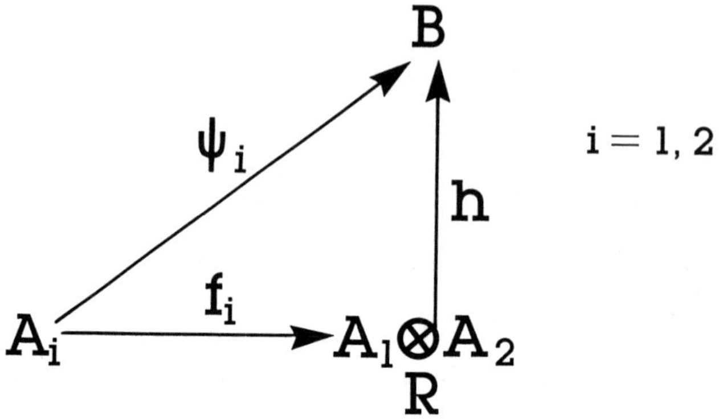

(f_i is the canonical homomorphism)

Proof. The map $A_1 \times A_2 \longrightarrow B$, $(a_1, a_2) \longmapsto \psi_1(a_1)\psi_2(a_2)$ is bilinear and so the map $h : A_1 \otimes_R A_2 \longrightarrow B$ defined by $h(a_1 \otimes a_2) = \psi_1(a_1)\psi_2(a_2)$ is a homomorphism of R-modules. Furthermore, if $x = a_1 \otimes a_2$ and $y = a_1' \otimes a_2'$, then

$$\begin{aligned} h(xy) &= h(a_1a_1' \otimes a_2a_2') = \psi_1(a_1)\psi_1(a_1')\psi_2(a_2)\psi_2(a_2') \\ &= [\psi_1(a_1)\psi_2(a_2)]\,[\psi_1(a_1')\psi_2(a_2')] \\ &= h(x)h(y) \end{aligned}$$

so that h is a homomorphism of R-algebras. Since $h(f_i(a_i)) = \psi_i(a_i)$, $a_i \in A_i$, the above diagram is commutative. Finally, the uniqueness of h is a consequence of the fact any R-algebra homomorphism $A_1 \otimes_R A_2 \longrightarrow B$ is uniquely determined by its restriction to $\{a_1 \otimes a_2 \mid a_i \in A_i, i = 1,2\}$. ■

There is a simple criterion for an algebra over a field to be a tensor product which is often useful. Let A be an algebra over a field F and let U, V be subspaces of A. Then U and V are said to be *linearly disjoint* over F if for any linearly independent elements u_i in U and v_j in V, the elements u_iv_j in A are linearly independent over F. Clearly this just means that the natural mapping

$$\begin{cases} U \otimes V \longrightarrow A \\ u \otimes v \longmapsto uv \end{cases}$$

is an injective F-homomorphism.

4.4. PROPOSITION. Let A be an algebra over a field F. Given subalgebras B,C of A, if (i) B and C are linearly disjoint over F, (ii) $A = BC$ and (iii) B and C commute elementwise, then $A \cong B \otimes_F C$ and $B \cap C = F$.

Proof. By Proposition 4.3, the map $B \otimes C \longrightarrow A$, $b \otimes c \longmapsto bc$ is a homomorphism of F-algebras. This is injective by (i) and surjective by (ii), hence $A \cong B \otimes_F C$. Assume that $\lambda \in B \cap C$. If $\lambda \notin F$, then $1,\lambda$ are F-linearly independent, hence by (i), $1,\lambda,\lambda,\lambda^2$ are F-linearly independent, which is absurd. Thus $\lambda \in F$ and the result follows. ■

4.5. PROPOSITION. Let A be a finite-dimensional algebra over a field F and let B,C be subalgebras which commute elementwise. Then B and C are linearly disjoint over F if and only if

$$\dim_F BC = (\dim_F B)(\dim_F C)$$

Proof. The image BC of the natural map $B \otimes_F C \longrightarrow A$, $b \otimes c \longmapsto bc$ has the same dimension as $B \otimes_F C$ if and only if the map is injective, as required. ■

5. MODULE-THEORETIC PREREQUISITES

Let R be an arbitrary ring and let M be a (left) R-module. We say that M is *noetherian* (artinian) if every ascending (descending) chain of submodules of M stops. The ring R is said to be (left) *noetherian* (artinian) if the regular module ${}_RR$ is noetherian (artinian); in other words, R is a noetherian (artinian) ring if and only if every ascending (descending) chain of left ideals of R stops.

5.1. PROPOSITION. Let M be an R-module. Then the following conditions are equivalents

(i) M is noetherian

(ii) Every submodule of M is finitely generated

(iii) Every nonempty collection of submodules of M contains a maximal element, that is, a submodule which is not properly contained in any other submodule in the collection.

Proof. (i) ⇒ (ii): Assume that (ii) is false, and let N be a submodule of M which is not finitely generated. Then, by induction, we can construct an infinite sequence of elements $a_1, a_2, \ldots$ in N such that if $M_k = Ra_1 + \ldots + Ra_k$, then for every k, M_k is properly contained in M_{k+1}. This contradicts (i) and so (i) ⇒ (ii) is true.

(ii) ⇒ (iii): Assume that (iii) does not hold. Then there exists a nonempty collection X of submodules which has no maximal element. By induction, we can find submodules $M_1, M_2, \ldots$ in X such that for every k, M_k is properly contained in M_{k+1}. The union $M^* = \bigcup_{k=1}^{\infty} M_k$ is clearly a submodule of M although not necessarily a member of X. We show that M^* is not finitely generated. Assume the contrary, say

$$M^* = Rm_1 + \ldots + Rm_s$$

for some s. Then each $m_i \in M_{k_i}$ for some k_i, and, choosing the largest of the modules M_{k_i}, we deduce that $M^* = M_{k_i}$ for some index k_i and hence that $M_{k_i+1} = M_{k_i}$. But this contradicts the way the submodules M_k were constructed, so the implication is true.

(iii) ⇒ (i): Deny the statement. Then there exists an infinite properly ascending chain of submodules, and it is clear that this collection of submodules can have no maximal element. The proof is therefore complete. ■

5.2. PROPOSITION. Let U be a submodule of an R-module V. Then V is noetherian if and only if both U and V/U are noetherian. Furthermore, if R is noetherian and if V is finitely generated, then V is a noetherian R-module.

Proof. If V is noetherian, then, clearly, so are U and V/U. Conversely, assume that U and V/U are noetherian, and let

$$A_1 \subseteq A_2 \subseteq \ldots \subseteq A_n \subseteq \ldots$$

be an ascending chain of R-submodules of V. Then both sequences $\{A_n \cap U\}$ and $\{(A_n+U)/U\}$ eventually stabilize, and hence for some integer t we have

$$A_n \cap U = A_t \cap U \quad \text{and} \quad A_n + U = A_t + U \qquad \text{for all} \quad n \geqslant t$$

Accordingly, since $A_n \supseteq A_t$ we infer from the modular law that

$$\begin{aligned} A_n &= A_n \cap (A_n + U) = A_n \cap (A_t + U) \\ &= A_t + (A_n \cap U) = A_t + (A_t \cap U) \\ &= A_t \end{aligned}$$

Hence the sequence $\{A_n\}$ stabilizes, and V is noetherian.

Finally, assume that R is noetherian. Then every one-generator R-module is a homomorphic image of ${}_RR$ and hence is noetherian. It therefore follows from the preceding, by induction on the number of generators, that every finitely generated R-module is also noetherian. ■

A submodule W of V is said to be *maximal*, if $W \neq V$ and there is no proper submodule of V strictly containing W. A finitely generated module over any ring has an important maximality property, which is a consequence of Zorn's lemma.

5.3. PROPOSITION. Let R be any ring and V a finitely generated R-module. Then every proper submodule of V is contained in a maximal submodule.

Proof. Write $V = Rv_1 + \ldots + Rv_n$. Given a proper submodule W of V, consider the set X of proper submodules of V which contain W. This is inductive: it contains W and if (W_λ) is a chain in X, then $\cup W_\lambda$ is a submodule. Assume $\cup W_\lambda = V$. Then each $v_i \in W_{\lambda_i}$ for some λ_i. Let W_μ be the largest of the modules $W_{\lambda_1}, \ldots, W_{\lambda_n}$. Then W_μ contains each v_i and so $V = W_\mu$, which contradicts the definition of X. Therefore $\cup W_\lambda$ is proper and this shows X to be inductive. We can therefore apply Zorn's lemma and deduce that X has a maximal element; this is a proper submodule of V containing W, as we had to construct. ■

5.4. COROLLARY. Let R be any ring. Then any proper left ideal of R is contained in a maximal left ideal of R. In particular, R has maximal left ideals.

Proof. The module ${}_RR$ is generated by the single element 1. Now apply Proposition 5.3. ■

The most important source of noetherian rings is the following classical result:

5.5. THEOREM. (Hilbert basis theorem). If R is a commutative noetherian ring, then so is $R[X_1,\ldots,X_n]$ for any integer $n \geq 1$.

Proof. By induction on n, it suffices to consider the polynomial ring $R[X]$. Let I be an ideal in $R[X]$ and denote by A_n the set of leading coefficients of polynomials of degree $\leq n$ in I. It is clear that A_n is an ideal in R and

$$A_0 \subseteq A_1 \subseteq \ldots$$

Now $A = \cup A_n$ is an ideal of R, finitely generated by virtue of Proposition 5.1. Let $f_1,\ldots,f_r$ be polynomials in I whose leading coefficients generate A. If f is any polynomial in I then we can reduce the degree of f by subtracting multiples of the f_i's, unless $\deg f < \deg f_i$ for some i. Hence, if

$$k = \max\{\deg f_1,\ldots,\deg f_r\},$$

then by repeated subtractions we obtain a polynomial g of degree $< k$:

$$f \equiv g(\mathrm{mod}(f_1,\ldots,f_r)) \qquad \deg g < k$$

and since $f_i \in I$, $g \in I$. Now let $g_1,\ldots,g_s$ be polynomials in I such that the leading coefficients of the g_t of degree $\leq i$ generate A_i, $i = 0,1,\ldots,k-1$. Then g can be written as a linear combination of the g_i's, hence I is generated by $f_1,\ldots,f_r,g_1,\ldots,g_s$. The desired conclusion now follows by virtue of Proposition 5.1. ■

5.6. COROLLARY. Let R be a noetherian subring of a commutative ring S. If S is finitely generated over R (as a ring), then S is noetherian.

Proof. By hypothesis, S is a homomorphic image of $R[X_1,\ldots,X_n]$ for some $n \geq 1$. Now apply Theorem 5.5. ■

5.7. PROPOSITION. Let V be a finitely generated free module over a commutative ring R. Then all bases of V are finite, with the same number of elements (called the rank of V).

Proof. By hypothesis, $V = Ru_1 + \ldots + Ru_n$ for some $n \geq 1$ and some $u_i \in V$, and also V has a basis (e_i), say. If we express each u_k in terms of the e_i, only finitely many e_i occur in the expression for any given u_k. Hence V

is generated by a finite subfamily of the e_i, and since the e_i's form a basis, the whole family (e_i) must be finite, say $V = Re_1 \oplus \ldots \oplus Re_m$. Let P be any maximal ideal of R, and set $\bar{R} = R/P$, $\bar{V} = V/PV$. Then $\bar{R}$ is a field and $\bar{V} = \bar{R} \oplus \ldots \oplus \bar{R}$ (m summands). Therefore the dimension $(\bar{V}:\bar{R}) = m$, and, since the dimension of a vector space over a field is uniquely determined, the result is proved. ∎

5.8. PROPOSITION. Let R be a principal ideal domain and let V be a free R-module of finite rank s. Then every nonzero submodule of V is free of rank $r \leqslant s$.

Proof. The case $s = 0$ being vacuously true, we argue by induction on s. Assume that the result is true for modules of rank $\leqslant s-1$. Let $v_1,\ldots,v_s$ be a basis of V, $s \geqslant 1$ and let $W \neq 0$ be a submodule of V. Each $x \in W$ can be expressed uniquely in the form

$$x = r_1v_1 + \ldots + r_sv_s \qquad (r_i \in R)$$

The set of all coefficients r_1 which occur as x ranges over all elements of W is an ideal I of R. Since R is a PID, there exists an element $y \in I$ such that $I = Ry$. Then y occurs as the coefficient of v_1 in some element

$$x_1 = yv_1 + r_2v_2 + \ldots + r_sv_s \in W$$

If we put $V' = Rv_2 \oplus \ldots \oplus Rv_s$, then $V' \cap W$ is a submodule of V' and has a basis of $\leqslant s-1$ elements $x_2,\ldots,x_t$, $t \leqslant s$, by the induction hypothesis. It follows that $W = Rx_1 + (V' \cap W)$ and that $x_1,x_2,\ldots,x_t$ is a basis of W (unless $y = 0$, in which case x_1 should be omitted). This completes the proof. ∎

6. TOPOLOGICAL PREREQUISITES

It is assumed that the reader has some familiarity with the basic concepts of general topology. However, for convenience we recall briefly those that are relevant for our subsequent investigations.

A set X becomes endowed with a *topology* as soon as there is given a collection Ω of subsets of X having the following properties:

(a) both X and its empty subset $\emptyset$ belong to Ω

(b) Ω is closed under arbitrary unions and finite intersections.

The members of Ω are then referred to as the *open sets* of the topology and the pair (X,Ω), or simply X, is called a topological space.

Let X be a topological space. A subset Y of X is called a *closed subset* if its complement with respect to X is open. The union of a finite number of closed sets is necessarily a closed set and so too is the intersection of an arbitrary family of closed sets. Of course it can happen that a set is both open and closed.

A bijection f of a topological space X onto another topological space Y such that $f(A)$ is open in Y if and only if A is open in X, is called a *homeomorphism*. We say that X and Y are *homeomorphic* if there exists a homeomorphism of X onto Y. The *discrete topology* on X is the topology for which every subset of X is open; the *trivial topology* on X is the topology whose only open sets are $\emptyset$ and X.

Suppose that Y is a subset of a topological space X. The intersection of all the closed sets containing Y is itself a closed set which will be denoted by $\bar{Y}$. This is called the closure of Y in X. If $x \in X$, then it is easy to show that x belongs to $\bar{Y}$ if and only if every open set containing x contains at least one element of Y. If it happens that $\bar{Y} = X$, then one says that Y is *dense* in X or that Y is a *dense subset* of Y.

Again, assume that Y is a subset of a topological space X. It is then possible to define a topology on Y whose open sets are precisely the intersections of Y with the open subsets of X. This is known as the *induced topology*. When Y is endowed with the induced topology, we say that Y is a *subspace* of X.

The elements of a topological space X are often referred to as *points*. By a *neighbourhood* of a point $x \in X$, we understand any set which contains an open set to which x belongs. More generally, a *neighbourhood of a subset* A of X is any subset of X which contains an open set containing A. A point $x \in X$ is said to be an *interior* point of a subset A of X if A is a neighbourhood of x. The set of interior points of A is called the *interior* of A. Thus A is open if and only if A coincides with its interior.

Assume that X has the following property: whenever x and x' are distinct points there exists a neighbourhood Y of x and a neighbourhood Y' of x' such that $Y \cap Y' = \emptyset$. We then say that X is a *Hausdorff space*.

Let X be a topological space. A collection C of subsets of X is called a *base for the topology* of X if (i) every set in C is an open set of X, and (ii) every open set of X is a union of a (possibly empty) family of sets belonging to C.

Let X and X' be two topological spaces and suppose that C and C' are bases for their respective topologies. If now $Y \in C$ and $Y' \in C'$, then $Y \times Y'$ is a subset of $X \times X'$. It is easy to see that there is a unique topology on $X \times X'$ which has the subsets $Y \times Y'$ as a base. Moreover, this topology is unaltered if we replace C and C' by other bases for the open sets of X and X', respectively. The topological space $X \times X'$ so obtained is called the *product* of X and X'. The topology on the product space is referred to as the *product topology*.

Let X and X' be topological spaces and $f : X \longrightarrow X'$ a mapping. Then f is said to be *continuous at the point* $x \in X$ if for each neighbourhood Y' of $f(x)$ there exists a neighbourhood Y of x such that $f(Y) \subseteq Y'$. When f is continuous at every point of X, we say simply that f is *continuous*. It is well known that f is continuous if and only if for every open set Y' in X', $f^{-1}(Y')$ is open in X.

Given two topologies Ω_1 and Ω_2 on the same set X, we say that Ω_1 is *finer* than Ω_2 (and that Ω_2 is *coarser* than Ω_1), if every subset of X which is open in Ω_2 is open in Ω_1 (equivalently, every subset of X which is closed in Ω_2 is closed in Ω_1).

Let X be a topological space, R an equivalence relation on X, X/R the quotient set of X with respect to the relation R, $\lambda : X \longrightarrow X/R$ the canonical mapping. The finest topology on X/R for which λ is continuous is called the *quotient* of the topology of X by the relation R; under this topology X/R is the *quotient space* of X by R. It follows from the definition that a subset A of X/R is open (respectively closed) if and only if $\lambda^{-1}(A)$ is open (respectively closed) in X. Note that if Y is a topological space, then a mapping $f : X/R \longrightarrow Y$ is continuous if and only if $f\circ\lambda : X \longrightarrow Y$ is continuous. Thus there is a one-to-one correspondence between the continuous mappings $X/R \rightarrow Y$ and the continuous maps $X \rightarrow Y$ which are constant on each equivalence class mod$R$.

Let X be a set. A set Ω of subsets of X is called a *filter* on X if it satisfies the axioms:

(i) $\Omega \neq \emptyset$ and $\emptyset \notin \Omega$

(ii) $Y \in \Omega$ and $Y \subseteq Y_1 \subseteq X$ implies $Y_1 \in \Omega$

(iii) $Y_1, Y_2 \in \Omega$ implies $Y_1 \cap Y_2 \in \Omega$

A set Γ of subsets of X is called a *filter base* if (a) $\Gamma \neq \emptyset$ and $\emptyset \notin \Gamma$ and (b) if $Y_1, Y_2 \in \Gamma$ then $Y_3 \subseteq Y_1 \cap Y_2$ for some $Y_3 \in \Gamma$. Every filter base Γ generates a unique filter Ω on X such that $Y \in \Omega$ if and only if $Y_1 \subseteq Y$ for at least one $Y_1 \in \Gamma$; Γ is called a *base* of the filter Ω. The set of all filters on a nonempty set X is inductively ordered by the relation $\Omega_1 \subseteq \Omega_2$; $\Omega_1 \subseteq \Omega_2$ is expressed by saying that Ω_1 is *coarser* than Ω_2, or that Ω_2 is *finer* than Ω_1. Every filter on X which is maximal with respect to this ordering, is called an *ultrafilter* on X; by Zorn's lemma, for each filter Ω on X there exists an ultrafilter finer than Ω.

Let X be a topological space. An open covering of X is a family $\{X_i | i \in I\}$ of open subsets of X such that $X = \bigcup_{i\in I} X_i$. We say that X is *compact* if for each open covering of X there is a finite subcovering.

A topological space X is said to be *connected* if it is not the union of two disjoint nonempty open sets. A subset A of a topological space X is said to be *connected* if the subspace A of X is connected.

Given a point x of a topological space X, the union of the connected subsets of X which contain x is connected; it is therefore the largest connected subset of X containing x. The *component* (or *connected component*) of a point of a topological space X is defined to be the largest connected subset of X which contains this point. The space X is said to be *totally disconnected* if the component of each point of X consists of the point alone. A subset A of X is said to be *totally disconnected set* if the subspace A of X is totally disconnected.

Let G be a group. We say that G is a *topological group* if G is also a topological space such that

(i) The mapping $(x,y) \longmapsto xy$ of $G \times G$ into G is continuous

(ii) The mapping $x \longmapsto x^{-1}$ of G into G is continuous

A group structure and a topology on a set G are said to be *compatible* if they satisfy (i) and (ii) above.

Let R be a ring. We say that R is a *topological ring* if R is also a topological space such that:

(i) The mapping $(x,y) \longmapsto x + y$ of $A \times A$ into A is continuous

(ii) The mapping $x \longmapsto -x$ of A into A is continuous

(iii) The mapping $(x,y) \longmapsto xy$ of $A \times A$ into A is continuous

Let G be a topological group and let Ω be the neighbourhood filter of the identity element 1 of G. If $g \in G$, then the mappings $x \longmapsto gx$ and $x \longrightarrow xg$ are homeomorphisms. It follows that the neighbourhood filter of g is the family $g\Omega$ of sets gX, $X \in \Omega$ and also the family Ωg of sets Xg, $X \in \Omega$. We close this section by quoting the following standard properties (see Bourbaki, 1966)).

6.1. PROPOSITION. Let G be a topological group and let Ω be the neighbourhood filter of 1. Then

(i) Given any $X \in \Omega$, there exists $Y \in \Omega$ such that $Y \cdot Y \subseteq X$

(ii) Given any $X \in \Omega$, we have $X^{-1} \in \Omega$

(iii) For all $g \in G$ and all $X \in \Omega$, we have $gXg^{-1} \in \Omega$

Conversely, if G is a group and Ω a filter on G satisfying (i) - (iii), then there is a unique topology on G compatible with the group structure of G, for which Ω is the neighbourhood filter of 1.

6.2. PROPOSITION. Let G be a topological group. Then the following conditions are equivalent:

(i) G is Hausdorff

(ii) The set $\{1\}$ is closed

(iii) The intersection of the neighbourhoods of 1 consists only of the point 1.

6.3. PROPOSITION. A subgroup of a topological group is open if and only if it has an interior point. Every open subgroup is closed.

Let X_i, $i \in I$, be an arbitrary collection of topological spaces. Then, by definition, the *product topology* on $\prod_{i\in I} X_i$ has a basis consisting of the sets of the form $\prod_{i\in I} U_i$ with U_i an open subset of X_i for every $i \in I$, and $U_i = X_i$ for all but finitely many $i \in I$.

6.4. PROPOSITION. A subspace of a Hausdorff space is Hausdorff. Every product of Hausdorff spaces is Hausdorff. Conversely, if a product of nonempty spaces is Hausdorff, then each factor is a Hausdorff space.

A collection Y of subsets of a topological space X is said to have the *finite intersection property* if every finite subcollection Y_1 of Y satisfies $\bigcap_{S\in Y_1} S \neq \emptyset$. For the proof of the following results we refer to Pontryagin (1977).

6.5. PROPOSITION. A topological space X is compact if and only if every collection of closed subsets of X possessing the finite intersection property has nonempty intersection.

6.6. PROPOSITION. (i) A closed subset of a compact topological space is compact.

(ii) A compact subset of a Hausdorff topological space is closed.

(iii) A continuous image of a compact topological space is compact.

(iv) Any continuous bijection $X \longrightarrow Y$ of a compact topological space X onto a Hausdorff space Y is a homeomorphism.

6.7. PROPOSITION. Let G be a compact totally disconnected topological group and let U be an arbitrary neighbourhood of the identity. Then there is an open normal subgroup N of G such that $N \subseteq U$.

6.8. PROPOSITION. (Tychonoff's Theorem). The product of an arbitrary collection of compact topological spaces is compact.

2
Classical topics in field theory

Our aim here is to introduce classical foundations of the theory of fields, with an emphasis on refinements and extensions achieved in scope of relatively recent developments. Among these, we exhibit canonical fundamental units of certain classes of pure cubic fields, prove Kneser's theorem on torsion groups of separable field extensions and establish a theorem of Schinzel which provides necessary and sufficient conditions for the Galois group of a binomial $X^n - a$ to be abelian. A separate section is devoted to a result of Isaacs concerning the degrees of sums in a separable field extension. If G is the Galois group of a finite Galois extension E/F, an interpretation of a distinguished subgroup of $H^3(G,E^*)$ is provided by using the methods of Eilenberg and MacLane. We also include some recent results pertaining to radical extensions, especially a cogalois correspondence discovered by Greither and Harrison(1986).

1. ALGEBRAIC EXTENSIONS

Let E be a field and F a subfield; we write E/F for the field E considered as an extension of F. We can view E as an F-algebra and hence a vector space over F; its dimension is either a positive integer or an infinite cardinal and is called the *degree* of E over F. In what follows we write $(E:F)$ for the degree of E over F. We say that E/F is a *finite* or an *infinite extension* accordingly as $(E:F)$ is finite or infinite.

If S is a subset of E, we write $F(S)$ (respectively, $F[S]$) for the smallest subfield (respectively, subring) containing S and F. In case $S = \{s_1, s_2, \ldots, s_n\}$, we also write $F(s_1, \ldots, s_n)$ and $[F\ s_1, \ldots, s_n]$ instead of $F(S)$ and $F[S]$, respectively. It is clear that $F(s_1, \ldots, s_n)$ is the quotient field of $F[s_1,\dots,s_n]$ and that $F[s_1,\dots,s_n]$ consists of all polynomials in $s_1,\dots,s_n$ over F. We shall refer to $F(s_1,\dots,s_n)$ as the *field generated over* F by $s_1,\dots,s_n$ or the field obtained by *adjoining* to F the elements $s_1,\dots,s_n$.

The field extension E/F is said to be *finitely generated* if $E = F(s_1,\dots,s_n)$ for suitable finitely many elements $s_1,\dots,s_n$ of E. The extension E/F is called *simple* if there is an element α in E, called a *primitive element* over F, such that $E = F(\alpha)$.

When we have a chain, also called a *tower* of fields:

$$F \subseteq E \subseteq K$$

we can regard K as an extension of E or F. Thus we have three extensions E/F, K/F and K/E; their degrees are related by the important product formula:

1.1. PROPOSITION. Let $F \subseteq E \subseteq K$ be a chain of fields. If A and B are bases for E/F and K/E, respectively, then $AB = \{ab \mid a \in A, b \in B\}$ is a basis for K/F. In particular, $(K:F) = (E:F)(K:E)$.

Proof. If λ is an element of K, then $\lambda = \sum_{j=1}^{m} \mu_j b_j$ for some $m \geq 1$ and some $\mu_j \in E, b_j \in B$. Furthermore, we have

$$\mu_j = \sum_{i=1}^{n} \alpha_{ij} a_i \qquad (\alpha_{ij} \in F,\ a_i \in A)$$

for some $n \geq 1$. Hence $\lambda = \sum_{i=1}^{n} \sum_{j=1}^{m} \alpha_{ij} a_i b_j$ which shows that K is spanned by AB over F. It remains to show that the elements of AB are linearly independent over F. Let $\sum_{i=1}^{n} \sum_{j=1}^{m} \lambda_{ij} a_i b_j = 0$ with $\lambda_{ij} \in F$. We set $c_j = \sum_{i=1}^{n} \lambda_{ij} a_i$. Then $\sum_{j=1}^{m} c_j b_j = 0$, $c_j \in E$, and since the b_j are E-linearly independent, we must have $c_j = 0$, $1 \leq j \leq m$. The latter in turn implies that the λ_{ij} are zero, since the a_i are F-linearly independent. This completes the proof. ■

1.2. COROLLARY. Let $F \subseteq E \subseteq K$ be a chain of fields. Then K/F is finite if and only if both E/F and K/E are finite.

Proof. Direct consequence of Proposition 1.1. ■

Let E/F be a field extension. An element α of E is said to be *algebraic* over F if α is a root of a nonzero polynomial over F. If $f(X)$ is a nonzero polynomial over F with $f(\alpha) = 0$ and if λ is the leading coefficient of $f(X)$, then $g(X) = \lambda^{-1}f(X)$ is a monic polynomial with $g(\alpha) = 0$. Thus α is algebraic over F if and only if α is integral over F. We say that the extension E/F is *algebraic* if all elements of E are algebraic over F.

1.3. PROPOSITION. Let E/F be a field extension, let $\alpha \in E$ be algebraic over F and let $f(X) \in F[X]$ be of least degree such that $f(\alpha) = 0$. Then

(i) $f(X)$ is irreducible

(ii) $f(X)$ divides each $g(X) \in F[X]$ with $g(\alpha) = 0$

(iii) $F(\alpha) = F[\alpha]$ and if $n = \deg f(X)$ then $1,\alpha,\ldots,\alpha^{n-1}$ is an F-basis for $F(\alpha)$. In particular, $(F(\alpha):F) = n$.

Proof. (i) Suppose that $f(X) = f_1(X)f_2(X)$, $f_i(X) \in F[X]$. Then

$$f_1(\alpha)f_2(\alpha) = 0,$$

and since F is a field, we have $f_1(\alpha) = 0$ or $f_2(\alpha) = 0$. Assume for definiteness that $f_1(\alpha) = 0$. Since $\deg f_1(X) \leqslant \deg f(X)$, we must have

$$\deg f_1(X) = \deg f(X)$$

which forces $\deg f_2(X) = 0$. This shows that $f(X)$ is irreducible.

(ii) Let $g(X) \in F[X]$ be such that $g(\alpha) = 0$. Owing to Corollary 1.2.7, we have

$$g(X) = q(X)f(X) + r(X) \qquad (q(X),r(X) \in F[X])$$

where either $r(X) = 0$ or $\deg r(X) < \deg f(X)$. Since $r(\alpha) = 0$, we must have $r(X) = 0$, as required.

(iii) By Proposition 1.3.6, $F[\alpha]$ is finite-dimensional over F and hence artinian. Since $F[\alpha]$ is an integral domain, we have $F(\alpha) = F[\alpha]$ by virtue of Proposition 1.1.3. It is clear that each element of $F[\alpha]$ belongs to the F-linear span of $\{1,\alpha,\ldots,\alpha^{n-1}\}$. Furthermore, if $\lambda_0 + \lambda_1\alpha + \ldots + \lambda_{n-1}\alpha^{n-1} = 0$ with $\lambda_i \in F$ and not all λ_i equal to 0, then α is a root of a polynomial over F of degree $< n$ which is impossible. Thus $1,\alpha,\ldots,\alpha^{n-1}$ are F-linearly independent and the result follows. ■

Let E/F be a field extension and let $\alpha \in E$ be algebraic over F. By

Proposition 1.3(ii), there exists a unique monic polynomial $f(X) \in F[X]$ of least degree for which α is a root. We shall refer to $f(X)$ as the *minimal polynomial* for α over F, and to $\deg f(X)$ as the *degree* of α over F. Alternatively, the minimal polynomial for α over F can be defined as the unique irreducible monic polynomial in $F[X]$ for which α is a root.

1.4. PROPOSITION. Let E/F be a field extension.

(i) The elements of E algebraic over F form a subfield of E containing F

(ii) If $E = F(S)$ for some subset S of E, then E/F is algebraic if and only if each element of S is algebraic over F

(iii) E/F is finite if and only if E/F is algebraic and finitely generated

(iv) If S is a subset of E such that all elements of S are algebraic over F, then $F(S) = F[S]$.

Proof. (i) Let K consist of all elements of E that are algebraic over F. Then K is a ring, by Corollary 1.3.2. If $0 \neq \alpha \in K$, then $F[\alpha]$ is a field by Proposition 1.3(iii). Hence $\alpha^{-1} \in F[\alpha] \subseteq K$, proving that K is a field. Since F is obviously contained in K, the assertion follows.

(ii) Direct consequence of (i).

(iii) Direct consequence of Proposition 1.3.6.

(iv) By (ii), $F(S)/F$ is algebraic. If $0 \neq \alpha \in F[S]$, then α is algebraic over F, hence $F[\alpha]$ is a field by Proposition 1.3(iii). Therefore

$$\alpha^{-1} \in F[\alpha] \subseteq F[S]$$

as required. ■

Let E/F be a field extension. By the *algebraic closure* of F in E we understand the subfield of E consisting of all elements that are algebraic over F.

1.5. PROPOSITION. Let $F \subseteq E \subseteq K$ be a chain of fields, let $\bar{F}$ be the algebraic closure of F in E and let $\lambda \in K$.

(i) If λ is algebraic over $\bar{F}$, then λ is algebraic over F.

(ii) If λ is algebraic over E and E/F is algebraic, then λ is algebraic over F.

Proof. Property (ii) is a particular case of (i) in which $\bar{F} = E$, while (i) is a consequence of Proposition 1.3.7. ■

Let E,K be extensions of a field F. If E and K are contained in some field L, then we denote by EK the smallest subfield of L containing both E and K, and call it the *composite* of E and K in L. Thus whenever we write EK we automatically assume that E and K are contained in some field L.

1.6. COROLLARY. (i) If $F \subseteq E \subseteq K$ is a chain of fields and both E/F and K/E are algebraic, then K/F is also algebraic.

(ii) If E,K are algebraic extensions of a field F, then EK/F is also algebraic.

Proof. (i) Direct consequence of Proposition 1.5(ii).

(ii) Put $S = E \cup K$. Then $EK = F(S)$ and each element of S is algebraic over F. Now apply Proposition 1.4(ii). ■

1.7. PROPOSITION. Let E,K be extensions of a field F.

(i) If both E/F and K/F are finitely generated, then so is EK/F

(ii) If both E/F and K/F are finite, then so is EK/F

Proof. (i) Write $E = F(\alpha_1,\ldots,\alpha_n)$ and $K = F(\beta_1,\ldots,\beta_m)$. Then

$$EK = F(\alpha_1,\ldots,\alpha_n,\beta_1,\ldots,\beta_m)$$

and so EK/F is finitely generated.

(ii) By Proposition 1.4(iii), both E/F and K/F are finitely generated and algebraic, so by (i), EK/F is finitely generated and, by Corollary 1.5(ii), EK/F is algebraic. Now apply Proposition 1.4(iii). ■

1.8. PROPOSITION. Let E/F and E'/F be two extensions of F and let α and α' be elements of E and E', respectively, which are algebraic over F. Then the following conditions are equivalent:

(i) α and α' have the same minimal polynomial $f(X)$ in $F[X]$

(ii) There exists an F-isomorphism $F(\alpha) \longrightarrow F(\alpha')$ which carries α into α'.

Proof. Assume that (i) holds. The kernel of the surjective homomorphism $F[X] \longrightarrow F(\alpha)$ which sends X to α is $(f(X))$. Hence

$F(\alpha) \longrightarrow F[X]/(f(X))$, $\alpha \longmapsto X + (f(X))$ is an F-isomorphism. On the other hand, $F[X]/(f(X)) \longrightarrow F(\alpha')$, $X + (f(X)) \longmapsto \alpha'$ is also an F-isomorphism. Taking the composite of these isomorphisms, (ii) follows. Conversely, assume that (ii) holds. If $f(X) = X^n + a_1 X^{n-1} + \dots + a_n$ is the minimal polynomial for α, then $\alpha^n + a_1\alpha^{n-1} + \dots + a_n = 0$ and hence $(\alpha')^n + a_1(\alpha')^{n-1} + \dots + a_n = 0$. Thus α' is a root of $f(X)$ and so $f(X)$ is divisible by the minimal polynomial $g(X)$ for α', by Proposition 1.3(ii). A similar argument shows that $g(X)$ is divisible by $f(X)$, hence the result. ■

Let E/F be a field extension. Two elements $\alpha,\beta \in E$ are said to be *F-conjugate* if they are algebraic over F and have the same minimal polynomial over F.

1.9. COROLLARY. Let E/F be a field extension and let $\alpha \in E$ be algebraic over F. If the minimal polynomial for α over F is of degree n, then the number of F-conjugates of α is at most n. Moreover, if $\beta \in E$ is conjugate to α, then $F(\alpha)$ and $F(\beta)$ are F-isomorphic.

Proof. Apply Proposition 1.8 and Corollary 1.2.9. ■

1.10. PROPOSITION. Let f be any nonconstant polynomial over a field F. Then there exists a finite field extension E/F such that f has a root in E.

Proof. Let I be a maximal ideal of $F[X]$ containing (f). Then $E = F[X]/I$ is a field containing an isomorphic copy $(F+I)/I$ of F. Hence we may regard E as a field extension of F and as such it is obviously finite. Since $\alpha = X + I \in E$ is a root of $f(X)$, the required assertion is proved. ■

1.11. COROLLARY. Let F be a field and let $f_1,\dots,f_n$ be polynomials in $F[X]$ of degrees $\geqslant 1$. Then there exists a finite field extension E/F in which each f_i has a root.

Proof. By Proposition 1.10, we may find a finite extension E_1/F in which f_1 has a root. We may view f_2 as a polynomial over E_1. Let E_2/E_1 be a finite extension in which f_2 has a root. Proceeding inductively, the result follows. ■

For future use, we next record the following elementary observations.

1.12 PROPOSITION. Let E/F be an algebraic extension and let $\sigma : E \longrightarrow E$ be an F-homomorphism. Then σ is an automorphism.

Proof. It suffices to verify that σ is surjective. To this end, fix $\alpha \in E$ and denote by $f(X)$ its minimal polynomial over F. If E' is the subfield of E generated by all the roots of $f(X)$ which lie in E, then E'/F is finitely generated and so, by Proposition 1.4(iii), is finite. Moreover, σ maps a root of $f(X)$ to a root of $f(X)$ and therefore maps E' into itself. Since $(E':F) = (\sigma(E'):F)$, we must have $\sigma(E') = E'$. Because $\alpha \in E'$, it follows that $\alpha \in \sigma(E')$, as required. ■

We define a field F to be *algebraically closed* if every polynomial in $F[X]$ of degree $\geqslant 1$ has a root in F. We observe that if F is an algebraically closed field, and $f(X) \in F[X]$ has degree $n \geqslant 1$, then there exists $\lambda \in F$ and $\alpha_1,\ldots,\alpha_n \in F$ such that

$$f(X) = \lambda(X-\alpha_1)\ \ldots\ (X-\alpha_n)$$

Indeed, $f(X)$ has a root α_1 in F, so there exists $g(X) \in F[X]$ such that

$$f(X) = (X-\alpha_1)g(X)$$

If $\deg g \geqslant 1$, we can repeat this argument inductively, and express $f(X)$ in the desired form. Note that λ is the leading coefficient of $f(X)$. Hence if the coefficients of $f(X)$ lie in a subfield K of F, then $\lambda \in K$.

The following result, whose proof is due to Artin, guarantees the existence of algebraically closed fields.

1.13. THEOREM. Let F be a field. Then there exists an algebraically closed field E containing F as a subfield.

Proof. We first construct an extension E_1/F in which every polynomial in $F[X]$ of degree $\geqslant 1$ has a root. To this end, put $S = \{X_f \mid f \in F[X], \deg f \geqslant 1\}$. We form the polynomial ring $F[S]$ and denote by I the ideal generated by all the polynomials $f(X_f)$ in $F[S]$. We claim that $I \neq F[S]$. Indeed, otherwise there is a finite combination of elements in I which is equal to 1:

$$g_1 f_1(X_{f_1}) + \ldots + g_n f_n(X_{f_n}) = 1$$

with $g_i \in F[S]$. Put $X_i = X_{f_i}$ and observe that the polynomials g_i involve actually only a finite number of variables, say $X_1, X_2, \ldots, X_m$, with $m \geq n$. Hence we may rewrite our relation as:

$$\sum_{i=1}^{n} g_i(X_1,\ldots,X_m) f_i(X_i) = 1.$$

Owing to Corollary 1.11, we may choose a finite extension K/F in which each polynomial f_i, $1 \leq i \leq n$, has a root, say α_i. Put $\alpha_i = 0$ for $n < i \leq m$. Substituting α_i for X_i, $1 \leq i \leq m$, in our relation, we get $0 = 1$, a contradiction.

Let M be a maximal ideal of $F[S]$ containing I. Then $E_1 = F[S]/M$ is a field containing an isomorphic copy $(F+M)/M$ of F. Hence we may regard E_1 as a field extension of F. Furthermore, for every polynomial $f \in F[X]$ of degree ≥ 1, $X_f + M \in E_1$ is a root of f.

Inductively, we can form a chain of fields

$$E_1 \subseteq E_2 \subseteq \ldots \subseteq E_n \subseteq \ldots$$

such that each polynomial in $E_n[X]$ of degree ≥ 1 has a root in E_{n+1}. Let E be the union of all fields E_n, $n = 1,2,\ldots$. Then E is obviously a field and every polynomial $f \in E[X]$ has its coefficients in some field E_n. But then f has a root in $E_{n+1} \subseteq E$, as asserted. ■

1.14. COROLLARY. Let F be a field. Then there exists an algebraic extension E/F such that E is algebraically closed.

Proof. By Theorem 1.13, there exists a field extension K/F such that K is algebraically closed. Let E be the algebraic closure of F in K. Then E/F is an algebraic extension. Moreover, if $f(X) \in E[X]$ is a polynomial of degree ≥ 1, then $f(X)$ has a root, say α, in K. Hence α is algebraic over E and therefore, by Proposition 1.5(ii), α is algebraic over F. Thus $\alpha \in E$ and the result follows. ■

Let F be a field. By an *algebraic closure* of F we understand any field E containing F such that E/F is algebraic and E is algebraically closed. Owing to Corollary 1.14, such E always exists. It will be shown below that E

is determined uniquely, up to an F-isomorphism. The following preliminary result will clear our path.

1.15. LEMMA. Let F be a field and let $\sigma : F \longrightarrow L$ be a homomorphism of F into an algebraically closed field L. Let α be an algebraic element in a field extension of F, and let $f(X) \in F[X]$ be the minimal polynomial for α. If t is the number of distinct roots of $f(X)$ (in the algebraic closure of F), then there exist exactly t extensions of σ to a homomorphism $F(\alpha) \longrightarrow L$.

Proof. Let $f^{\sigma}(X) \in L[X]$ be obtained from $f(X)$ by applying σ to the coefficients of $f(X)$. Since L is algebraically closed, there exists a root β of $f^{\sigma}(X)$ in L. Given an element of $F(\alpha) = F[\alpha]$, we can write it in the form $\psi(\alpha)$ with some polynomial $\psi(X) \in F[X]$. We define an extension of σ by mapping

$$\psi(\alpha) \longmapsto \psi^{\sigma}(\beta)$$

This is well defined, i.e. independent of the choice of the polynomial $\psi(X)$ used to express our element in $F[\alpha]$. Indeed, if $\psi_1(X) \in F[X]$ is such that

$$\psi(\alpha) = \psi_1(\alpha),$$

then $(\psi-\psi_1)(\alpha) = 0$, whence $f(X)$ divides $\psi(X) - \psi_1(X)$. Hence $f^{\sigma}(X)$ divides $\psi^{\sigma}(X) - \psi_1^{\sigma}(X)$, and thus $\psi^{\sigma}(\beta) = \psi_1^{\sigma}(\beta)$. It is now clear that our map is a homomorphism $F(\alpha) \longrightarrow L$ inducing σ on F. Now the number of distinct roots of $f(X)$ in the algebraic closure of F is equal to the number of distinct roots of $f^{\sigma}(X) \in L[X]$. Since the image of α under any extension $F(\alpha) \longrightarrow L$ of σ is a root of $f^{\sigma}(X)$, the result follows. ■

1.16. PROPOSITION. Let E/F be an algebraic extension and let $\sigma : F \longrightarrow L$ be a homomorphism into an algebraically closed field L.

(i) There exists an extension σ' of σ to a homomorphism $E \longrightarrow L$

(ii) If E is algebraically closed and L is algebraic over $\sigma(F)$, then σ' is an isomorphism.

Proof. Let us show first that (ii) is a consequence of (i). Indeed, if E is algebraically closed, and L is algebraic over $\sigma(F)$, then $\sigma'(E)$ is algebraically closed and L is algebraic over $\sigma'(E)$, hence $L = \sigma'(E)$.

To prove (i), denote by S the set of all pairs (K,λ) where K is a subfield of E containing F, and λ is an extension of σ to a homomorphism $K \longrightarrow L$. If (K,λ) and (K',λ') are in S, we write $(K,\lambda) \leqslant (K',\lambda')$ if $K \subseteq K'$ and $\lambda'|K = \lambda$. Note that $(F,\sigma) \in S$, so $S \neq \emptyset$, and S is inductively ordered: If $\{(K_i,\lambda_i)\}$ is a totally ordered subset, we let $K = \cup K_i$ and define λ on K to be equal to λ_i on each K_i. Then (K,λ) is an upper bound for the totally ordered subset. Owing to Zorn's lemma, we may therefore choose a maximal element, say (T,μ), in S. Then μ is an extension of σ, and we claim that $T = E$. Otherwise, there exists $\alpha \in E-T$ and, by Lemma 1.15, μ has an extension to $T(\alpha)$, thereby contradicting the maximality of (T,μ). This proves that there exists an extension of σ to E, as required. ■

1.17. COROLLARY. Let F be a field. Then any two algebraic closures of F are F-isomorphic.

Proof. Let E and E' be algebraic closures of F and let $\sigma : F \longrightarrow E'$ be the inclusion map. Then, by Proposition 1.16(ii), σ extends to an isomorphism $E \longrightarrow E'$, as required. ■

1.18. COROLLARY. Let E/F be an algebraic field extension and let $\bar{F}$ be the algebraic closure of F containing E. Then any F-homomorphism $E \longrightarrow \bar{F}$ is extendible to an automorphism of $\bar{F}$.

Proof. Since $\bar{F}/E$ is an algebraic extension and $\bar{F}$ is algebraically closed, the result follows by virtue of Propositions 1.16(i) and 1.12. ■

We close this section by proving the following observation.

1.19 PROPOSITION. Let E/F be an algebraic field extension and let F be infinite. Then $|E| = |F|$.

Proof. Let S be the subset of $F[X]$ consisting of monic polynomials of positive degree and, for any $n \geqslant 0$, let $S(n)$ be the subset of S consisting of all polynomials of degree $n+1$. The elements of $S(n)$ have the form $X^{n+1} + \lambda_1 X^n + \dots + \lambda_n$, $\lambda_i \in F$, so

$$|S(n)| = |\underbrace{F \times \ldots \times F}_{\leftarrow n \text{ factors} \rightarrow}| = |F|$$

Also

$$|S| = |\bigcup_{n \geqslant 0} S(n)| = |F|$$

We now map each $f(X) \in S$ into the finite set K_f (possibly empty) of its roots in E. Because every element of E is algebraic,

$$E = \bigcup_{f \in S} K_f$$

Since each K_f is finite the cardinality of the collection $\{K_f\}$ is the same as $|E|$. Therefore $|E| = |\{R_f\}| \leqslant |S| = |F|$ which clearly implies that $|E| = |F|$.■

2. NORMAL EXTENSIONS

Let F be a field and let $\{f_i | i \in I\}$ be a family of polynomials in $F[X]$ of degree $\geqslant 1$. By a *splitting field* for this family we understand an extension E of F such that every f_i splits into linear factors in $E[X]$, and E is generated by all the roots of all the polynomials $f_i, i \in I$. We note that any family of polynomials in $F[X]$ always has a splitting field, namely the field generated by the roots of the given family in any algebraic closure of F. It turns out that the splitting field of $\{f_i | i \in I\}$ is determined uniquely up to an F-isomorphism, as we shall see below.

2.1. LEMMA. Let $f(X) \in F[X]$ and let E, K be two splitting fields of $f(X)$.

(i) E and K are F-isomorphic

(ii) If $F \subseteq E \subseteq \bar{F}$ where $\bar{F}$ is an algebraic closure of F, then any F-homomorphism $K \longrightarrow \bar{F}$ is an isomorphism of K onto E.

Proof. (i) Let $\bar{E}$ be an algebraic closure of E. Then $\bar{E}$ is algebraic over F, hence is an algebraic closure of F. Owing to Proposition 1.16, there exists an F-homomorphism $\sigma : K \longrightarrow \bar{E}$. We are therefore left to verify that

$$\sigma(K) = E \tag{1}$$

We have a factorization

$$f(X) = \lambda(X-\beta_1)\ldots(X-\beta_n)$$

with $\beta_i \in K$, $1 \leqslant i \leqslant n$, $\lambda \in F$. Then

$$f(X) = f^{\sigma}(X) = \lambda(X-\sigma(\beta_1))\ldots(X-\sigma(\beta_n))$$

Hence $\alpha_1,\ldots,\alpha_n$ with $\alpha_i = \sigma(\beta_i)$, $1 \leqslant i \leqslant n$, are all roots of $f(X)$ in E. Since $K = F(\beta_1,\ldots,\beta_n)$ and $E = F(\alpha_1,\ldots,\alpha_n)$, we deduce that $\sigma(K) = E$, as required.

(ii) Since $\bar{F}$ is an algebraic closure of E, it suffices to put $\bar{E} = \bar{F}$ and apply (1). ■

2.2. PROPOSITION. Let $\{f_i | i \in I\}$ be a family of polynomials in $F[X]$ and let K,E be two splitting fields for this family. Then K and E are F-isomorphic.

Proof. Let $\bar{K}$ be an algebric closure of K. Then, by Proposition 1.16, there is an injective F-homomorphism $\sigma : E \longrightarrow \bar{K}$. Note that E contains a unique splitting field E_i of f_i and K contains a unique splitting field K_i of f_i. By Lemma 2.1(ii), $\sigma(E_i) = K_i, i \in I$, and since K is generated by all the K_i, we have $\sigma(E) = K$. ■

We next present two applications of splitting fields of polynomials. Let R be an integral domain with quotient field F. The procedure for determining whether or not an element in a field extension of F is integral over R might generally be a lengthy one since there is no clear way to select the polynomial which expresses integral dependence. The following result provides a circumstance where this procedure is simplified.

2.3. PROPOSITION. Let R be a subring of a field F, let E/F be a field extension and let $\lambda \in E$ be algebraic over F with minimal polynomial

$$f(X) \in F[X]$$

(i) If λ is integral over R, then the coefficients of $f(X)$ are integral over R

(ii) If F is the quotient field of R and R is integrally closed, then λ is integral over R if and only if $f(X) \in R[X]$.

Proof. It is clear that (ii) is a consequence of (i). To prove (i), let

K denote a splitting field of $f(X)$. Let $\lambda_1 = \lambda, \lambda_2, \ldots, \lambda_n$ be all roots of $f(X)$ in K and let $g(X) \in R[X]$ be a monic polynomial with $g(\lambda) = 0$. Since $f(X)$ is irreducible and has a root in common with $g(X)$, it follows from Proposition 1.3(ii) that $f(X)$ divides $g(X)$ in $F[X]$. Hence the roots $\lambda_1, \ldots, \lambda_n$ of $f(X)$ are also roots of $g(X)$. So $\lambda_1, \ldots, \lambda_n$ are integral over R. It follows that the coefficients of $f(X) = (X-\lambda_1)\ldots(X-\lambda_n)$ are integral over R. ∎

2.4. PROPOSITION. Let R be an integrally closed integral domain with quotient field F. Let $f(X) \in R[X]$ be a monic polynomial, and suppose that

$$f(X) = g(X)h(X),$$

where $g(X)$ and $h(X)$ are monic polynomials in $F[X]$. Then both $g(X)$ and $h(X)$ lie in $R[X]$.

Proof. Let E be a finite extension of F which is a splitting field for $f(X)$, and write

$$f(X) = \prod_{i=1}^{n}(X-\alpha_i) \qquad \alpha_i \in E$$

Let S denote the integral closure of R in E. Since $f(X) \in R[X]$ is monic, and $f(\alpha_i) = 0$ for each i, it follows that each $\alpha_i \in S$.

In $E[X]$ we have

$$g(X)h(X) = \prod_{i=1}^{n}(X-\alpha_i)$$

whence both $g(X)$ and $h(X)$ are expressible as partial products $\prod'(X-\alpha_i)$. Thus both $g(X)$ and $h(X)$ have coefficients in S. It therefore follows that the coefficients of $g(X)$ and $h(X)$ are in $S \cap F$. But $S \cap F = R$, since R is integrally closed. Hence both $g(X)$ and $h(X)$ lie in $R[X]$, as asserted. ∎

We next provide a useful characterization of splitting fields of families of polynomials.

2.5. PROPOSITION. Let E be an algebraic extension of F contained in an algebraic closure $\bar{F}$ of F. Then the following conditions are equivalent:

(i) Every F-homomorphism $\sigma : E \longrightarrow \bar{F}$ is an automorphism of E.

(ii) E is the splitting field of a family of polynomials in $F[X]$

(iii) Every irreducible polynomial of $F[X]$ which has a root in E splits into linear factors in $E[X]$

(iv) Every conjugate of an element of E is in E.

Proof. The equivalence of (iii) and (iv) is a consequence of the definition of conjugacy. We now show that (i) ⇒ (ii) and (i) ⇒ (iii). Indeed, assume that (i) holds. Given an element $\lambda \in E$, let $f_\lambda(X)$ be its minimal polynomial over F. Let μ be a root of $f_\lambda(X)$ in $\bar{F}$. There exists an F-isomorphism

$$F(\lambda) \longrightarrow F(\mu)$$

sending λ to μ. By Proposition 1.16(i), we may extend this isomorphism to an F-homomorphism $E \longrightarrow \bar{F}$. This extension is an automorphism σ of E by hypothesis, hence $\sigma(\lambda) = \mu$ lies in E. Therefore every root of $f_\lambda(X)$ lies in E, and $f_\lambda(X)$ splits into linear factors in $E[X]$. Thus E is a splitting field of the family $\{f_\lambda(X)\}$, $\lambda \in E$, which implies (ii) and (iii).

Assume that (ii) holds and let $\{f_i(X) | i \in I\}$ be the family of polynomials of which E is a splitting field. If α is a root of some f_i in E, then for any F-homomorphism $\sigma : E \longrightarrow \bar{F}$, $\sigma(\alpha)$ is also a root of f_i. Because E is generated by the roots of all the polynomials f_i, it follows that σ maps E into itself. This proves (i), by applying Proposition 1.12.

Finally, assume that (iii) holds. Let $\sigma : E \longrightarrow \bar{F}$ be an F-homomorphism, let $\alpha \in E$ and let $f(X)$ be its minimal polynomial over F. Then $\sigma(\alpha)$ is a root of $f(X)$, hence $\sigma(\alpha) \in E$. Therefore σ maps E into itself. By Proposition 1.12, it follows that σ is an automorphism. ■

2.6. REMARK. If E/F is a finite extension, then condition (ii) of Proposition 2.5 is equivalent to the requirement that E is a splitting field of a single polynomial in $F[X]$. Indeed, if E is a splitting field of $\{f_i | i \in I\}$ and, for each finite product g of the f_i,$F(g)$ is the splitting field for g contained E, then

$$F \subseteq F(f_1) \subseteq F(f_1 f_2) \subseteq \cdots \subseteq F(f_1 \cdots f_n) \subseteq \cdots$$

is an ascending chain of subfields of E. Hence $E = F(f_1 \cdots f_n)$ for some $n \geqslant 1$. ■

Let E/F be an algebraic field extension. We say that E is *normal* over F if it satisfies the equivalent conditions of Proposition 2.4.

Note that if $F \subseteq E \subseteq K$ is a chain of fields such that both E/F and K/E are normal, then K/F need not be normal. For example, in the chain

$$\mathbb{Q} \subseteq \mathbb{Q}(\sqrt{2}) \subseteq \mathbb{Q}(\sqrt[4]{2})$$

both $\mathbb{Q}(\sqrt{2})/\mathbb{Q}$ and $\mathbb{Q}(\sqrt[4]{2})/\mathbb{Q}(\sqrt{2})$ are normal, but $\mathbb{Q}(\sqrt[4]{2})/\mathbb{Q}$ is not normal since the complex roots of $X^4 - 2$ are not in $\mathbb{Q}(\sqrt[4]{2})$. Nevertheless we do have the following properties:

2.7. PROPOSITION. (i) If $F \subseteq E \subseteq K$ and K/F is normal, then K/E is also normal

(ii) If K_1/F and K_2/F are normal and K_1, K_2 are subfields of a field E, then both K_1K_2 and $K_1 \cap K_2$ are normal over F.

Proof. (i) Let $\bar{E}$ be an algebraic closure of E containing K and let $\sigma : K \longrightarrow \bar{E}$ be an E-homomorphism. Then σ is an F-homomorphism and $\bar{E}$ is an algebraic closure of F containing K. Hence, by Proposition 2.5(i), σ is an automorphism of K. Thus K/E is normal, again by Proposition 2.5(i).

(ii) Let $\bar{F}$ be an algebraic closure of F containing K_1K_2, and let

$$\sigma : K_1K_2 \longrightarrow \bar{F}$$

be an F-homomorphism. Then, by Proposition 2.5(i), σ restricts to an automorphism of K_i, $i = 1,2$, hence to automorphism of K_1K_2 since

$$\sigma(K_1K_2) = \sigma(K_1)\sigma(K_2).$$

Thus, by Proposition 2.5(i), K_1K_2/F is normal. The fact that $K_1 \cap K_2$ is normal over F is an immediate consequence of Proposition 2.5(iii). ■

Let E/F be an algebraic field extension. We define a *normal closure* of E/F as a normal extension of F containing E and having the property that no proper subfield containing E is normal over F.

2.8. PROPOSITION. Let E/F be an algebraic field extension. A normal closure of E/F exists and is unique, up to an F-isomorphism. Furthermore, if E/F is finite, then so are the normal closures of E/F.

Proof. For each $\lambda \in E$, let $f_\lambda(X) \in F[X]$ be the minimal polynomial of λ and let K be a splitting field of the family $\{f_\lambda | \lambda \in E\}$. Then $F \subseteq E \subseteq K$ and K/F is normal, by Proposition 2.5. Furthermore, if K_1 is such that

$$F \subseteq E \subseteq K_1 \subseteq K$$

and K_1/F is normal, then by Proposition 2.5, $K_1 = K$. This shows that K is a normal closure of E/F.

Let $F \subseteq E \subseteq S$ where S is a normal closure of E/F. Then obviously S is a splitting field of the family $\{f_\lambda | \lambda \in E\}$. Hence S and K are F-isomorphic, by Proposition 2.2.

Finally, if E/F is finite, then $E = F(\alpha_1, \ldots, \alpha_n)$ for some $n \geqslant 1$ and some $\alpha_1, \ldots, \alpha_n$ in E. Adjoining to E all conjugates of α_i, $1 \leqslant i \leqslant n$, we obtain a normal closure of E/F which is finite-dimensional over F. ∎

The following simple observation is often useful.

2.9. PROPOSITION. Let $F \subseteq K \subseteq E$ be a chain of fields, where E/F is normal, and let $\bar{F}$ be an algebraic closure of F containing E. Then every F-homomorphism of K into $\bar{F}$ is extendible to an F-automorphism of E. In particular every F-automorphism of K extends to an F-automorphism of E.

Proof. By Corollary 1.18, every F-homomorphism $K \longrightarrow \bar{F}$ is extendible to an automorphism of $\bar{F}$; its restriction to E provides, by virtue of Proposition 2.5, the desired F-automorphism of E. ∎

2.10. PROPOSITION. Let $F \subseteq K \subseteq E \subseteq L$ be a chain of fields, where L/F is a finite normal extension. If the number of K-homomorphisms $E \longrightarrow L$ is equal to n, then every F-homomorphism $K \longrightarrow L$ has exactly n extensions to homomorphisms $E \longrightarrow L$.

Proof. Let G be the group of all F-automorphisms of L and let $G(K)$ (respectively, $G(E)$) be the subgroup of G consisting of those automorphisms of L which leave fixed every element of K (respectively, of E). It is clear that $G(E)$ is a subgroup of $G(K)$. Let

$$G(K) = \bigcup_{i=1}^{n} \phi_i G(E) \quad \text{and} \quad G = \bigcup_{j=1}^{m} \psi_j G(K)$$

be cosets decompositions of $G(K)$ and G. Then

$$G = \bigcup_{i,j} \psi_j\phi_i G(E)$$

is a coset decomposition of G with respect to $G(E)$. It is clear that the m automorphisms ψ_j have distinct restrictions to K and that the restriction of any element ψ of G to K coincides with the restriction of one of the ψ_j. Since by Proposition 2.9, every F-homomorphism of K into L is the restriction of some automorphism of L, it follows that K has exactly m F-homomorphisms into L and that these are given by the restrictions of $\psi_1,\ldots,\psi_m$ to K.

A similar argument shows that E has exactly mn F-homomorphisms into L and that these are given by the restrictions of the mn products $\psi_j\phi_i$ to E. Now each ϕ_i restricts to the identity on K and ψ_j and ψ_k have distinct restrictions to K if $j \neq k$. Thus each F-homomorphism of K into L, say the one represented by the restriction of ψ_j, has exactly n extensions to E which are F-homomorphisms of E into L, namely the restrictions of

$$\psi_j\phi_1, \psi_j\phi_2, \ldots, \psi_j\phi_n$$

to E. ∎

3. SEPARABLE, PURELY INSEPARABLE AND SIMPLE EXTENSIONS

Let F be a field and f a polynomial in $F[X]$. If λ is a root of f in F, then

$$f(X) = (X-\lambda)^m g(X)$$

for some $g(X) \in F[X]$ with $g(\lambda) \neq 0$. We call m the *multiplicity* of λ in F, and say that λ is a *multiple root* if $m > 1$ and a *simple root* if $m = 1$. Our aim is to be able to test whether f has multiple roots in some field containing F, without having to go outside F. To this end, we first define derivatives formally.

Given any polynomial $f(X) = a_0 + a_1X + \ldots + a_nX^n$ over a commutative ring R, we define its *derivative* Df or f' by the rule

$$f'(X) = a_1 + 2a_2X + \ldots + na_nX^{n-1}$$

The mapping

$$\begin{cases} R[X] \longrightarrow R[X] \\ f \longmapsto f' \end{cases}$$

enjoys the following properties

(i) $(f+g)' = f' + g'$ $\qquad (f,g \in R[X])$

(ii) $(rf)' = rf'$ $\qquad (r \in R)$

(iii) $(fg)' = f'g + fg'$

(iv) $X' = 1$.

Conversely, f' is completely determined by (i) - (iv) and this provides an alternative definition.

3.1. PROPOSITION. Let F be a field, let $f \in F[X]$ and let λ be any root of f in some field extension of F. Then λ is a multiple root if and only if $f'(\lambda) = 0$.

Proof. Dividing f by $(X-\lambda)^2$ we obtain

$$f(X) = (X-\lambda)^2 g(X) + h(X)$$

where $h(X) = 0$ or $\deg h(X) \leqslant 1$. Putting $X = \lambda$, we see that $f(\lambda) = h(\lambda) = 0$. Differentiating and putting $X = \lambda$, we find that $h'(\lambda) = f'(\lambda)$. Hence

$$f(X) = (X-\lambda)^2 g(X) + f'(\lambda)(X-\lambda)$$

which shows that $(X-\lambda)^2 | f(X)$ if and only if $f'(\lambda) = 0$. ■

3.2. COROLLARY. Let $f \in F[X]$ with $\deg f \geqslant 1$. Then all the roots of f (in splitting field of f) are simple if and only if f is prime to its derivative f'. In particular, if f is irreducible, then all the roots of f are simple if and only if $f' \neq 0$.

Proof. By Proposition 3.1, f has a multiple root if and only if f and f' have a common factor. ■

3.3. PROPOSITION. Let F be a field and let $f(X) \in F[X]$ be of degree $\geqslant 1$.

(i) If $\mathrm{char} F = 0$, then $f' \neq 0$

(ii) If $\mathrm{char} F = p > 0$, then $f' = 0$ if and only if f has the form $g(X^p)$.

Proof. (i) Write $f(X) = a_0 + a_1X + \dots + a_nX^n$ with $a_n \neq 0$, $n \geqslant 1$. If $f' = 0$, then $na_n = 0$, which is impossible.

(ii) If f has the form $g(X^p)$, then obviously $f' = 0$. Conversely, assume that $f' = 0$. Then $ia_i = 0$, $1 \leqslant i \leqslant n$, and hence $a_i = 0$ if $p \nmid i$. Thus f has the form

$$f(X) = a_0 + a_pX^p + a_{2p}X^{2p} + \dots + a_{rp}X^{rp},$$

in other words, f is now a polynomial in X^p. ■

In what follows by the roots of a polynomial we mean those lying in a splitting field of that polynomial.

3.4. COROLLARY. Let f be an irreducible polynomial over a field F.

(i) If $\text{char}F = 0$, then f has only simple roots

(ii) If $\text{char}F = p > 0$, then f has multiple roots if and only if f has the form $g(X^p)$.

Proof. (i) Apply Corollary 3.2 and Proposition 3.3(i).

(ii) Apply Corollary 3.2 and Proposition 3.3(ii). ■

The next observation is a useful companion to Corollary 3.4(ii).

3.5. PROPOSITION. Let F be a field of characteristic $p > 0$ and let $f(X) \in F[X]$ be an irreducible polynomial. Choose the nonnegative integer e such that

$$f(X) \in F[X^{p^e}] \quad \text{but} \quad f(X) \notin F[X^{p^{e+1}}]$$

(i) If $f(X) = \psi(X^{p^e})$, then ψ is irreducible and has only simple roots.

(ii) All roots of $f(X)$ have the same multiplicity equal to p^e and $\deg\psi$ is equal to the number of distinct roots of $f(X)$. In particular,

$$\deg f = (\deg\psi)p^e$$

Proof. (i) It is clear that ψ is irreducible. Moreover, $\psi' \neq 0$, since otherwise $\psi(y)$ is of the form $g(y^p)$ and hence $f(X) \in F[X^{p^{e+1}}]$, contrary to the assumption. Hence, by Corollary 3.2, ψ has only simple roots.

(ii) By (i), we may split $\psi(y)$ into distinct linear factors (in a suitable extension of F):

$$\psi(y) = \prod_{i=1}^{m}(y-\beta_i)$$

Then

$$f(X) = \prod_{i=1}^{m}(X^{p^e}-\beta_i)$$

Let α_i be a root of $X^{p^e} - \beta_i$. Then

$$\alpha_i^{p^e} = \beta_i \qquad (\alpha_1,\ldots,\alpha_m \text{ are distinct })$$

$$X^{p^e} - \beta_i = X^{p^e} - \alpha_i^{p^e} = (X-\alpha_i)^{p^e}$$

Hence

$$f(X) = \prod_{i=1}^{m}(X-\alpha_i)^{p^e}$$

as required. ∎

The degree of the polynomial ψ (in the notation of Proposition 3.5) is called the *separable degree* of $f(X)$, while e is called the *exponent of inseparability* of $f(X)$. If α is an algebraic element over F with minimal polynomial $f(X)$, then the *separable degree* of α over F (respectively, the *exponent of inseparability* of α over F) is defined to be the separable degree of $f(X)$ (respectively, the exponent of inseparability of $f(X)$). In the case where char$F = 0$, the separable degree of α is defined to be the degree of α. Thus, in all cases, the separable degree of α is equal to the number of distinct roots of the minimal polynomial of α.

Let E/F be a finite extension and let K be an algebraic closure of E. The *separable degree* $(E:F)_s$ of E/F is defined to be the number of F-homomorphisms of E into K. This is obviously independent of the choice of K.

3.6. LEMMA. Let $F(\alpha)/F$ be a finite extension. Then $(F(\alpha):F)_s$ is equal to the separable degree of α over F and is also equal to the number of F-homomorphisms $F(\alpha) \longrightarrow E$ where E is any normal extension of F containing α.

Proof. The first assertion is a particular case of Lemma 1.15 in which σ is equal to the inclusion map. To prove the second assertion, let K be an algebraic closure of E. Then K is obviously an algebraic closure of $F(\alpha)$. Since E/F is normal, the image of any F-homomorphism $F(\alpha) \longrightarrow K$ is contained in E, as required. ■

We now introduce a property of field extensions which will play an important role in our subsequent investigations.

Let F be a field and let $f \in F[X]$. We say that f is *separable over* F if all irreducible factors of f over F have only simple roots (in a splitting field of f). In particular, an irreducible polynomial is separable if and only if its roots are simple. An algebraic element α of a field extension E of F is called *separable* over F if its minimal polynomial over F is separable. If E/F is an algebraic extension such that all elements of E are separable over F, we call E/F a *separable extension*. It is an immediate consequence of the definition that α is separable over F if and only if the degree and the separable degree of α over F coincide.

3.7. PROPOSITION. Let F be a field of characteristic 0. Then all polynomials over F are separable. In particular, any algebraic extension of F is separable.

Proof. Apply Corollary 3.4(i). ■

3.8. PROPOSITION. Let F be a field of characteristic p, let α be an algebraic element in a field extension of F and let e be the exponent of inseparability of α. Then

(i) $(F(\alpha):F) = p^e(F(\alpha):F)_s$

(ii) α^{p^e} is separable over F

(iii) α is separable over F if and only if $F(\alpha) = F(\alpha^p)$.

Proof. (i) Let $f(X) \in F[X]$ be the minimal polynomial of α and let m be the separable degree of α. Then

$$(F(\alpha):F) = \deg f(X)$$

and so $(F(\alpha):F) = p^e m$, by Proposition 3.5(ii). Now apply Lemma 3.6.

(ii) In the notation of the proof of Proposition 3.5, α^{p^e} is a root of the irreducible polynomial ψ which has only simple roots. Hence α^{p^e} is separable over F.

(iii) Assume that α is not separable. Then, by Proposition 3.5(ii),

$f(X) = g(X^p)$ and α^p is a root of $g(X)$. Hence

$$(F(\alpha^p):F) \leqslant \deg g(X) < \deg f(X) = (F(\alpha):F)$$

which shows that $F(\alpha^p) \neq F(\alpha)$.

Now suppose α is separable so that $f(X)$ has distinct roots. Let $g(X)$ be the minimal polynomial of α over $F(\alpha^p)$. Then $g(X)|f(X)$, so $g(X)$ has distinct roots. Also α is a root of the polynomial $X^p-\alpha^p$ over $F(\alpha^p)$, so $g(X)$ divides $X^p-\alpha^p = (X-\alpha)^p$. Since $g(X)$ has distinct roots, this implies that $g(X) = X-\alpha$. Hence $\alpha \in F(\alpha^p)$ and therefore $F(\alpha) = F(\alpha^p)$. ■

The next observation will enable us to take full advantage of the results so far obtained.

3.9. LEMMA. Let $F \subseteq K \subseteq E$ be a chain of fields, where E is an algebraic closure of F and let $\lambda : K \longrightarrow E$ be an F-homomorphism. Then, for any $\alpha \in E$, there exist exactly $(K(\alpha):K)_s$ F-homomorphisms $K(\alpha) \longrightarrow E$ extending λ.

Proof. Let $f(X)$ be the minimal polynomial of α over K. By Lemma 3.6, $(K(\alpha):K)_s$ is equal to the separable degree of α over K, hence to the number of distinct roots of $f(X)$ in E. Now apply Lemma 1.15. ■

As an application of the above result, we now prove

3.10. PROPOSITION. Let K/F be a finite field extension, say $K = F(\alpha_1,\ldots,\alpha_r)$.

(i) $(K:F)_s = \prod_{i=1}^{r} (F(\alpha_1,\ldots,\alpha_i) : F(\alpha_1,\ldots,\alpha_{i-1}))_s$

(ii) $(K:F)_s = (K:F)$ if and only if each α_i is separable over $F(\alpha_1,\ldots,\alpha_{i-1})$

(iii) If $\operatorname{char} F = p > 0$ and e_i is the exponent of inseparability of α_i over $F(\alpha_1,\ldots,\alpha_{i-1})$, then

$$(K:F) = (K:F)_s p^{e_1+\cdots+e_r}$$

Proof. (i) Put $n_i = (F(\alpha_1,\ldots,\alpha_i) : F(\alpha_1,\ldots,\alpha_{i-1}))_s$ and

$$m_i = (F(\alpha_1,\ldots,\alpha_i) : F(\alpha_1,\ldots,\alpha_{i-1})), \quad 1 \leqslant i \leqslant r.$$

Then, by Lemma 3.6, n_i is equal to the separable degree of α_i over $F(\alpha_1,\ldots,\alpha_{i-1})$. This, together with Proposition 1.1, ensures that

$$(K:F) = \prod_{i=1}^{r} m_i \quad \text{and} \quad n_i \leqslant m_i \quad (1 \leqslant i \leqslant r) \tag{1}$$

The case $r = 1$ being obvious, we argue by induction on r. If $K_1 = F(\alpha_1,\ldots,\alpha_{r-1})$ and E is an algebraic closure of K(hence of K_1), then by induction there exist exactly $n_1 n_2 \ldots n_{r-1}$ F-homomorphisms $K_1 \longrightarrow E$. Since $K = K_1(\alpha_r)$, it follows from Lemma 3.9 that each F-homomorphism $K_1 \longrightarrow E$ has exactly n_r extensions to F-homomorphisms $K \longrightarrow E$. Because the restriction of an F-homomorphism $K \longrightarrow E$ to K_1 is an F-homomorphism $K_1 \longrightarrow E$, it follows that $(K:F)_s = n_1 \ldots n_r$, as required.

(ii) Direct consequence of Lemma 3.6, (i) and (1).

(iii) By Proposition 3.8(i), $m_i = p^{e_i} n_i$, hence by (1) and (i),

$$(K:F) = \prod_{i=1}^{r} m_i = \left(\prod_{i=1}^{r} n_i\right) p^{e_1+\ldots+e_r} = (K:F)_s p^{e_1+\ldots+e_r}$$

as required. ■

The result above has a number of important applications. Before putting it to use, we make the following observation:

If $F \subseteq E \subseteq K$ is a chain of fields and $\alpha \in K$ is separable over F, then α is separable over E (2)

Indeed, if $f(X) \in F[X]$ is the minimal polynomial of α and $f(X) = f_1(X)\ldots f_n(X)$ with irreducible $f_i(X) \in E[X]$, then α is a root of some $f_i(X)$ and $f_i(X)$ must have only simple roots since so does $f(X)$. Thus α is separable over E.

3.11. COROLLARY. Let E/F be a field extension. Then the set of all elements of E which are separable over F is a subfield of E containing F. In particular, if E is generated over F by a family of separable elements, then E/F is separable.

Proof. Assume that α_1, α_2 are separable over F and let

$$\lambda \in \{\alpha_1\alpha_2, \alpha_1+\alpha_2, \alpha_1^{-1}\}, \alpha_1 \neq 0.$$

Then $F(\alpha_1,\alpha_2) = F(\lambda,\alpha_1,\alpha_2)$ and α_2 is separable over $F(\alpha_1)$ by (2). Hence, by Proposition 3.10(ii), λ is separable over F. Since all elements of F are obviously separable over F, the result follows. ■

Let E/F be a field extension. By the *separable closure* of F in E we understand the subfield of E consisting of all elements which are separable over

F. If $\bar{F}$ is an algebraic closure of F, then the separable closure of F in $\bar{F}$ is called the *separable closure* of F. Due to Corollary 1.17, the separable closure of F is uniquely determined up to an F-isomorphism.

Let K/F be a finite extension. We define the inseparable degree $(K:F)_i$ of K over F by

$$(K:F)_i = (K:F)/(K:F)_s$$

If $\text{char}F = 0$, then by Propositions 3.10(ii) and 3.7, $(K:F)_i = 1$, while if $\text{char}F = p > 0$, then by Proposition 3.10(iii), $(K:F)_i$ is a power of p.

3.12. COROLLARY. (i) If K/F is a finite extension, then $(K:F)_s = (K:F)$ if and only if K/F is separable

(ii) If $F \subseteq E \subseteq K$ is a chain of fields and K/F is finite, then

$$(K:F)_s = (K:E)_s(E:F)_s \quad \text{and} \quad (K:F)_i = (K:E)_i(E:F)_i$$

Proof. (i) Write $K = F(\alpha_1,\ldots,\alpha_r)$ where each α_i is algebraic over F. Then, by Proposition 3.10(ii), the condition that each α_i is separable over $F(\alpha_1,\ldots,\alpha_{i-1})$ is equivalent to the condition that each α_i is separable over F (simply rewrite $F(\alpha_1,\ldots,\alpha_r)$ as $F(\alpha_i,\alpha_1,\ldots,\alpha_{i-1},\alpha_{i+1},\ldots,\alpha_r)$. The latter condition, in view of Corollary 3.11, is equivalent to K/F being separable. Now apply Proposition 3.10(ii).

(ii) Apply Proposition 3.10(i),(iii). ■

3.13. COROLLARY. If F is a field of characteristic $p > 0$ and E a finite field extension of F with $p \nmid (E:F)$, then E/F is separable.

Proof. Owing to Proposition 3.10(iii), the assumption that $p \nmid (E:F)$ forces $(E:F)_s = (E:F)$. Now apply Corollary 3.12. ■

3.14. COROLLARY. (i) If $F \subseteq E \subseteq K$ is a chain of fields, then K/F is separable if and only if both E/F and K/E are separable.

(ii) If E/F is separable and K/F is any extension with both E and K contained in some field, then KE/K is separable.

Proof. (i) If K/F is separable, then obviously both E/F and K/E are separable. Assume that E/F and K/E are separable. If K/F is finite, then the result is true by virtue of the second equality in Corollary 3.12(ii). If K/F is infinite, let $\alpha \in K$. Then α is a root of a separable polynomial $f(X) \in E[X]$. Let $a_0, a_1, \ldots, a_n$ be the coefficients of $f(X)$ and let

$$E_0 = F(a_0, \ldots, a_n).$$

Then we have $F \subseteq E_0 \subseteq E_0(\alpha)$ where $E_0(\alpha)/F$ is finite and both E_0/F and $E_0(\alpha)/E_0$ are separable. Hence $E_0(\alpha)/F$ is separable and therefore α is separable over F, proving that K/F is separable.

(ii) Every element of E is separable over F, hence separable over K. Since KE is generated over K by the elements of E, it follows that KE/K is separable by Corollary 3.11. ∎

If $F \subseteq K \subseteq E$ is a chain of fields with E/F normal, then E/K is normal but K/F need not be normal. The following result provides a condition under which K/F is normal.

3.15. PROPOSITION. Let E/F be a normal field extension and let F_s be the separable closure of F in E. Then F_s/F is normal.

Proof. Let $\bar{E}$ be an algebraic closure of E and let $\sigma : F_s \longrightarrow \bar{E}$ be an F-homomorphism. By Proposition 1.16, we may extend σ to an F-homomorphism $E \longrightarrow \bar{E}$. Since E/F is normal, we see that $\sigma(E) = E$, so σ is an F-automorphism of E. Furthermore, $\sigma(F_s)$ is separable over F, hence is contained in F_s. Thus $\sigma(F_s) = F_s$ and, by Proposition 2.5, F_s/F is normal. ∎

We next provide a class of fields over which all extensions are separable.

Let F be a field of characteristic $p > 0$ and let $F^p = \{x^p \mid x \in F\}$. Then the mapping

$$\begin{cases} F \longrightarrow F \\ x \longmapsto x^p \end{cases}$$

is obviously an injective homomorphism whose image is the subfield F^p of F. If it is also surjective (i.e. if $F = F^p$), and hence an automorphism, the field F is said to be *perfect*. Thus F is perfect if and only if every element is a p-th power. In addition, every field of characteristic 0 is perfect by definition.

Let F be a field of characteristic $p > 0$ and let $\bar{F}$ be an algebraic closure of F. For each $n \geq 1$, define

$$F^{p^{-n}} = \{x \in \bar{F} | x^{p^n} \in F\}$$

Then $F^{p^{-n}}$ is obviously a subfield of $\bar{F}$ and we have a chain

$$F \subseteq F^{p^{-1}} \subseteq F^{p^{-2}} \subseteq \cdots \subseteq F^{p^{-n}} \subseteq \cdots$$

of subfields of $\bar{F}$. Put

$$F^{p^{-\infty}} = \bigcup_{n=1}^{\infty} F^{p^{-n}}$$

3.16. LEMMA. With the notation above, $F^{p^{-\infty}}$ is the smallest perfect subfield of $\bar{F}$ containing F. Furthermore, if $F \neq F^{p^{-\infty}}$, then $F^{p^{-\infty}}/F$ is an infinite extension.

Proof. Let K be a perfect subfield of $\bar{F}$ containing F. If $x \in F^{p^{-n}}$ then $x^{p^n} \in F \subseteq K$, so $x^{p^n} = \lambda^{p^n}$ for some $\lambda \in K$ since K is perfect. Hence $x = \lambda \in K$ and therefore $F^{p^{-n}} \subseteq K$, proving that $F^{p^{-\infty}} \subseteq K$. The fact that $F^{p^{-\infty}}$ is perfect is a consequence of the equality

$$(F^{p^{-n}})^p = F^{p^{-(n-1)}} \qquad (n \geq 1)$$

Finally, assume that $F \neq F^{p^{-\infty}}$. If $F^{p^{-\infty}}/F$ is finite, then $F^{p^{-\infty}} = F^{p^{-n}}$ for some $n \geq 1$, hence $F = (F^{p^{-n}})^{p^n} = F^{p^{-n}}$, a contradiction. ∎

We shall refer to the field $F^{p^{-\infty}}$ as the *perfect closure* of F. By Corollary 1.17, $F^{p^{-\infty}}$ is determined uniquely, up to an F-isomorphism.

3.17. PROPOSITION. Every finite field is perfect.

Proof. Direct consequence of the fact that an injection of a finite set into itself is a bijection. ∎

The simplest example of an imperfect field is the field $F(X)$ of rational functions in X over a field F of characteristic $p > 0$. This is so since X cannot be a p-th power in $F(X)$. We now provide some criteria for a field to be perfect.

3.18. PROPOSITION. The following conditions are equivalent:

(i) The field F is perfect

(ii) Any finite extension of F is separable

(iii) Any algebraic extension of F is separable

(iv) Any polynomial in $F[X]$ is separable.

Proof. For characteristic 0 there is nothing to prove: in that case every field is perfect and, by Proposition 3.7, (ii), (iii) and (iv) always hold. Assume that $\text{char}F = p > 0$. It is obvious that (iii) and (iv) are equivalent and that (iii) implies (ii). Furthermore, (ii) implies (iii) by Corollary 3.11. We are therefore left to verify that (i) is equivalent to (ii).

Let F be a perfect field and let f be any irreducible polynomial over F. If f is not separable, then by Corollary 3.2, $f' = 0$. Hence, by Proposition 3.3(ii), $f(X) = g(X^p)$, say

$$f(X) = a_0X^{pr} + a_1X^{p(r-1)} + \dots + a_r$$

Since F is perfect, $a_i = b_i^p$ for some $b_i \in F$ and hence

$$\begin{aligned} f(X) &= b_0^pX^{pr} + b_1^pX^{p(r-1)} + \dots + b_r^p \\ &= (b_0X^r + b_1X^{r-1} + \dots + b_r)^p \end{aligned}$$

But this contradicts the irreducibility of f. Thus every finite extension of F is separable.

Conversely, assume that every finite extension of F is separable. Fix $\lambda \in F$ and consider the splitting field of $X^p - \lambda$. If μ is a root, then

$$X^p - \lambda = (X-\mu)^p$$

and so the minimal polynomial of μ over F is of degree 1. Hence $\mu \in F$ and $\lambda = \mu^p$, as required. ■

Let F be a field of characteristic $p > 0$ and let E/F be an algebraic extension. We say that E/F is *purely inseparable* if the only elements of E that are separable over F are the elements of F. Expressed otherwise, E/F is purely inseparable if and only if the separable closure of F in E is F. It is clear that E/F can be inseparable without being purely inseparable.

An algebraic element α of a field containing F is said to be *purely inseparable* over F if $\alpha^{p^n} \in F$ for some $n \geqslant 0$. If E/F is an algebraic field extension, then the set of purely inseparable elements of E over F is obviously a subfield of E containing F. We shall refer to this subfield as the *pure inseparable closure* of F in E. The next result shows that it is a purely inseparable field extension of F containing any other such extension K/F with $K \subseteq E$.

3.19. PROPOSITION. Let F be a field of characteristic $p > 0$ and let E/F be an algebraic extension. Then the following conditions are equivalent:

(i) E/F is purely inseparable

(ii) All elements of E are purely inseparable over F

(iii) E is generated over F by a family of purely inseparable elements

(iv) There exists exactly one F-homomorphism of E into an algebraic closure of E.

Proof. Since the elements of E which are purely inseparable over F form a subfield of E containing F, (ii) and (iii) are equivalent.

(i) $\Rightarrow$ (ii): If $\alpha \in E$, then α^{p^e} is separable over F for some $e \geqslant 0$ (Proposition 3.8(ii)), hence $\alpha^{p^e} \in F$. Thus α is purely inseparable over F.

(ii) $\Rightarrow$ (iv): Let S be a set of purely inseparable elements of E such that $E = F(S)$ and let $\bar{E}$ be an algebraic closure of E. Each element $\alpha \in S$ is a root of the polynomial $X^{p^n} - \lambda \in F[X]$ for some $n \geqslant 0$ and some $\lambda \in F$. Since all the roots of this polynomial are the same, the minimal polynomial $f(X) \in F[X]$ of α has α as the only root. Now the image of α under any F-homomorphism $E \longrightarrow \bar{E}$ is a root of $f(X)$, hence $\alpha \longmapsto \alpha$ proving (iv).

(iv) $\Rightarrow$ (i): Assume by way of contradiction that $\alpha \in E$, $\alpha \notin F$ and α is separable. Then $(F(\alpha):F) = (F(\alpha):F)_s > 1$ and hence there are more than one F-homomorphisms $F(\alpha) \longrightarrow \bar{E}$. But each such homomorphism is extendible to a homomorphism $E \longrightarrow \bar{E}$ by Proposition 2.9. This provides the desired contradiction and completes the proof. ■

3.20. COROLLARY. Let F be a field of characteristic $p > 0$ and let $F \subseteq E \subseteq K$ be a chain of fields. Then K/F is purely inseparable if and only if both E/F

and K/E are purely inseparable.

Proof. Direct consequence of Proposition 3.19(ii). ■

3.21. PROPOSITION. Let F be a field of characteristic $p > 0$ and let E/F be an algebraic field extension. If F_s is the separable closure of F in E, then F_s/F is separable and E/F_s is purely inseparable.

Proof. That F_s/F is separable is a consequence of the definition of F_s. Let K be the separable closure of F_s in E. Then K/F is separable by Corollary 3.14(i). Hence $K = F_s$ and therefore E/F_s is purely inseparable. ■

3.22. COROLLARY. Let F be a field of characteristic $p > 0$. Then every algebraic extension E/F may be obtained by taking a separable extension followed by a purely inseparable extension.

Proof. Direct consequence of Proposition 3.21. ■

The following result provides a criterion under which the order of the extensions in Corollary 3.22 can be reversed.

3.23. PROPOSITION. Let E/F be an algebraic extension of fields of characteristic $p > 0$ and let K and F_s be respectively the pure inseparable and separable closure of F in E. Then E/K is separable if and only if $E = KF_s$.

Proof. Assume that E/K is separable. Then E is separable over KF_s and, by Corollary 3.20 and Proposition 3.21, E is purely inseparable over KF_s. Hence $E = KF_s$. Conversley, if $E = KF_s$ then E/K is separable by Corollary 3.14(ii). ■

3.24. PROPOSITION. Let F be a field of characteristic $p > 0$ and let E/F be a finite extension.

(i) E/F is purely inseparable if and only if $(E:F)_s = 1$. In particular, by Proposition 3.10(iii), if E/F is purely inseparable, then $(E:F)$ is a power of p.

(ii) $(E:F)_s = (F_s:F)$, where F_s is the separable closure of F in E.

Proof. (i) Direct consequence of Proposition 3.19(iv).

(ii) By Proposition 3.21, E/F_s is purely inseparable, hence by (i), $(E:F_s)_s = 1$.

Now

$$(E:F)_s = (F_s:F)_s(E:F_s)_s$$

by Corollary 3.12(ii), and $(F_s:F)_s = (F_s:F)$ by Corollary 3.12(i). Hence

$$(E:F)_s = (F_s:F)$$

as required. ■

3.25. PROPOSITION. Let F be a field of characteristic $p > 0$ and let E be a finite field extension of F. Then E/F is separable if and only if $E = E^pF$.

Proof. Assume that E/F is separable. Then E/E^pF is both separable and purely inseparable, hence $E = E^pF$. Conversely, assume that $E = E^pF$. Let F_s be the separable closure of F in E. Since E/F is finite, we may write $E = F(\alpha_1,\ldots,\alpha_r)$. Now E/F_s is purely inseparable (Proposition 3.21) and hence, by Proposition 3.19, $\alpha_i^{p^{m_i}} \in F_s$ for some $m_i \geq 0$, $1 \leq i \leq r$. Setting

$$m = \max\{m_i\},$$

we have $\alpha_i^{p^m} \in F_s$ for all $i \in \{1,2,\ldots,r\}$. Hence $E^{p^m} \subseteq F_s$. But $E = E^pF$ obviously implies $E = E^{p^m}F$, hence $E = F_s$ is separable over F. ■

3.26. COROLLARY. Let E/F be a finite separable extension of fields of characteristic $p > 0$. If $u_1,\ldots,u_n$ is a basis for E over F, then so is

$$u_1^{p^m},\ldots,u_n^{p^m}$$

for any $m \geq 1$.

Proof. By Proposition 3.25, $E = E^pF$. Therefore $u_1^p,\ldots,u_n^p$ again spans E over F, and so is a basis. Now the result follows by induction on m. ■

As a preliminary to the next result, we record the following observation:

3.27. LEMMA. Let A be a finite-dimensional commutative algebra over a field F. If A has no nonzero nilpotent elements and no nontrivial idempotents, then A is a field.

Proof. Since A has no nonzero nilpotent elements, the Jacobson radical $J(A)$ of A is 0. Hence A is a finite direct product of fields. But A has no nontrivial idempotents, hence A is a field. ■

3.28. PROPOSITION. Let F be a field of characteristic $p > 0$ and let E/F and K/F be respectively a separable and purely inseparable field extensions. Then $E \otimes_F K$ is a field.

Proof. Given $0 \neq x \in E \otimes_F K$, we may, by Proposition 1.4.1, write

$$x = \sum_{i=1}^{n} \alpha_i \otimes \beta_i$$

for some F-linearly independent $\alpha_i \in E$ and some nonzero $\beta_i \in K$, $1 \leq i \leq n$. Since E/F is separable, so is $F(\alpha_1,\ldots,\alpha_n)/F$ and hence $\alpha_1^{p^m},\ldots,\alpha_n^{p^m}$ are F-linearly independent for all $m \geq 1$, by Corollary 3.26. Since K/F is purely inseparable, we may choose $m \geq 1$ such that $\beta_i^{p^m} \in F$ for all i. Then

$$x^{p^m} = \sum_{i=1}^{n} \alpha_i^{p^m} \otimes \beta_i^{p^m} = \sum_{i=1}^{n} \beta_i^{p^m} (\alpha_i^{p^m} \otimes 1) \in E \otimes 1$$

and $x^{p^m} \neq 0$. Since $E \otimes 1$ is an isomorphic copy of E, we conclude that $E \otimes_F K$ has no nonzero nilpotent elements and no nontrivial idempotents. Hence, by Lemma 3.27, for any $\alpha \in E$, $F(\alpha) \otimes_F K$ is a field. Therefore $E \otimes_F K$ is also a field, as required. ■

3.29. PROPOSITION. Let $F \subseteq K \subseteq E$ be a chain of fields of characteristic $p > 0$ such that K/F is purely inseparable and E/K is separable. If F_s is the separable closure of F in E, then $E = F_sK$ and F_s, K are linearly disjoint over F. In particular, by Proposition 1.4.4, $E \cong F_s \otimes_F K$.

Proof. Since K is obviously the pure inseparable closure of F in E, we have $E = F_sK$ by Proposition 3.23. Hence the natural homomorphism $F_s \otimes_F K \longrightarrow E$ is surjective. But $F_s \otimes_F K$ is a field, by Proposition 3.28, hence the result. ■

We close this section by providing circumstances under which a given extension E/F is simple, i.e. $E = F(\alpha)$ for some $\alpha \in E$.

3.30. PROPOSITION. Let F be an infinite field and let $E = F(\alpha_1,\ldots,\alpha_n,\beta)$ where the α_i are separable and β is algebraic over F. Then E/F is a simple extension.

Proof. By induction on n, it suffices to treat the case $n = 1$. Let $f(X)$ and $g(X)$ be the minimal polynomials over F of α_1 and β_1, respectively. If K is a splitting field of $f(X)g(X)$, then $K \supseteq E$. Let

$$\gamma_1 = \alpha_1, \gamma_2, \ldots, \gamma_m$$

be the distinct roots of $f(X)$ and $\delta_1 = \beta, \delta_2, \ldots, \delta_r$ those of $g(X)$. Since α_1 is separable over F, $\deg f(X) = m$. We may therefore assume that $m > 1$, since otherwise, $\alpha_1 \in F$ and $E = F(\beta)$. Consider one of the linear equations

$$x\gamma_1 + \delta_1 = x\gamma_i + \delta_j \qquad (2 \leqslant i \leqslant m,\ 1 \leqslant j \leqslant r)$$

This has at most one solution in F. Bearing in mind that F is infinite, we may therefore choose $x = \gamma \in F$ such that

$$\gamma\gamma_1 + \delta_1 \neq \gamma\gamma_i + \delta_j \qquad (2 \leqslant i \leqslant m,\ 1 \leqslant j \leqslant r) \quad (3)$$

It will next be shown that $\theta = \gamma\gamma_1 + \delta_1$ is a primitive element of E. To this end, consider the polynomial $g(\theta - \gamma X)$ which obviously belongs to $F(\theta)[X]$. Then $g(\theta - \gamma\gamma_1) = g(\delta_1) = 0$ and, by (3), $g(\theta - \gamma\gamma_i) \neq 0$, $2 \leqslant i \leqslant m$. Therefore the greatest common divisor of $g(\theta - \gamma X)$ and $f(X) = \prod_{i=1}^{m} (X - \gamma_i)$ is $X - \gamma_1 = X - \alpha_1$. But then there exist $g_1(X), g_2(X) \in F(\theta)[X]$ such that

$$X - \alpha_1 = g_1(X)f(X) + g_2(X)g(\theta - \gamma X)$$

which implies that $\alpha_1 \in F(\theta)$. Hence $\beta = \theta - \gamma\alpha_1 \in F(\theta)$ and the result follows.■

To treat finite fields, we need the following simple observation.

3.31. PROPOSITION. Any finite subgroup of the multiplicative group of a field is cyclic.

Proof. Let F^* be the multiplicative group of a field F and let G be a finite subgroup of F^* of order m. Let n be the exponent of G. Then $\lambda^n = 1$ for all $\lambda \in G$. On the other hand, the polynomial $X^n - 1$ has at most n roots in F. Since $n | m$, we have $n = m$. Moreover, if $\mu \in G$ is an element of order m, then the order of the cyclic group $\langle\mu\rangle$ generated by μ is m. Hence $G = \langle\mu\rangle$ as required. ■

3.32. COROLLARY. Every finite-dimensional separable field extension is simple.

Proof. Let E/F be a finite-dimensional separable field extension. If F is finite, then so is E and the result follows by virtue of Proposition 3.31. Assume that F is infinite. Then, by Proposition 1.4, $E = F(\alpha_1, \ldots, \alpha_r)$ for some $\alpha_i \in E$ and some $r \geqslant 1$. Since E/F is separable, each α_i is separable over F. Now apply Proposition 3.30. ∎

The result above can also be derived as a consequence of the following general assertion due to Artin.

3.33. PROPOSITION. Let E/F be a finite-dimensional field extension, where F is an infinite field. Then E/F is a simple extension if and only if there are only a finite number of intermediate fields between E and F.

Proof. Assume that E/F is simple, say $E = F(\theta)$ and let K be an intermediate field. Let $f(X)$ be the minimal polynomial of θ over K and let K' be the field generated by F and the coefficients of $f(X)$. Then $K' \subseteq K$ and $f(X)$ is also the minimal polynomial of θ over K'. Therefore

$$(E:K') = \deg f(X) = (E:K)$$

and thus $K = K'$ is generated by the coefficients of $f(X)$. Note also that $f(X)$ divides the minimal polynomial $g(X)$ of θ over F and both $f(X), g(X) \in E[X]$. Because $g(X)$ has only a finite number of distinct factors in $E[X]$ with leading coefficients 1, the number of intermediate fields K is finite.

Conversely, assume that there are only a finite number of intermediate fields between E and F. Given $\alpha, \beta \in E$, we need only verify that $F(\alpha,\beta)/F$ is simple. To this end, let $\gamma \in F$ and consider the subfield $E_\gamma = F(\alpha+\gamma\beta)$. We have an infinite number of $\gamma \in F$ and a finite number of E_γ. Therefore, there exist γ, δ in $F, \gamma \neq \delta$, such that $E_\gamma = E_\delta$. Then

$$\beta = (\gamma-\delta)^{-1}(\alpha+\gamma\beta-\alpha-\delta\beta) \in E_\gamma$$

and therefore $\alpha = \alpha + \gamma\beta - \gamma\beta \in E_\gamma$. Thus $F(\alpha,\beta) = F(\alpha+\gamma\beta)$, as required. ∎

4. GALOIS EXTENSIONS

Let F be a field and let G be a group of automorphisms of F. We denote by F^G the subfield of F consisting of all elements $x \in F$ such that

$$\sigma(x) = x \qquad \text{for all } \sigma \in G$$

We shall refer to F^G as the *fixed field* of G.

Let E/F be a field extension. The group of all F-automorphisms of E is called the *Galois group* of the extension and is written $\mathrm{Gal}(E/F)$. We say that E/F is a *Galois extension* if E is algebraic over F and F is the fixed field of the group $\mathrm{Gal}(E/F)$.

4.1. PROPOSITION. Let E/F be an algebraic extension. Then the following conditions are equivalent:

(i) E/F is Galois

(ii) E/F is normal and separable

(iii) E is the splitting field of a family of separable polynomials over F

Proof. (i) ⇒ (ii): Let $\alpha \in E$ and let f be the minimal polynomial of α over f. We must show that all roots of f are distinct and lie in E. Let $\bar{E}$ be an algebraic closure of E and let $\alpha_1 = \alpha, \alpha_2, \ldots, \alpha_r$ be all distince roots of f in $\bar{E}$. By Corollary 1.18, each $\sigma \in \mathrm{Gal}(E/F)$ is extendible to an automorphism of $\bar{E}$. In this way each $\sigma \in \mathrm{Gal}(E/F)$ permutes the α_i and hence fixes $g = (X-\alpha_1)(X-\alpha_2) \ldots (X-\alpha_r)$. Since E/F is Galois, we deduce that $g \in F[X]$. Hence $f|g$ and, by the definition of g, $g|f$ proving that $f = g$. Since g is irreducible, the group $\mathrm{Gal}(E/F)$ permutes the α_i transitively, hence each $\alpha_i \in E$, as required.

(ii) ⇒ (iii): Since E/F is separable, E is generated over F by a family $\{\alpha_i\}$ of separable elements. We denote by f_i the minimal polynomial of α_i over F. Then each f_i is a separable polynomial over F. Since E/F is normal, f_i splits over E, hence E is the splitting field of the family $\{f_i\}$ of separable polynomials over F.

(iii) ⇒ (i): Assume that E is the splitting field of a family $\{f_i | i \in I\}$ of separable polynomials over F. Then E/F is normal and separable. Fix $\alpha \in E - F$. Since α is separable, it follows from Lemma 3.6 that there is an F-homomorphism $\sigma : F(\alpha) \longrightarrow E$ for which $\sigma(\alpha) \neq \alpha$. On the other hand, by Proposition 2.9, σ is extendible to an automorphism of E, proving that E/F is Galois. ■

Owing to Corollary 3.22, any algebraic extension may be obtained by taking a separable extension followed by a purely inseparable extension. Usually, one cannot reverse the order of the tower. However, there is an important case when it can be done.

4.2. PROPOSITION. Let E/F be a normal extension and let $G = \mathrm{Gal}(E/F)$. Then
(i) E^G/F is purely inseparable and E/E^G is separable
(ii) If F_s is the separable closure of F in E, then $E = F_sE^G$ and F_s, E^G are linearly disjoint over F. In particular,

$$E \cong F_s \otimes_F E^G \quad \text{and} \quad F_s \cap E^G = F$$

(iii) If $\lambda \in E^G - F$, then $\lambda^{p^n} \in F$ for some prime p and some $n \geqslant 1$.

Proof. If char$F = 0$, then $E = F_s$ and hence, by Proposition 4.1, $E^G = F$ which clearly yields the result. Assume that char$F = p > 0$. Owing to Propositions 3.29 and 3.19, we need only verify (i).

Since $\mathrm{Gal}(E/F) = \mathrm{Gal}(E/E^G)$, it follows that if $\lambda \in E$ is such that $\sigma(\lambda) = \lambda$ for all $\sigma \in \mathrm{Gal}(E/E^G)$, then $\lambda \in E^G$. Hence E/E^G is Galois and therefore, by Proposition 4.1, is separable. Let $\bar{E}$ be an algebraic closure of E and let $\sigma : E^G \longrightarrow \bar{E}$ be an F-homomorphism. By Corollary 1.18, σ is extendible to an automorphism of $\bar{E}$. Its restriction to E is an F-automorphism of E since E/F is normal. Hence $\sigma(x) = x$ for all $x \in E^G$ and the result follows by virtue of Proposition 3.19. ■

We next provide a useful characterization of finite Galois extensions.

4.3. PROPOSITION. Let E/F be a finite field extension. Then

$$|\mathrm{Gal}(E/F)| \leqslant (E:F)$$

and

$$|\mathrm{Gal}(E/F)| = (E:F)$$

if and only if E/F is Galois.

Proof. The first inequality follows from $\mathrm{Gal}(E/F) \leqslant (E:F)_s \leqslant (E:F)$. Assume that E/F is Galois. Then, by Proposition 4.1, E/F is both normal and separable. Since E/F is separable, we have $(E:F)_s = (E:F)$ (Corollary 3.12), while since E/F is normal we have $(E:F)_s = |\mathrm{Gal}(E/F)|$. Hence $|\mathrm{Gal}(E/F)| = (E:F)$.

Conversely, assume that $|\mathrm{Gal}(E/F)| = (E:F)$. Then $|\mathrm{Gal}(E/F)| = (E:F)_s = (E:F)$. Since $(E:F)_s = (E:F)$, E/F is separable by Corollary 3.12. Since

$$|\mathrm{Gal}(E/F)| = (E:F)_s,$$

E/F must also be normal. Thus E/F is Galois, by Proposition 4.1. ■

As a preliminary to the next result, we prove

4.4. LEMMA. Let E/F be a separable extension. Assume that there is an integer $n \geqslant 1$ such that every element α of E is of degree $\leqslant n$ over F. Then E/F is finite and $(E:F) \leqslant n$.

Proof. Let $\alpha \in E$ be such that the degree $(F(\alpha):F)$ is maximal, say $m \leqslant n$. We claim that $E = F(\alpha)$. Deny the claim. Then there exists an element $\beta \in E$ such that $\beta \notin F(\alpha)$, and by Corollary 3.32, $F(\alpha,\beta) = F(\gamma)$ for some $\gamma \in F(\alpha,\beta)$. But from the chain

$$F \subseteq F(\alpha) \subseteq F(\alpha,\beta)$$

we see that $(F(\alpha,\beta):F) > m$ whence γ has degree $> m$ over F, a contradiction. ■

We are now ready to prove the following classical result.

4.5. PROPOSITION. (Artin). Let F be a field and let G be a finite group of automorphisms of F. Then F/F^G is a finite Galois extension, $G = \mathrm{Gal}(F/F^G)$ and

$$|G| = (F:F^G)$$

Proof. In view of Proposition 4.3, it suffices to verify that F/F^G is a finite Galois extension and $(F:F^G) \leqslant |G|$. Let $\alpha \in F$ and let $\sigma_1,\ldots,\sigma_r$ be a maximal set of elements of G such that $\sigma_1(\alpha),\ldots,\sigma_r(\alpha)$ are distinct. If $\sigma \in G$ then $(\sigma\sigma_1(\alpha),\ldots,\sigma\sigma_r(\alpha))$ differs from $(\sigma_1(\alpha),\ldots,\sigma_r(\alpha))$ by a permutation, since σ is injective and each $\sigma\sigma_i(\alpha) \in \{\sigma_1(\alpha),\ldots,\sigma_r(\alpha)\}$. Therefore α is a root of the polynomial

$$f(X) = \prod_{i=1}^{r}(X-\sigma_i(\alpha))$$

and for any $\sigma \in G$, $f^\sigma = f$. Thus the coefficients of f lie in F^G. Moreover,

f is separable. It follows that every element α of F is a root of a separable polynomial over F^G of degree $\leqslant |G|$. Furthermore, this polynomial splits into linear factors in F. Thus F/F^G is both separable and normal, hence F/F^G is Galois, by Proposition 4.1. By Lemma 4.4, we have $(F:F^G) \leqslant |G|$ and the result follows. ∎

4.6. COROLLARY. Let E/F be a finite Galois extension and let G be its Galois group. Then, for any subgroup H of G, there exists an intermediate field K such that E/K is Galois and $H = \mathrm{Gal}(E/K)$.

Proof. Put $K = E^H$ and apply Proposition 4.5. ∎

We close by examining some properties of automorphisms of field extensions. Let S be a set and F a field. The maps $f_i : S \longrightarrow F$, $1 \leqslant i \leqslant n$, are said to be *linearly independent* over F if whenever we have a relation

$$\lambda_1 f_1 + \dots + \lambda_n f_n = 0 \qquad (\lambda_i \in F)$$

then all $\lambda_i = 0$. Recall that a *monoid* is a set G, with an associative binary operation and having an identity element.

4.7. PROPOSITION. (Artin). Let $\chi_1, \chi_2, \dots, \chi_n$ be distinct homomorphisms from a monoid G to the multiplicative group F^* of a field F. Then $\chi_1, \dots, \chi_n$ are linearly independent over F.

Proof. If $\chi_1, \chi_2, \dots, \chi_n$ are linearly dependent, then we may harmlessly assume that $\chi_1 = \sum_{i=2}^{n} \lambda_i \chi_i$ for some $\lambda_i \in F$ and that $\chi_2, \dots, \chi_n$ are linearly independent. Then

$$\chi_1(g) = \sum_{i=2}^{n} \lambda_i \chi_i(g) \qquad \text{for all } g \in G \qquad (1)$$

Replace g by gx in (1):

$$\chi_1(g)\chi_1(x) = \sum_{i=2}^{n} \lambda_i \chi_i(g)\chi_i(x) \qquad \text{for all } g,x \in G \qquad (2)$$

Multiply (1) by $\chi_1(x)$ and subtract the result from (2):

$$\sum_{i=2}^{n} \lambda_i(\chi_i(x) - \chi_1(x))\chi_i(g) = 0 \qquad \text{for all } g,x \in G$$

Hence $\sum_{i=2}^{n} \lambda_i(\chi_i(x) - \chi_1(x))\chi_i = 0$ and so $\lambda_i(\chi_i(x) - \chi_1(x)) = 0$, $2 \leqslant i \leqslant n$, since $\chi_2, \dots, \chi_n$ are linearly independent. Since $\chi_i \neq \chi_1$ for $i \neq 1$, it follows that $\lambda_i = 0$ for $i \neq 1$ and (1) reduces to the form $\chi_1(g) = 0$ which is impossible. ■

4.8. COROLLARY. Let $f_1, f_2, \ldots, f_n$ be distinct homomorphisms of a field E into another field F. Then $f_1, \ldots, f_n$ are linearly independent over F.

Proof. Put $G = E^*$ and denote by χ_i the restriction of f_i to G. Then each χ_i is a homomorphism of G into F^* and the χ_i are obviously distinct. Since $\sum_{i=1}^{n} \lambda_i f_i = 0$ obviously implies $\sum_{i=1}^{n} \lambda_i \chi_i = 0$, the result follows by virtue of Proposition 4.7. ■

4.9. COROLLARY. Let E/F and K/F be two field extensions and assume that $(E:F) = n < \infty$. Then there are at most n F-homomorphisms from E to K.

Proof. Let $u_1, \ldots, u_n$ be an F-basis of E, and suppose that there are $n + 1$ distinct F-homomorphisms $\sigma_0, \sigma_1, \ldots, \sigma_n$ from E to K. Then the n equations in the $n + 1$ unknowns λ_i:

$$\sum_{i=0}^{n} \lambda_i \sigma_i(u_j) = 0 \qquad (1 \leq j \leq n)$$

have a nonzero solution in K. Any $\mu \in E$ has the form $\sum_{j=1}^{n} \mu_j u_j, \mu_j \in F$ and

$$\sum_{i=0}^{n} \lambda_i \sigma_i(\mu) = \sum_{i=0}^{n} \sum_{j=1}^{n} \lambda_i \mu_j \sigma_i(u_j) = 0$$

contrary to Corollary 4.8. ■

4.10. COROLLARY. Let $\lambda_1, \ldots, \lambda_n$ be distinct nonzero elements of a field F. If $\mu_1, \ldots, \mu_n$ are elements of F such that for all $k \in \mathbb{Z}$ we have

$$\mu_1 \lambda_1^k + \ldots + \mu_n \lambda_n^k = 0$$

then all $\mu_i = 0$.

Proof. Apply Proposition 4.7 to the distinct homomorphisms

$$\chi_i : k \longmapsto \lambda_i^k$$

of $\mathbb{Z}$ into F^*. ■

4.11. COROLLARY. Let E/F be a finite separable field extension, let $\sigma_1, \ldots, \sigma_n$ be all F-homomorphisms of E into an algebraic closure $\bar{E}$ of E, and let $u_1, \ldots, u_n$ be a basis of E over F. Then the vectors

$$v_1 = (\sigma_1(u_1),\ldots,\sigma_1(u_n))$$

$$- - - - - - - - - - - - -$$

$$v_n = (\sigma_n(u_1),\ldots,\sigma_n(u_n))$$

are linearly independent over E.

Proof. Let $\lambda_1,\ldots,\lambda_n$ be elements of E such that

$$\lambda_1 v_1 + \ldots + \lambda_n v_n = 0$$

Then we have

$$\left(\sum_{i=1}^{n} \lambda_i \sigma_i\right)(u_j) = 0 \qquad \text{for all } j \in \{1,2,\ldots,n\}$$

But $\sigma_1,\ldots,\sigma_n$ are linearly independent as homomorphisms from E^* to $\bar{E}^*$, by virtue of Proposition 4.7. Hence $\lambda_i = 0$ for all $i \in \{1,\ldots,n\}$ and the result follows. ■

5. FINITE FIELDS, ROOTS OF UNITY AND CYCLOTOMIC EXTENSIONS

A field with a finite number of elements is called a *finite field*, or also a *Galois field*.

Let F be a finite field. Then char$F = p > 0$ is a prime and the image of $\mathbb{Z}$ in F, denoted by $\mathbb{F}_p$, is isomorphic to the field $\mathbb{Z}/p\mathbb{Z}$ consisting of p elements. It is clear that $\mathbb{F}_p$ is the prime subfield of F, i.e. the intersection of all subfields of F.

5.1. PROPOSITION. (i) If F is a finite field of characteristic p and $n = (F:\mathbb{F}_p)$, then $|F| = p^n$ and F^* is cyclic of order $p^n - 1$.

(ii) For each prime p and each $n \geqslant 1$ there is exactly one field of $q = p^n$ elements (up to isomorphism), namely the splitting field of $X^q - X$ over $\mathbb{F}_p$, and its elements are the roots of $X^q - X$. Furthermore, every finite field is isomorphic to exactly one of these fields.

Proof. (i) Since F is finite, n is obviously finite and hence $|F| = p^n$. The second assertion is a consequence of Proposition 3.31.

(ii) Let F be a field of $q = p^n$ elements. Since F^* has $q - 1$ elements, we have

$$x^{q-1} = 1 \qquad \text{for all } x \in F^*$$

and hence

$$x^q = x \qquad \text{for all } x \in F$$

Since this equation has at most q roots, its roots are precisely the elements of F. We can therefore write

$$X^q - X = \prod_{\lambda \in F}(X-\lambda)$$

and F is the splitting field of $X^q - X$ over $\mathbb{F}_p$.

Conversely, let $q = p^n$ where p is a prime. Denote by F the splitting field of $X^q - X$ over $\mathbb{F}_p$. To show that this splitting field has exactly q elements, we observe that the roots of $X^q - X$ are distinct since

$$(X^q-X)' = qX^{q-1} - 1 = -1 \neq 0$$

(hence there are exactly q of them) and that they form a field : if $a^q = a$, $b^q = b$, then $(a-b)^q = a^q - b^q = a - b$, $(ab)^q = a^q b^q = ab$, and if $b \neq 0, (b^{-1})^q = (b^q)^{-1} = b^{-1}$. ■

5.2. PROPOSITION. Let F be a field of $q = p^m$ elements. Then, for any $n \geqslant 1$, there exists a field extension K of F of dimension n, namely the splitting field of

$$X^{q^n} - X \text{ over } F$$

Moreover, any two field extensions of F of dimension n are F-isomorphic.

Proof. Assume that E is a field extension of F of dimension n. Then E has q^n elements and, by Proposition 5.1(ii), the elements of E are precisely the roots of $X^{q^n} - X$ in an algebraic closure of $\mathbb{F}_p$. Therefore E is a splitting field of $X^{q^n} - X$ over F. This shows that any two field extensions of F of dimension n are F-isomorphic. Conversely, the splitting field of $X^{q^n} - X$ over F is the same as the splitting field of $X^{q^n} - X$ over $\mathbb{F}_p$, since all elements of F are roots of $X^{q^n} - X$. Hence it consists of precisely q^n elements and therefore has degree n over F. ■

The field of q elements is denoted by $\mathbb{F}_q$. For $q = p$ this agrees with the notation $\mathbb{F}_p$ introduced earlier.

5.3. PROPOSITION. If $q = p^n$, where p is a prime and $n \geq 1$, then $\mathbb{F}_q/\mathbb{F}_p$ is a Galois extension with cyclic Galois group of order n, generated by the Frobenius automorphism $x \longmapsto x^p$ of $\mathbb{F}_q$.

Proof. The extension $\mathbb{F}_q/\mathbb{F}_p$ being a finite extension of the perfect field $\mathbb{F}_p$ must be separable (Proposition 3.18). On the other hand, $\mathbb{F}_q/\mathbb{F}_p$ is normal by Proposition 5.1(ii) and hence is Galois by Proposition 4.1. Therefore, by Proposition 4.3, $\mathrm{Gal}(\mathbb{F}_q/\mathbb{F}_p)$ is of order n. Since $x^p = x$ for all $x \in \mathbb{F}_p$, the Frobenius automorphism α of $\mathbb{F}_q$ is an $\mathbb{F}_p$-automorphism of $\mathbb{F}_q$. Finally, since

$$\alpha^r(x) = x^{p^r} \qquad \text{for all } x \in \mathbb{F}_q$$

and this is the identity for $r = n$ and not the identity for $1 \leq r < n$, we see that α has order n. Thus $\mathrm{Gal}(\mathbb{F}_q/\mathbb{F}_p) = \langle\alpha\rangle$ as required. ∎

5.4. PROPOSITION. All subfields of $\mathbb{F}_{p^m}$ are of the form $\mathbb{F}_{p^n}$ where $n|m$ and, for any $n|m$, there is exactly one subfield of $\mathbb{F}_{p^m}$ of order p^n. Furthermore, if $n|m$, then $\mathbb{F}_{p^m}/\mathbb{F}_{p^n}$ is a Galois extension with cyclic Galois group of order m/n, generated by α^n, where α is the Frobenius automorphism of $\mathbb{F}_{p^m}$.

Proof. Assume that $n|m$. Then any root of $X^{p^n} - X$ is also a root of $X^{p^m} - X$. Hence, by Proposition 5.1(ii), all the roots of $X^{p^n} - X$ form a unique subfield of $\mathbb{F}_{p^m}$ of order p^n. Conversely, if F is a subfield of $\mathbb{F}_{p^m}$ and $|F| = p^n$, then

$$m = (\mathbb{F}_{p^m}:\mathbb{F}_p) = (\mathbb{F}_{p^m}:F)(F:\mathbb{F}_p) = (\mathbb{F}_{p^m}:F)n$$

proving that $n|m$.

By the above $(\mathbb{F}_{p^m}:\mathbb{F}_{p^n}) = m/n$ and, by Proposition 5.3, $\mathbb{F}_{p^m}/\mathbb{F}_{p^n}$ is Galois hence its Galois group is of order m/n. Since α^n is of order m/n and α^n fixes $\mathbb{F}_{p^n}$ elementwise, the result follows. ∎

We now turn our attention to roots of unity. Let F be a field. An element $\lambda \in F$ is said to be an *n-th root* of 1 (or of *unity*) if $\lambda^n = 1$. If $\mathrm{char}F = p > 0$ and $n = p^m s$ with $(s,p) = 1$, then $\lambda^n = 1$ implies $(\lambda^s)^{p^m} = 1$

or $\lambda^s = 1$. Hence, if λ is an n-th root of unity then λ is an s-th root of unity. We shall therefore assume in what follows that the characteristic of the field does not divide n.

5.5. PROPOSITION. Let F be a field and let $n \geq 1$ be such that $\mathrm{char}F \nmid n$.
(i) The n-th roots of unity in F form a cyclic group whose order divides n.
(ii) If $X^n - 1$ splits into linear factors over F, then the n-th roots of unity in F form a cyclic group of order n.

Proof. It is clear that (i) is a consequence of (ii). To prove (ii), note that the n-th roots of unity in F certainly form a group, say G. Since $(X^n-1)' = nX^{n-1} \neq 0$, it follows from Proposition 3.1, that all roots of $X^n - 1$ are simple. Hence F contains exactly n n-th roots of unity, i.e. $|G| = n$. Let m be the exponent of G. Then $m \leq n$ and all elements of G are roots of $X^m - 1$, hence $|G| \leq m$. This shows that $m = n$ and therefore G is cyclic. ■

Let F be a field. We say that $\lambda \in F$ is a *primitive n-th root of unity* if the order of λ in F^* is precisely n. Given $n \geq 1$, by a *primitive n-th root of unity over* F we mean a primitive n-th root of unity in the splitting field of $X^n - 1$ over F. By Proposition 5.5, if $\mathrm{char}F \nmid n$, then a primitive n-th root of unity over F always exists. Furthermore, if ε_n is a primitive n-th root of unity over F, then $F(\varepsilon_n)$ is obviously the splitting field of $X^n - 1$ over F. The field $F(\varepsilon_n)$ is called *the n-th cyclotomic extension* of F. By the *maximal cyclotomic extension* of F we understand the field obtained from F by adjoining n-th roots of unity for all $n \geq 1$.

5.6. COROLLARY. Let F be a finite field of q elements, let $m \geq 1$ be an integer coprime to q and let ε be a primitive m-th root of unity over F. Then

(i) $(F(\varepsilon):F)$ is equal to the order of q modulo m

(ii) $\mathrm{Gal}(F(\varepsilon)/F)$ is cyclic generated by $\varepsilon \longmapsto \varepsilon^q$

Proof. It is clear that (i) is a consequence of (ii). Let α be the Frobenius automorphism of F. If $q = p^n$, p prime, then $\alpha^n(\varepsilon) = \varepsilon^{p^n} = \varepsilon^q$.

Therefore (ii) is a consequence of Proposition 5.4. ■

5.7. PROPOSITION. Let F be a field, let $n \geq 1$ be such that char$F \nmid n$ and let E and K be respectively the n-th cyclotomic and maximal cyclotomic extensions of F. Then both E/F and K/F are Galois extensions.

Proof. Since char$F \nmid n$, $X^n - 1$ has no multiple roots. It remains to notice that E is the splitting field of $X^n - 1$ over F and K is the splitting field of the family of all such polynomials $X^n - 1$ with char$F \nmid n$ (see Proposition 4.1). ■

The following observation is often useful.

5.8. LEMMA. Let E/F be a Galois extension and let $\mu_1, \mu_2, \ldots, \mu_n$ be elements of E such that each $\sigma \in \mathrm{Gal}(E/F)$ permutes the $\mu_1, \ldots, \mu_n$. Then

$$(X-\mu_1)(X-\mu_2) \ldots (X-\mu_n)$$

is a monic polynomial over F. If, furthermore, F is the quotient field of an integrally closed integral domain R and each μ_i is integral over R, then $(X-\mu_1) \ldots (X-\mu_s)$ is a monic polynomial over R.

Proof. Put $f(X) = (X-\mu_1) \ldots (X-\mu_n)$. Then

$$f(X) = X^n - s_1 X^{n-1} + s_2 X^{n-2} + \ldots + (-1)^k s_k X^{n-k} + \ldots + (-1)^n s_n$$

where

$$s_1 = \mu_1 + \mu_2 + \ldots + \mu_n$$
$$s_2 = \mu_1\mu_2 + \mu_1\mu_3 + \ldots + \mu_{n-1}\mu_n$$
$$\text{- - - - - - - - - - - - - - - - - -}$$
$$s_k = \sum_{i_1 < i_2 < \ldots < i_k} \mu_{i_1}\mu_{i_2} \ldots \mu_{i_k}$$
$$\text{- - - - - - - - - - - - - - - - - -}$$
$$s_n = \mu_1\mu_2 \ldots \mu_n$$

Hence $\sigma(s_i) = s_i$ for all $\sigma \in \mathrm{Gal}(E/F)$ and all $i \in \{1,2,\ldots,n\}$. Since E/F is Galois, we deduce that each $s_i \in F$, hence $f(X) \in F[X]$. Under the additional hypotheses, each s_i is integral over R, hence belongs to R since R is integrally closed. ■

In what follows we write $(\mathbb{Z}/n\mathbb{Z})^*$ for the unit group of the ring $\mathbb{Z}/n\mathbb{Z}$. The order of this group is denoted by $\phi(n)$, where ϕ is known as the *Eiler function*. Note that $(\mathbb{Z}/n\mathbb{Z})^* = \{\mu+n\mathbb{Z} \mid 1 \leqslant \mu < n,\ (\mu,n) = 1\}$ and that $\phi(n)$ coincides with the number of generators of a cyclic group of order n. In particular, the number of primitive n-th roots of unity is equal to $\phi(n)$.

Let F be a field, let $n \geqslant 1$ be such that $\mathrm{char}F \nmid n$ and let $\varepsilon_1,\ldots,\varepsilon_r$, $r = \phi(n)$, be all primitive n-th roots of unity over F. Then, by Proposition 5.7 and Lemma 5.8,

$$\Phi_n(X) = \prod_{i=1}^{r}(X-\varepsilon_i)$$

is a monic polynomial over the prime subfield of F and hence over F. Furthermore, if $\mathrm{char}F = 0$, then by Lemma 5.8,

$$\Phi_n(X) \in \mathbb{Z}[X]$$

We shall refer to $\Phi_n(X)$ as the *n-th cyclotomic polynomial over* F.

5.9. PROPOSITION. Let F be a field, let $n \geqslant 1$ be such that $\mathrm{char}F \nmid n$ and let F_n be the n-th cyclotomic extension of F. Then $\mathrm{Gal}(F_n/F)$ is isomorphic to a subgroup of $(\mathbb{Z}/n\mathbb{Z})^*$. Furthermore, the following conditions are equivalent:

(i) $\mathrm{Gal}(F_n/F) \cong (\mathbb{Z}/n\mathbb{Z})^*$

(ii) $\Phi_n(X)$ is irreducible over F

(iii) $(F_n:F) = \phi(n)$

Proof. Let ε be a primitive n-th root of unity so that $F_n = F(\varepsilon)$. If $\sigma \in \mathrm{Gal}(F_n/F)$, then $\sigma(\varepsilon) = \varepsilon^{\mu}$ for some $1 \leqslant \mu < n$ with $(\mu,n) = 1$. Furthermore, σ is uniquely determined by μ modulo n and may be denoted by σ_μ. Then the map

$$\begin{cases} \mathrm{Gal}(F_n/F) \longrightarrow (\mathbb{Z}/n\mathbb{Z})^* \\ \sigma_\mu \longmapsto \mu + n\mathbb{Z} \end{cases}$$

is obviously an injective homomorphism. The second assertion follows from the above and Propositions 4.3 and 5.7. ∎

Let E/F be a field extension. We say that E/F is *abelain* (respectively, *cyclic*) if E/F is Galois and $\mathrm{Gal}(E/F)$ is abelian (respectively, cyclic).

5.10. PROPOSITION. Let F be a field and let E be the maximal cyclotomic extension of F. Then E/F is abelian.

Proof. Let $\sigma_1,\sigma_2 \in \mathrm{Gal}(E/F)$. Since E is generated over F by roots of unity, it suffices to show that $(\sigma_1\sigma_2)(\varepsilon) = (\sigma_2\sigma_1)(\varepsilon)$ where ε is a root of unity in E. But if n is the order of ε, $\sigma_i(\varepsilon)$ is also of order n, hence $\sigma_i(\varepsilon) = \varepsilon^{\mu_i}$ with $(\mu_i,n) = 1$. Therefore $\sigma_1\sigma_2(\varepsilon) = \sigma_2\sigma_1(\varepsilon) = \varepsilon^{\mu_1\mu_2}$, as required. ■

Let $n = p_1^{n_1}p_2^{n_2}\dots p_k^{n_k}$ be the canonical decomposition of n. Then

$$\mathbb{Z}/n\mathbb{Z} \cong \mathbb{Z}/p_1^{n_1}\mathbb{Z} \times \dots \times \mathbb{Z}/p_k^{n_k}\mathbb{Z}$$

and therefore

$$(\mathbb{Z}/n\mathbb{Z})^* \cong (\mathbb{Z}/p_1^{n_1}\mathbb{Z})^* \times \dots \times (\mathbb{Z}/p_k^{n_k}\mathbb{Z})^*$$

For this reason, to determine the isomorphism class of $(\mathbb{Z}/n\mathbb{Z})^*$, we may harmlessly assume that n is a power of a prime.

5.11. PROPOSITION. (i) If p is an odd prime, then $(\mathbb{Z}/p^n\mathbb{Z})^*$ is cyclic of order $p^{n-1}(p-1)$

(ii) Both $(\mathbb{Z}/2\mathbb{Z})^*$ and $(\mathbb{Z}/4\mathbb{Z})^*$ are cyclic, and if $n \geqslant 3$, then

$$(\mathbb{Z}/2^n\mathbb{Z})^* \cong \mathbb{Z}_2 \times \mathbb{Z}_{2^{n-2}}$$

Proof. (i) Since $\phi(p^n) = p^{n-1}(p-1)$ (for any prime p), $(\mathbb{Z}/p^n\mathbb{Z})^*$ is of order $p^{n-1}(p-1)$. Hence $(\mathbb{Z}/p^n\mathbb{Z})^*$ is a direct product of its subgroup H of order p^{n-1} consisting of the elements which satisfy $x^{p^{n-1}} = 1$ and the subgroup K of order $p - 1$ of the elements satisfying $x^{p-1} = 1$. Since the orders of H and K are coprime, it suffices to verify that both H and K are cyclic.

If $n = 1$, then $(\mathbb{Z}/p\mathbb{Z})^* = K$ and this is cyclic by Proposition 5.1(i). Hence we can choose an integer m such that $m + p\mathbb{Z}, m^2 + p\mathbb{Z}, \dots ,m^{p-1} + p\mathbb{Z}$ are distinct in $\mathbb{Z}/p\mathbb{Z}$. Set $k = m^{p^{n-1}}$. Because $(m,p) = 1, (k,p^n) = 1$ and $k + p^n\mathbb{Z}$ and $m + p^n\mathbb{Z} \in (\mathbb{Z}/p^n\mathbb{Z})^*$. Also

$$k^{p-1} = (m^{p^{n-1}})^{p-1} = m^{\phi(p^n)} \equiv 1 \pmod{p^n}$$

which implies $k + p^n\mathbb{Z} \in K$. Since

$$k = m^{p^{n-1}} \equiv m \pmod p$$

the elements $k + p\mathbb{Z}, k^2 + p\mathbb{Z}, \ldots, k^{p-1} + p\mathbb{Z}$ are distinct. Hence also

$$k + p^n\mathbb{Z}, k^2 + p^n\mathbb{Z}, \ldots, k^{p-1} + p^n\mathbb{Z}$$

are distinct. This implies that the order of $k + p^n\mathbb{Z}$ is precisely $p - 1$. But the order of K is $p - 1$, hence K is cyclic generated by $k + p^n\mathbb{Z}$.

To prove that H is cyclic, we may assume that $n \geqslant 2$ since otherwise $H = 1$ Then H is a direct product of $t \geqslant 1$ cyclic groups of order p^{n_i}, $n_i \geqslant 1$. Hence the number of solutions of the equation $x^p = 1$, $x \in H$ is p^t. It therefore suffices to show that the number of integers s, $0 < s < p^n$, satisfying $s^p \equiv 1 \pmod{p^n}$ does not exceed p. Now if s satisfies these conditions, then, since $s^p \equiv s \pmod p$, we have $s \equiv 1 \pmod p$. Then if $s \neq 1$, we may write

$$s = 1 + yp^f + zp^{f+1} \quad (1 \leqslant f \leqslant n-1, 0 < y < p,\ z \geqslant 0)$$

in which case

$$\begin{aligned} s^p &= 1 + \binom{p}{1}(y+zp)p^f + \binom{p}{2}(y+zp)^2p^{2f} \\ &\quad + \ldots + (y+zp)^p p^{pf} \\ &\equiv 1 + yp^{f+1} \pmod{p^{f+2}} \end{aligned}$$

If $s^p \equiv 1 \pmod{p^n}$ and $f < n-1$, this gives

$$yp^{f+1} \equiv 0 \pmod{p^{f+2}}$$

so $y \equiv 0 \pmod p$ contrary to $0 < y < p$. Thus, if $1 < s < p^n$ satisfies $s^p \equiv 1 \pmod{p^n}$, then $s = 1 + yp^{n-1}$, $0 < y < p$. This gives altogether at most p solutions including 1 and proves that H is cyclic.

(ii) The order of $(\mathbb{Z}/2^n\mathbb{Z})^*$ is $\phi(2^n) = 2^{n-1}$. If $n = 1$ or 2, then these orders are 1 and 2, respectively, and there is nothing to prove. So assume that $n \geqslant 3$. We first claim that there are four distinct elements $x \in (\mathbb{Z}/2^n\mathbb{Z})^*$ satisfying $x^2 = 1$; if sustained, it will follow that $(\mathbb{Z}/2^n\mathbb{Z})^*$ is a direct product of at least two distinct cyclic groups $\neq 1$. Put $a_1 = 1$, $a_2 = -1$, $a_3 = 1 + 2^{n-1}$, $a_4 = -1 + 2^{n-1}$, $x_i = a_i + 2^n\mathbb{Z}$. Then the x_i are distinct and satisfy $x_i^2 = 1$, which substantiates our claim.

Since $(\mathbb{Z}/2^n\mathbb{Z})^*$ is a direct product of at least two cyclic groups $\neq 1$ and the order of $(\mathbb{Z}/2^n\mathbb{Z})^*$ is 2^{n-1}, we see that, if $x \in (\mathbb{Z}/2^n\mathbb{Z})^*$, then $x^{2^{n-2}} = 1$ or, what is the same thing, if m is an odd integer, then $m^{2^{n-2}} \equiv 1 \pmod{2^n}$. To complete the proof, we are therefore left to exhibit an x such that $x^{2^{n-3}} \neq 1$.

Put $x = 5 + 2^n\mathbb{Z}$. Observe first that, if $n = 3$, then $5^{2^{n-3}} = 5 \not\equiv 1 \pmod{2^n}$ but $5^{2^{n-3}} \equiv 1 \pmod{2^{n-1}}$. Now let $f \geq 3$ and let $k(f)$ be the largest integer k such that

$$5^{2^{f-3}} \equiv 1 \pmod{2^k}$$

Then we have $k(3) = 2$. Also for any $f \geq 3$ we have $5^{2^{f-3}} = 1 + y2^{k(f)}$ where y is odd. This implies that

$$5^{2(f+1)-3} = (5^{2^{f-3}})^2 = 1 + y2^{k(f)+1} + y2^{2k(f)}$$

whence $k(f+1) \geq k(f)$, so $k(f) \geq 2$ if $f \geq 3$. Then the relation shows that

$$5^{2^{(f+1)-3}} = 1 + z2^{k(f)+1}$$

where $z = y + 2^{k(f)-1}y^2$ is odd. Thus $k(f+1) = k(f) + 1$. Since $k(3) = 2$, we deduce that $k(f) = f - 1$ for all $f \geq 3$. Hence

$$5^{2^{n-3}} \not\equiv 1 \pmod{2^n} \qquad \text{for } n \geq 3$$

and the result follows. ■

It will be shown below (Proposition 5.13) that $\mathrm{Gal}(\mathbb{Q}_n/\mathbb{Q}) \cong (\mathbb{Z}/n\mathbb{Z})^*$. With this in mind, we now derive

5.12. COROLLARY. Let $n = p^k$, p a prime, let F be a field with $\mathrm{char}F \neq p$ and let F_n be the n-th cyclotomic extension of F. Then F_n/F is cyclic unless $p = 2$ and $k \geq 3$, in which case F_n/F is either cyclic or $\mathrm{Gal}(F_n/F)$ is a direct product of a cyclic group of order 2 and one of order 2^{t-2} where $t \leq k$. Furthermore

$$\mathrm{Gal}(\mathbb{Q}_n/\mathbb{Q}) \simeq \begin{cases} \mathbb{Z}_{p^{k-1}(p-1)} & \text{if } p \text{ is odd} \\ \mathbb{Z}_2 \times \mathbb{Z}_{2^{k-2}} & \text{if } p = 2 \end{cases}$$

Proof. Direct consequence of Propositions 5.9 and 5.11. ■

We now turn our attention to the case where $F = \mathbb{Q}$.

5.13. PROPOSITION. Let $\mathbb{Q}_n$ be the n-th cyclotomic extension of $\mathbb{Q}$ and let $\Phi_n(X)$ be the n-th cyclotomic polynomial over $\mathbb{Q}$. Then

(i) $(\mathbb{Q}_n:\mathbb{Q}) = \phi(n)$

(ii) $\Phi_n(X)$ is irreducible

(iii) $\mathrm{Gal}(\mathbb{Q}_n/\mathbb{Q}) \cong (\mathbb{Z}/n\mathbb{Z})^*$

Proof. In view of Proposition 5.9, only (i) deserves a proof. Let ε be a primitive n-th root of unity in $\mathbb{Q}_n$ and let $f(X)$ be the minimal polynomial of ε over $\mathbb{Q}$. We assert that for any prime $p \nmid n$, ε^p is a root of f. Since any primitive n-th root of unity can be obtained by raising ε to a succession of prime powers with primes not dividing n, this will prove that $\deg f \geqslant \phi(n)$ and hence that $(\mathbb{Q}_n:\mathbb{Q}) = \phi(n)$.

To prove the required assertion, we first observe that $f(X)$ divides $X^n - 1$, say $X^n - 1 = f(X)h(X)$ where f and h are monic polynomials in $\mathbb{Q}[X]$. By Theorem 1.2.21(iii) we must have $f(X), h(X) \in \mathbb{Z}[X]$. Assume by way of contradiction that ε^p is not a root of f. Then ε^p is a root of h and hence ε is a root of $h(X^p)$. We may therefore write $h(X^p) = f(X)g(X)$ for some

$$g(X) \in \mathbb{Q}[X]$$

Since both $h(X^p)$ and $f(X)$ are monic, so is $g(X)$, hence $g(X) \in \mathbb{Z}[X]$ by Theorem 1.2.21(iii). Now $\ell^p \equiv \ell \pmod p$ for any integer ℓ, so

$$h(X^p) \equiv h(X)^p \pmod p$$

and thus

$$h(X)^p \equiv f(X)g(X) \pmod p$$

It follows that the images $\bar{f}$ and $\bar{h}$ of f and h in $\mathbb{F}_p[X]$ have a nonconstant factor in common. On the other hand $X^n - 1 = \bar{f}(X)\bar{g}(X)$ and $X^n - 1$ has no multiple roots since $p \nmid n$. This provides the desired contradiction. ■

5.14. COROLLARY. Let n, m be relatively prime integers $\geqslant 1$ and let ε_n and ε_m be primitive n-th and m-th roots of unity over $\mathbb{Q}$, respectively. Then

$$\mathbb{Q}(\varepsilon_n) \cap \mathbb{Q}(\varepsilon_m) = \mathbb{Q}$$

Proof. Note that, since $(m,n) = 1$, $\varepsilon = \varepsilon_n \varepsilon_m$ is a primitive mn-th root of unity and that ε^n and ε^m are primitive m-th and n-th roots of unity, respectively. Hence

$$\mathbb{Q}(\varepsilon_n)\mathbb{Q}(\varepsilon_m) = \mathbb{Q}(\varepsilon)$$

On the other hand, $\phi(mn) = \phi(m)\phi(n)$ since $(m,n) = 1$ and thus, by Proposition 5.13,

$$(\mathbb{Q}(\varepsilon_n):\mathbb{Q})(\mathbb{Q}(\varepsilon_m):\mathbb{Q}) = (\mathbb{Q}(\varepsilon):\mathbb{Q})$$

Hence $\mathbb{Q}(\varepsilon_n)$ and $\mathbb{Q}(\varepsilon_m)$ are linearly disjoint over $\mathbb{Q}$ and the result follows by virtue of Proposition 1.4.4. ■

We now examine cyclotomic polynomials in detail. We begin by defining the Möbius μ-function on the positive rational integers. Let $p_1,\ldots,p_r$ denote primes and set

$$\mu(1) = 1,\ \mu(p_1\cdots p_r) = \begin{cases} (-1)^r, & \text{if the } p_i \text{ are distinct primes} \\ 0, & \text{otherwise} \end{cases}$$

It is easily verified that

$$\mu(ab) = \mu(a)\,\mu(b) \quad \text{if } (a,b) = 1$$

and that

$$\sum_{d|n} \mu(d) = \begin{cases} 1, & \text{if } n = 1 \\ 0, & \text{if } n > 1 \end{cases} \tag{1}$$

Given a function f, we define its "Möbius transform" g by

$$g(n) = \sum_{d|n} \mu(\tfrac{n}{d})f(d) = \sum_{d|n} \mu(d)f(\tfrac{n}{d}) \tag{2}$$

5.15. LEMMA. With the notation above,

$$f(t) = \sum_{n|t} g(n) \tag{3}$$

In particular,

(a) The Euler function ϕ is the Möbius transform of the identity function

(b) $t = \sum_{n|t} \phi(n)$

Proof. We have

$$\sum_{n|t} g(n) = \sum_{n|t} \sum_{d|n} \mu(\tfrac{n}{d}) f(d) = \sum_{d|t} \{\sum_{n} \mu(\tfrac{n}{d})\} f(d)$$

where, in the sum in braces, n ranges over all divisors of t which are multiples of d. In that case, n/d ranges over all divisors of t/d and so, by (1), the sum in braces vanishes except when $d = t$, in which case it is 1. This proves (2).

In particular, let ϕ be the Euler function. Using the fact that

$$\phi(ab) = \phi(a)\phi(b)$$

if $(a,b) = 1$, one easily obtains the well-known formula

$$\phi(n) = n \prod_{p|n} (1 - \tfrac{1}{p}) = n \sum_{d|n} \frac{\mu(d)}{d}$$
$$= \sum_{d|n} \mu(d) \cdot \frac{n}{d}$$

where the first product is taken over the distinct primes dividing n. This proves (a) and by (2), also (b). ■

Formulas (2) and (3) are called the *Möbius inversion formulas*, Formula (3) expresses f in terms of its Möbius transform g. Conversely, from (3) it is easy to deduce that g is given by (2).

5.16. PROPOSITION. Let $n \geq 1$ and let $\Phi_n(X)$ be the n-th cyclotomic polynomial over $\mathbb{Q}$. Then

(i) $X^n - 1 = \prod_{d|n} \Phi_d(X)$

(ii) $\Phi_n(X) = \prod_{d|n} (X^d - 1)^{\mu(n/d)}$

(iii) If p is a prime, then

$$\Phi_p(X) = X^{p-1} + X^{p-2} + \dots + 1$$

and, for any integer $r \geq 1$,

$$\Phi_{p^r}(X) = \Phi_p(X^{p^{r-1}})$$

(iv) If n is odd, then

$$\Phi_{2n}(X) = \Phi_n(-X)$$

(v) If p is a prime not dividing n, then

$$\Phi_{pn}(X) = \frac{\Phi_n(X^p)}{\Phi_n(X)}$$

(vi) If $n = p_1^{r_1} p_2^{r_2} \cdots p_s^{r_s}$ is the canonical decomposition of n, then

$$\Phi_n(X) = \Phi_{p_1 p_2 \cdots p_s}(X^{p_1^{r_1-1} \cdots p_s^{r_s-1}})$$

Proof. (i) This is derived by collecting together all terms $(X-\varepsilon)$ of $\Phi_n(X)$ belonging to roots of unity of the same order

(ii) Apply (i) and the Möbius inversion formula (2), written multiplicatively.

(iii) We have

$$X^p - 1 = (X-1)(X^{p-1} + X^{p-2} + \dots + 1)$$

Any primitive p-th root of unity is a root of the second factor on the right of this equation. Since there are exactly $p - 1$ primitive p-th roots of unity, we conclude that these roots are precisely the roots of $X^{p-1} + X^{p-2} + \dots + 1$. Hence

$$\Phi_p(X) = X^{p-1} + X^{p-2} + \dots + 1.$$

The second assertion follows from the fact that $\deg\Phi_p(X^{p^{r-1}}) = \deg\Phi_{p^r}(X)$ and that any primitive p^r-th root of unity ε is a root of $\Phi_p(X^{p^{r-1}})$. The latter can be seen by observing that $\varepsilon^{p^{r-1}}$ is a primitive p-th root of unity, hence a root of $\Phi_p(X)$.

(iv) Since n is odd, $\phi(2n) = \phi(n)$ and a typical $2n$-th root of unity is of the form $-\varepsilon$ where ε is a primitive n-th root of unity. But $-\varepsilon$ is a root of $\Phi_n(-X)$, hence the assertion.

(v) Put $\psi(X) = \Phi_n(X^p)/\Phi_{pn}(X)$. If ε is a primitive pn-th root of unity, then ε^p is a primitive n-th root of unity, hence ε is a root of $\Phi_n(X^p)$. Thus $\Phi_{pn}(X) | \Phi_n(X^p)$ and so $\psi(X)$ is a monic polynomial in $\mathbb{Q}[X]$ of degree $\phi(n)$. If δ is a primitive n-th root of unity, then so is δ^p since $(p,n) = 1$. Hence δ is a root of $\Phi_n(X^p)$ and therefore of $\psi(X)$. This proves that $\psi(X) = \Phi_n(X)$,

Deg $\Phi_{p_1 \cdots p_s}(x) = (p_1 - 1) \cdots (p_s - 1)$

hence the assertion.

(vi) The polynomial $\Phi_{p_1 p_2 \cdots p_s}(X^{p_1^{r_1 - 1} \cdots p_s^{r_s - 1}})$ is obviously of degree $\phi(n)$. On the other hand, if ε is a primitive n-th root of unity, then $\varepsilon^{p_1^{r_1-1} \cdots p_s^{r_s-1}}$ is a primitive $p_1 p_2 \cdots p_s$-th root of unity. Hence ε is a root of

$$\Phi_{p_1 p_2 \cdots p_s}(X^{p_1^{r_1 - 1} \cdots p_s^{r_s - 1}}),$$

which shows that it must be equal to $\Phi_n(X)$. ■

The rest of this section is devoted to exhibiting some general properties of cyclotomic extensions. The following two preliminary observations will clear our path.

5.17. LEMMA. Let $F(\theta)/F$ be a separable field extension and let E/F be an arbitrary field extension. Let $f = f_1 \cdots f_r$ be the decomposition of the minimal polynomial f of θ over F into irreducible factors over E. Then

$$E \otimes_F F(\theta) \cong E(\alpha_1) \times E(\alpha_2) \times \cdots \times E(\alpha_r)$$

where α_i is a root of f_i, $1 \leqslant i \leqslant r$. In particular, if K/F is a finite Galois extension, then for any field extension E/F, $E \otimes_F K$ is isomorphic to a direct product of $(K:F)$ copies of E.

Proof. Since $F(\theta) \cong F[X]/(f)$ we have an exact sequence

$$0 \longrightarrow (f) \longrightarrow F[X] \longrightarrow F(\theta) \longrightarrow 0$$

Tensoring with E, we obtain

$$0 \longrightarrow (f) \longrightarrow E[X] \longrightarrow E \otimes_F F(\theta) \longrightarrow 0$$

Thus

$$\begin{aligned} E \otimes_F F(\theta) &\cong E[X]/(f) \cong E[X]/(f_1) \times \cdots \times E[X]/(f_r) \\ &\cong E(\alpha_1) \times \cdots \times E(\alpha_r) \end{aligned}$$

Suppose now that K/F is a finite Galois extension. Then writing $K = F(\theta)$ it follows that f splits into $r = (K:F)$ linear factors over K. Hence for any field extension E/F, $E \otimes_F K$ is isomorphic to a direct product of r copies of E. ■

5.18. LEMMA. Let p be a prime, let F be a field of characteristic distinct from p, and for each integer $n \geqslant 1$ let δ_n be a primitive p^n-th root of unity over F. Then, for any $n \geqslant 2$,

$$F(\delta_n) \supset F(\delta_{n-1}) \quad \text{implies} \quad (F(\delta_n):F(\delta_{n-1})) = p$$

In particular, if $\delta_1 \in F$ then $(F(\delta_n):F)$ is a nonnegative power of p.

Proof. We first note that $\delta_n^p \in \langle\delta_{n-1}\rangle \subseteq F(\delta_{n-1})$, so that δ_n is a root of the polynomial

$$f(X) = X^p - \delta_n^p \in F(\delta_{n-1})[X]$$

Next observe that because $n \geqslant 2$, any root of $f(X)$ is a primitive p^n-th root of unity and hence generates $F(\delta_n)$ over $F(\delta_{n-1})$. Consequently, it may be inferred that all irreducible factors of f have the same degree, namely,

$$r = (F(\delta_n):F(\delta_{n-1}))$$

But then r divides $p = \deg f$ and $r \neq 1$; thus we must have $r = p$.

Suppose now that $\delta_1 \in F$. Then

$$F = F(\delta_1) \subseteq \cdots \subseteq F(\delta_{i-1}) \subseteq F(\delta_i) \subseteq \cdots \subseteq F(\delta_n)$$

and, by the above, $(F(\delta_i):F(\delta_{i-1})) = 1$ or p for all $i \geqslant 2$. This proves the lemma. ■

We are now ready to prove the following result in which ε_n denotes a primitive n-th root of unity over F.

5.19. PROPOSITION. Let p be a prime, let r be a natural number such that $p \nmid \phi(r)$, and let F be a field such that $\varepsilon_p \in F$ and $\operatorname{char}F \nmid r$. Then for any $j \geqslant 1$,

(i) $(F(\varepsilon_{p^j}):F) = (F(\varepsilon_{p^j},\varepsilon_r):F(\varepsilon_r))$

(ii) $F(\varepsilon_r) \otimes_F F(\varepsilon_{p^j}) \cong F(\varepsilon_r,\varepsilon_{p^j})$

Proof. (i) It follows from Lemma 5.18 that there exist nonnegative integers k,s such that

$$(F(\varepsilon_{p^j}):F) = p^k \quad \text{and} \quad (F(\varepsilon_{p^j},\varepsilon_r):F(\varepsilon_r)) = p^s$$

Note also that because $p \nmid \phi(r)$, p divides neither $(F(\varepsilon_{p^j},\varepsilon_r):F(\varepsilon_{p^j}))$ nor

$[F(\varepsilon_r):F] \mid \varphi(r)$

$(F(\varepsilon_r):F)$. Denoting the above dimensions by m and n, respectively, we deduce that

$$(F(\varepsilon_{p^j},\varepsilon_r):F) = p^k m = p^s n$$

where $p \nmid m$ and $p \nmid n$. Thus $k = s$, as asserted.

(ii) If f is a minimal polynomial of ε_{p^s} over F, then by (i), f stays irreducible over $F(\varepsilon_r)$. The desired conclusion now follows from Lemma 5.17 applied for the case $E = F(\varepsilon_r)$ and $\theta = \varepsilon_{p^j}$. ■

5.20. PROPOSITION. Let p be a prime, let F be a field of characteristic distinct from p, and for each integer $m \geqslant 1$, let ε_m be a primitive p^m-th root of unity over F. Let the number q be defined by

$$q = \begin{cases} 1 & \text{if } p \neq 2 \\ 2 & \text{if } p = 2 \end{cases}$$

Then exactly one of the following two properties hold:

(i) For all $m \geqslant q$, $F(\varepsilon_q) = F(\varepsilon_m)$

(ii) There exists an integer $n \geqslant q$ such that

$$F(\varepsilon_q) = F(\varepsilon_{q+1}) = \dots = F(\varepsilon_n) \subset F(\varepsilon_{n+1}) \subset \dots$$

Proof. Assume that (i) does not hold. Then by choosing the least integer m violating (i) and setting $n = m - 1$, we derive

$$F(\varepsilon_q) = F(\varepsilon_{q+1}) = \dots = F(\varepsilon_n) \neq F(\varepsilon_{n+1})$$

Because $F(\varepsilon_i) \subseteq F(\varepsilon_{i+1})$ for all i, it suffices to show that if $t > n$, then $F(\varepsilon_t) \neq F(\varepsilon_{t+1})$. Assume the contrary. Then choosing the minimal $t > n$ with $F(\varepsilon_t) = F(\varepsilon_{t+1})$, we must have $F(\varepsilon_{t-1}) \subset F(\varepsilon_t)$. Writing $K = F(\varepsilon_{t-1})$ for convenience, we have $\varepsilon_{t+1}^{p^2} \in K$ so that ε_{t+1} is a root of the polynomial

$$g(X) = X^{p^2} - \varepsilon_{t+1}^{p^2} \in K[X]$$

Owing to Lemma 5.18, $K(\varepsilon_{t+1}) = F(\varepsilon_{t+1})$ has degree p over K, so the irreducible polynomial $f(X) \in K[X]$ satisfied by ε_{t+1} has degree p and divides $g(X)$. However all the roots of $g(X)$ are of the form $\varepsilon_{t+1}\delta$ where δ is a p^2-th root of unity. Therefore, there exist exactly p p^2-th roots of unity, say $\delta_1,\ldots,\delta_p$, such that $\varepsilon_{t+1}\delta_1, \varepsilon_{t+1}\delta_2,\ldots,\varepsilon_{t+1}\delta_p$ are all roots of $f(X)$. Now taking the product of the roots of $f(X)$ we have

$$\varepsilon_{t+1}^p(\delta_1\delta_2\ldots\delta_p) \in K$$

If $t - 1 \geqslant 2$, then $\varepsilon_2 \in K$, so $\delta_i \in K$ and therefore $\varepsilon_{t+1}^p \in K$. Since the latter contradicts the fact that $F(\varepsilon_{t-1}) \subset F(\varepsilon_t)$, we conclude that $t = 2$. Hence

$$F(\varepsilon_1) \subset F(\varepsilon_2) = F(\varepsilon_3)$$

and $p \neq 2$.

We now show that the latter configuration is impossible. Because $\varepsilon_3^{p^2}$ and ε_3^p are, respectively a p-th and a p^2-th root of unity, we may assume that $\varepsilon_1 = \varepsilon_3^{p^2}$ and $\varepsilon_2 = \varepsilon_3^p$. By Lemma 5.18, $(F(\varepsilon_2):F(\varepsilon_1)) = p$ and so the Galois group of $F(\varepsilon_3)$ over $F(\varepsilon_1)$ has a generator ψ of order p. With $\psi(\varepsilon_3) = \varepsilon_3^\mu$ for some $\mu \in \mathbb{Z}$, it follows that $\mu^p \equiv 1 \pmod{p^3}$. Bearing in mind that

$$\psi(\varepsilon_1) = \varepsilon_3^{\mu p^2},$$

we must also have $\mu p^2 \equiv p^2 \pmod{p^3}$ so that $\mu \equiv 1 \pmod{p}$. Furthermore, $\mu \not\equiv 1 \pmod{p^2}$, since otherwise $\psi(\varepsilon_2) = \varepsilon_2$. Therefore, there is an integer k such that $\mu = 1 + kp$ and $p \nmid k$. It follows that

$$\begin{aligned}\mu^p &= (1+kp)^p \\ &= 1 + \binom{p}{1}(kp) + \binom{p}{2}(kp)^2 + \ldots + (kp)^p \\ &\equiv 1 + kp^2 \not\equiv 1 \pmod{p^3},\end{aligned}$$

a contradiction. ■

The following simple observation will be needed for the proof of the final result of this section.

5.21. LEMMA. Let F be a finite field of q elements and let $n \geqslant 1$ be an integer. Then

$$\sum_{\alpha\in F} \alpha^n = \begin{cases} 0, & \text{if } n \not\equiv 0 \pmod{q-1} \\ q-1, & \text{if } n \equiv 0 \pmod{q-1} \end{cases}$$

Proof. Let ε be a generator of F^* and let $\lambda = \sum_{\alpha\in F} \alpha^n$. Then

$$\lambda = \sum_{\alpha\in F^*} \alpha^n = \sum_{i=0}^{q-2} \varepsilon^{ni}$$

which is the sum of a geometric progression. Therefore, if $\varepsilon^n \neq 1$, i.e. if $n \not\equiv 0 \pmod{q-1}$, then

$$\lambda = \frac{\varepsilon^{n(q-1)} - 1}{\varepsilon^n - 1} = 0$$

since $\varepsilon^{q-1} = 1$. If $\varepsilon^n = 1$, i.e. if $n \equiv 0 \pmod{q-1}$, then $\lambda = q-1$, as required. ∎

Let F be a field and let $f \in F[X_1,\ldots,X_n]$. By a *zero* of f in

$$F^n = F \times \ldots \times F$$

($n$times) one means an element $(\alpha_1,\ldots,\alpha_n) \in F^n$ such that

$$f(\alpha_1,\ldots,\alpha_n) = 0$$

By a *nontrivial zero* we mean a zero $(\alpha_1,\ldots,\alpha_n)$ such that some $\alpha_i \neq 0$. Given a field F and an integer $j \geqslant 0$, one says that F is a C_j*-field* if every homogeneous polynomial over F of degree d in $n > d^j$ variables, has a nontrivial zero in F^n.

5.22. PROPOSITION. (Chevalley). Every finite field is a C_1-field.

Proof. Let F be a finite field of q elements, let $f(X_1,\ldots,X_n)$ be a homogeneous polynomial of degree $d < n$ over F and let $p = \text{char}F$. Let $V \subseteq F^n$ be the set of zeros of F and write α for a point $(\alpha_1,\ldots,\alpha_n) \in F^n$. Then

$$f(\alpha)^{q-1} = 0 \quad \text{for all} \quad \alpha \in V$$

and, by Proposition 5.1(i),

$$f(\alpha)^{q-1} = 1 \qquad \text{for all} \quad \alpha \in F^n - V$$

It follows that the polynomial $g(X_1,\ldots,X_n) = f(X_1,\ldots,X_n)^{q-1}$ is the characteristic function of $F^n - V$ with values in $\mathbb{F}_p$. The number modulo p of points of F^n-V is therefore equal to $\sum_{\alpha\in F^n} g(\alpha)$. We claim that this sum is zero in $\mathbb{F}_p$; if sustained, it will follow that $|F^n-V|$ is divisible by p. Since $p \mid |F^n|$, we deduce that $|V|$ is divisible by p. This will prove the result since $|V| \geqslant 1$.

To substantiate our claim, note that g is a linear combination of monomials $M_t(X) = X_1^{t_1} \dots X_n^{t_n}$. Since

$$\sum_{\alpha\in F^n} M_t(\alpha) = \sum_{\alpha\in F^n} \alpha_1^{t_1} \dots \alpha_n^{t_n} = \Big(\sum_{\alpha_1\in F} \alpha_1^{t_1}\Big) \dots \Big(\sum_{\alpha_n\in F} \alpha_n^{t_n}\Big)$$

it follows from Lemma 5.21 that $\sum_{\alpha\in F^n} \alpha_1^{t_1} \dots \alpha_n^{t_n}$ vanishes unless all the t_i's are nonzero and multiples of $q-1$. In this case, the degree $t_1 + \dots + t_n$ of the monomial is at least $(q-1)n$. But, since $g = f^{q-1}$, g has degree $(q-1)d$ and $(q-1)d < (q-1)n$ by assumption. The proof is therefore complete. ∎

6. NORMS, TRACES AND THEIR APPLICATIONS

Let R be a commutative ring and let A be an R-algebra. We say that A is *R-free* if it is free as an R-module.

Let A be an R-algebra which is R-free of finite rank. By an *R-representation* of A (or simply a *representation* of A) we understand any homomorphism

$$A \longrightarrow \operatorname{End}_R(V)$$

of R-algebras, where V is a finitely generated free R-module. If n is the rank of V, then $\operatorname{End}_R(V)$ is identifiable with the R-algebra $M_n(R)$ of all $n \times n$-matrices with entries in R. Thus the given representation of A defines a homomorphism

$$A \longrightarrow M_n(R)$$

We shall refer to any such homomorphism as a *matrix representation* of A. The special case in which $V = A$ and the image ℓ_a of $a \in A$ in $\operatorname{End}_R(V)$ is given by

$$\ell_a(v) = av \qquad (a, v \in A)$$

is of particular importance. The representation

$$\begin{cases} A \longrightarrow \underset{R}{\mathrm{End}}(A) \\ a \longmapsto \ell_a \end{cases}$$

is called the *regular representation* of A.

Let R be a commutative ring and let V be a free R-module of finite rank with basis $v_1,\dots,v_n$. If $f \in \underset{R}{\mathrm{End}}(V)$, then with respect to this basis f can be represented as an $n \times n$-matrix (a_{ij}) over R. The *trace, determinant*, and the characteristic polynomial of f are respectively,

$$Tr(f) = \sum_{i=1}^{n} a_{ii}\ , \ \det(f) = \det(a_{ij}) \text{ and } \det(X\cdot I_n - (a_{ij})),$$

where I_n is the identity $n \times n$-matrix. These quantities are independent of the choice of the basis. Note also that

$$\det(X\cdot I_n-(a_{ij})) = X^n - (Tr(f))X^{n-1} + \dots + (-1)^n\det f \tag{1}$$

Assume that R is a subring of a commutative ring S and that S is R-free of finite rank. Let $S \longrightarrow \underset{R}{\mathrm{End}}(S)$, $s \longmapsto \ell_s$ be the regular representation of S. By the *trace* (*norm*, *characteristic polynomial*) of $s \in S$ we understand the trace (determinant, characteristic polynomial of ℓ_s). The trace and norm of s will be denoted by

$$Tr_{S/R}(s) \text{ and } N_{S/R}(s)$$

respectively.

Now assume that E/F is a finite field extension. Then E is a finite-dimensional F-algebra and so for any $\lambda \in E$, we may define the trace (norm, characteristic polynomial) as above.

6.1. LEMMA. Let E/F be a finite field extension, let $\lambda \in E$ and let f be the characteristic polynomial of λ.

(i) λ is a root of f and, if $E = F(\lambda)$, then f is the minimal polynomial of λ over F.

(ii) If K/F is any field extension in which f splits into linear factors, $f = \prod_{i=1}^{n}(X-\lambda_i)$, then

$$N_{E/F}(\lambda) = \prod_{i=1}^{n}\lambda_i \text{ and } Tr_{E/F}(\lambda) = \sum_{i=1}^{n}\lambda_i$$

Proof. (i) Let M be the matrix for multiplication by λ in E with respect to a given F-basis of E. We know from matrix theory that $f(M) = 0$. This implies that $f(\lambda) = 0$ because $f(M)$ multiplies elements of E by $f(\lambda)$. If $E = F(\lambda)$, then $f(\lambda) = 0$ and f is a monic polynomial of degree equal to the degree of the minimal polynomial of λ over F, hence the assertion.

(ii) This is a direct consequence of the equality (1). ■

6.2. LEMMA. Let E/F be a finite field extension, let K be an intermediate field and let $\lambda \in K$. Let $g(X)$ be the characteristic polynomial of λ over F, when λ is regarded as an element of K, and let $f(X)$ be the characteristic polynomial of λ over F, when λ is regarded as an element of E. Then

(i) $f(X) = g(X)^m$, where $m = (E:K)$

(ii) $N_{E/F}(\lambda) = [N_{K/F}(\lambda)]^m$

(iii) $Tr_{E/F}(\lambda) = m[Tr_{K/F}(\lambda)]$

Proof. Properties (ii) and (iii) follow from (i) and Lemma 6.1(ii). To prove (i), let $\{y_i | 1 \leqslant i \leqslant k\}$ be a basis of K over F and let $\{z_j | 1 \leqslant j \leqslant m\}$ be a basis of E over K. Then, by Proposition 1.1, $\{y_i z_j | 1 \leqslant i \leqslant k, 1 \leqslant j \leqslant m\}$ is a basis of E over F. Let $M = (a_{ih})$ be the matrix for multiplication by λ in K with respect to the basis $\{y_i\}$. Then $\lambda y_i = \sum_h a_{ih} y_h$ and therefore

$$\lambda(y_i z_j) = (\sum_h a_{ih} y_h) z_j = \sum_h a_{ih}(y_h z_j)$$

Ordering lexicographically the basis $\{y_i z_j\}$ of E over F, we see that the matrix M_1 for multiplication by λ in E with respect to this basis is given by

$$M_1 = \begin{bmatrix} M & 0 & \dots & 0 \\ 0 & M & \dots & 0 \\ - & - & - & - \\ 0 & 0 & \dots & M \end{bmatrix}$$

Hence the matrix $X \cdot I_n - M_1$, $n = (E:F)$, consists of m diagonal blocks, each of the form $X \cdot I_k - M$. This shows that

$$f(X) = \det(X \cdot I_n - M_1) = (\det(X \cdot I_k - M))^m = g(X)^m$$

as required. ∎

6.3. COROLLARY. Let E/F be a finite field extension, let $\lambda \in E$ and let $f(X)$ and $g(X)$ be the characteristic and minimal polynomials of λ over F, respectively. Then

$$f(X) = g(X)^n \quad \text{where } n = (E:F(\lambda)).$$

In particular, $f(X) = g(X)$ if and only if $E = F(\lambda)$.

Proof. Put $K = F(\lambda)$ and apply Lemmas 6.1(i) and 6.2(i). ∎

We are now ready to prove the following important result.

6.4. PROPOSITION. Let E/F be a finite field extension, let $\bar{E}$ be an algebraic closure of E and let $f(X)$ be the characteristic polynomial of $\lambda \in E$. Let $\sigma_1, \sigma_2, \ldots, \sigma_n$, $n = (E:F)_s$, be all distinct F-homomorphisms of E into $\bar{E}$ and let $p^e = (E:F)_i$. Then

$$f(X) = \prod_{i=1}^{n} (X - \sigma_i(\lambda))^{p^e}$$

and $\sigma_1(\lambda), \ldots, \sigma_n(\lambda)$ are all conjugates of λ in $\bar{E}$, each repeated $(E:F(\lambda))_s$ times. In particular, by Lemma 6.1(ii),

$$N_{E/F}(\lambda) = \left(\prod_{i=1}^{n} \sigma_i(\lambda)\right)^{p^e}$$

$$Tr_{E/F}(\lambda) = p^e \sum_{i=1}^{n} \sigma_i(\lambda)$$

Proof. Put $t = (F(\lambda):F)_s$, $k = (E:F(\lambda))_s$, $r = (F(\lambda):F)_i$ and $a = (E:F(\lambda))$. Then, by Corollary 3.12(ii), $p^e k = ar$. By Lemma 3.6, there exist exactly t, say $\lambda = \lambda_1, \ldots, \lambda_t$ of conjugates of λ in $\bar{E}$. Denote by L the normal closure of E/F. Then $L \subseteq \bar{E}$ and any F-homomorphism of $F(\lambda)$ into $\bar{E}$ is in fact an F-homomorphism of $F(\lambda)$ into L. Hence, by Proposition 2.10 and Corollary 3.12(ii), each of the t F-homomorphisms of $F(\lambda)$ into L has exactly

$$(E:F)_s/(F(\lambda):F)_s = (E:F(\lambda))_s$$

extensions among the σ_i. Therefore, each of the conjugates λ_j of λ occurs $(E:F(\lambda))_s$ times in the set $\{\sigma_1(\lambda), \ldots, \sigma_n(\lambda)\}$. It follows that

$$\prod_{i=1}^{n} (X - \sigma_i(\lambda))^{p^e} = \prod_{i=1}^{t} (X - \lambda_i)^{p^e k} = \prod_{i=1}^{t} (X - \lambda_i)^{ar}$$

Let $g(X)$ be the minimal polynomial of λ over F. By Proposition 3.5(ii), all the roots $\lambda_1,\ldots,\lambda_t$ of $g(X)$ have the same multiplicity equal to r. Hence

$$g(X) = \prod_{i=1}^{t}(X-\lambda_i)^r$$

and, applying Corollary 6.3, we derive

$$f(X) = \prod_{i=1}^{t}(X-\lambda_i)^{ar}$$

as required. ■

6.5. COROLLARY. Let E/F be a finite separable extension of degree n, let $\sigma_1,\ldots,\sigma_n$ be all distinct F-homomorphisms of E into its algebraic closure $\bar{E}$ and let $f(X)$ be the characteristic polynomial of $\lambda \in E$. Then

(i) $f(X) = \prod_{i=1}^{n}(X-\sigma_i(\lambda))$

and $\sigma_1(\lambda),\ldots,\sigma_n(\lambda)$ are all conjugates of λ in $\bar{E}$, each repeated $(E:F(\lambda))$ times

(ii) $N_{E/F}(\lambda) = \prod_{i=1}^{n}\sigma_i(\lambda)$

(iii) $Tr_{E/F}(\lambda) = \sum_{i=1}^{n}\sigma_i(\lambda)$

Proof. Since E/F is separable, we have $(E:F)_s = (E:F)$ and $(E:F)_i = 1$. Now apply Proposition 6.4. ■

6.6. PROPOSITION. Let E/F be a field extension of degree n.

(i) $Tr_{E/F}(x+y) = Tr_{E/F}(x) + Tr_{E/F}(y)$ $(x,y \in E)$

(ii) $Tr_{E/F}(\lambda x) = \lambda Tr_{E/F}(x)$ $(x \in E, \lambda \in F)$

(iii) $N_{E/F}(xy) = N_{E/F}(x)N_{E/F}(y)$ $(x,y \in E)$

(iv) $N_{E/F}(\lambda x) = \lambda^n N_{E/F}(x)$ $(x \in E, \lambda \in F)$

(v) If $F \subseteq K \subseteq E$ is a chain of fields, then

(a) $Tr_{E/F}(x) = Tr_{K/F}(Tr_{E/K}(x))$ $(x \in E)$

(b) $N_{E/F}(x) = N_{K/F}(N_{E/K}(x))$ $(x \in E)$

Proof. The proof of (i) - (iv) is obvious. Let $a_1,\ldots,a_k$ be a basis of K over F and $b_1,\ldots,b_m$ a basis of E over K. For x in E and y in K, let

$$xb_i = \Sigma\beta_{ij}(x)b_j \ , \ ya_i = \Sigma\alpha_{ij}(y)a_j$$

Then

$$Tr_{K/F}(y) = \Sigma\alpha_{ii}(y),\ Tr_{E/K}(x) = \Sigma\beta_{ii}(x)$$

It follows that

$$Tr_{K/F}(Tr_{E/K}(x)) = \Sigma\Sigma\alpha_{ii}(\beta_{jj}(x))$$

The products $a_i b_j$ give a basis of E over F and

$$\begin{aligned} xa_s b_t &= \sum_j a_s \beta_{tj}(x) b_j \\ &= \sum_j \sum_i \alpha_{si}(\beta_{tj}(x)) a_i b_j \end{aligned}$$

Thus

$$Tr_{E/F}(x) = \Sigma\Sigma\alpha_{ii}(\beta_{jj}(x)),$$

proving (a).

We employ the notation of the proof of Proposition 2.10. We may assume that ψ_1 is the identity automorphism of L. We have by Proposition 6.4.

$$N_{E/K}(x) = \Big(\prod_{i=1}^{n}\phi_i(x)\Big)^{p^a},\quad p^a = (E:K)_i$$

$$\begin{aligned} N_{K/F}(N_{E/K}(x)) &= \Big(\prod_{j=1}^{m}\psi_j\, N_{E/K}(x)\Big)^{p^\beta},\quad p^\beta = (K:F)_i \\ &= \Big(\prod_{j=1}^{m}\prod_{i=1}^{n}\psi_j\phi_i(x)\Big)^{p^{\alpha+\beta}} \end{aligned}$$

Now, we know from the proof of Proposition 2.10, that the restrictions of the products $\psi_j\phi_i$ to E are distinct and give all F-homomorphisms of E into L. Furthermore, by Corollary 3.12(ii), $p^{\alpha+\beta} = (E:F)_i$ the desired conclusion is therefore a consequence of Proposition 6.4. ■

Let R be a commutative ring, and let M,N be R-modules. An *R-bilinear form* on $M \times N$ is a map

$$f : M \times N \longrightarrow R$$

having the following properties: For each $x \in M$, the map

$$y \longmapsto f(x,y)$$

is R-linear, and for each $y \in N$, the map

$$x \longmapsto f(x,y)$$

is R-linear. We shall omit the subscript R and write $<x,y>$ instead of $f(x,y)$.

The bilinear form $f : M \times N \longrightarrow R$ is said to be *nonsingular* if

(i) $<x,b> = 0$ for all $x \in M$ implies $b = 0$, and

(ii) $<a,y> = 0$ for all $y \in N$ implies $a = 0$

We shall refer to a bilinear form $f : M \times M \longrightarrow R$ as being *symmetric* if

$$<x,y> = <y,x> \qquad \text{for all } x,y \in M$$

For future use, we next record the following standard result.

6.7. PROPOSITION. Given two finite-dimensional vector spaces U and V over a field F, let $U \times V \longrightarrow F$ be a nonsingular bilinear form. Then

$$\dim U = \dim V = n,$$

say, and to any basis $u_1,\ldots,u_n$ of U there corresponds just one family of elements $v_1,\ldots,v_n$ of V such that

$$<u_i,v_j> = \delta_{ij} \qquad (\delta_{ij} = 0 \text{ if } i \neq j,\ \delta_{ii} = 1).$$

Moreover, $v_1,\ldots,v_n$ is a basis of V

Proof. The rule

$$y \longmapsto (<u_1,y>,\ldots,<u_n,y>) \tag{1}$$

defines a mapping $V \longrightarrow F^n$ which is easily seen to be linear. The kernel consists of all $y \in V$ such that $<u_1,y> = \ldots = <u_n,y> = 0$. This clearly implies that $y = 0$ since the given form is nonsingular. Thus the mapping (1) is injective, whence $\dim V \leqslant n = \dim U$. By symmetry we also have the opposite inequality, and combining the two we find $\dim U = \dim V$.

The map (1) is an injective F-homomorphism from one n-dimensional space to another and hence is an isomorphism. Take the standard basis $e_1,\ldots,e_n$ in F^n and let v_i be the inverse image of e_i under the mapping (1). Then $<u_i,v_j> = \delta_{ij}$, as claimed.

Finally, if $v'_1,\ldots,v'_n$ is another family with this property, then each $v_i-v'_i$ lies in the kernel of the mapping (6) and so must be zero. Hence there is only one such family. Now the fact that the v_i's form a basis follows because they are the inverse image, under an F-isomorphism, of a basis of F^n. ∎

Let U and V be two finite-dimensional vector spaces of dimension n over a field F and let $U \times V \longrightarrow F$ be a nonsingular bilinear form. Any two bases $u_1,\ldots,u_n$ and $v_1,\ldots,v_n$ of U and V, respectively, satisfying

$$<u_i,v_j> = \delta_{ij}$$

are said to be *dual* to one another with respect to the given form.

Let E/F be a finite field extension. Then the trace map may be used to define a symmetric bilinear form $E \times E \longrightarrow F$ by the rule

$$<x,y> = Tr_{E/F}(xy)$$

The bilinear properties are easily verified and also the symmetric property $<x,y> = <y,x>$.

6.8. PROPOSITION. Let E/F be a finite field extension. Then the following conditions are equivalent:

(i) The bilinear form $<x,y> = Tr_{E/F}(xy)$ is nonsingular

(ii) $Tr_{E/F}(x) \neq 0$ for some $x \in E$

(iii) The extension E/F is separable.

Proof. (i) ⇒ (ii): Obvious

(ii) ⇒ (iii): Direct consequence of Proposition 6.4.

(iii) ⇒ (i): Assume E/F is separable. Then, by Corollary 3.32, $E = F(\theta)$ for some $\theta \in E$. It follows that $1,\theta,\ldots,\theta^{n-1}$ is a basis of E over F for some $n \geqslant 1$. Thus

$$<x,\ a_i\theta^i> = \Sigma a_i<x,\theta^i>$$

and so $<x,E> = 0$ if and only if $<x,\theta^i> = 0$ for all $i \in \{0,1,\ldots,n-1\}$. Our task is to prove that $x = 0$ under these conditions.

Let $x = \Sigma b_i\theta^i$. Then

$$<x,\theta^j> = \Sigma b_i<\theta^i,\theta^j> \tag{2}$$

Let D denote the matrix (d_{ij}) with $d_{ij} = <\theta^{i-1},\theta^{j-1}>$. The equation (2) and the assumption $\langle x,E\rangle = 0$ ensure that

$$(b_0, b_1, \ldots, b_{n-1}) D = (0,0,\ldots,0)$$

It will be shown that D is nonsingular which will imply that each $b_i = 0$ and hence $x = 0$.

Let $f(X)$ be the minimal polynomial of θ over F and let $\bar{E}$ be an algebraic closure of E. Since E/F is separable, $f(X)$ has n distinct roots, say $\theta_1, \ldots, \theta_n$ in $\bar{E}$. Hence, by Lemma 6.1,

$$Tr_{E/F}(\theta) = \sum_{i=1}^{n} \theta_i \qquad (\theta = \theta_1)$$

We wish to find a similar formula for $Tr_{E/F}(\theta^k)$. The linear transformation ℓ_θ has the distinct characteristic roots $\theta_1, \ldots, \theta_n$ in $\bar{E}$. Hence the matrix for ℓ_θ (with respect to some basis of E over F) can be diagonalized over $\bar{E}$. So there exists U such that

$$U^{-1} \ell_\theta U = \text{diag}\{\theta_1, \ldots, \theta_n\}$$

Raising this to the k-th power we deduce that

$$U^{-1} \ell_{\theta^k} U = \text{diag}\{\theta_1^k, \ldots, \theta_n^k\}$$

Taking traces we get

$$Tr_{E/F}(\theta^k) = \sum_{i=1}^{n} \theta_i^k \tag{3}$$

Now let $V = V(\theta_1, \ldots, \theta_n)$ denote the matrix

$$\begin{bmatrix} 1 & 1 & \cdots & 1 \\ \theta_1 & \theta_2 & \cdots & \theta_n \\ \theta_1^2 & \theta_2^2 & \cdots & \theta_n^2 \\ \cdot & \cdot & & \cdot \\ \cdot & \cdot & & \cdot \\ \cdot & \cdot & & \cdot \\ \theta_1^{n-1} & \theta_2^{n-1} & & \theta_n^{n-1} \end{bmatrix}$$

and V^t the transposed matrix. Applying (3), we see that the (i,j)-th entry of VV^t is

$$\sum_k \theta_k^{i-1} \theta_k^{j-1} = \sum_k \theta_k^{i+j-2} = Tr_{E/F}(\theta^{i-1}\theta^{j-1}) = d_{ij}$$

It therefore follows that $VV^t = D$. Because V and V^t have the same determinant, we conclude that

$$\det D = (\det V)^2$$

On the other hand, the matrix V is a van der Monde matrix and

$$\det V = \prod_{i>j} (\theta_i - \theta_j)$$

The θ_i are distinct, so $\det V \neq 0$ and this proves that D is nonsingular, as we wished to show. ■

6.9. COROLLARY. Let E/F be a finite separable field extension and let $u_1,\ldots,u_n$ be a basis of E over F. Then there exists a second basis $v_1,\ldots,v_n$ with the property

$$Tr_{E/F}(u_i v_j) = \delta_{ij}$$

Proof. Apply Propositions 6.7 and 6.8. ■

A related property to the above corollary is given by

6.10. PROPOSITION. Let E/F be a finite separable extension, let $\alpha \in E$ be such that $E = F(\alpha)$ and let $f(X)$ be the minimal polynomial of α over F. Let

$$\frac{f(X)}{X-\alpha} = \beta_0 + \beta_1 X + \ldots + \beta_{n-1} X^{n-1} \qquad (\beta_i \in E)$$

Then the dual basis of $1, \alpha,\ldots,\alpha^{n-1}$ is

$$\frac{\beta_0}{f'(\alpha)}, \ldots, \frac{\beta_{n-1}}{f'(\alpha)}$$

Proof. Let $\alpha_1,\ldots,\alpha_n$ be the distinct roots of f. We claim that

$$\sum_{i=1}^{n} \frac{f(X)}{(X-\alpha_i)} \frac{\alpha_i^r}{f'(\alpha_i)} = X^r \qquad (0 \leqslant r \leqslant n-1) \qquad (4)$$

Indeed, denote by $g(X)$ the difference of the left and right-hand side of this equality. Then $\deg g \leqslant n-1$ and g has n roots $\alpha_1,\ldots,\alpha_n$. Hence g is identically zero, as claimed.

The polynomials

$$\frac{f(X)}{(X-\alpha_i)} \frac{\alpha_i^r}{f'(\alpha_i)}$$

are all conjugate to each other. Define the trace of a polynomial in $E[X]$ to be the polynomial obtained by applying $Tr_{E/F}$ to the coefficients. Then, by (4),

$$Tr\left[\frac{f(X)}{(X-\alpha)}\ \frac{\alpha^r}{f'(\alpha)}\right] = X^r$$

Comparing the coefficients of each power of X in this equation, we derive

$$Tr_{E/F}(\alpha^i\ \frac{\beta_j}{f'(\alpha)}) = \delta_{ij}$$

as required. ■

Let E/F be a finite field extension and let $f(X)$ be the characteristic polynomial of $\lambda \in E$ over F. It will next be shown that $f(X)$ can itself be interpreted as a norm. In what follows, we write $F(X)$ for the field of all rational functions in the indeterminate X. The observation below will clear our path.

6.11. LEMMA. Let E/F be a finite extension of degree n with F-basis $\lambda_1,\ldots,\lambda_n$. Then $\lambda_1,\ldots,\lambda_n$ is also a basis of $E(X)$ over $F(X)$.

Proof. We first observe that the algebraic closure of $F(X)$ in $E(X)$ contains E (since E/F is an algebraic extension) and X (since $X \in F(X)$), hence coincides with $E(X)$. In other words, $E(X)$ is an algebraic extension of $F(X)$. Moreover, since

$$E = \sum_{i=1}^{n} F\lambda_i = F(\lambda_1,\ldots,\lambda_n),$$

we have

$$E(X) = F(X)(\lambda_1,\ldots,\lambda_n)$$

Therefore, by Proposition 1.1.3,

$$E(X) = F(X)[\lambda_1,\ldots,\lambda_n] = F(X)E = \sum_{i=1}^{n} F(X)\lambda_i$$

and so we need only verify that the λ_i's are linearly independent over $F(X)$. But this follows immediately from the linear independence of the λ_i's over F and from the fact that X is transcendental over $F(\lambda_1,\ldots,\lambda_n)$. So the lemma is true. ■

We are now ready to prove

6.12. PROPOSITION. Let E/F be a finite field extension and let $f(X)$ be the characteristic polynomial of $\lambda \in E$ over F. Then

$$N_{E(X)/F(X)}(X-\lambda) = f(X)$$

Proof. Let $\{u_1,\ldots,u_n\}$ be an F-basis of E. Then, by Lemma 6.11, $\{u_1,\ldots,u_n\}$ is an $F(X)$-basis of $E(X)$. Write

$$\lambda u_j = \sum_{i=1}^{n} a_{ij}u_i \qquad (a_{ij} \in F)$$

or, in matrix notation, $\lambda(u_1,\ldots,u_n) = (u_1,\ldots,u_n)A$, where $A = (a_{ij})$. It follows that

$$\begin{aligned}(X-\lambda)(u_1,\ldots,u_n) &= X(u_1,\ldots,u_n) - \lambda(u_1,\ldots,u_n)\\ &= (u_1,\ldots,u_n)(X\cdot I_n - A)\end{aligned}$$

Hence

$$N_{E(X)/F(X)}(X-\lambda) = \det(X\cdot I_n - A) = f(X)$$

as asserted. ∎

The rest of this section is devoted to some applications of norms and traces.

By an *algebraic number field*, we understand a finite field extension of $\mathbb{Q}$. Let F be an algebraic number field. The integral closure of $\mathbb{Z}$ in F is a subring of F, called the *ring of integers* of F.

6.13. PROPOSITION. Let R be a subring of a field F and let S be the integral closure of R in a finite field extension E of F.

(i) $N_{E/F}(S) \subseteq S$ and $Tr_{E/F}(S) \subseteq S$

(ii) If F is the quotient field of R and R is integrally closed, then

$$N_{E/F}(S) \subseteq R \text{ and } Tr_{E/F}(S) \subseteq R$$

In particular, if S is the ring of integers of an algebraic number field E, then

$$N_{E/\mathbb{Q}}(S) \subseteq \mathbb{Z} \text{ and } Tr_{E/\mathbb{Q}}(S) \subseteq \mathbb{Z}$$

Proof. (i) Fix $s \in S$ and let $g(X)$ and $f(X)$ be the minimal and characteristic polynomials of s over F, respectively. Since s is integral over

R, Proposition 2.3 tells us that $g(X) \in S[X]$. Hence, by Corollary 6.3, $f(X) \in S[X]$. Since $N_{E/F}(s)$ and $Tr_{E/F}(s)$ differ from certain coefficients of $f(X)$ by a factor ± 1, the assertion follows.

(ii) By hypothesis, $S \cap F = R$. Since $f(X) \in S[X] \cap F[X] = R[X]$, the result is established. ■

We are now ready to characterize the unit group of algebraic integers via the norm map.

6.14. PROPOSITION. Let R be the ring of integers of an algebraic number field F and let $r \in R$. Then the following conditions are equivalent:

(i) $r \in U(R)$

(ii) $N_{F/\mathbb{Q}}(r) = \pm 1$

(iii) There is a monic polynomial $f(X) \in \mathbb{Z}[X]$ with $f(0) = \pm 1$ and $f(r) = 0$

(iv) There is a polynomial $g(X) = a_n X^n + \dots + a_0 \in R[X]$ such that $g(r) = 0$ and $a_0 \in U(R)$.

Proof. (i) ⇒ (ii): By hypothesis $rr_1 = 1$ for some $r_1 \in R$. Hence

$$1 = N_{F/\mathbb{Q}}(rr_1) = N_{F/\mathbb{Q}}(r)N_{F/\mathbb{Q}}(r_1)$$

Since by Proposition 6.13(ii), $N_{F/\mathbb{Q}}(r) \in \mathbb{Z}$ we conclude that $N_{F/\mathbb{Q}}(r) = \pm 1$

(ii) ⇒ (iii): Let $g(X)$ be the minimal polynomial for r over F. Then, by Proposition 2.3(ii), $g(X) \in \mathbb{Z}[X]$. Hence, by Corollary 6.3, the characteristic polynomial $f(X)$ of r is also in $\mathbb{Z}[X]$. Since $N_{F/\mathbb{Q}}(r) = \pm f(0)$, the assertion follows.

(iii) ⇒ (iv): Obvious

(iv) ⇒ (i): Put $\psi(X) = a_0^{-1}X^n g(X^{-1})$. Then $\psi(X)$ is a monic polynomial in $R[X]$ and $\psi(r^{-1}) = 0$. Hence r^{-1} is integral over R and therefore $r^{-1} \in R$, as required. ■

As an application of the trace map, we now prove

6.15. PROPOSITION. Let R be an integrally closed integral domain with quotient field F, let E/F be a finite separable field extension of degree n and let S be the integral closure of R in E. Then S is an R-submodule of a free

R-module of rank n and S contains a basis of E over F.

Proof. Let $x_1,\dots,x_n$ be a basis of E over F. Each x_i is algebraic over F, so for any $i \in \{1,2,\dots,n\}$, we have an equation of the form

$$a_n x_i^n + a_{n-1}x_i^{n-1} + \dots + a_0 = 0 \qquad (a_j \in R \text{ for all } j)$$

We may assume that $a_n \neq 0$. Multiplying through by a_n^{n-1}, we see that $a_n x_i$ is integral over R. Put $x_i' = a_n x_i$, $1 \leqslant i \leqslant n$. Then $x_1',\dots,x_n'$ is a basis of E over F contained in S.

Owing to Corollary 6.9, there is another basis $y_1,\dots,y_n$ of E over F such that

$$Tr_{E/F}(x_i'y_j) = \delta_{ij} \qquad (1 \leqslant i,j \leqslant n)$$

Fix $s \in S$. Since $y_1,\dots,y_n$ is a basis of E over F, we may write

$$s = \sum_{j=1}^{n} b_j y_j$$

with $b_j \in F$. For any $i \in \{1,2,\dots,n\}$, we have $x_i's \in S$ and therefore, by Proposition 6.13(ii), $Tr_{E/F}(x_i's) \in R$. It follows that

$$Tr_{E/F}(x_i's) = Tr_{E/F}(\sum_{j=1}^{n} b_j x_i' y_j) = \sum_{j=1}^{n} b_j Tr(x_i'y_j) = b_i$$

which shows that each $b_i \in R$. Thus S is a submodule of the free R-module $\sum_{j=1}^{n} Ry_j$. ■

6.16. COROLLARY. Further to the assumptions of Proposition 6.15, assume that R is noetherian. Then S is a finitely generated R-module and, in particular, S is noetherian.

Proof. Apply Propositions 6.15 and 1.5.2. ■

6.17. COROLLARY. Let R be a *PID* with quotient field F, let E/F be a finite separable field extension of degree n and let S be the integral closure of R in E. The S is a free R-module of rank n.

Proof. Owing to Propositions 6.15 and 1.5.8, S is a free R-module of rank $\leqslant n$. On the other hand, by Proposition 6.15, S contains a basis of E over F. Therefore it is of rank n. ■

If F is an algebraic number field, then the degree $(F:\mathbb{Q})$ is called the *degree* of F.

6.18. COROLLARY. Let R be the ring of integers of an algebraic number field F of degree n. Then R is a free $\mathbb{Z}$-module of rank n.

Proof. Apply Corollary 6.17 for the case $R = \mathbb{Z}$. ∎

7. DISCRIMINANTS AND INTEGRAL BASES

Let R be a subring of a commutative ring S such that S is a free R-module of finite rank n. Given $(s_1,\ldots,s_n) \in S^n$, the *discriminant* of $(s_1,\ldots,s_n)$, written $D_{S/R}(s_1,\ldots,s_n)$ is defined by

$$D_{S/R}(s_1,\ldots,s_n) = \det(Tr_{S/R}(s_i s_j))$$

7.1. LEMMA. If $(s_1',\ldots,s_n') \in S^n$ and $s_i' = \sum_{j=1}^{n} r_{ij}s_j$ with $r_{ij} \in R$, then

$$D_{S/R}(s_1',\ldots,s_n') = (\det(r_{ij}))^2 D_{S/R}(s_1,\ldots,s_n)$$

In particular, the discriminants of any two bases of S over R differ by a square of a unit (hence generate the same prinicpal ideal of R).

Proof. We have

$$Tr_{S/R}(s_p's_q') = Tr_{S/R}\Big(\sum_{i,j} r_{pi}r_{qj}s_is_j\Big)$$
$$= \sum_{i,j} r_{pi}r_{qj} \quad T_{S/R}(s_i s_j)$$

This gives the matrix equation

$$(Tr_{S/R}(s_p's_q')) = (r_{pi})(Tr_{S/R}(s_is_j))\,{}^t(r_{qj})$$

where ${}^t(r_{qj})$ is the transpose of (r_{qj}). Taking the determinants, the result follows. ∎

Owing to Lemma 7.1, we may define the *discriminant ideal* of S over R, written $D_{S/R}$, as the principal ideal of R generated by the discriminant of any basis of S over R.

7.2. LEMMA. Assume that $D_{S/R}$ contains an element which is not a zero divisor. Then the elements $s_1,\ldots,s_n$ of S form a basis of S over R if and only if

$D_{S/R}$ is generated by $D_{S/R}(s_1,\dots,s_n)$.

Proof. It clearly suffices to prove the "if" part. Assume that $d = D_{S/R}(s_1,\dots,s_n)$ generates $D_{S/R}$. Fix a basis $e_1,\dots,e_n$ of S over R and put $d' = D_{S/R}(e_1,\dots,e_n)$. We may write

$$s_i = \sum_{j=1}^{n} r_{ij}e_j \qquad (r_{ij} \in R)$$

in which case $d = \det(r_{ij})^2 d'$, by Lemma 7.1. By hypothesis, we have $Rd = Rd'$. Therefore $d' = rd$ for some $r \in R$. It follows that

$$d(1-r\det(r_{ij})^2) = 0$$

Furthermore, d is not a zero divisor, since otherwise every element of $Rd = D_{S/R}$ would be a zero divisor. Accordingly,

$$1 - r\det(r_{ij})^2 = 0$$

which shows that $\det(r_{ij})$ is a unit of R. But then the matrix (r_{ij}) is invertibel and thus $s_1,\dots,s_n$ is a basis of S over R, as asserted. ■

Turning our attention to field extensions, we next record a number of useful observations.

7.3. LEMMA. Let R be a subring of a field F, let S be the integral closure of R in a finite field extension E of F of degree n, and let

$$\lambda_1,\dots,\lambda_n \in S.$$

Then

(i) $D_{E/F}(\lambda_1,\dots,\lambda_n) \in S$

(ii) $D_{E/F}(\lambda_1,\dots,\lambda_n) \in R$ if F is the quotient field of R and R is integrally closed

(iii) $D_{E/\mathbb{Q}}(\lambda_1,\dots,\lambda_n) \in \mathbb{Z}$ if S is the ring of integers of an algebraic number field E.

Proof. Direct consequence of Proposition 6.13. ■

7.4. PROPOSITION. Let E/F be a finite field extension of degree n and let $\lambda_1,\dots,\lambda_n$ be a basis of E over F.

(i) $D_{E/F}(\lambda_1,\dots,\lambda_n) = 0$ if and only if $Tr_{E/F}(x) = 0$ for all $x \in E$

(ii) E/F is separable if and only if $D_{E/F}(\lambda_1,\dots,\lambda_n) \neq 0$

Proof. (i) The "if" part is obvious. Assume that $D_{E/F}(\lambda_1,\dots,\lambda_n) = 0$. Then there exist n elements $\mu_1,\dots,\mu_n$ in F, not all zero, such that

$$\sum_j \mu_j Tr_{E/F}(\lambda_i\lambda_j) = 0 \qquad (1 \leq i \leq n)$$

We set $z = \sum_j \mu_j\lambda_j$. Then $z \neq 0$, and we have $Tr_{E/F}(\lambda_i z) = 0$, $1 \leq i \leq n$. It therefore follows that $Tr_{E/F}(yz) = 0$ for all y in E. If $x \in E$, we take $y = x/z$ and we find $Tr_{E/F}(x) = 0$, as required.

(ii) Apply (i) and Proposition 6.8. ∎

7.5. COROLLARY. Let E/F be a finite separable field extension of degree n and let $\lambda_1,\dots,\lambda_n \in E$. Then $\lambda_1,\dots,\lambda_n$ is a basis of E over F if and only if $D_{E/F}(\lambda_1,\dots,\lambda_n) \neq 0$.

Proof. Apply Lemma 7.2 and Proposition 7.4(ii). ∎

As an application of discriminants, we now prove

7.6. PROPOSITION. Let R be an integrally closed integral domain with quotient field F, let E/F be a finite separable field extension of degree n and let S be the integral closure of R in E. Denote by $\lambda_1,\dots,\lambda_n$ any basis of E over F contained in S (by Proposition 6.15, such $\lambda_1,\dots,\lambda_n$ always exist). If $d = D_{E/F}(\lambda_1,\dots,\lambda_n)$, then d is a nonzero element of R such that

$$\sum_{i=1}^{n} R\lambda_i \subseteq S \subseteq d^{-1}\left(\sum_{i=1}^{n} R\lambda_i\right)$$

In particular, if d is a unit of R, then S is a free R-module with $\lambda_1,\dots,\lambda_n$ as a basis.

Proof. The fact that d is a nonzero element of R is a consequence of Lemma 7.3(ii) and Corollary 7.5. Owing to Corollary 6.9, we may choose another basis $\mu_1,\dots,\mu_n$ of E over F such that

$$Tr_{E/F}(\lambda_i\mu_j) = \delta_{ij}$$

Let x be an arbitrary element of S. Then $x = \sum_{i=1}^{n} a_i\mu_i$ with $a_i \in F$. Since for any $i \in \{1,2,\dots,n\}$, $a_i = Tr_{E/F}(x\lambda_i)$ it follows from Proposition 6.13(ii) that each a_i is in R. This shows that

$$S \subseteq \sum_{i=1}^{n} R\mu_i \qquad (1)$$

Now write

$$\mu_j = \sum_{i=1}^{n} \alpha_{ji}\lambda_i \qquad (\alpha_{ji} \in F)$$

Then

$$\sum_{i=1}^{n} \alpha_{ji}\, Tr_{E/F}(\lambda_j\lambda_k) = \delta_{jk} \qquad \text{for all } j,k$$

Applying Cramer's rule, we infer that each α_{ji} is in $d^{-1}R$, whence

$$\sum_{i=1}^{n} R\mu_i \subseteq d^{-1}\left(\sum_{i=1}^{n} R\lambda_i\right) \qquad (2)$$

Comparing (1) and (2), we find $S \subseteq d^{-1}(\sum_{i=1}^{n} R\lambda_i)$. The containment $\sum_{i=1}^{n} R\lambda_i \subseteq S$ being obvious, the result follows. ■

We now provide a useful characterization of discriminants of finite separable field extensions.

7.7. PROPOSITION. Let E/F be a finite separable field extension of degree n and let $\sigma_1,\ldots,\sigma_n$ be the n distinct F-homomorphisms of E into an algebraic closure of E. If $\lambda_1,\ldots,\lambda_n$ are any n elements of E, then

$$D_{E/F}(\lambda_1,\ldots,\lambda_n) = \det(\sigma_i(\lambda_j))^2$$

Proof. Since

$$\begin{aligned} D_{E/F}(\lambda_1,\ldots,\lambda_n) &= \det(Tr_{E/F}(\lambda_i\lambda_j)) = \det(\textstyle\sum_k \sigma_k(\lambda_i\lambda_j)) \\ &= \det(\textstyle\sum_k \sigma_k(\lambda_i)\sigma_k(\lambda_j)) = \det(\sigma_k(\lambda_i))\det(\sigma_k(\lambda_i)) \\ &= \det(\sigma_i(\lambda_j))^2, \end{aligned}$$

the result follows. ■

Let E/F be a finite separable field extension of degree n. If λ is an arbitrary element of E, then the *discriminant* of this element with respect to F, denoted by $D_{E/F}(\lambda)$ is defined by

$$D_{E/F}(\lambda) = D_{E/F}(1,\lambda,\ldots,\lambda^{n-1})$$

Note that, by Corollary 7.5, $D_{E/F}(\lambda) \neq 0$ if and only if $E = F(\lambda)$.

7.8. PROPOSITION. Let E/F be a finite separable field extension of degree n, let $\lambda \in E$ be such that $E = F(\lambda)$, and let $\sigma_1,\ldots,\sigma_n$ be the n distinct F-homomorphisims of E into an algebraic closure of E. Then

$$D_{E/F}(\lambda) = (-1)^{n(n-1)/2} N_{E/F}(f'(\lambda)) = (-1)^{n(n-1)/2} \det(a_{sk})$$

where $f(X)$ is the minimal polynomial of λ over F and

$$\lambda^k f'(\lambda) = \sum_{s=0}^{n-1} a_{sk}\lambda^s \qquad (a_{sk} \in F,\ 0 \leqslant k \leqslant n-1)$$

Proof. Put $c = (-1)^{n(n-1)/2}$, $\lambda_i = \sigma_i(\lambda)$ and $b_i = \sigma_i(f'(\lambda))$, $1 \leqslant i \leqslant n$. Then

$$\begin{aligned}
D_{E/F}(\lambda) &= \det(\lambda_i^{j-1})^2 && (1 \leqslant i,j \leqslant n, \text{ by Proposition 7.7}) \\
&= \Big(\prod_{j<i}(\lambda_i-\lambda_j)\Big)^2 && \text{(Vandermonde)} \qquad (3) \\
&= c\prod_{i\neq j}(\lambda_i-\lambda_j) \\
&= c\prod_{i}\Big(\prod_{j\neq i}(\lambda_i-\lambda_j)\Big) \\
&= c\prod_{i} f'(\lambda_i) = c(b_1\ldots b_n) && (\text{since } b_i = f'(\lambda_i)) \\
&= cN_{E/F}(f'(\lambda))
\end{aligned}$$

Moreover,

$$\begin{aligned}
b_1\ldots b_n \det(\lambda_i^{j-1}) &= \det(\lambda_i^{j-1} b_i) \\
&= \det\Big(\sum_{s=0}^{n-1} \lambda_i^s a_{s,j-1}\Big) \\
&= \det(\lambda_i^{j-1})\det(a_{sk})
\end{aligned}$$

and thus

$$b_1\ldots b_n = \det(a_{sk}),$$

as required. ■

7.9. COROLLARY. Let E/F be a finite separable field extension. Then

$$D_{E/F}(\lambda) = D_{E/F}(\lambda+a) \qquad \text{for all } \lambda \in E,\ a \in F$$

Proof. Let $\sigma_1,\ldots,\sigma_n$ be all distinct F-homomorphisms of E into an algebraic closure of E. If $E \neq F(\lambda)$, then $E \neq F(\lambda+a)$ and so

$$D_{E/F}(\lambda) = D_{E/F}(\lambda+a) = 0.$$

We may therefore assume that $E = F(\lambda)$ (and hence that $E = F(\lambda+a)$). Since

$$\sigma_i(\lambda) - \sigma_j(\lambda) = \sigma_i(\lambda+a) - \sigma_j(\lambda+a)$$

the desired conclusion follows by virtue of (3). ■

Our next aim is to provide some applications of discriminants to the study of algebraic number fields.

Let F be an algebraic number field of degree n and let R be the ring of integers of F. We know, from Corollary 6.18, that R is a free $\mathbb{Z}$-module of rank n. By an *integral basis* of the field F, we understand any $\mathbb{Z}$-basis of R. Expressed otherwise, a system $r_1,\dots,r_n$ of elements of R is an integral basis of F if and only if $r_1,\dots,r_n$ are $\mathbb{Q}$-linearly independent and generate R as a $\mathbb{Z}$-module. This is so since $r_1,\dots,r_n$ are $\mathbb{Q}$-linearly independent if and only if $r_1,\dots,r_n$ are $\mathbb{Z}$-linearly independent. The following result provides a constructive method of obtaining integral bases of algebraic number fields.

7.10. PROPOSITION. Let F be an algebraic number field of degree n, let R be the ring of integers of F and let $a_1,\dots,a_n$ be $\mathbb{Q}$-linearly independent elements of R. For each $i \in \{1,2,\dots,n\}$, define d_{ii} to be the least natural number such that

$$\omega_i = D_{F/\mathbb{Q}}(a_1,\dots,a_n)^{-1}(d_{i1}a_1 + \dots + d_{i,i-1}a_{i-1} + d_{ii}a_i) \in R$$

for suitable $d_{ik} \in \mathbb{Z}$, $1 \leqslant k \leqslant i-1$. Then $\omega_1,\dots,\omega_n$ is an integral basis of F.

Proof. Put $d_{ij} = 0$ for $j > i$ and note that, since

$$\det(d_{ij}) = d_{11} \dots d_{nn} \neq 0,$$

the elements $\omega_1,\dots,\omega_n$ are $\mathbb{Q}$-linearly independent. We are therefore left to verify that $R = \sum_{i=1}^{n} \mathbb{Z}\omega_i$. We claim that if $r \in R$ can be written in the form

$$r = D_{F/\mathbb{Q}}(a_1,\dots,a_n)^{-1}(z_1a_1 + \dots + z_ja_j) \qquad (z_j \in \mathbb{Z},\ 1 \leqslant i \leqslant j) \quad (4)$$

then

$$r \in \sum_{i=1}^{j} \mathbb{Z}\omega_i \qquad (5)$$

Since, by Proposition 7.6, each r can be written in this form with $j = n$, the result will follow.

To prove (5), we first note that $d_{jj}|z_j$ for all $j \in \{1,\ldots,n\}$. Indeed, if $z_j = sd_{jj} + t$ with $0 < t < d_{jj}$, then $r - s\omega_j \in R$ and so

$$r - s\omega_j = D_{F/\mathbb{Q}}(a_1,\ldots,a_n)^{-1}((z_1-d_{j1})a_1 + \ldots + ta_j) \in R$$

contrary to our choice of d_{jj}. We now argue by induction on j. If $j = 1$, then $r = D_{F/\mathbb{Q}}(a_1,\ldots,a_n)^{-1}(z_1a_1) = (z_1/d_{11})\omega_1 \in \mathbb{Z}\omega_1$. Assume that (4) is true for $j-1$ and let r be as in (4). Since $d_{jj}|z_j$ we have $z_j = ad_{jj}$ for some $a \in \mathbb{Z}$, and so $r - a\omega_j \in R$. By induction hypothesis, $r - a\omega_j \in \sum_{i=1}^{j-1} \mathbb{Z}\omega_i$ and hence $r \in \sum_{i=1}^{j} \mathbb{Z}\omega_i$, as asserted. ∎

7.11. COROLLARY. Let R be the ring of integers of an algebraic number field F of degree n and let $a_1,\ldots,a_n \in R$ be $\mathbb{Q}$-linearly independent. Then there exists an integral basis $\omega_1,\ldots,\omega_n$ of F such that

$$a_j \in \sum_{i=1}^{j} \mathbb{Z}\omega_i \quad \text{for all } j \in \{1,2,\ldots,n\}.$$

Proof. In the notation of the proof of Proposition 7.10, each a_j can be written in the form (3). Hence $a_j \in \sum_{i=1}^{j} \mathbb{Z}\omega_i$, as asserted. ∎

Let F be an algebraic number field. By the *discriminant* of F, written $d(F)$, we understand the discriminant of any integral basis of F. Note that, by Lemma 7.1, the value of $d(F)$ is independent of the choice of an integral basis. Furthermore, by Lemma 7.3 and Corollary 7.5,

$$d(F) \in \mathbb{Z} \text{ and } d(F) \neq 0$$

As a preliminary to the next result, note that if A is a free $\mathbb{Z}$-module of rank n and B a submodule generated by n $\mathbb{Z}$-linearly independent elements of A, then the index $(A:B)$ is finite. This is so since the abelian group A/B is both torsion and finitely generated.

7.12. PROPOSITION. Let F be an algebraic number field of degree n, let R be the ring of integers of F and let $a_1,\ldots,a_n \in R$ be $\mathbb{Q}$-linearly independent. Put

$$M = \sum_{i=1}^{n} \mathbb{Z}a_i$$

Then

$$D_{F/\mathbb{Q}}(a_1,\ldots,a_n) = (R:M)^2 d(F)$$

In particular, if $D_{F/\mathbb{Q}}(a_1,\ldots,a_n) = d(F)$, then $a_1,\ldots,a_n$ is an integral basis of F.

Proof. Let $\omega_1,\ldots,\omega_n$ be any integral basis of F, and choose $b_i \in M$ such that

$$b_i = \sum_{k=1}^{i} c_{ik}\omega_k \qquad (1 \leqslant i \leqslant n, c_{ik} \in \mathbb{Z}) \qquad (6)$$

and c_{ii} is positive of smallest possible value. By the argument in the proof of Proposition 7.10, we see that $b_1,\ldots,b_n$ is a $\mathbb{Z}$-basis of M and that $t_1\omega_1 + \ldots + t_n\omega_n, t_i \in \mathbb{Z}$, can lie in M only if $c_{ii}|t_i$. This shows that the numbers

$$a_1\omega_1 + \ldots + a_j\omega_j \qquad (0 \leqslant a_i < c_{jj}, 1 \leqslant j \leqslant n)$$

are all incongruent modulo M and clearly there are $c_{11} \ldots c_{nn}$ of them. We now claim that all elements in R/M are represented by them; if sustained, it will follow that $(R:M) = c_{11}\ldots c_{nn}$ and hence, by (6) and Lemma 7.1, that

$$D_{F/\mathbb{Q}}(a_1,\ldots,a_n) = D_{F/\mathbb{Q}}(b_1,\ldots,b_n) = (c_{11}\ldots c_{nn})^2 d(F) = (R:M)^2 d(F)$$

as required.

To substantiate our claim, let

$$x = \sum_{k=1}^{n} \lambda_k\omega_k \qquad (\lambda_k \in \mathbb{Z})$$

be an arbitrary element from R and denote by μ_n the least nonnegative residue of $\lambda_n \pmod{c_{nn}}$. Then we have

$$x = \frac{\lambda_n-\mu_n}{c_{nn}} b_n + \mu_n\omega_n + \sum_{k=1}^{n-1} (\lambda_k - c_{nk})\omega_k$$

Now let μ_{n-1} denote the least nonnegative residue of $\lambda_{n-1} - c_{n,n-1} \pmod{c_{n-1,n-1}}$. Then in the same way we obtain

$$x = \frac{\lambda_n-\mu_n}{c_{nn}} b_n + \frac{\lambda_{n-1} - c_{n,n-1} - \mu_{n-1}}{c_{n-1,n-1}} b_{n-1} + \mu_n\omega_n + \mu_{n-1}\omega_{n-1} +$$

$$+ \sum_{k=1}^{n-2} (\lambda_k - c_{n,k} - c_{n-1,k})\omega_k$$

Continuing this process, we finally derive the equality of the form

$$x = \sum_{k=1}^{n} a_k b_k + \sum_{k=1}^{n} \mu_k \omega_k \qquad (0 \leqslant \mu_j < c_{jj}; \mu_j, a_j \in \mathbb{Z})$$

Therefore

$$x \equiv \sum_{k=1}^{n} \mu_k \omega_k \pmod{M}$$

as required. ■

The following result shows that not every nonzero rational integer is a discriminant of a suitable algebraic number field.

7.13. PROPOSITION. (Stickelberger's criterion). Let F be an algebraic number field of degree n and let $\alpha_1, \ldots, \alpha_n$ be algebraic integers in F. Then

$$D_{F/\mathbb{Q}}(\alpha_1, \ldots, \alpha_n) \equiv 0 \text{ or } 1 \pmod{4}$$

In particular, $d(F) \equiv 0$ or $1 \pmod{4}$.

Proof. Let $\sigma_1, \ldots, \sigma_n$ be the n distinct $\mathbb{Q}$-homomorphisms of F into its algebraic closure. Then, by Proposition 7.7,

$$D_{F/\mathbb{Q}}(\alpha_1, \ldots, \alpha_n) = \det(\sigma_i(\alpha_j))^2$$

The determinant $\det(\sigma_i(\alpha_j))$ has an expansion as a sum of terms with plus and minus signs in front of them. Let a be the sum of terms with plus signs, and b the sum of terms with minus signs, so that

$$D_{F/\mathbb{Q}}(\alpha_1, \ldots, \alpha_n) = (a-b)^2 = (a+b)^2 - 4ab$$

But $a + b$ and ab are both invariant under any σ_i and hence are rational integers. The desired conclusion follows at once. ■

Our next result determines the sign of the discriminant.

7.14. PROPOSITION. Let $F = \mathbb{Q}(\theta)$ be an algebraic number field and let r be the number of pairs of complex conjugate roots of the minimal polynomial of θ over $\mathbb{Q}$. Then $d(F)$ is positive if and only if r is even.

Proof. Let $\omega_1, \ldots, \omega_n$ be an integral basis of F and let $\sigma_1, \ldots, \sigma_n$ be all $\mathbb{Q}$-homomorphisms $F \longrightarrow \mathbb{C}$. Write $\det(\sigma_k(\omega_j)) = d_1 + id_2$ with $d_1, d_2 \in \mathbb{R}$. Because the change of i into $-i$ in this determinant is equivalent to the interchange of r pairs of rows, we have

$$d_1 - id_2 = (-1)^r \det(\sigma_k(\omega_j))$$

and so

$$(-1)^r(d_1+id_2) = d_1 - id_2$$

If r is even, this implies $d_2 = 0$ and $d(F) = d_1^2$ is positive. If r is odd, then we have $d_1 = 0$ and $d(F) = (id_2)^2 = -d_2^2$ is negative. ■

We now relate discriminants in a chain of algebraic number fields. The following simple observation will be useful.

7.15. LEMMA. Let $R \subseteq S$ be the rings of integers of algebraic number fields $F \subseteq E$. Then there is an integral basis of E which contains an integral basis of F.

Proof. Put $m = (F:\mathbb{Q}), n = (E:F)$ and fix an integral basis $a_1,\ldots,a_m$ of F. We may find $a_{m+1},\ldots,a_{mn} \in E$ such that $a_1,\ldots,a_{mn}$ are $\mathbb{Q}$-linearly independent. Moreover, multiplying each of the $a_{m+1},\ldots,a_{mn}$ by a suitable integer, if necessary, we may assume that

$$a_1,\ldots,a_m,a_{m+1},\ldots,a_{mn} \in S$$

Owing to Corollary 7.11, we may find an integral basis, say $\omega_1,\ldots,\omega_{mn}$, of E such that

$$a_j \in \sum_{i=1}^{j} \mathbb{Z}\omega_i \qquad (1 \leqslant j \leqslant mn)$$

The elements $\omega_1,\ldots,\omega_m$ of this basis lie in $F \cap S = R$. Hence

$$R = \sum_{i=1}^{m} \mathbb{Z}a_i \subseteq \sum_{i=1}^{m} \mathbb{Z}\omega_i \subseteq R$$

proving that $R = \sum_{i=1}^{m} \mathbb{Z}\omega_i$. Thus $\omega_1,\ldots,\omega_n$ is an integral basis of F, as required. ■

7.16. PROPOSITION. Let $F \subseteq E$ be algebraic number fields. Then $d(F)$ divides $d(E)$.

Proof. Put $m = (F:\mathbb{Q})$ and $n = (E:F)$. By Lemma 7.4, we may find an integral basis $\omega_1,\ldots,\omega_{mn}$ of E such that $\omega_1,\ldots,\omega_m$ is an integral basis of F. By Lemma 3.9, each $\mathbb{Q}$-homomorphism $F \longrightarrow \mathbb{C}$ has exactly n extensions to $\mathbb{Q}$-homomorphisms $E \longrightarrow \mathbb{C}$. Hence, if $\sigma_1,\ldots,\sigma_m$ are all distinct $\mathbb{Q}$-homomorphisms of F into $\mathbb{C}$, we may choose the mn distinct $\mathbb{Q}$-homomorphisms

$$\mu_1,\ldots,\mu_m,\mu_{m+1},\ldots,\mu_{mn}$$

of E into $\mathbb{C}$ such that

$$\mu_k(\omega_1) = \sigma_k(\omega_1) \qquad (1 \leqslant k \leqslant m)$$

$$\mu_i(x) = \mu_j(x) \qquad \text{if} \quad i \equiv j (\text{mod } m) \qquad (\text{for all} \quad x \in F)$$

Setting $\mu_i(\omega_j) = \omega_j^{(i)}$, $1 \leqslant i,j \leqslant mn$, it follows from Proposition 7.7 that

$$d(E) = [\det(\omega_j^{(i)})]^2$$

$$= \begin{vmatrix} \omega_1^{(1)} \ldots \omega_1^{(m)} \omega_1^{(1)} \ldots \omega_1^{(m)} \ldots \omega_1^{(1)} \ldots \omega_1^{(m)} \\ - - - - - - - - - - - - - - - \\ \omega_m^{(1)} \ldots \omega_m^{(m)} \omega_m^{(1)} \ldots \omega_m^{(m)} \ldots \omega_m^{(1)} \ldots \omega_m^{(m)} \\ \omega_{m+1}^{(1)} \ldots \omega_{m+1}^{(m)} \omega_{m+1}^{(m+1)} \ldots\ldots\ldots\ldots\ldots \omega_{m+1}^{(mn)} \\ - - - - - - - - - - - - - - - \\ \omega_{mn}^{(1)} \ldots \omega_{mn}^{(m)} \omega_{mn}^{(m+1)} \ldots\ldots\ldots\ldots\ldots \omega_{mn}^{(mn)} \end{vmatrix}^2$$

$$= \begin{vmatrix} \omega_1^{(1)} \ldots \omega_1^{(m)} & \\ - - - - - - & 0 \\ \omega_m^{(1)} \ldots \omega_m^{(m)} & \\ - - - - - - & - - - - - - - - - - \\ \omega_{m+1}^{(1)} \ldots \omega_{m+1}^{(m)} & \omega_{m+1}^{(m+1)} - \omega_{m+1}^{(1)} \ldots \omega_{m+1}^{(mn)} - \omega_{m+1}^{(m)} \\ - - - - - - & - - - - - - - - - - \\ \omega_{mn}^{(1)} \ldots \omega_{mn}^{(m)} & \omega_{mn}^{(m+1)} - \omega_{mn}^{(1)} \ldots \omega_{mn}^{(mn)} - \omega_{mn}^{(m)} \end{vmatrix}^2$$

$$= d(F)a,$$

where a is an algebraic integer. But $a = d(E)/d(F) \in \mathbb{Q}$, hence $a \in \mathbb{Z}$ and the result follows. ∎

The next observation is often used for determining the discriminant.

7.17. PROPOSITION. Let R be the ring of integers of an algebraic number field F of degree n, let $\lambda \in R$ be such that $F = \mathbb{Q}(\lambda)$ and assume that the minimal polynomial of λ over $\mathbb{Q}$ is of the form

$$X^n + a_{n-1}X^{n-1} + \dots + a_0 \qquad (a_i \in \mathbb{Z})$$

where all the a_i are divisible by some prime p and $p^2 \nmid a_0$. Then

$$p \nmid (R: \sum_{i=0}^{n-1} \mathbb{Z}\lambda^i)$$

In particular, by Proposition 7.12, the highest powers of p dividing $D_{F/\mathbb{Q}}(\lambda)$ and $d(F)$ are the same.

Proof. By hypothesis, $\lambda^n/p \in R$ and $N_{F/\mathbb{Q}}(\lambda) = (-1)^n a_0$ is not divisible by p^2. Assume by way of contradiction that p divides $(R: \sum_{i=0}^{n-1} \mathbb{Z}\lambda^i)$. Then there exists $r \in R$ such that

$$r = (b_0 + b_1\lambda + \dots + b_{n-1}\lambda^{n-1})p^{-1} \qquad (b_i \in \mathbb{Z})$$

and not all the b_i's are divisible by p. Let j be the minimal index for which $p \nmid b_j$. Then the number

$$s = (b_j\lambda^j + \dots + b_{n-1}\lambda^{n-1})p^{-1} = r - ((b_0/p) + (b_1/p)\lambda + \dots + (b_{j-1}/p)\lambda^{j-1})$$

is an algebraic integer and so is also

$$s_1 = b_j\lambda^{n-1}p^{-1} = s\lambda^{n-j-1} - \lambda^n(b_{j+1} + b_{j+2}\lambda + \dots + b_{n-1}\lambda^{n-j-2})p^{-1}$$

It follows that

$$p^n N_{F/\mathbb{Q}}(s_1) = N_{F/\mathbb{Q}}(ps_1) = N_{F/\mathbb{Q}}(b_j\lambda^{n-1}) = b_j^n N_{F/\mathbb{Q}}(\lambda)^{n-1}$$

Since $p^2 \nmid N_{F/\mathbb{Q}}(\lambda)$, the latter implies that $p|b_j$, a contradiction. ∎

We close this section by providing a number of examples of the calculation of a discriminant.

7.18. PROPOSITION. Let E/F be a finite separable field extension of degree $n \geqslant 2$, let $\lambda \in E$ be such that $E = F(\lambda)$ and assume that the minimal polynomial of λ over F is

$$f(X) = X^n + aX + b$$

Then

$$D_{E/F}(\lambda) = (-1)^{n(n-1)/2}(n^n b^{n-1} + (-1)^{n-1}(n-1)^{n-1}a^n)$$

Proof. By Proposition 7.8, $D_{E/F}(\lambda) = (-1)^{n(n-1)/2}N_{E/F}(f'(\lambda))$. Putting $\mu = f'(\lambda)$, we have

$$\mu = n\lambda^{n-1} + a = -(n-1)a - nb\lambda^{-1}$$

since $\lambda^n + a\lambda + b = 0$ implies $n\lambda^{n-1} = -na - nb\lambda^{-1}$. It follows from this that

$$\lambda = -nb(\mu + (n-1)a)^{-1}$$

The minimal polynomial $g(X)$ of μ over F is therefore the numerator of

$$b^{-1}f(-nb(X+(n-1)a)^{-1})$$

since $F(\lambda) = F(\mu)$
deg $g(x)$ = deg $f(x)$
$g(\mu) = 0$

Hence we must have

$$g(X) = (X + (n-1)a)^n - na(X + (n-1)a)^{n-1} + (-1)^n b^{n-1}$$

The norm of μ is $(-1)^n$ times the constant term of this polynomial, i.e.

$$N_{E/F}(\mu) = n^n b^{n-1} + (-1)^{n-1}(n-1)^{n-1}a^n$$

which clearly yields the result. ■

We now turn our attention to *quadratic fields*, i.e. algebraic number fields of degree two. An integer $d \in \mathbb{Z}$ is said to be *square-free* if $d \neq 1$ and d is not divisible by any square except 1. Thus d is square-free if and only if $d = -1$ or $|d|$ is a product of distinct primes. Every integer n which is not a square can be uniquely written in the form $n = m^2 s$ where s is square-free. We refer to s as the *square-free part* of n.

For convenience, we assume that all quadratic fields are embedded in $\mathbb{C}$. In what follows by $\sqrt{x}$, $x \in \mathbb{Q}$, we mean one of the two elements of $\mathbb{C}$ whose square is x.

7.19. LEMMA. (i) Every quadratic field is of the form $\mathbb{Q}(\sqrt{d})$, where d is a square-free integer.

(ii) If $n,m \in \mathbb{Z}$ are nonsquares, then $\mathbb{Q}(\sqrt{n}) = \mathbb{Q}(\sqrt{m})$ if and only if the square-free parts of n and m are the same.

Proof. (i) Let F be a quadratic field. Then any element $x \in F - \mathbb{Q}$ is of degree two over $\mathbb{Q}$. Let $f(X) = X^2 + bX + c$ $(b,c \in \mathbb{Q})$ be the minimal polynomial of such an element $x \in F$. Solving the quadratic equation $x^2 + bx + c = 0$ gives

$$2x = -b \pm \sqrt{b^2-4ac}$$

and thus $F = \mathbb{Q}(\sqrt{b^2-4c})$. Now $b^2-4c \in \mathbb{Q}$ so that $b^2-4c = m/n = mn/n^2$ with $m,n \in \mathbb{Z}$, $n \neq 0$. Hence $F = \mathbb{Q}(\sqrt{mn})$ with $m,n \in \mathbb{Z}$. Let d be the square-free part of mn, say $mn = da^2$ for some $a \in \mathbb{Z}$. Then $F = \mathbb{Q}(\sqrt{d})$ as required.

(ii) We may harmlessly assume that both n and m are square-free, in which case we need only show that $\mathbb{Q}(\sqrt{n}) = \mathbb{Q}(\sqrt{m})$ implies $n = m$. If $\mathbb{Q}(\sqrt{n}) = \mathbb{Q}(\sqrt{m})$, then

$$\sqrt{m} = x + y\sqrt{n} \qquad \text{for some } x,y \in \mathbb{Q}$$

whence

$$m = x^2 + ny^2 + 2xy\sqrt{n}$$

and thus

$$m = x^2 + ny^2, \quad 2xy = 0$$

If $y = 0$, then $m = x^2$ which is impossible. If $x = 0$, then $m = ny^2$ and hence $m = n$, as asserted. ∎

7.20. PROPOSITION. Let $F = \mathbb{Q}(\sqrt{d})$ be a quadratic field with $d \in \mathbb{Z}$ square-free (so $d \not\equiv 0 \pmod 4$). Then

(i) If $d \equiv 1 \pmod 4$, then $1, (1/2)(1+\sqrt{d})$ is an integral basis of F and $d(F) = d$.

(ii) If $d \equiv 2 \pmod 4$ or $d \equiv 3 \pmod 4$, then $1, \sqrt{d}$ is an integral basis of F and $d(F) = 4d$.

Proof. The field F has precisely two $\mathbb{Q}$-homomorphisms into $\mathbb{C}$, namely $\sigma_i : \mathbb{Q}(\sqrt{d}) \longrightarrow \mathbb{C}$, $i = 1,2$, given by

$$\sigma_1(a+b\sqrt{d}) = a + b\sqrt{d} \quad \text{and} \quad \sigma_2(a+b\sqrt{d}) = a - b\sqrt{d}$$

Hence $N_{F/\mathbb{Q}}(a+b\sqrt{d}) = (a+b\sqrt{d})(a-b\sqrt{d}) = a^2 - db^2$. If $d \equiv 1 \pmod 4$, then the number $\lambda = (1/2)(1+\sqrt{d})$ is integral, since it is a root of the monic polynomial

$$f(X) = X^2 - X + (1-d)/4 \in \mathbb{Z}[X]$$

Moreover, since $f(X)$ is obviously the minimal polynomial and $f'(\lambda) = 2\lambda - 1 = \sqrt{d}$, it follows from Proposition 7.8 that

$$D_{F/\mathbb{Q}}(\lambda) = -N_{F/\mathbb{Q}}(\sqrt{d}) = d$$

Since d is square-free, it follows from Proposition 7.12 that $d(F) = d$ and that $1, \lambda$ is an integral basis of F.

Assume that $d \equiv 2 (\mathrm{mod}\ 4)$ or $d \equiv 3 (\mathrm{mod}\ 4)$. Let $f(X) = X^2 - d$ and $g(X) = X^2 + 2X + 1 - d$. Since $f(X)$ is the minimal polynomial of $\sqrt{d}$ and $f'(\sqrt{d}) = 2\sqrt{d}$, we have

$$D_{F/\mathbb{Q}}(\sqrt{d}) = -N_{F/\mathbb{Q}}(2\sqrt{d}) = 4d$$

Let p be any prime dividing d. Then $f(X)$ satisfies the hypotheses of Proposition 7.17 with respect to p, hence the highest powers of p dividing $4d$ and $d(F)$ are the same. If $d \equiv 2 (\mathrm{mod}\ 4)$, then d is even, hence $d(F) = 4d$ by Proposition 7.12. If $d \equiv 3 (\mathrm{mod}\ 4)$, then $g(X)$ satisfies the hypotheses of Proposition 7.17 for $p = 2$. Since $\sqrt{d} - 1$ is a root of $g(X)$ and

$$D_{F/\mathbb{Q}}(\sqrt{d}) = D_{F/\mathbb{Q}}(\sqrt{d}-1) = 4d$$

by Corollary 7.9, we again deduce that $d(F) = 4d$. The desired assertion is now a consequence of Proposition 7.12. ■

We now concentrate on pure cubic fields. A rational integer $d > 1$ is said to be *cube-free* if d is not divisible by a cube of a prime. The field $F = \mathbb{Q}(\sqrt[3]{d})$, in which $d > 1$ is a cube-free rational integer and $\sqrt[3]{d}$ is real, is called a *pure cubic field*.

Let $F = \mathbb{Q}(\sqrt[3]{d})$ be a pure cubic field. Since d is cube-free, we can write

$$d = ab^2$$

where ab is square-free. Furthermore, $X^3 - d$ is the minimal polynomial of $\sqrt[3]{d}$ over $\mathbb{Q}$. Taking into account that $\sqrt[3]{d^2} = b\sqrt[3]{a^2b}$, we see that

$$\{1, \sqrt[3]{ab^2}, \sqrt[3]{a^2b}\}$$

is a $\mathbb{Q}$-basis of F. Following Dedekind, we say that F is of the *first* or *second kind*, according as 9 does not, or does, divide $a^2 - b^2$. It can be easily verified that $9 \mid a^2 - b^2$ if and only if $ab^2 \equiv \pm 1 (\mathrm{mod}\ 9)$. Thus $\mathbb{Q}(\sqrt[3]{d})$ is of the first or second kind, according as $d \not\equiv \pm 1 (\mathrm{mod}\ 9)$ or $d \equiv \pm 1 (\mathrm{mod}\ 9)$. Note also that, by Proposition 7.18,

$$D_{F/\mathbb{Q}}(\sqrt[3]{d}) = -27d^2 \tag{7}$$

We are now ready to prove

7.21. PROPOSITION. (Dedekind(1900)). Let $F = \mathbb{Q}(\sqrt[3]{d})$ be a pure cubic field and let $d = ab^2$ where ab is square-free.

(i) If F is of the first kind, then $d(F) = -27(ab)^2$ and the numbers

$$1, \sqrt[3]{ab^2}, \sqrt[3]{a^2b}$$

form an integral basis of F.

(ii) If F is of the second kind, then $d(F) = -3(ab)^2$ and the numbers

$$(1/3)(1+a\sqrt[3]{ab^2} + b\sqrt[3]{a^2b}), \sqrt[3]{ab^2}, \sqrt[3]{a^2b}$$

form an integral basis of F.

Proof. Since the $\mathbb{Z}$-linear span of $\{1,\sqrt[3]{ab^2}, \sqrt[3]{a^2b}\}$ contains the $\mathbb{Z}$-linear span of $\{1,\sqrt[3]{d},\sqrt[3]{d^2}\}$ as a subgroup of index b, it follows from (7) and Proposition 7.12 that $d(F) < 0$ and that $d(F)$ divides $27a^2b^2$. The minimal polynomial for $\sqrt[3]{d}$ is $f(X) = X^3-d$ and $f(X)$ satisfies the hypotheses of Proposition 7.17 for every prime p dividing a. It follows from (7) that if $a \equiv 0 \pmod 3$, then $d(F)$ is divisible by $27a^2$, and if $a \not\equiv 0 \pmod 3$, then $d(F)$ is divisible by a^2. The number $\sqrt[3]{a^2b}$ is a root of $g(X) = X^3 - a^2b$ and $g(X)$ satisfies the hypotheses of Proposition 7.7 for every prime p dividing b. Applying once more the foregoing argument, we see that $d(F)$ is divisible by 3^nb^2, where $n = 3$ if $b \equiv 0 \pmod 3$ and $n = 0$ otherwise. Putting this together, it follows that $d(F) = -3^n(ab)^2, n$ being equal to 3 if $d \equiv 0 \pmod 3$, and equal to 1 or 3 otherwise.

Assume that $d \not\equiv \pm 1 \pmod 9$. Then $d^3 \not\equiv d \pmod 9$ and $d^3 \equiv d \pmod 3$, whence the polynomial

$$f(X) = (X+d)^3-d = X^3 + 3dX^2 + 3d^2X + (d^3-d)$$

satisfies the hypotheses of Proposition 7.17 with respect to $p = 3$ and $\lambda = \sqrt[3]{d}-d$. Since $D_{F/\mathbb{Q}}(\sqrt[3]{d}-d) = D_{F/\mathbb{Q}}(\sqrt[3]{d}) = -27d^2$, it follows from Proposition 7.17 that $27|d(F)$ and hence $d(F) = -27(ab)^2$

If $d \equiv 1 \pmod 9$, then the number

$$x = (1+\sqrt[3]{d}+\sqrt[3]{d^2})/3$$

is integral. Indeed, its trace equals 1, since $T_{F/\mathbb{Q}}(\sqrt[3]{d}) = Tr_{F/\mathbb{Q}}(\sqrt[3]{d^2}) = 0$, its norm equals

$$\frac{1}{27} N_{F/\mathbb{Q}}(1+\sqrt[3]{d}+\sqrt[3]{d^2}) = \frac{1}{27} \frac{N_{F/\mathbb{Q}}(d-1)}{N_{F/\mathbb{Q}}(\sqrt[3]{d}-1)} = \frac{1}{27}(d-1)^2$$

since the minimal polynomial for $\sqrt[3]{d}-1$ is $X^3 + 3X^2 + 3X + (1-d)$, and finally the remaining coefficient of the minimal polynomial for x equals $(d-1)/3$. This shows that the index of $\sum_{i=0}^{2} \mathbb{Z}x^i$ in the additive group of the ring of integers of F is divisible by 3. Hence, by Proposition 7.12, $d(F) = -3(ab)^2$. The same argument applies also to the case where $d \equiv -1 \pmod 9$ by considering the number $(1-\sqrt[3]{d}+\sqrt[3]{d^2})/3$ instead of x.

Let σ_1 = identity, σ_2, σ_3 be all $\mathbb{Q}$-homomorphisms of F into $\mathbb{C}$. Then we may write

$$\sigma_2(x_1+x_2\sqrt[3]{ab^2}+x_3\sqrt[3]{a^2b}) = x_1 + \varepsilon x_2\sqrt[3]{ab^2} + \varepsilon^2 x_3\sqrt[3]{a^2b} \qquad (x_i \in \mathbb{Q})$$

$$\sigma_3(x_1+x_2\sqrt[3]{ab}+x_3\sqrt[3]{a^2b}) = x_1 + \varepsilon^2 x_2\sqrt[3]{ab^2} + \varepsilon x_3\sqrt[3]{a^2b}$$

where ε is a primitive cube root of unity. Hence

$$D_{F/\mathbb{Q}}(1,\sqrt[3]{ab^2},\sqrt[3]{a^2b}) = \det \begin{bmatrix} 1 & \sqrt[3]{ab^2} & \sqrt[3]{a^2b} \\ 1 & \varepsilon\sqrt[3]{ab^2} & \varepsilon^2\sqrt[3]{a^2b} \\ 1 & \varepsilon^2\sqrt[3]{ab^2} & \varepsilon\sqrt[3]{a^2b} \end{bmatrix}^2$$

$$= [3ab(\varepsilon^2-\varepsilon)]^2$$

$$= -27(ab)^2$$

and

$$D_{F/\mathbb{Q}}((1/3)(1+a\sqrt[3]{ab^2}+b\sqrt[3]{a^2b}),\sqrt[3]{ab^2},\sqrt[3]{a^2b})$$

$$= \det \begin{bmatrix} (1/3)(1+a\sqrt[3]{ab^2}+b\sqrt[3]{a^2b}) & \sqrt[3]{ab^2} & \sqrt[3]{a^2b} \\ (1/3)(1+\varepsilon a\sqrt[3]{ab^2}+\varepsilon^2 b\sqrt[3]{a^2b}) & \varepsilon\sqrt[3]{ab^2} & \varepsilon^2\sqrt[3]{a^2b} \\ (1/3)(1+\varepsilon^2 a\sqrt[3]{ab^2}+\varepsilon b\sqrt[3]{a^2b}) & \varepsilon^2\sqrt[3]{ab^2} & \varepsilon\sqrt[3]{a^2b} \end{bmatrix}^2$$

$$= (ab(\varepsilon^2-\varepsilon))^2$$
$$= -3(ab)^2$$

The desired conclusion now follows by virtue of Proposition 7.12. ■

Turning to cyclotomic fields, we now prove

7.22. PROPOSITION. Let $n = p^k > 2$, p prime, and let ε be a primitive n-th root of unity over $\mathbb{Q}$. Then

$$d(\mathbb{Q}(\varepsilon)) = (-1)^{\phi(p^k)/2}p^{p^{k-1}(k(p-1)-1)}$$

and the numbers $1,\varepsilon,\varepsilon^2,\dots,\varepsilon^{\phi(n)-1}$ form an integral basis of $\mathbb{Q}(\varepsilon)$.

Proof. Put $F = \mathbb{Q}(\varepsilon)$ and $\mu = \phi(p^k)$. If $\Phi_n(X)$ is the n-th cyclotomic polynomial, then by Proposition 5.13, $\Phi_n(X)$ is the minimal polynomial of ε over $\mathbb{Q}$. Hence, by Proposition 7.8, we have

$$D_{F/\mathbb{Q}}(\varepsilon) = (-1)^{\mu(\mu-1)/2}N_{F/\mathbb{Q}}(\Phi_n'(\varepsilon))$$

Since $X^n-1 = (X^{n/p}-1)\Phi_n(X)$, we have

$$nX^{n-1} = (n/p)X^{(n/p)-1}\Phi_n(X) + (X^{n/p}-1)\Phi_n'(X)$$

and thus

$$n\,\varepsilon^{n-1} = (\varepsilon^{n/p}-1)\Phi_n{}'(\varepsilon) \tag{8}$$

Let $P(n)$ denote the set of all nonzero residue classes $\bar{a} = a + n\mathbb{Z}$ modulo n with $(a,n) = 1$. Then considering the conjucates of (8) and taking their product, we arrive at

$$n^{\mu}\Big(\prod_{\bar{a}\in P(n)} \varepsilon^{a}\Big)^{n-1} = \prod_{\bar{a}\in P(n)} (\varepsilon^{a(n/p)}-1)\prod_{\bar{a}\in P(n)} \Phi_n'(\varepsilon^{a})$$

But

$$\prod_{\bar{a}\in P(n)} \varepsilon^{a} = N_{F/\mathbb{Q}}(\varepsilon) = (-1)^{\mu}\,1 = (-1)^{\mu},$$

$$\prod_{\bar{a}\in P(n)} (\varepsilon^{a(n/p)}-1) = (-1)^{\mu}\prod_{\bar{a}\in P(n)} (1-\varepsilon^{ap^{k-1}}) = (-1)^{\mu}p^{p^{k-1}}$$

since $\varepsilon^{p^{k-1}}$ is a primitive p-th root of unity, $\prod_{\bar{c}\in P(p)} (1-\varepsilon^{c}) = \Phi_p(1) = p$ and a system of representatives of the prime residue classes modulo $n = p^k$ gives rise to p^{k-1} systems of representatives of prime residue classes modulo p. Hence

$$n^{\mu}(-1)^{\mu(n-1)} = (-1)^{\mu}p^{p^{k-1}} N_{F/\mathbb{Q}}(\Phi_n'(\varepsilon)),$$

and so

$$D_{F/\mathbb{Q}}(\varepsilon) = (-1)^{(1/2)\mu(\mu-1)+\mu(n-2)}p^{p^{k-1}(k(p-1)-1)}$$

Since $\mu(n-2)$ is always even and $(1/2)\mu(\mu-1) \equiv (1/2)\mu (\mathrm{mod}\ 2)$, then

$$D_{F/\mathbb{Q}}(\varepsilon) = (-1)^{(1/2)\mu}p^{p^{k-1}(k(p-1)-1)}$$

Consider the polynomial $f(X) = \Phi_n(X+1)$. Since

$$\Phi_n(X) = (X^{p^k}-1)/(X^{p^{k-1}}-1) = X^{p^{k-1}(p-1)} + X^{p^{k-1}(p-2)} + \dots + 1,$$

one immediately verifies that $f(X)$ satisfies the hypotheses of Proposition 7.17 for $\lambda = \varepsilon-1$ and with respect to the prime p. Hence

$$D_{F/\mathbb{Q}}(\varepsilon) = D_{F/\mathbb{Q}}(\varepsilon-1) = d(F)$$

which, by Proposition 7.12, also shows that the numbers $1, \varepsilon, \dots, \varepsilon^{\phi(n)-1}$ form an integral basis of F. ■

8. UNITS IN QUADRATIC FIELDS

Throughout this section, d denotes a square-free integer and R the ring of integers of the quadratic field $\mathbb{Q}(\sqrt{d})$. Our aim is to investigate the group $U(R)$ of all invertible elements of the ring R. We shall use here and in the next section the Dirichlet Unit Theorem (to be proved later) which asserts that if F is an algebraic number field, then the unit group of the ring of integers of F is finitely generated of torsion-free rank $r_1 + r_2 - 1$, where r_1 and r_2 are respectively the number of real and the number of pairs of complex conjugate roots of the minimial polynomial of a primitive element of F.

8.1. LEMMA. (i) If $d \equiv 2 (\mathrm{mod}\ 4)$ or $d \equiv 3 (\mathrm{mod}\ 4)$, then

$$U(R) = \{a + b\sqrt{d} \mid a,b \in \mathbb{Z},\ a^2-db^2 = \pm 1\}$$

(ii) If $d \equiv 1 (\mathrm{mod}\ 4)$, then

$$U(R) = \{(1/2)(a + b\sqrt{d}) \mid a,b \in \mathbb{Z},\ a^2-db^2 = \pm 4\}$$

Proof. If $x = a + b\sqrt{d}$, $a,b \in \mathbb{Q}$, then

$$N_{F/\mathbb{Q}}(x) = (a + b\sqrt{d})(a-b\sqrt{d}) = a^2 - db^2 \qquad (F = \mathbb{Q}(\sqrt{d}))$$

Note also that if $a^2 - db^2 = \pm 4$ with $a,b \in \mathbb{Z}$, then since d is square-free a and b are of the same parity. It is a consequence of Proposition 7.20 that if $d \equiv 1 (\text{mod } 4)$, then

$$R = \{(1/2)(a+b\sqrt{d}) \mid a,b \in \mathbb{Z} \text{ of the same parity}\}$$

while if $d \equiv 2(\text{mod } 4)$ or $d \equiv 3(\text{mod } 4)$, then

$$R = \{a+b\sqrt{d} \mid a,b \in \mathbb{Z}\}$$

The desired conclusion is therefore a consequence of Proposition 6.14. ∎

8.2. REMARK. The formula

$$U(R) = \{(1/2)(a+b\sqrt{d}) \mid a,b \in \mathbb{Z},\ a^2 - db^2 = \pm 4\}$$

is true for any square-free d.

Indeed, if $a \equiv b \equiv 1(\text{mod } 2)$, then $a^2 \equiv b^2 \equiv 1(\text{mod } 4)$ and hence

$$a^2 - db^2 = \pm 4 \text{ implies } d \equiv 1(\text{mod } 4)$$

Now apply Lemma 8.1. ∎

If $d > 0$, $\mathbb{Q}(\sqrt{d})$ is called a *real quadratic field*, while if $d < 0$ then $\mathbb{Q}(\sqrt{d})$ is said to be an *imaginary quadratic field*.

8.3. PROPOSITION. Let R be the ring of integers of an imaginary quadratic field F. Then $U(R) = \{-1,1\}$, except in the following two cases:

(i) If $F = \mathbb{Q}(i)$, $i^2 = -1$, then $U(R) = \langle i \rangle$

(ii) If $F = \mathbb{Q}(\sqrt{-3})$, then $U(R) = \langle \varepsilon \rangle$ where ε is a primitive 6-th root of unity.

Proof. Write $R = \mathbb{Q}(\sqrt{-m})$, where $m = 1$ or m is a square-free positive integer.

(i) If $m \equiv 1$ or $2(\text{mod } 4)$, then we may apply Lemma 8.1(i). For $x = a+b\sqrt{-m}$, $a,b \in \mathbb{Z}$, to be a unit of R we must have $a^2 + mb^2 = 1$. If $m \geqslant 2$, this implies $b = 0$ and $a = \pm 1$, so $x = \pm 1$. If $m = 1$, besides the solution $x = \pm 1$, there are the solutions $a = 0$, $b = \pm 1$, i.e. $x = \pm i$ with $i^2 = -1$.

(ii) If $m \equiv 3(\text{mod } 4)$, then we may apply Lemma 8.1(ii). For $x = (1/2)(a+b\sqrt{-m})$,

$a,b \in \mathbb{Z}$, to be a unit of R, we must have $a^2 + mb^2 = 4$. If $m \geqslant 7$, this implies $b = 0$ so $a = \pm 2$ and $x = \pm 1$. If $m = 3$ and $b \neq 0$, then the relations $a = \pm 1$, $b = \pm 1$ entail the additional solutions $x = (1/2)(\pm 1 \pm \sqrt{-3})$. So the proposition is true. ■

Let us now concentrate on the more challenging case where F is a real quadratic field. As a preliminary observation, we first prove

8.4. LEMMA. Let R be the ring of integers of a real quadratic field F.

(i) There is a unit $\varepsilon > 1$ in R such that $U(R) = \{\pm 1\} \times \langle\varepsilon\rangle$

(ii) If $F = \mathbb{Q}(\sqrt{d})$, where $d \geqslant 2$ is a square-free integer, and $u > 1$ is a unit of R, then

$$u = a + b\sqrt{d} \qquad \text{for some } a > 0,\ b > 0 \text{ in } \mathbb{Q}$$

Proof. (i) Let $t(F^*)$ be the torsion subgroup of F^*. Since $F \subseteq R$, we have $t(F^*) = \{\pm 1\}$. Now apply the Dirichlet Unit Theorem.

(ii) Let $u = a + b\sqrt{d}$ $(a,b \in \mathbb{Q})$ be a unit of R. The numbers $u, u^{-1}, -u$, and $-u^{-1}$ are units of R and, because

$$N(u) = (a+b\sqrt{d})(a-b\sqrt{d}) = \pm 1$$

these four numbers are $\pm a \pm b\sqrt{d}$. For $u \neq \pm 1$ only one of the four numbers $u, u^{-1}, -u, -u^{-1}$ is greater than one, and it is the largest of the four. Hence $u > 1$ implies $a > 0, b > 0$ as required. ■

In what follows we refer to the unit ε in Lemma 8.4(i) as the *canonical fundamental unit* of R (or, by abuse of language, of F). Hence a knowledge of $U(R)$ is equivalent to a knowledge of the canonical fundamental unit ε, and for this reason we henceforth restrict ourselves to the computation of ε.

For the rest of this section, we fix the following notation:

$d \geqslant 2$ is a square-free integer

R is the ring of integers of $\mathbb{Q}(\sqrt{d})$

ε_d is the canonical fundamental unit of R

$N(\varepsilon_d)$ is the norm of ε_d

In view of Lemma 8.1, it is important to investigate the Diophantine equation

$$x^2 - dy^2 = m \tag{1}$$

in the special case where $m = \pm 1, \pm 4$. The equation (1), commonly known as *Pell's equation*, was actually never considered by Pell; it was because of a mistake on Euler's part that his name has been attached to it. If $x = a$ and $y = b$ are integers which satisfy (1) we say, for simplicity, that the number

$$a + b\sqrt{d}$$

is a solution of equation (1). It will be called *positive* if a and b are positive integers. The positive solutions will be ordered by the size of $a + b\sqrt{d}$. If $a_1 + b_1\sqrt{d}$ and $a_2 + b_2\sqrt{d}$ are two solutions of (1), then

$$a_1^2 - a_2^2 = d(b_1^2 - b_2^2).$$

Therefore, if both $a_1 + b_1\sqrt{d}$ and $a_2 + b_2\sqrt{d}$ are positive solutions of (1), then

$$a_1 + b_1\sqrt{d} > a_2 + b_2\sqrt{d} \Leftrightarrow a_1 > a_2 \Leftrightarrow b_1 > b_2 \tag{2}$$

We warn the reader that if $a_1 + b_1\sqrt{d}$ and $a_2 + b_2\sqrt{d}$ are positive solutions of the double equation $x^2 - dy^2 = \pm m$, then $a_1 + b_1\sqrt{d} > a_2 + b_2\sqrt{d}$ need not imply $b_1 > b_2$. Indeed, consider the double equation $x^2 - 5y^2 = \pm 4$. Then $3 + \sqrt{5}$ and $1 + \sqrt{5}$ are positive solutions with $3 + \sqrt{5} > 1 + \sqrt{5}$ but $b_1 = b_2 = 1$.

Our next result provides a useful characterization of ε_d. Although it is possible to treat the cases $d \equiv 2$ or $3 \pmod 4$ and $d \equiv 1 \pmod 4$ simultaneously by applying Remark 8.2, we consider them separably for practical purposes.

8.5. PROPOSITION. (i) If $d \equiv 2$ or $3 \pmod 4$, then

$$\varepsilon_d = a + b\sqrt{d}$$

where $a + b\sqrt{d}$ is the minimal positive solution of $x^2 - dy^2 = \pm 1$.

(ii) If $d \equiv 1 \pmod 4$, then

$$\varepsilon_d = (1/2)(a + b\sqrt{d})$$

where $a + b\sqrt{d}$ is the minimal positive solution of $x^2 - dy^2 = \pm 4$.

Proof. (i) Owing to Lemma 8.1(i), we may write $\varepsilon_d = a + b\sqrt{d}$ with $a, b \in \mathbb{Z}$ satisfying

$$a^2 - db^2 = \pm 1$$

Since $\varepsilon_d > 1$, Lemma 8.4(ii) implies that $a > 0$ and $b > 0$. Hence $a + b\sqrt{d}$ is a positive solution of $x^2 - dy^2 = \pm 1$. Let $a_1 + b_1\sqrt{d}$ be any positive solution of $x^2 - dy^2 = \pm 1$. Then $a_1 + b_1\sqrt{d} > 1$ and, by Lemma 8.1(i),

$$a + b\sqrt{d} \in U(R).$$

Therefore, by Lemma 8.4(i),

$$a_1 + b_1\sqrt{d} = \varepsilon_d^n \qquad \text{for some } n \geqslant 1$$

Because $\varepsilon_d > 1$, this shows that $a_1 + b_1\sqrt{d} \geqslant a + b\sqrt{d}$, as required.

(ii) Apply arguments of (i) with Lemma 8.1(i) replaced by Lemma 8.1(ii). ∎

Observe that if $N(\varepsilon_d) = 1$, then $N(u) = 1$ for all units u of R. Hence, if $d \equiv 2$ or $3 (\text{mod } 4)$ and $\varepsilon_d = a + b\sqrt{d}$, then by Proposition 8.5(i), the following holds:

(a) $a + b\sqrt{d}$ is the minimal positive solution of $x^2 - dy^2 = 1$ or $x^2 - dy^2 = -1$ according to whether $N(\varepsilon_d) = 1$ or $N(\varepsilon_d) = -1$. In particular, $N(\varepsilon_d) = 1$ if and only if the equation

$$x^2 - dy^2 = -1$$

has no solution in $\mathbb{Z}$.

Similarly, if $d \equiv 1 (\text{mod } 4)$ and $\varepsilon_d = (1/2)(a+b\sqrt{d})$, then by Proposition 8.5(ii), we have

(b) $a + b\sqrt{d}$ is the minimal positive solution of $x^2 - dy^2 = 4$ or $x^2 - dy^2 = -4$ according to whether $N(\varepsilon_d) = 1$ or $N(\varepsilon_d) = -1$.

In particular, $N(\varepsilon_d) = 1$ if and only if the equation

$$x^2 - dy^2 = -4$$

has no solution in $\mathbb{Z}$.

8.6. PROPOSITION. (i) Assume that $d \equiv 2$ or $3 (\text{mod } 4)$ and let $\varepsilon_d = a_1 + b_1\sqrt{d}$ with $a_1, b_1 \in \mathbb{N}$. Then the sequence

$$a_n + b_n\sqrt{d} = (a_1 + b_1\sqrt{d})^n \qquad (n \geqslant 1)$$

gives all positive solutions of $x^2 - dy^2 = \pm 1$.

(ii) Assume that $d \equiv 1 \pmod 4$ and let $\varepsilon_d = (1/2)(a_1+b_1\sqrt{d})$ with $a_1, b_1 \in \mathbb{N}$. Then the sequence

$$a_n + b_n\sqrt{d} = 2^{1-n}(a_1+b_1\sqrt{d})^n \qquad (n \geqslant 1)$$

gives all positive solutions of $x^2 - dy^2 = \pm 4$.

Proof. (i) Because ε_d is of infinite order, all the $a_n + b_n\sqrt{d}$, $n \geqslant 1$, are distinct. Moreover, by Lemma 8.1(i) and 8.4(ii), they are positive solutions of $x^2 - dy^2 = \pm 1$. If $z = a + b\sqrt{d}$ is any positive solution of $x^2 - dy^2 = \pm 1$, then $z \in U(R)$ and $z > 1$ so that by Lemma 8.4(i), $z = \varepsilon_d^n = a_n + b_n\sqrt{d}$ for some $n \geqslant 1$.

(ii) Apply arguments of (i) with Lemma 8.1(i) replaced by Lemma 8.1(ii). ∎

We now apply the result above to provide a rather crude method for calculating the canonical fundamental unit ε_d of R. It will be convenient to treat the cases $d \equiv 2$ or $3 \pmod 4$ and $d \equiv 1 \pmod 4$ separately.

8.7. PROPOSITION. Assume that $d \equiv 2$ or $3 \pmod 4$. In order to calculate

$$\varepsilon_d = a_1 + b_1\sqrt{d} \qquad (a_1, b_1 \in \mathbb{N})$$

it suffices to write down the sequence db^2, $b = 1,2,\ldots$, and to stop at the first number db_1^2 of this sequence which differs from ± 1 by a square a_1^2.

Proof. Let $a_n+b_n\sqrt{d}$ be the sequence defined in Proposition 8.6(i). It suffices to verify that

$$b_{n+1} > b_n \qquad \text{for all } n \geqslant 1$$

Indeed, in this case ε_d can be characterized as the unique positive solution $a + b\sqrt{d}$ of $x^2 - dy^2 = \pm 1$ with smallest possible b. Because

$$\begin{aligned} a_{n+1} + b_{n+1}\sqrt{d} &= (a_1+b_1\sqrt{d})(a_n+b_n\sqrt{d}) \\ &= (a_1a_n+b_1b_nd) + (b_1a_n+a_1b_n)\sqrt{d} \end{aligned}$$

we have

$$b_{n+1} = b_1a_n + a_1b_n > b_n$$

as required. ∎

To illustrate the case $d \equiv 2$ or $3 \pmod 4$, let $d \in \{7,6,2\}$. If $d = 7$, the sequence db^2, $b = 1,2,\ldots$ is $7, 28, 7.3^2 = 8^2 - 1$, so taking $b_1 = 3$ and $a_1 = 8$, we see that $8 + 3\sqrt{7}$ is the canonical fundamental unit of $\mathbb{Q}(\sqrt{7})$. A similar argument shows that $5 + 2\sqrt{6}$ is the canonical fundamental unit of $\mathbb{Q}(\sqrt{6})$. Note that both units $8 + 3\sqrt{7}$ and $5 + 2\sqrt{6}$ have norm 1. By taking $d = 2$, and hence $\varepsilon_d = 1 + \sqrt{2}$, we obtain an example of a canonical fundamental unit of norm -1.

Turning to the case $d \equiv 1 \pmod 4$, observe that it is not true in general that $b_{n+1} > b_n$ for all $n > 1$. Indeed, it will be shown below that $(1/2)(1+\sqrt{5})$ is the canonical fundamental unit of $\mathbb{Q}(\sqrt{5})$. Hence, if

$$a_n + b_n\sqrt{5} = 2^{1-n}(1+\sqrt{5})^n$$

then $b_1 = b_2 = 1$. This remark explains why we have to modify slightly the method of Proposition 8.7 in order to treat the case $d \equiv 1 \pmod 4$.

8.8. PROPOSITION. Assume that $d \equiv 1 \pmod 4$. In order to calculate

$$\varepsilon_d = (1/2)(a_1+b_1\sqrt{d}) \qquad (a_1, b_1 \in \mathbb{N})$$

it suffices to write down the sequence $db^2, b = 1,2,\ldots,$ and to stop at the first term db_1^2 of this sequence which differs from ± 4 by a square a_1^2 (if there are two such squares, then as a_1^2 choose the minimal one).

Proof. Let $a + b\sqrt{d}$ be the positive solution of $x^2 - dy^2 = \pm 4$ with smallest possible b and such that if there are two positive solutions of the form $c_1 + b\sqrt{d}$ and $d_1 + b\sqrt{d}$, then $a = \min\{c_1, d_1\}$. The assertion of the proposition amounts to saying that

$$\varepsilon_d = (1/2)(a+b\sqrt{d})$$

Put $a_n + b_n = 2^{1-n}(a_1+b_1\sqrt{d})$ so that, by Proposition 8.6(ii), the $a_n + b_n\sqrt{d}$, $n \geqslant 1$, are all positive solutions of $x^2 - dy^2 = \pm 4$. We distinguish two cases.

First assume that $N(\varepsilon_d) = 1$. Then $x^2 - dy^2 = \pm 4$ implies $x^2 - dy^2 = 4$. Because

$$a_{n+1} + b_{n+1}\sqrt{d} = 2\varepsilon_d^{n+1} > 2\varepsilon_d^n = a_n + b_n\sqrt{d}$$

it follows from (2) that $b_{n+1} > b_n$ for all n. Hence $\varepsilon_d = a + b\sqrt{d}$.

Next assume that $N(\varepsilon_d) = -1$. Then

$$a_1 + b_1\sqrt{d} < a_3 + b_3\sqrt{d} < a_5 + b_5\sqrt{d} < \dots$$

are solutions of $x^2 - dy^2 = -4$, while

$$a_2 + b_2\sqrt{d} < a_4 + b_4\sqrt{d} < a_6 + b_6\sqrt{d} < \dots$$

are solutions of $x^2 - dy^2 = 4$. Hence, by (2),

$$b_1 < b_3 < b_5 < \dots \text{ and } b_2 < b_4 < b_6 < \dots$$

This ensures that $a + b\sqrt{d} = a_1 + b_1\sqrt{d}$ or $a_2 + b_2\sqrt{d}$. If $b_1 = b_2$, then $a = \min\{a_1,a_2\} = a_1$ so that $a + b\sqrt{d} = a_1 + b_1\sqrt{d}$. We are therefore left to verify that $b_1 \leqslant b_2$. However,

$$b_{n+1} = (1/2)(a_1b_n + b_1a_n) \qquad (n \geqslant 1)$$

and so $b_2 = a_1b_1 \geqslant b_1$, as required. ∎

8.9. REMARK. Assume that $d \equiv 1 \pmod 4$ and that $d \neq 5$. If

$$\varepsilon_d = (1/2)(a_1 + b_1\sqrt{d})$$

then $a_1 + b_1\sqrt{d}$ is the unique positive solution of $x^2 - dy^2 = \pm 4$ with smallest possible b_1. Indeed, if $a + b\sqrt{d}$ is a positive solution of $x^2 - dy^2 = \pm 4$ with smallest possible b, then by the above

$$a + b\sqrt{d} = a_1 + b_1\sqrt{d} \text{ or } a + b\sqrt{d} = a_2 + b_2\sqrt{d}$$

Hence, if $a + b\sqrt{d} = a_2 + b_2\sqrt{d}$, then $b_1 = b_2 = a_1b_1$ in which case $a_1 = 1$. Accordingly, $1 - db_1^2 = -4$ and so $db_1^2 = 5$. Because $d \geqslant 2$, we infer that $d = 5$, as claimed. ∎

The remark above together with Proposition 8.7 shows that for any square-free integer $d \geqslant 2$ distinct from 5, $\varepsilon_d = (1/2)(a + b\sqrt{d})$ where $a + b\sqrt{d}$ is a unique positive solution of $x^2 - dy^2 = \pm 4$ with smallest possible b.

To illustrate the case $d \equiv 1 \pmod 4$, let $d \in \{5,13,17,69\}$. If $d = 5$, the first term of the sequence $db^2, b = 1,2,\dots,$ is $5 = 1^2 + 4 = 3^2 - 4$. Hence $b_1 = 1$ and $a_1 = \min\{1,3\} = 1$. Therefore $(1/2)(1+\sqrt{5})$ is the canonical fundamental unit of $\mathbb{Q}(\sqrt{5})$.

If $d = 13$, then the first term of the sequence db^2, $b = 1,2,\ldots$, is $13 = 3^2 + 4$. Thus $b_1 = 1$ and $a_1 = 3$ so that $(1/2)(3+\sqrt{13})$ is the canonical fundamental unit of $\mathbb{Q}(\sqrt{13})$. If $d = 17$, then the sequence db^2, $b = 1,2,\ldots$, is $17, 68 = 8^2 + 4$ and so $(1/2)(8 + 2\sqrt{17}) = 4 + \sqrt{17}$ is the canonical fundamental unit of $\mathbb{Q}(\sqrt{17})$. Note that all the above three units have norm -1.

Finally, if $d = 69$, the sequence db^2, $b = 1,2,\ldots$ is

$$69, 276, 621 = 25^2 - 4$$

so that $a_1 = 25$ and $b_1 = 3$. Thus $(1/2)(25+3\sqrt{69})$ is the canonical fundamental unit of $\mathbb{Q}(\sqrt{69})$ with norm 1.

The above calculations become very awkward for large d. Note, however, that using the theory of continuous fractions, one can find other, more rapid procedures for calculating the canonical fundamental unit.

We now turn our attention to the investigation of the sign of $N(\varepsilon_d)$. Let a and $m > 1$ be coprime integers. We say that a is a *quadratic residue modulo* m if $x^2 \equiv a \pmod{m}$ for some $x \in \mathbb{Z}$. Otherwise, we say that a is a *quadratic nonresidue modulo* m.

For each $n \in \mathbb{Z}$, let $\bar{n}$ be its image in $\mathbb{Z}/m\mathbb{Z}$ and let $U(m)$ be the unit group of $\mathbb{Z}/m\mathbb{Z}$. Then a is a quadratic residue modulo m if and only if $\bar{a}$ is a square in $U(m)$.

Let p be an odd prime and let a be an integer coprime to p. We define the *Legendre* symbol $(a|p)$ of a, relative to p, as follows:

$$(a|p) = \begin{cases} 1, & \text{if } a \text{ is a quadratic residue modulo } p \\ -1, & \text{if } a \text{ is a quadratic nonresidue modulo } p \end{cases}$$

A straightforward verification shows that

(i) If $a \equiv b \pmod{p}$, then $(a|p) = (b|p)$

(ii) $(ab|p) = (a|p)(b|p)$ (in particular, $(a^2|p) = 1$)

8.10. LEMMA. Let p be an odd prime and let a be an integer coprime to p.

Then

$$(a|p) \equiv a^{(p-1)/2} \pmod p$$

and, in particular,

$$(-1|p) = (-a^2|p) = (-1)^{(p-1)/2}$$

Proof. Let $a \equiv \omega^t \pmod p$, where ω is a primitive root modulo p and $0 \leqslant t < p - 1$. Because $\bar{\omega}$ is not a square in $U(p)$ and

$$\omega^{(p-1)/2} \equiv -1 \pmod p$$

we have

$$(a|p) = (\omega^t|p) = (\omega|p)^t = (-1)^t = (\omega^{(p-1)/2})^t$$

$$= (\omega^t)^{(p-1)/2} \equiv a^{(p-1)/2} \pmod p$$

as required. ∎

The above lemma will be used in the proof of the following useful observation

8.11. LEMMA. If $a,b \in \mathbb{N}$ satisfy $a^2 - db^2 = -4$, then all odd prime factors of d and b are congruent to 1 modulo 4 and $b \not\equiv 0 \pmod 4$.

Proof. Because $a^2 - db^2 = -4$, we have $a^2 \equiv -4 \pmod{db^2}$. Therefore, for any odd prime factor p of db^2, we have $a^2 \equiv -4 \pmod p$. Accordingly, by Lemma 8.10,

$$1 = (-4|p) = (-1)^{(p-1)/2}$$

which implies $p \equiv 1 \pmod 4$.

If $b \equiv 0 \pmod 4$, then $a^2 - db^2 = -4$ implies $a \equiv 0 \pmod 2$, and hence we may put $b = 4b_0, a = 2a_0$, in which case $a_0^2 - 4db_0^2 = -1$. Thus $a_0^2 \equiv -1 \pmod 4$, which is a contradiction. ∎

We are now ready to provide a necessary condition for $N(\varepsilon_d) = -1$.

8.12. PROPOSITION. If $N(\varepsilon_d) = -1$, then all odd prime factors of d are congruent to 1 modulo 4.

Proof. By Remark 8.2, $\varepsilon_d = (1/2)(a+b\sqrt{d})$ where $a,b \in \mathbb{N}$ satisfy

$$a^2 - db^2 = -4$$

Now apply Lemma 8.11. ■

The converse of the above result is decidedly false. Indeed, let $d = 34$ so that the only odd prime factor of 34 is $17 \equiv 1 (\mathrm{mod}\ 4)$. Applying Proposition 8.7, we see that $\varepsilon_d = 35 + 6\sqrt{34}$ and hence $N(\varepsilon_d) = 1$. Our next result provides a partial converse of Proposition 8.12.

8.13. PROPOSITION. If d is a prime congruent to 1 modulo 4, then

$$N(\varepsilon_d) = -1$$

Proof. It suffices to show that there is at least one unit of norm -1. The latter will hold provided we prove that the equation $x^2 - dy^2 = -1$ is solvable in $\mathbb{Z}$.

Let $x_1 + y_1\sqrt{d}$ be the minimal positive solution of $x^2 - dy^2 = 1$. Then

$$x_1^2 - 1 = dy_1^2 \tag{3}$$

Consequently, x_1 cannot be even, for in this case we should have $-1 \equiv d (\mathrm{mod}\ 4)$. Thus x_1 is odd, in which case the numbers $x_1 - 1$ and $x_1 + 1$ have the greatest common divisor 2. It follows from (3) that

$$x_1 \pm 1 = 2a^2\ ,\ x_1 \mp 1 = 2db^2$$

where a and b are natural numbers and $y_1 = 2ab$. By elimination of x_1, we get

$$\pm 1 = a^2 - db^2$$

Because $b < y_1$ we cannot have the upper sign. Therefore the lower sign must be taken and the result follows. ■

In order to provide information on those $\mathbb{Q}(\sqrt{d})$ for which $N(\varepsilon_d) = -1$, we first give a generating function for all such fields. All the remaining results of this section are based on a work of Yokoi(1968).

8.14. LEMMA. A unit ε of R has norm -1 if and only if

$$\varepsilon = (t \pm \sqrt{t^2+4})/2$$

for some $t \in \mathbb{Z}$ with $\mathbb{Q}(\sqrt{d}) = \mathbb{Q}(\sqrt{t^2+4})$. In particular, $N(\varepsilon_d) = -1$ if and only if $\mathbb{Q}(\sqrt{d}) = \mathbb{Q}(\sqrt{n^2+4})$ for some $n \in \mathbb{N}$ or, equivalently, if and only if d

is the square-free part of $n^2 + 4$ for some $n \in \mathbb{N}$.

Proof. If ε is a unit of R with norm -1, then by Remark 8.2,

$$\varepsilon = (t+u\sqrt{d})/2$$

where $t,u \in \mathbb{Z}$ satisfy

$$t^2 - du^2 = -4 \quad (4)$$

By (4), we have $t^2 + 4 = du^2$ so that $\mathbb{Q}(\sqrt{d}) = \mathbb{Q}(\sqrt{t^2+4})$ and $u\sqrt{d} = \pm \sqrt{t^2+4}$. Thus $\varepsilon = (t \pm \sqrt{t^2+4})/2$.

Conversely, suppose that $\varepsilon = (t \pm \sqrt{t^2+4})/2$ where $t \in \mathbb{Z}$ is such that $\mathbb{Q}(\sqrt{d}) = \mathbb{Q}(\sqrt{t^2+4})$. Then, by Lemma 7.19(ii), $t^2 + 4 = du^2$ for some $u \in \mathbb{N}$ and hence $\varepsilon = (t \pm u\sqrt{d})/2$. Because $t^2 - du^2 = -4$, we conclude, by Remark 8.2, that ε is a unit of R with norm -1. ∎

We have now accumulated all the information necessary to provide a global picture of canonical fundamental units of norm -1.

8.15. PROPOSITION. The sequence

$$(n + \sqrt{n^2+4})/2$$

with $n \in \mathbb{N}$ such that $n < m$ whenever $n \neq m$ and $n^2 + 4, m^2 + 4$ have the same square-free part, gives all canonical fundamental units of norm -1. In fact, for any such n,

$$(n + \sqrt{n^2+4})/2 = \varepsilon_d$$

where d is the square-free part of $n^2 + 4$.

Proof. Fix $n \in \mathbb{N}$ such that $n < m$ if $n \neq m$ and $n^2 + 4, m^2 + 4$ have the same square-free part. We claim that $\varepsilon_d = (n + \sqrt{n^2+4})/2$ and $N(\varepsilon_d) = -1$, where d is the square-free part of $n^2 + 4$. Indeed, by Lemma 8.14, $(n + \sqrt{n^2+4})/2$ is a unit of R with norm -1. Therefore $N(\varepsilon_d) = -1$ and so, by Lemma 8.14,

$$\varepsilon_d = (m + \sqrt{m^2+4})/2$$

for some $m \in \mathbb{N}$ such that d is the square-free part of $m^2 + 4$. If $n \neq m$, then $n < m$ by hypothesis and so $(n + \sqrt{n^2+4})/2 < \varepsilon_d$, a contradiction. Consequently, $m = n$ and $\varepsilon_d = (n + \sqrt{n^2+4})/2$.

Conversely, suppose that $N(\varepsilon_d) = -1$. Then, by Lemma 8.14, there exists $n \in \mathbb{N}$ such that d is the square-free part of $n^2 + 4$ and $\varepsilon_d = (n + \sqrt{n^2+4})/2$. If $n \neq m$ and d is the square-free part of $m^2 + 4$, then $\varepsilon = (m + \sqrt{m^2+4})/2$ is a unit of R, again by Lemma 8.14. If $n > m$, then $\varepsilon_d > \varepsilon$ which is impossible. Thus $n < m$ and the result follows. ∎

It is now easy to derive the following properties:

8.16. PROPOSITION. (i) If $d = n^2 + 4$ for some $n \in \mathbb{N}$, then

$$\varepsilon_d = (n + \sqrt{n^2+4})/2 \quad \text{and} \quad N(\varepsilon_d) = -1$$

(ii) If $d = m^2 + 1$ with $2 \neq m \in \mathbb{N}$, then

$$\varepsilon_d = m + \sqrt{m^2+1} \quad \text{and} \quad N(\varepsilon_d) = -1$$

Proof. (i) Suppose that $m \in \mathbb{N}$ is such that the square-free parts of $m^2 + 4$ and $n^2 + 4$ are the same. Because d is square-free, if $m \neq n$, then

$$n^2 + 4 < m^2 + 4$$

and so $n < m$. Now apply Proposition 8.15.

(ii) Put $\varepsilon = m + \sqrt{m^2+1}$ so that $\varepsilon = (1/2)(2m + \sqrt{(2m)^2+4})$. By Proposition 8.15, it suffices to prove that if $k \neq 2m$ and d is the square-free part of $k^2 + 4$, then $2m < k$. Now

$$4m^2 + 4 = 2^2 d \quad \text{and} \quad k^2 + 4 = a^2 d$$

for $a \geqslant 1$. Hence we need only verify that $a \neq 1$. Assume by way of contradiction that $k^2 + 4 = d$. Then $k^2 + 4 = m^2 + 1$ which implies $(k-m)(k+m) = -3$. Because the latter is possible only in the case $m = 2, k = 1$, we derive a desired contradiction. ∎

As an application of the result above, we prove

8.17. COROLLARY. If $\varepsilon_d = (1/2)(t+u\sqrt{d})$, then the following conditions are equivalent:

(i) $N(\varepsilon_d) = -1$ and $u \in \{1,2\}$

(ii) $d = n^2 + 4$ or $d = n^2 + 1$ for some $n \in \mathbb{N}$

Proof. (i) ⇒ (ii): Because $N(\varepsilon_d) = -1$, we have $t^2 - du^2 = -4$. Bearing in mind that $u = 1$ or 2, we obtain respectively $d = t^2 + 4$ or $d = t_0^2 + 1$ for $t_0 = t/2$.

(ii) $\Rightarrow$ (i): If $d = n^2 + 4$, then by Proposition 8.16, $\varepsilon_d = (n + \sqrt{d})/2$ (so $u = 1$) and $N(\varepsilon_d) = -1$. If $d = n^2 + 1$ with $n \neq 2$, then by Proposition 8.16, $\varepsilon_d = n + \sqrt{n^2+1}$ (so $u = 2$) and $N(\varepsilon_d) = -1$. Finally, if $d = n^2 + 1$ and $n = 2$, then $d = 1^2 + 4$ and we may apply the first case. ∎

We close this section by remarking that the following problem is still outstanding.

Problem. Given a square-free integer $d \geqslant 2$, determine ε_d exclusively in terms of d.

In this vein we quote the following result due to Richaud (1866) and Degert (1958).

8.18. PROPOSITION. Let $d \geqslant 2$ be a square-free integer and write $d = n^2 + r$, $-n < r \leqslant n$. If r divides $4n$, then

$$\varepsilon_d = \begin{cases} n + \sqrt{d} \text{ with } N(\varepsilon_d) = -\operatorname{sgn} r & \text{for } |r| = 1 \text{ (except for } d = 5, n = 2, r = 1) \\ (n + \sqrt{d})/2 \text{ with } N(\varepsilon_d) = -\operatorname{sgn} r & \text{for } |r| = 4 \\ [(2n^2+r) + 2n\sqrt{d}]/r \text{ with } N(\varepsilon_d) = 1 & \text{for } |r| \neq 1,4 \end{cases}$$

9. UNITS IN PURE CUBIC FIELDS

A rational integer $d > 1$ is said to be *cube-free* if d is not divisible by a cube of a prime. The field $F = \mathbb{Q}(\sqrt[3]{d})$, in which $d > 1$ is a cube-free rational integer and $\sqrt[3]{d}$ is real, is called a *pure cubic field*. Throughout R denotes the ring of integers of the pure cubic field $F = \mathbb{Q}(\sqrt[3]{d})$. Our aim is to investigate the structure of the unit group of R.

9.1. PROPOSITION. Let $1 \neq d \in \mathbb{N}$ be cube-free and let $d = ab^2$ where ab is square-free.

$9 \nmid a^2 - b^2$

(i) If $\mathbb{Q}(\sqrt[3]{d})$ is of the first kind, then $U(R)$ consists precisely of those elements

$$x + y\sqrt[3]{ab^2} + z\sqrt[3]{a^2b} \qquad (x,y,z \in \mathbb{Z})$$

for which

$$x^3 + ab^2y^3 + a^2bz^3 - 3abxyz = \pm 1 \tag{1}$$

(ii) If $\mathbb{Q}(\sqrt[3]{d})$ is of the second kind, then $U(R)$ consists precisely of those elements

$$(u/3)(1+a\sqrt[3]{ab^2} + b\sqrt[3]{a^2b}) + v\sqrt[3]{ab^2} + \omega\sqrt[3]{a^2b}+ \qquad (u,v,\omega \in \mathbb{Z})$$

for which

$$x^3 + ab^2y^3 + a^2bz^3 - 2abxyz = \pm 27 \tag{2}$$

where $u = x$, $au + 3v = y$, and $bu + 3\omega = z$.

Proof. (i) Let $\omega = x + y\sqrt[3]{ab^2} + z\sqrt[3]{a^2b}$ and let ε be a primitive cube root of unity. Then $N(\omega) = \omega\omega'\omega''$, where

$$\omega' = x + \varepsilon y\sqrt[3]{ab^2} + \varepsilon^2 z\sqrt[3]{a^2b} \quad \text{and} \quad \omega'' = x + \varepsilon^2 y\sqrt[3]{ab^2} + \varepsilon z\sqrt[3]{a^2b}$$

An easy calculation shows that

$$N(\omega) = x^3 + ab^2y^3 + a^2bz^3 - 3abxyz \tag{3}$$

and the required assertion follows by Proposition 7.21.

(ii) Apply the same argument as in (i). ∎

9.2. PROPOSITION. There is a unique unit ε_d of R with $0 < \varepsilon_d < 1$ such that

$$U(R) = \{\pm 1\} \times \langle\varepsilon_d\rangle$$

Moreover,

(i) For any unit ω of R, $N(\omega) = 1$ if and only if $\omega > 0$.

(ii) If $\omega = x + y\sqrt[3]{ab^2}$ or $\omega = x + y\sqrt[3]{a^2b}$, $x \in \mathbb{Z}$, $0 \neq y \in \mathbb{Z}$, and ω is a positive unit of R, then $\omega < 1$. In particular, $\omega = \varepsilon_d^n$ for some $n > 0$.

Proof. The field F has the property that each of its elements is either rational or of degree 3. It follows that ± 1 are the only roots of unity in F.

Since $\sqrt[3]{d}$, $\varepsilon\sqrt[3]{d}, \varepsilon^2\sqrt[3]{d}$ are the roots of X^3-d (ε is a primitive cube root of unity) and $\varepsilon\sqrt[3]{d}$, $\varepsilon^2\sqrt[3]{d}$ are complex conjugates, it follows from the Dirichlet Unit Theorem that $U(R) = \{\pm 1\} \times \langle\delta\rangle$ for some $\delta > 0$ in R. If $\delta' \neq \delta$ satisfies $\delta' > 0$ and $U(R) = \{\pm 1\} \times \langle\delta'\rangle$, then $\delta' = \delta^{-1}$. Hence $\varepsilon_d = \min\{\delta, \delta^{-1}\}$ is a unique unit of R with $0 < \varepsilon_d < 1$ and with

$$U(R) = \{\pm 1\} \times \langle\varepsilon_d\rangle$$

If ω', ω'' are conjugates of ω, then $N(\omega) = \omega\omega'\omega''$. But ω', ω'' are complex conjugates, so $\omega > 0$ if and only if $N(\omega) > 0$. In particular, for $\omega \in U(R)$, $N(\omega) = 1$ if and only if $\omega > 0$.

Suppose that $\omega = x + y\sqrt[3]{ab^2}$, $x, y \in \mathbb{Z}$, is a positive unit of R and $y \neq 0$. Then $N(\omega) = 1$ and by (3),

$$x^3 + ab^2y^3 = 1$$

which shows that $xy < 0$. We conclude therefore that for $\alpha = \sqrt[3]{ab^2}$

$$\omega^{-1} = \omega'\omega'' = x^2 - xy\alpha + y^2\alpha^2 \geqslant 1 + \alpha + \alpha^2 > 1$$

and hence that $\omega < 1$. A similar argument can be applied to the case $\omega = x + y\sqrt[3]{a^2b}$. ■

From now on, we refer to the unit ε_d in Proposition 9.2 as the *canonical fundamental unit*. Thus, by Proposition 9.2, effective knowledge of $U(R)$ ultimately depends upon our ability to express ε_d exclusively in terms of d. To solve this problem, however, seems to pose almost insurmountable difficulties. Indeed, write $d = ab^2$ where ab is square-free. Then if we could express ε_d in terms of d, it would also be possible to provide all solutions of the Diophantine equation

$$x^3 + ab^2y^3 + a^2bz^3 - 3abxyz = 1$$

The latter, needless to say, is an exceptionally difficult task. It is nevertheless possible, for a given numerical value of d, to calculate ε_d by means of algorithms. For example, with the aid of the computer, Wada(1970) calculated

$$\varepsilon_d \quad \text{for all} \quad 2 \leqslant d < 250$$

Another possibility is to put a certain constraint on d and then try to find a formula for ε_d. It is in this direction that most of the successful results have been achieved. The following two cases are among the significant ones:

(i) $d = D^3 + k$, where D, k are rational integers with $D > 0$, $|k| > 1$, and

$k|3D^2$

(ii) $d = (n^3+1)(n^3+2)$ where $n \in \mathbb{N}$

In case (i), extending a result of Stender(1969), it was proved by Rudman(1973) that

$$\varepsilon_d = (\sqrt[3]{d}-D)^3/k$$

except for

$$(D,k) = (2,-6),(1,3),(2,2),(3.1),(5,-25),(2,-4)$$

In case (ii), a noteworthy result of Bernstein(1975) (established by ingenious calculations) asserts that

$$\varepsilon_d = 1 - 3n(n^3+1)\sqrt[3]{d} + 3n^2\sqrt[3]{d^2}$$

The proofs of these facts are very computational in nature and therefore are not suitable to be presented here. Instead, we decided after discussing some general facts to prove a beautiful result known as the Delaunay-Nagell theorem which asserts that any unit $u \neq 1$ of R which is of norm 1 and of the form

$$x + y\sqrt[3]{d} \qquad (x,y \in \mathbb{Z}, y \neq 0)$$

is either the canonical fundamental unit or its square. As an application, we disclose ε_d for infinitely many d. Namely, we prove that for

$$d = n(u^6n^2+3u^3n+3) \qquad (n,u \in \mathbb{N})$$

where $nu \equiv 0(\text{mod}3)$ or $n \equiv u(\text{mod}3)$,

$$\varepsilon_d = (1+u^3n) - u\sqrt[3]{d}$$

Applying a similar argument, we also show that if

$$d = n(u^6n^2-3u^3n+3) \qquad (n,u \in \mathbb{N})$$

where $nu \equiv 0(\text{mod}3)$ or $n + u \equiv 0(\text{mod}3)$, then

$$\varepsilon_d = (1-u^3n) + u\sqrt[3]{d}$$

Finally, we provide a number of numerical examples pertaining to the above results.

We begin by proving two technical lemmas to be applied later. For simplicity in notation, we define the binomial coefficient $\binom{m}{k}$ to be zero for $k > m$.

9.3. LEMMA. Let m be a positive integer. Then

$$\binom{m}{0} + \binom{m}{3} + \binom{m}{6} + \dots \not\equiv 0 \pmod{3}$$

Proof. Put

$$s_0 = \binom{m}{0} + \binom{m}{3} + \binom{m}{6} + \dots$$
$$s_1 = \binom{m}{1} + \binom{m}{4} + \binom{m}{7} + \dots$$
$$s_2 = \binom{m}{2} + \binom{m}{5} + \binom{m}{8} + \dots$$

Then

$$s_0 + s_1 + s_2 = 2^m \equiv (-1)^m \pmod{3}$$

and

$$s_2 = \binom{m}{1}(m-1)/2 + \binom{m}{4}(m-4)/5 + \dots \equiv -ms_1 + s_1 \pmod{3}$$
$$s_1 = \binom{m}{0}(m/1) + \binom{m}{3}(m-3)/4 + \dots \equiv ms_0 \pmod{3}$$

Therefore

$$(1+2m-m^2)s_0 \equiv (-1)^m \pmod{3}$$

and the result follows. ■

9.4. LEMMA. Assume that x and y are integers such that $(x,dy) = 1$ and let

$$(x+y\sqrt[3]{d})^n = X + Y\sqrt[3]{d} + Z(\sqrt[3]{d})^2$$

where $X,Y,Z \in \mathbb{Q}$ and $n > 1$. Then $XYZ \neq 0$ except in the following cases:

$$(\sqrt[3]{10}-1)^5 = 99 - 45\sqrt[3]{10}$$
$$(\sqrt[3]{4}-1)^4 = -15 + 12\sqrt[3]{2}$$

Proof. We first observe that, since $(x,d) = 1$, we have $X \neq 0$. Suppose that $Z = 0$, so that

$$\binom{n}{2}x^{n-2}y^2 + \binom{n}{5}x^{n-5}y^5d + \binom{n}{8}x^{n-8}y^8d^2 + \dots = 0 \tag{4}$$

Dividing by $\binom{n}{2}y^2$, we obtain

$$-x^{n-2} = \sum_{k\geqslant 1} \binom{n-2}{3k} \frac{2x^{n-3k-2}y^{3k}d^k}{(3k+1)(3k+2)} \tag{5}$$

Let p be a prime divisor of y. Then since $p^{3k} \geqslant 2^{3k} > 3k + 2$ for $k \geqslant 1$, each term in the last sum is divisible by p, which is impossible since $(x,y) = 1$. Thus $y = \pm 1$.

When $n \equiv 0 (\mathrm{mod} 3)$, equation (5) can be written in the form

$$-y^{n-3} d^{(1/3)(n-3)} = \sum_{k \geq 1} \binom{n-1}{3k} \frac{x^{3k} y^{n-3k-3} d^{(1/3)(n-3)-k}}{3k+1}$$

when $n \equiv 1 (\mathrm{mod} 3)$,

$$-y^{n-4} d^{(1/3)(n-4)} = \sum_{k \geq 1} \binom{n-2}{3k} \frac{2x^{3k} y^{n-3k-4} d^{(1/3)(n-4)-k}}{(3k+1)(3k+2)}$$

and when $n \equiv 2 (\mathrm{mod} 3)$,

$$-y^{n-2} d^{(1/3)(n-2)} = \sum_{k \geq 1} \binom{n}{3k} x^{3k} y^{n-3k-2} d^{(1/3)(n-2)-k}$$

A similar argument now shows that $x = \pm 1$ and since it is clear from (4) that $xy < 0$, we have $x = -y$.

Now assume that p is a prime divisor of d with $p^{\alpha} | d$ but $p^{\alpha+1} \nmid d$. If $p^{\alpha} > 5$, then $p^{\alpha k} > 5^k \geq 3k + 2$ for $k \geq 1$, so each term in the sum (5) is divisible by p, which is impossible since $(x,d) = 1$. If $p = 3$, then since $3 \nmid (3k+1)(3k+2)$ we reach the same contradiction. Therefore $p^{\alpha} = 2$ or 5, and $d \in \{2,5,10\}$.

The information obtained so far shows that

$$1 - \binom{n-2}{3} \frac{2d}{4\cdot 5} + \binom{n-2}{6} \frac{2d^2}{7\cdot 8} - \dots = 0 \qquad (6)$$

If $d = 10$, this becomes

$$\binom{n-2}{3} - 1 = \frac{1}{6}(n-5)(n^2-4n+6)$$

$$= \sum_{k \geq 2} (-1)^k \binom{n-2}{3k} \frac{2 \cdot 10^k}{(3k+1)(3k+2)}$$

This equation is true for $n = 5$, and leads to the first of the exceptions mentioned in the lemma. For other values of n, we may divide through by (n-5)/6 and obtain

$$n^2 - 4n + 6$$

$$= - \sum_{k\geq 1} (-1)^k \binom{n-6}{3k-1} \frac{(n-2)(n-3)(n-4)\cdot 12\cdot 10^{k+1}}{3k(3k+1)(3k+2)(3k+3)(3k+4)(3k+5)}$$

The highest power of 5 which divides the denominator of a term in the sum is obviously at most $5(3k+5)$, and because $5^{k+1} > 5(3k+5)$ for $k \geqslant 2$, we have

$$\begin{aligned} n^2 - 4n + 6 &= (n-2)^2 + 2 \\ &\equiv \binom{n-6}{2} \frac{(n-2)(n-3)(n-4)\cdot 12\cdot 10}{3\cdot 4\cdot 5\cdot 6\cdot 7\cdot 8} \\ &\equiv 0 \pmod 5 \end{aligned}$$

which is impossible since -2 is a quadratic nonresidue of 5.

When $d = 2$ or 5, equation (6) implies the congruence

$$1 + \binom{n-2}{3} + \binom{n-2}{6} + \dots \equiv 0 \pmod 3$$

which is again impossible, by Lemma 9.3.

We are therefore left to examine the possibility $Y = 0$. The proof that this happens only in the case of the second exception mentioned in the lemma is completely similar to what has just been done for the case $Z = 0$ (the only variation lies in the fact that d may now have the sole prime divisor 2, so that $d = 2$ or 4). The lemma is therefore established. ■

Let $d > 1$ be a cube-free integer and R the ring of integers of $\mathbb{Q}(\sqrt[3]{d})$. As before, we write

$$d = ab^2 \ , \quad \alpha = \sqrt[3]{ab^2} \quad \text{and} \quad \beta = \sqrt[3]{a^2b}$$

where ab is square-free. Then

$$u = x + y\alpha + z\beta \qquad (x, y, z \in \mathbb{Z})$$

is a unit of norm 1 if and only if

$$x^3 + ab^2y^3 + a^2bz^3 - 3abxyz = 1 \tag{7}$$

Finding such u supplies infinitely many solutions of (7), by taking all powers of u; finding one nontrivial solution of (7) means having a unit $u \neq 1$, which again supplies infinitely many solutions of (7). Of course, if we know a fundamental unit u satisfying (7), then we have all solutions of (7). Therefore there is an intimate connection between units of R and Diophantine equation (7). It is probably a hopeless task to try to solve (7) for any a,b. Instead, it is reasonable to specify a certain value for x,y,z and a natural choice would fall on $z = 0$. Hence we are confronted with the following equation:

$$x^3 + dy^3 = 1 \qquad (x,y \in \mathbb{Z}) \qquad (8)$$

This equation was first considered by Delaunay (for a detailed account of this work refer to Delaunay and Fadeev(1964)). His method was later refined by Nagell and it is our immediate purpose to prove Nagell's result. In the context of units Nagell's theorem asserts that any $u \neq 1$ of R which is of norm 1 and of the form

$$x + y\sqrt[3]{d} \qquad (x,y \in \mathbb{Z}, y \neq 0)$$

is either the canonical fundamental unit ε_d or ε_d^2. Whether for a given d equation (8) has a nontrivial solution is a very delicate problem. Some partial results can be found in Bernstein(1971) and J.H.E. Cohn(1967). Here we only mention that for values $d \leqslant 100$ such solutions exist if $d \in \{2,7,9,17,19,20,26,28,37,43,63,65,91\}$ (see J.H.E. Cohn(1967)).

9.5. LEMMA. The only solutions in $\mathbb{Z}$ of the equation

$$x^2 + 2 = y^3 \qquad (9)$$

are $x = \pm 5$, $y = 3$.

Proof. By Proposition 7.20(ii), the integers of the field $\mathbb{Q}(\sqrt{-2})$ are of the form $a + b\sqrt{-2}$, where $a,b \in \mathbb{Z}$. A routine calculation shows that they form a Euclidean domain: given $a,b,c,k \in \mathbb{Z}$, with $ck \neq 0$, there are $e,f,g,h \in \mathbb{Z}$ such that

$$a + b\sqrt{-2} = (c+k\sqrt{-2})(e+f\sqrt{-2}) + (g+h\sqrt{-2})$$
$$g^2 + 2h^2 < c^2 + 2k^2$$

It follows that $\mathbb{Z}[\sqrt{-2}]$ is a unique factorization domain.

We now prove that if x and y satisfy (9), then $x + \sqrt{-2}$ and $x - \sqrt{-2}$ are relatively prime. It is clear that

$$(x+\sqrt{-2}, x-\sqrt{-2}) \mid -2\sqrt{-2}$$

and since $-2\sqrt{-2} = (\sqrt{-2})^3$ and $\sqrt{-2}$ is prime in $\mathbb{Z}[\sqrt{-2}]$, we must have

$$(x+\sqrt{-2}, x-\sqrt{-2}) = (\sqrt{-2})^m \qquad (0 \leqslant m \leqslant 3)$$

But if $x + \sqrt{-2} = (a+b\sqrt{-2})\sqrt{-2}$, then $x = -2b$, whence by (9)

$$4b^2 + 2 = y^3$$

$$y^3 \equiv 2(\text{mod} 4)$$

which is impossible.

By Proposition 8.3, the only units of $\mathbb{Z}[\sqrt{-2}]$ are ± 1. It follows from (9) that

$$x + \sqrt{-2} = (a+b\sqrt{-2})^3$$

for some $a,b \in \mathbb{Z}$, and equating real and imaginary parts gives

$$a^3 - 6ab^2 = x \qquad (10)$$

$$3a^2b - 2b^3 = 1 \qquad (11)$$

From (ii) it follows that $b = \pm 1$ and therefore $3a^2 - 2 = \pm 1$, or $a = \pm 1$. From (10), we therefore derive $x = \pm 1 \mp 6 = \pm 5$, thus completing the proof. ■

The next four lemmas carry most of the burden of the proof of Nagell's theorem

9.6. LEMMA. The square of a positive rational unit of R of the form

$$\lambda = x + y\alpha + z\beta \qquad (x,y,z \in \mathbb{Z})$$

is itself of the form $X + Y\alpha$ only if $d = 20$ and

$$\lambda = 1 + \sqrt[3]{20} - \sqrt[3]{50}$$

Proof. Let $\lambda = x + y\alpha + z\beta$ be a positive unit of R so that, by (3) and Proposition 9.2(i),

$$x^3 + ab^2y^3 + a^2bz^3 - 3abxyz = 1 \qquad (12)$$

and

$$\lambda^2 = (x^2+2abyz) + (2xy+az^2)\alpha + (2xz+by^2)\beta$$

If $2xz+by^2 = 0$, then

$$z = -\frac{by^2}{2x}$$

and substituting this into (12) we have

$$x^3 + dy^3 - d^2\frac{y^6}{8x^3} + 3d\frac{y^3}{2} = 1,$$

or

$$d^2y^6 - 20x^3dy^3 - 8(x^6-x^3) = 0$$

whence

$$dy^3 = 10x^3 \pm 2x\sqrt{27x^4-2x} \tag{13}$$

Thus the number $27x^4-2x$ must be a square:

$$(27x^3-2)x = t^2 \tag{14}$$

If x is even, then $(27x^3-2,x) = 2$, so that

$$27x^3 - 2 = \pm 2u^2, \quad x = \pm 2v^2$$

Since -1 is a quadratic nonresidue of 3, we must choose the lower sign, and eliminating x we derive

$$108v^6 + 1 = u^2$$

$$(u-1)(u+1) = 108v^6$$

Since $(u-1,u+1) = 2$, this implies that

$$u \pm 1 = 54r^6, \quad u \mp 1 = 2s^6$$

whence

$$27r^6 - s^6 = (3r^2)^3 - (s^2)^3 = \pm 1$$

From the truth of Fermat's conjecture for $n = 3$, we infer that $r = 0$, which gives $v = 0$ and $x = 0$. It follows also that $y = 0$, by (13), which is impossible because $z\beta$ is not a unit.

Now assume that x is odd. Then (14) gives

$$27x^3-2 = \pm u^2, \quad x = \pm v^2$$

Here the upper sign must be chosen, and we obtain

$$(3x)^3 = u^2 + 2$$

which by Lemma 9.5 has the sole solution $x = 1$, $u = \pm 5$. By (13), $dy^3 = 10 \pm 10$, so that $d = 20, y = 1$ (if $y = 0$, then $z = 0$, and λ is rational). The sole solution is therefore

$$(1 + \sqrt[3]{20} - \sqrt[3]{50})^2 = -19 + 7\sqrt[3]{20}$$

as asserted. ∎

9.7. LEMMA. If the square of a positive unit of R of the form

$$\lambda = (1/3)(x+y\alpha+z\beta) \qquad 3 \nmid xyz \quad (x,y,z \in \mathbb{Z})$$

is itself of the form $X + Y\alpha$, then either $x = 2, y = 2, z = -1$ and $d = 19$ or x is odd and $x = -v^2$ for some $v \in \mathbb{Z}$.

Proof. By (2), we have

$$x^3 + ab^2y^3 + a^2bz^3 - 3abxyz = 27 \tag{15}$$

and

$$9\lambda^2 = (x^2+2abyz) + (2xy+az^2)\alpha + (2xz+by^2)\beta \tag{16}$$

If the coefficient of β in the expression for λ^2 is 0, we have

$$z = -\frac{by^2}{2x}$$

Substituting this into (15), it follows that

$$dy^3 = 10x^3 + 6x\sqrt{3x^4-6x} \tag{17}$$

so that

$$3x^4 - 6x = t^2 \tag{18}$$

If x is even, the fact that $3 \nmid x$ implies that

$$x^3 - 2 = \pm 6u^2, \quad x = \pm 2v^2$$

whence

$$\pm 4v^6 - 1 = \pm 3u^2$$

Since $3 \nmid (4v^6+1)$, we must choose the upper sign; the last equation can then be written as

$$(u+1)^3 - (u-1)^3 = (2v^2)^3$$

so that $|u| = 1$. Hence $x = 2$, and by (17), $dy^3 = 80 \pm 72$. The lower sign yields $d = 1$ or 8, both of which are excluded since $d > 1$ is cube-free. Therefore $d = 19$, $y = 2$, and $z = -1$. The only solution in this case is

$$[(1/3)(2+2\sqrt[3]{19}-(\sqrt[3]{19})^2)]^2 = -8 + 3\sqrt[3]{19}$$

If x is odd, (18) implies that

$$x^3 - 2 = \pm 3u^2, \ x = \pm v^2$$

Hence the lower sign must be chosen and we have $x = -v^2$. ■

9.8. LEMMA. The fourth power of a positive irrational unit of R is never of the form $X + Y\alpha$.

Proof. Let σ be such a unit,

$$\sigma = (1/3)(x_1+y_1\alpha+z_1\beta)$$

and assume that

$$\sigma^4 = X + Y\alpha$$

Then, since the coefficient of β in σ^4 is 0, we obtain

$$6bx_1^2y_1^2 + 4x_1^3z_1 + 4ab^2y_1^3z_1 + 12abx_1y_1z_1^2 + a^2bz_1^4 = 0 \tag{19}$$

If we put

$$\lambda = \sigma^2 = (1/3)(x+y\alpha+z\beta),$$

then

$$x = (1/3)(x_1^2+2aby_1z_1)$$
$$y = (1/3)(2x_1y_1+az_1^2)$$
$$z = (1/3)(2x_1z_1+by_1^2)$$

Since $\lambda^2 = X + Y\alpha$, we may apply Lemmas 9.6 and 9.7. The cases

$$d = 20, \ x = y = -z = 3$$
$$d = 19, \ x = y = 2, \ z = -1$$

are impossible, since in the first the above equation for z becomes

$$-9 = 2x_1z_1 + 2y_1^2,$$

while in the second the system is easily seen to be inconsistent for all choices of signs of x_1, y_1, z_1. Therefore, we must have $x = -v^2$, where v is odd, so that

$$3v^2 + x_1^2 = -2aby_1z_1$$

Because v is odd, so is x_1, so that $3v^2 + x_1^2 \equiv 4 \pmod 8$. Thus three of the numbers a, b, y_1, z_1 are odd, and the fourth is even. Owing to (19), $a^2bz_1^4$ is even, so y_1 is odd. If either a or z_1 is even, (19) implies that

$$6bx_1^2y_1^2 \equiv 0 \pmod 4,$$

which is impossible. If b is even, (19) ensures that $a^2bz_1^4 \equiv 0 \pmod 4$, which is again impossible since b is square-free. This completes the proof of the lemma. ■

9.9. LEMMA. The cube of a positive irrational unit of R is never of the form $X + Y\alpha$.

Proof. If $\lambda = (1/3)(x+y\alpha+z\beta)$, $x, y, z \in \mathbb{Z}$, is a positive unit of R, then the coefficient of β in λ^3 is

$$(1/9)(bxy^2+x^2z+abyz^2)$$

From the equation

$$x^3 + ab^2y^3 + a^2bz^3 - 3abxyz = 27 \tag{20}$$

we infer that $(x,b) = 1$ and hence the equation

$$bxy^2 + x^2z + abyz^2 = 0 \tag{21}$$

implies that $b|z$. From (20) again, $\delta = (x,y,z) = 1$ or 3. Because $\lambda \neq \pm 1$, y and z are not both zero, and we can write

$$x = \delta d_1d_2x_1, \quad y = \delta d_2d_3y_1, \quad z = \delta bd_1d_3z_1 \tag{22}$$

where

$$(x/\delta, z/\delta b) = |d_1|, \ (x/\delta, y/\delta) = |d_2|, \ (y/\delta, z/\delta b) = |d_3|,$$

and $x_1 > 0$, $y_1 > 0$, $z_1 > 0$. The numbers d_1x_1, d_2y_1, d_3z_1 are pairwise relatively prime. Substituting the values from (22) into (20) and dividing by $\delta^3 bd_1d_2d_3$, we obtain

$$d_2^2 d_3 x_1 y_1^2 + d_1^2 d_2 x_1^2 z_1 + ab^2 d_1 d_3^2 y_1 z_1^2 = 0$$

It follows from this that $d_1|x_1, d_2|y_1$, and $d_3|z_1$. Setting

$$x_1 = d_1x_2, y_1 = d_2y_2, z_1 = d_3z_2,$$

substituting, and dividing by $d_1d_2d_3$, we obtain

$$d_2^3 x_2 y_2^2 + d_1^3 x_2^2 z_2 + ab^2 d_3^3 y_2 z_2^2 = 0$$

It follows that $x_2|ab^2d_3^3y_2z_2^2$, which in turn implies that $x_2 = 1$. Similarly, $y_2 = z_2 = 1$, so that

$$d_1^3 + d_2^3 + ab^2d_3^3 = 0 \tag{23}$$

and

$$x = \delta d_1^2 d_2,\ y = \delta d_2^2 d_3,\ z = \delta b d_1 d_3^2$$

Substituting these values into (20), we obtain

$$d_1^6 d_2^3 + ab^2 d_2^6 d_3^2 + a^2b^4 d_1^3 d_3^6 - 3ab^2 d_1^3 d_2^3 d_3^3 = 27/\delta^3$$

Eliminating $ab^2d_3^3$ between this equation and (23), we derive

$$d_1^9 + 6d_1^6d_2^3 + 3d_1^3d_2^6 - d_2^9 = (3/\delta)^3 \tag{24}$$

and putting $d_1^3 = u$, $d_2^3 = v$, $3/\delta = w$, this becomes

$$u^3 + 6u^2v + 3uv^2 - v^3 = w^3 \tag{25}$$

On the other hand, it is easily verified that

$$(u^3+6u^2v+3uv^2-v^3)U^3 = V^3 + W^3,$$

where

$$U = u^2 + uv + v^2,\ V = u^3 + 3u^2v - v^3,\ W = 3u^2v + 3uv^2$$

Since neither U nor V is zero for relatively prime u and v, (25) can hold with $w \neq 0$ only if $W = 0$, that is, if $u = -v$. In this case $w = v$. Because

$(d_1,d_2) = 1$, it follows that $d_1 = -1$, $d_2 = 1$, $\delta = 3$. But this leads to the values $x = 3$, $y = -3$, $z = 0$ for which the coefficient of β in λ^3 is not zero, a contradiction. ■

We have now accumulated all the information necessary to prove the following result.

9.10. THEOREM. (Nagell(1925)). Let $d > 1$ be a cube-free integer and R the ring of integers of $\mathbb{Q}(\sqrt[3]{d})$. Then the equation

$$x^3 + dy^3 = 1 \qquad (x \in \mathbb{Z},\ 0 \neq y \in \mathbb{Z})$$

has at most one solution. If x_0, y_0 with $y_0 \neq 0$ is a solution, then

$$x_0 + y_0\sqrt[3]{d}$$

is either the canonical fundamental unit of R or its square.

Proof. Assume that $x_0 \in \mathbb{Z}$, $0 \neq y_0 \in \mathbb{Z}$ are such that $x_0^3 + dy_0^3 = 1$. Then $\delta = x_0 + y_0\sqrt[3]{d}$ is a unit of norm 1, by virtue of (7). Hence, by Proposition 9.2, $\delta = \varepsilon_d^n$ for some $n > 0$ where ε_d is the canonical fundamental unit of R. It therefore suffices to show that $n = 1,2$. By Lemmas 9.8 and 9.9, the latter will follow provided we establish the following assertion:

If $p > 3$ is a prime and

$$\lambda = (1/3)(x+y\alpha+z\beta) \qquad (x,y,z \in \mathbb{Z})$$

is a positive unit smaller than 1, then λ^p is not of the form $X + Y\alpha$.

Assume that $z = 0$. Then $3|x$ and $3|y$ and

$$N(\lambda) = (x/3)^3 + d(y/3)^3 = 1$$

so that $(x/3, dy/3) = 1$; by Lemma 9.4, the coefficient of β in λ^p is not zero. Hence $z \neq 0$ and by the same reasoning (applied to the field $\mathbb{Q}(\beta) = \mathbb{Q}(\alpha)$) y cannot be zero.

It follows from the representation

$$r = x_1 + x_2\alpha + x_3\beta$$

of an arbitrary $r \in R$ that

$$\alpha(r+\varepsilon r'+\varepsilon^2 r'') = 3abx_3$$

where ε is a primitive cube root of unity and r', r'' are conjugates of r. Taking $r = \lambda^p$, we see that if the coefficient of β in λ^p is zero, then

$$[(1/3)(x+y\alpha+z\beta)]^p + \varepsilon[(1/3)(x+y\varepsilon d+z\varepsilon^2\beta)]^p$$
$$+ \varepsilon^2[(1/3)(x+y\varepsilon^2\alpha+z\varepsilon\beta)]^p = 0 \qquad (26)$$

Assume first that $p \equiv 1(\text{mod}3)$. Then since $\varepsilon^p = \varepsilon$, we can write (26) in the form

$$[(1/3)(x\varepsilon+y\varepsilon^2\alpha+z\beta)]^p + [(1/3)(x\varepsilon^2+y\varepsilon\alpha+z\beta)]^p = -[(1/3)(x+y\alpha+z\beta)]^p$$

Since p is odd, the left side is divisible by

$$(1/3)(x\varepsilon+y\varepsilon^2\alpha+z\beta) + (1/3)(x\varepsilon^2+y\varepsilon\alpha+z\beta) = (1/3)(-x-y\alpha+2z\beta);$$

this number is an integer, and since it divides λ^p, it is a unit. Thus

$$-x^3 - ab^2y^3 + 8a^2bz^3 - 6abxyz = \pm 27$$

Since λ is a positive unit, we have

$$x^3 + ab^2y^3 + a^2bz^3 - 3abxyz = 27$$

and by addition,

$$9a^2bz^3 - 9abxyz = 0 \text{ or } 54$$

In the first case $az^2 - xy$ must be zero. However, this number is the coefficient of α in $3\lambda^{-1}$ and, by Proposition 9.2(ii), it is not zero since $\lambda^{-1} > 1$.

In the second case we have

$$abz(az^2 - xy) = 6$$

But then x, y, z are not divisible by 3, so that $\mathbb{Q}(\sqrt[3]{d})$ is of the second kind. This is impossible, since if $ab|6$ then $a^2 - b^2 \not\equiv 0(\text{mod}9)$.

The case in which $p \equiv 2(\text{mod}3)$ proceeds similarly. Equation (26) can be written in the form

$$(1/3)(x\varepsilon^2+y\alpha+z\varepsilon\beta)^p + (1/3)(x\varepsilon+y\alpha+z\varepsilon^2\beta)^p = -[(1/3)(x+y\alpha+z\beta)]^p$$

which implies that the number

$$(1/3)(x\varepsilon^2+y\alpha+z\varepsilon\beta) + (1/3)(x\varepsilon+y\alpha+z\varepsilon^2\beta) = (1/3)(-x+2y\alpha-z\beta)$$

is a unit. As before,

$$9ab^2y^3 - 9abxyz = 0 \text{ or } 54$$

Since by^2-xz is the coefficient of β in $3\lambda^{-1}$, it is not zero. On the other hand, it is impossible that $(by^2-xz)|6$ and $ab|6$, because then $\mathbb{Q}(\sqrt[3]{d})$ must be both the first and second kinds. This completes the proof of the theorem. ■

9.11. COROLLARY. Let R be the ring of integers of $\mathbb{Q}(\sqrt[3]{d})$, where $d \neq 20$ and $\mathbb{Q}(\sqrt[3]{d})$ is of the first kind. If x_0, y_0 is an integer solution of the equation $x^3 + dy^3 = 1$ and $y_0 \neq 0$, then

$$x_0 + y_0\sqrt[3]{d}$$

is the canonical fundamental unit of R.

Proof. Because $\mathbb{Q}(\sqrt[3]{d})$ is of the first kind, the canonical fundamental unit ε_d of R is of the form

$$x + y\alpha + z\beta \qquad (x,y,z \in \mathbb{Z})$$

by Proposition 9.1. Therefore, if $\varepsilon_d^2 = x_0 + y_0\sqrt[3]{d}$, then by Lemma 9.6, $d = 20$. Invoking Theorem 9.10, we therefore derive $\varepsilon_d = x_0 + y_0\sqrt[3]{d}$. ■

9.12. REMARK. The above result need not be true for fields $\mathbb{Q}(\sqrt[3]{d})$ of the second kind. Indeed, let $d = 28$. Then $(x,y) = (-3,1)$ is a nontrivial solution of

$$x^3 + dy^3 = 1$$

Hence $\delta = -3 + \sqrt[3]{28}$ is either the canonical fundamental unit or its square (Theorem 9.10). Because

$$\delta = [(1/3)(-1 - \sqrt[3]{28} + \sqrt[3]{98})]^2$$

it follows that δ is the square of the canonical fundamental unit. ■

We now apply Corollary 9.11 to disclose ε_d for infinitely many d. In what follows, $d > 1$ is a cube-free integer, R is the ring of integers of $\mathbb{Q}(\sqrt[3]{d})$ and ε_d is the canonical fundamental unit of R.

9.13. THEOREM. (Yokoi(1974)). Assume that

$$d = n(u^6n^2+3u^3n+3) \qquad (n,u \in \mathbb{N})$$

If $x = 1 + u^3 n$, $d = n(x^2+x+1)$

where $nu \equiv 0 \pmod 3$ or $n \equiv u \pmod 3$. Then

$$\varepsilon_d = (1+u^3n) - u\sqrt[3]{d}$$

Proof. It is easily verified that $(x,y) = (1+u^3n,-u)$ is a nontrivial solution of the equation

$$x^3 + dy^3 = 1$$

Therefore, by Corollary 9.11, it suffices to verify that $d \neq 20$ and that $\mathbb{Q}(\sqrt[3]{d})$ is of the first kind. If $u \neq 1$ or $n \geqslant 2$, then obviously $d > 20$. On the other hand, if $u = n = 1$, then $d = 7$ and thus $d \neq 20$.

Put $x = 1 + u^3n$. If $x \equiv -1 \pmod 3$, we get $u^3n \equiv 1 \pmod 3$. Hence $n \equiv u \not\equiv 0 \pmod 3$ which ensures that $d \equiv \pm 2 \pmod 9$. In case $x \equiv 1 \pmod 3$, we get $u^3n \equiv 0 \pmod 3$ and therefore $n \equiv 0 \pmod 3$ or $n \not\equiv u \equiv 0 \pmod 3$. The latter implies $d \equiv 0$ or $\pm 3 \pmod 9$.

Finally, suppose that $x \equiv 0 \pmod 3$ so that $u^3n \equiv -1 \pmod 3$. Then $nu \not\equiv 0 \pmod 3$ and so, by hypothesis, $n \equiv u \pmod 3$. Hence $u^4 \equiv -1 \pmod 3$, which is impossible. Thus $\mathbb{Q}(\sqrt[3]{d})$ is of the first kind and the result follows. ∎

9.14. THEOREM. (Yokoi(1974)). Assume that

$$d = n(u^6n^2 - 3u^3n + 3) \qquad (n,u \in \mathbb{N})$$

where $nu \equiv 0 \pmod 3$ or $n + u \equiv 0 \pmod 3$. Then

$$\varepsilon_d = (1-u^3n) + u\sqrt[3]{d}$$

Proof. It is clear that $(x,y) = (1-u^3n,u)$ is a nontrivial solution of the equation

$$x^3 + dy^3 = 1$$

Again, by Corollary 9.11, we are left to verify that $d \neq 20$ and that $\mathbb{Q}(\sqrt[3]{d})$ is of the first kind. Now if $u \geqslant 2$, then obviously $d > 20$. If $u = 1$, then we have $d = n(n^2-n+3) \neq 20$.

Put $x = 1 - u^3n$. If $x \equiv 1 \pmod 3$, then we get $u^3n \equiv 0 \pmod 3$. Thus $n \equiv 0 \pmod 3$ or $n \not\equiv u \equiv 0 \pmod 3$, which implies that $d \equiv 0$ or $\pm 3 \pmod 9$. If $x \equiv -1 \pmod 3$, then we get $u^3n \equiv -1 \pmod 3$. Hence $nu \not\equiv 0 \pmod 3$ and so

$n \equiv -u \equiv \pm 1 \pmod 3$, by hypothesis. The latter easily implies that $d \equiv \pm 2 \pmod 9$.

Finally, suppose that $x \equiv 0 \pmod 3$ so that $u^3 n \equiv 1 \pmod 3$. Then $nu \not\equiv 0 \pmod 3$ and so $n \equiv -u \pmod 3$, in which case $u^4 \equiv -1 \pmod 3$, a contradiction. Thus $\mathbb{Q}(\sqrt[3]{d})$ is of the first kind and the result follows. ■

9.15. COROLLARY. (Yokoi(1974)). Assume that

$$d = n(u^6 n^2 \pm 3u^3 n + 3) \qquad (n,u \in \mathbb{N})$$

Then

$$\delta = (1 \pm u^3 n) \mp u\sqrt[3]{d}$$

is either the canonical fundamental unit of R or its square

Proof. As has been observed in the proofs of Theorems 9.13 and 9.14,

$$(x,y) = (1 \pm u^3 n, \mp u)$$

is a nontrivial solution of the equation

$$x^3 + dy^3 = 1$$

The desired conclusion is therefore a consequence of Theorem 9.10. ■

9.16. EXAMPLES for $d = n(u^6 n^2 + 3u^3 n + 3)$

(a) Assume that $u = n = 1$. Then $d = 7$ and so, by Theorem 9.13,

$$\varepsilon_7 = 2 - \sqrt[3]{7}$$

(b) Assume that $u = 1$, $n = 3$. Then $d = 63$ and so, by Theorem 9.13,

$$\varepsilon_{63} = 4 - \sqrt[3]{63}$$

(c) Assume that $u = n = 2$. Then $d = 614$ and thus, by Theorem 9.13,

$$\varepsilon_{614} = 17 - 2\sqrt[3]{614}$$

(d) Assume that $u = 3$, $n = 1$. Then $d = 813$ and therefore, by Theorem 9.13,

$$\varepsilon_{813} = 28 - 3\sqrt[3]{813}$$

9.17. EXAMPLES for $d = n(u^6 n^2 - 3u^3 n + 3)$

(a) Assume that $u = 1$, $n = 2$. Then $d = 2$ and thus, by Theorem 9.14,

$$\varepsilon_2 = -1 + \sqrt[3]{2}$$

(b) Assume that $u = 1, n = 3$. Then $d = 9$ and thus, by Theorem 9.14,

$$\varepsilon_9 = -2 + \sqrt[3]{9}$$

(c) Assume that $u = 2, n = 1$. Then $d = 43$ and hence, by Theorem 9.14,

$$\varepsilon_{43} = -7 + 2\sqrt[3]{43}$$

(d) Assume that $u = 3, n = 1$. Then $d = 651$ and thus, by Theorem 9.14,

$$\varepsilon_{651} = -26 + 3\sqrt[3]{651}$$

10. FINITE GALOIS THEORY

Throughout this section, given a field F and a group G of automorphisms of F, we write F^G for the fixed field of G. Our aim is to prove the main result of the Galois theory for finite Galois extensions. Since in the next section we shall deal with infinite extensions, some of the preparatory results presented are stated and proved for infinite extensions.

10.1. LEMMA. Let $F \subseteq K \subseteq E$ be a chain of fields, where E/F is a Galois extension.

(i) E/K is a Galois extension and so, if $H = \text{Gal}(E/K)$, then $K = E^H$

(ii) The map $K \longmapsto \text{Gal}(E/K)$ from the set of intermediate fields into the set of subgroups of $\text{Gal}(E/F)$ is injective. Furthermore, if E/F is a finite Galois extension, then this map is a bijection with the inverse map given by $H \longmapsto E^H$.

(iii) If K_1 and K_2 are two intermediate fields and $H_i = \text{Gal}(E/K_i), i = 1,2$, then

(a) $\text{Gal}(E/K_1K_2) = H_1 \cap H_2$

(b) $E^{<H_1,H_2>} = E^{H_1} \cap E^{H_2}$

(c) $K_1 \subseteq K_2$ if and only if $H_2 \subseteq H_1$

Proof. (i) By Proposition 4.1, E/F is both normal and separable. Hence, by Proposition 2.7 and Corollary 3.14, E/K is normal and separable. Therefore E/K is Galois, by Proposition 4.1.

(ii) Assume that $\text{Gal}(E/K_1) = \text{Gal}(E/K_2) = H$, say, for two intermediate fields K_1 and K_2. Then, by (i),

$$K_1 = E^H = K_2$$

proving the first assertion. The second assertion is a consequence of Corollary 9.6.

(iii) An automorphism ϕ of E fixes K_1K_2 elementwise if and only if ϕ fixes each element of K_1 and each element of K_2. This proves (a). If $\lambda \in E^{<H_1, H_2>}$, then λ is fixed by all elements of H_1 and H_2, hence $\lambda \in E^{H_1} \cap E^{H_2}$. Conversely, if $\lambda \in E^{H_1} \cap E^{H_2}$, then λ is fixed by all elements of H_1 and H_2, hence by $<H_1, H_2>$, proving (b). If $K_1 \subseteq K_2$, then every $\sigma \in \text{Gal}(E/F)$ which is the identity map on K_2 is also the identity map on K_1, i.e. $H_2 \subseteq H_1$. Conversely, if $H_2 \subseteq H_1$, then

$$K_1 = E^{H_1} \subseteq E^{H_2} = K_2$$

as asserted. ■

Let E/F be a Galois extension. We shall sometimes call the group $\text{Gal}(E/K)$ of an intermediate field K the group *associated* with K. We also say that a subgroup H of $\text{Gal}(E/F)$ *belongs* to an intermediate field K if $H = \text{Gal}(E/K)$. If K_1, K_2 are two intermediate fields, then we say that K_1 and K_2 are *conjugate under an automorphism* $\sigma \in \text{Gal}(E/F)$ if $\sigma(K_1) = K_2$.

10.2. LEMMA. Let E/F be a normal extension and let K be an intermediate field such that K/F is normal. Then the restriction map

$$\begin{cases} \text{Gal}(E/F) \longrightarrow \text{Gal}(K/F) \\ \qquad \sigma \longmapsto \sigma|K \end{cases}$$

induces an isomorphism

$$\text{Gal}(E/F)/\text{Gal}(E/K) \cong \text{Gal}(K/F)$$

Proof. Since K/F is normal, the restriction of any $\sigma \in \text{Gal}(E/F)$ to K is an automorphism of K, hence $\sigma|K \in \text{Gal}(K/F)$. The given map is obviously a homomorphism with kernel $\text{Gal}(E/K)$. Since $\sigma \longmapsto \sigma|K$ is surjective by Proposition 2.9, the result follows. ■

10.3. LEMMA. Let E/F be a Galois extension, let K_1, K_2 be two intermediate fields and let $H_i = \mathrm{Gal}(E/K_i), i = 1,2$.

(i) Then, for any given $\sigma \in \mathrm{Gal}(E/F)$, $\sigma(K_1) = K_2$ if and only if $H_2 = \sigma H_1 \sigma^{-1}$.

(ii) If K is any intermediate field, then K/F is normal if and only if $\mathrm{Gal}(E/K)$ is a normal subgroup of $\mathrm{Gal}(E/F)$ and when this is so,

$$\mathrm{Gal}(K/F) \cong \mathrm{Gal}(E/F)/\mathrm{Gal}(E/K)$$

Proof. (i) By definition, $\tau \in H_i$ if and only if $\tau(x) = x$ for all $x \in K_i$, $i = 1,2$. Assume that $K_2 = \sigma(K_1)$. Then, for any $y \in K_2$, $\sigma^{-1}(y) \in K_1$, hence $\tau\sigma^{-1}(y) = \sigma^{-1}(y)$ for all $\tau \in H_1$, i.e. $\sigma\tau\sigma^{-1}(y) = y$, which means that

$$\sigma\tau\sigma^{-1} \in H_2.$$

Thus $\sigma H_1 \sigma^{-1} \subseteq H_2$ and since $K_1 = \sigma^{-1}(K_2)$, it follows that $\sigma^{-1} H_2 \sigma \subseteq H_1$, i.e. $H_2 = \sigma H_1 \sigma^{-1}$. Retracing our steps, we obtain the converse.

(ii) Apply (i) and Lemma 10.2. ■

We are now ready to record our main result.

10.4. THEOREM. (Main theorem of finite Galois theory). Let E/F be a finite Galois extension with Galois group G.

(i) The map $K \longmapsto \mathrm{Gal}(E/K)$ from the set of intermediate fields into the set of subgroups of G is an inclusion-reversing bijection with the inverse map given by $H \longmapsto E^H$.

(ii) If H is a subgroup of G, then

$$|H| = (E:E^H) \quad \text{and} \quad (G:H) = (E^H:F)$$

(iii) The extension K/F is normal if and only if $\mathrm{Gal}(E/K) \lhd G$ and when this is so

$$\mathrm{Gal}(K/F) \cong G/\mathrm{Gal}(E/K)$$

Proof. (i) Direct consequence of Lemma 10.1.

(ii) By (i), $H = \mathrm{Gal}(E/E^H)$ and hence, by Proposition 4.3, $|H| = (E:E^H)$. Since $|G| = (E:F)$ by Proposition 4.3, we have $(G:H) = (E:F)/(E:E^H) = (E^H:F)$.

(iii) Direct consequence of Lemma 10.3(ii). ■

Every field F comes accompanied by a canonical Galois extension: the separable closure $\bar{F}/F$. Its Galois group is called the *absolute Galois group* of F. The extension $\bar{F}/F$ has infinite degree in almost all cases. We now show that Theorem 10.4 does not hold for the extension $\bar{F}/F$. More precisely, it is no longer true that every subgroup of $\mathrm{Gal}(\bar{F}/F)$ belongs to some subfield K such that $F \subseteq K \subseteq \bar{F}$.

10.5. EXAMPLE. Let $F = \mathbb{F}_p$ be the field of p elements, p prime, let $\bar{F}$ be the separable closure of F and let $G = \mathrm{Gal}(\bar{F}/F)$. Then G contains the Frobenius automorphism ϕ which is defined by

$$\phi(x) = x^p \qquad \text{for all} \quad x \in \bar{F}$$

and we put $H = \langle\phi\rangle$. We claim that there is no intermediate field K such that

$$H = \mathrm{Gal}(\bar{F}/K)$$

Since the fixed fields of H and G are the same, namely F, it suffices to show that $G \neq \langle\phi\rangle$. To this end, choose a sequence $\{a_n\}, n \in \mathbb{N}$, of integers such that

$$a_n \equiv a_m (\mathrm{mod}\, m) \qquad \text{if} \quad m|n,$$

but for no integer a the congruences

$$a_n \equiv a (\mathrm{mod}\, n)$$

hold for all $n \in \mathbb{N}$. The simplest example of such a sequence can be obtained as follows: write $n = n'p^{\nu(n)}$, $(n',p) = 1$ and $1 = n'x_n + p^{\nu(n)}y_n$, then $a_n = n'x_n$ is a required sequence. We now define ψ_n to be the restriction of ϕ^{a_n} to $\mathbb{F}_{p^n}$. Then, obviously, $\psi_n \in \mathrm{Gal}(\mathbb{F}_{p^n}/F)$. If $\mathbb{F}_{p^m} \subseteq \mathbb{F}_{p^n}$, then $m|n$, so $a_n \equiv a_m(\mathrm{mod}\, m)$, and thus

$$\psi_n|_{\mathbb{F}_{p^m}} = \phi^{a_n}|_{\mathbb{F}_{p^m}} = \phi^{a_m}|_{\mathbb{F}_{p^m}} = \psi_m$$

if we recall that $\phi|_{\mathbb{F}_{p^m}}$ has order m. Consequently, the ψ_n define an automorphism ψ of $\bar{F} = \bigcup_{n=1}^{\infty} \mathbb{F}_{p^n}$. If $\psi = \phi^a$ for some $a \in \mathbb{Z}$, then

$$\psi|_{\mathbb{F}_{p^n}} = \phi^{a_n}|_{\mathbb{F}_{p^n}} = \phi^{a}|_{\mathbb{F}_{p^n}}$$

which implies that

$$a_n \equiv a(\mathrm{mod}\, n) \qquad \text{for all} \quad n \in \mathbb{N}$$

a contradiction. ■

We close this section by providing an illustration of the main theorem of finite Galois theory.

10.6. PROPOSITION. The field $\mathbb{C}$ is algebraically closed.

Proof. We shall use the standard fact that every polynomial of odd degree in $\mathbb{R}[X]$ has a root in $\mathbb{R}$. Let i be a root of $X^2 + 1$. Every element of $\mathbb{R}(i)$ has a square root: if $a + bi \in \mathbb{R}(i)$, $a,b \in \mathbb{R}$, then the square root is given by $c + di$, where

$$c^2 = \frac{a+\sqrt{a^2+b^2}}{2} \qquad d^2 = \frac{-a+\sqrt{a^2+b^2}}{2}$$

Indeed, since both $a + \sqrt{a^2+b^2}$ and $-a + \sqrt{a^2+b^2}$ are nonnegative, they have a square root in $\mathbb{R}$. If $b > 0$, then choose the signs of c and d to be equal and if $b_1 < 0$, choose then to be opposite. Then it is clear that $(c+di)^2 = a+bi$.

Since $\mathbb{R}$ has characteristic 0, every finite extension of $\mathbb{R}$ is separable. Thus every finite extension of $\mathbb{R}(i)$ is contained in a field K such that $K/\mathbb{R}$ is a finite Galois extension. We are left to verify that $K = \mathbb{R}(i)$.

Let $G = \mathrm{Gal}(K/\mathbb{R})$ and let H be a Sylow 2-subgroup of G. By Theorem 10.4 (ii), $(G:H) = (K^H:\mathbb{R})$ and thus $(K^H:\mathbb{R})$ is odd. On the other hand, by Corollary 3.32, $K^H = \mathbb{R}(\alpha)$ for some $\alpha \in K^H$. Then α is a root of an irreducible polynomial in $\mathbb{R}[X]$ of odd degree. But every polynomial of odd degree in $\mathbb{R}[X]$ has a root in $\mathbb{R}$, hence $K^H = \mathbb{R}$ and therefore $G = H$ is a 2-group.

Since $G_1 = \mathrm{Gal}(K/\mathbb{R}(i))$ is a subgroup of G, we see that G_1 is a 2-group. Assume by way of contradiction that $K \neq \mathbb{R}(i)$. Then $G_1 \neq 1$ and so G_1 has a subgroup G_2 of index 2. Hence, by Theorem 10.4(ii), $(K^{G_2}:\mathbb{R}(i)) = 2$. But we say that every element of $\mathbb{R}(i)$ has a square root and hence $\mathbb{R}(i)$ has no extensions of degree 2, a contradiction. ■

11. PROFINITE GROUPS

The Galois group of an infinite (i.e. not necessarily finite) Galois extension E/F is a profinite group. For this reason, we devote this section to examining some general properties of profinite groups.

We begin by introducing the notion of a projective (or inverse) limit which stems from topology and which has become very useful in algebra too.

Let $\{G_i\} i \in I$ be a system of sets (groups, rings, etc) indexed by a partially ordered set I, which is directed in the sense that, given $i,j \in I$, there exists a $k \in I$ such that $i \leqslant k$ and $j \leqslant k$. Suppose that for every pair $i,j \in I$ with $i \leqslant j$, there is given a map (homomorphism)

$$f_{ij} : G_j \longrightarrow G_i$$

such that

(i) f_{ii} is the identity map of G_i for each $i \in I$

(ii) for all $i \leqslant j \leqslant k$ in I, we have $f_{ij} \circ f_{jk} = f_{ik}$

Then the system

$$\{G_i, f_{ij} | i,j \in I\}$$

is called a *projective* system. The *projective limit*

$$G = \varprojlim G_i$$

of this system is defined by

$$G = \{(g_i) \in \prod_{i \in I} G_i | f_{ij}(g_j) = g_i \quad \text{if} \quad i \leqslant j\}$$

It is clear that if each G_i is a group (ring), then G is a group (ring).

11.1. EXAMPLE. Let G be an arbitrary group (an arbitrary ring) and let $\{N_i | i \in I\}$ be a set of normal subgroups (ideals) of G of finite index which is closed under finite intersections. Write $i \leqslant j$ if $N_i \supseteq N_j$. Then I becomes a directed set. If $i \leqslant j$, let $f_{ij} : G/N_j \longrightarrow G/N_i$ be the natural projection, i.e.

$$f_{ij}(gN_j) = gN_i \qquad (f_{ij}(g+N_j) = g+N_i)$$

Then $\{G/N_i, f_{ij} | i,j \in I\}$ is a projective system of groups (rings) and the map

$$\begin{cases} G \longrightarrow \varprojlim G/N_i \\ g \longmapsto (gN_i) \end{cases} \qquad (g \longmapsto (g+N_i))$$

is a homomorphism with kernel $\bigcap_{i \in I} N_i$. ∎

If the sets G_i are topological spaces, we assume that the maps f_{ij} are *continuous*. Note that $\varprojlim G_i$ may be empty and that

$$\varprojlim G_i = \bigcap_{\substack{j,k \in I \\ j \leqslant k}} R_{kj}, \quad \text{where } R_{kj} = \{(g_i) \in \prod_{i \in I} G_i | f_{jk}(g_k) = g_j\}$$

Assume that each G_i is a topological space. Recall that the product topology on $\prod_{i \in I} G_i$ has a basis consisting of the sets of form $\prod_{i \in I} U_i$, with U_i an open subset of G_i for every $i \in I$ and $U_i = G_i$ for all but finitely many $i \in I$. Since $\varprojlim G_i$ is a subset of $\prod_{i \in I} G_i$, we may, and in future will, regard $\varprojlim G_i$ as a topological space. Note that if each G_i is a topological group (topological ring), then $\varprojlim G_i$ is a topological group (topological ring).

11.2. PROPOSITION. Let $\{G_i, f_{ij} | i,j \in I\}$ be a projective system of topological spaces and let $G = \varprojlim G_i$.

(i) If each G_i is a Hausdorff space, then G is a closed subset of $\prod G_i$

(ii) If each G_i is a nonempty compact Hausdorff space, then G is a nonempty compact Hausdorff space.

Proof. (i) Let $g = (g_i)$ be an element of $\prod G_i$ that does not belong to G. Then there exist i and j with $i \leqslant j$ such that $f_{ij}(g_j) \neq g_i$. Take open disjoint neighbourhoods U_i and U_i' of g_i and $f_{ij}(g_j)$, respectively. Then $U_i \times f_{ij}^{-1}(U_i') \times \prod_{k \neq i,j} G_k$ is an open neighbourhood of g that does not intersect G.

(ii) Assume that each G_i is a nonempty compact Hausdorff space. Since each G_i is compact, then $\prod G_i$ is also compact, by Tychonoff's theorem (Proposition 1.6.8). Thus, by (i) and Proposition 1.6.6, G is compact. By Proposition 1.6.4, G is also Hausdorff. We are therefore left to verify that G is nonempty.

The natural map $pr_k \times pr_j : G \longrightarrow G_k \times G_j$ is continuous. Since

$$T = \{(g_k, g_j) \in G_k \times G_j | f_{jk}(g_k) = g_j\}$$

is a closed subset of $G_k \times G_j$ (use the Hausdorff property), $R_{kj} = (pr_k \times pr_j)^{-1}(T)$ is a closed subset of $\prod_{i \in I} G_i$. Because $\prod G_i$ is compact, we have only to show that the intersection of finitely many of the R_{kj} is nonempty.

Let $j(1) \leqslant k(1), \ldots, j(n) \leqslant k(n)$ be n pairs in I. Choose $t \in I$ with $k(i) \leqslant t$, $1 \leqslant i \leqslant n$, and $g_t \in G_t$. Define

$$g_{j(i)} = f_{j(i),t}(g_t) \quad \text{and} \quad g_{k(i)} = f_{k(i),t}(g_t) \qquad (1 \leqslant i \leqslant n)$$

For each $r \in I - \{j(1), \ldots, j(n), k(1), \ldots, k(n)\}$, pick g_r in G_r. The element g thus defined belongs to $\bigcap_{i=1}^{n} R_{k(i),j(i)}$, as required. ■

It will be convenient at this stage to record the following elementary properties of topological groups.

11.3. PROPOSITION. Let G be a topological group

(i) Every closed subgroup of G of finite index is open.

(ii) If G is compact, then a subgroup of G is open if and only if it is closed and of finite index.

Proof. (i) Let H be a closed subgroup of G such that

$$G = H \cup g_2 H \cup \ldots \cup g_n H$$

(disjoint union). Since H is closed, so is $g_i H$, $2 \leqslant i \leqslant n$, hence $\bigcup_{i=2}^{n} g_i H$ is closed. Thus, H being the complement of $\bigcup_{i=2}^{n} g_i H$ in G, is open.

(ii) Let H be an open subgroup of G. Then H is closed by Proposition 1.6.3. Since the cosets of H provide an open covering of G and G is compact, H can have only finitely many cosets in G. Thus H is closed and of finite index. The converse being true by virtue of (i), the result follows. ■

We now return to the discussion of projective limits of groups (rings). Let

$$\{G_{ij}, f_{ij} | i,j \in I\}$$

be a projective system of groups (rings) and let

$$G = \varprojlim G_i$$

The following simple facts are worth mentioning:

(a) There exist homomorphisms

$$f_i : G \longrightarrow G_i \qquad (i \in I)$$

such that the diagrams

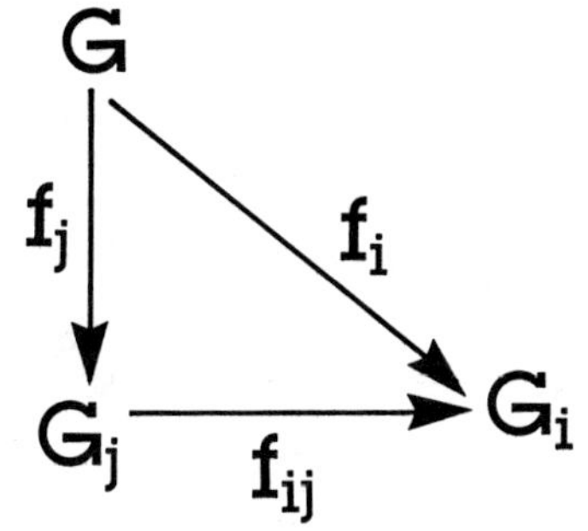

$(i \leqslant j)$

are all commutative.

In fact $f_i : g \longmapsto g_i$ (i.e. the restriction of the i-th coordinate projection of $\prod G_i$ to G) satisfies this condition. These f_i are called *canonical homomorphisms*.

(b) If every f_{ij} is injective, then so are all the f_i.
Assume that each f_{ij} is injective, and $f_i(x) = f_i(y)$ for some $x,y \in G$. Given $j \in I$, let $k \in I$ satisfy $i,j \leqslant k$. Now $f_{ik}f_k(x) = f_i(x) = f_i(y) = f_{ik}f_k(y)$, whence by our hypothesis, $f_k(x) = f_k(y)$. Therefore $f_j(x) = f_{jk}f_k(x) = f_{jk}f_k(y) = f_j(y)$ for every $j \in I$, and so $x = y$.
(c) If J is a *cofinal subset* of I (i.e. if for every $i \in I$ there exists $j \in J$ such that $i \leqslant j$), then

$$\varprojlim_{i \in I} G_i \cong \varprojlim_{j \in J} G_j$$

Indeed, given $g' \in \varprojlim_{j \in J} G_j \subseteq \prod_{j \in J} G_j$, there exists a unique $g \in \varprojlim_{i \in I} G_i$ such that, for every $j \in J$, the j-th coordinates of g' and g are equal. In fact, if we define $g = (g_i)$ with $g_i = f_{ij}(g_j') \; (i \leqslant j)$, then $g' \longmapsto g$ is a required isomorphism.

The following result provides a universal characterization of projective limits.

11.4. PROPOSITION. The projective limit $G = \varprojlim G_i$ of the projective system

$\{G_i, f_{ij} | i, j \in I\}$ of groups (rings) satisfies: if H is a group (ring) and if there are homomorphisms $\sigma_i : H \longrightarrow G_i$ with commutative diagrams:

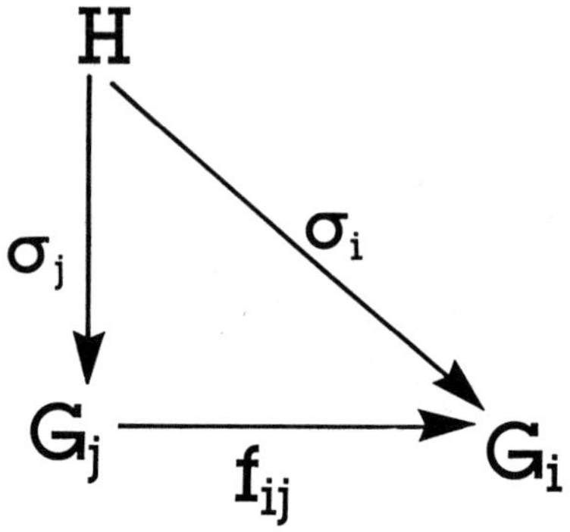

(1)

$(i \leqslant j)$

then there exists a unique homomorphism $\sigma : H \longrightarrow G$ for which all the daigrams

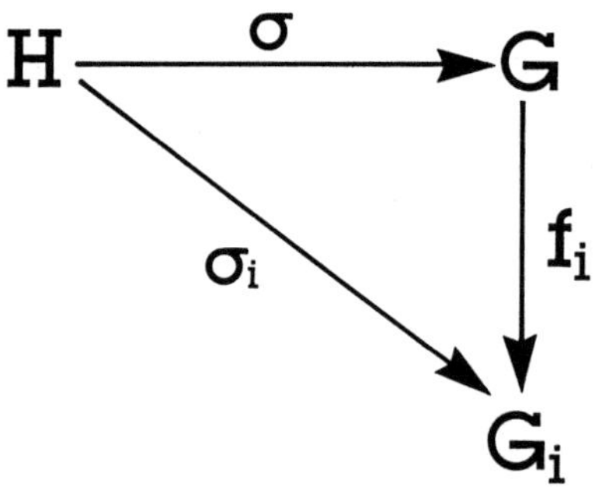

(2)

$(i \in I)$

are commutative (where f_i is the canonical homomorphism). This property characterizes G up to isomorphism.

Proof. Given $h \in H$, set $\sigma(h) = (\sigma_i(h)) \in \prod G_i$. Because of the commutativity of (1), $\sigma(h) \in G$. Thus $\sigma : H \longrightarrow G$ is a homomorphism satisfying $\sigma_i(h) = f_i(\sigma(h))$, whence the commutativity of (2) results. If $\sigma' : H \longrightarrow G$ also makes (2) commutative, then $f_i[\sigma(h)] = f_i[\sigma'(h)]$ for all $h \in H$, $i \in I$. Thus $\sigma = \sigma'$.

In order to establish the second assertion, assume H_0 and maps $\tau_i : H_0 \longrightarrow G_i$ have the property formulated in the first statement of the theorem. Then there exist unique homomorphisms $\sigma : H_0 \longrightarrow G$ and $\sigma_0 : G \longrightarrow H_0$ satisfying $f_i\sigma = \tau_i$

and $\tau_i\sigma_0 = f_i$. We infer $f_i = f_i\sigma\sigma_0$ and $\tau_i = \tau_i\sigma_0\sigma$ for all $i \in I$. Thus $\sigma\sigma_0$ and $\sigma_0\sigma$ are the identity maps of G and H_0, respectively, proving that σ is an isomorphism. ∎

We next introduce the notion of homomorphism for projective systems of topological groups (rings). Let

$$X = \{G_i, f_{ij} \mid i,j \in I,\ i \leqslant j\} \quad \text{and} \quad Y = \{H_i,\ \sigma_{ij} \mid i,j \in I,\ i \leqslant j\}$$

be two projective systems of topological groups (rings) indexed by the same directed set I. A *homomorphism* $\phi : X \longrightarrow Y$ is a set of continuous homomorphisms $\{\phi_i : G_i \longrightarrow H_i,\ i \in I\}$ subject to the condition that the diagrams

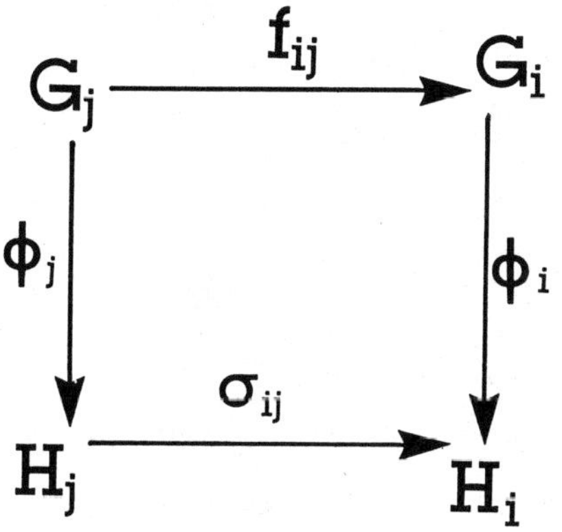

$(i \leqslant j)$

be commutative.

11.5. PROPOSITION. Let $X = \{G_i, f_{ij} \mid i,j \in I,\ i \leqslant j\}$ and $Y = \{H_i, \sigma_{ij} \mid i,j \in I,\ i \leqslant j\}$ be two projective systems of topological groups (rings) and let $\phi: X \longrightarrow Y$ be a homomorphism. Then, there is a unique homomorphism (which is necessarily continuous)

$$\phi^* : \varprojlim G_i \longrightarrow \varprojlim H_i$$

such that, for every $i \in I$, the diagram

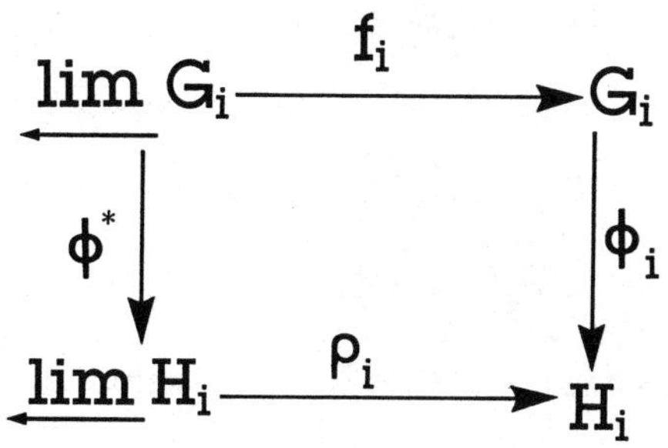

(3)

is commutative (f_i, ρ_i denote the canonical homomorphisms). In fact

$$\phi^*((g_i)) = (\phi_i(g_i)) \tag{4}$$

Furthermore, if every ϕ_i is injective, then ϕ^* is injective.

Proof. The homomorphisms $\phi_i, i \in I$, induce a homomorphism

$$\tilde{\phi} : \prod G_i \longrightarrow \prod H_i$$

given by $\tilde{\phi}((g_i)) = (\phi_i(g_i))$. The commutativity of (3) implies that if

$$g = (g_i) \in \varprojlim G_i \text{ ,}$$

then $\tilde{\phi}(g) \in \varprojlim H_i$; hence define ϕ^* to be the restriction of $\tilde{\phi}$ to $\varprojlim G_i$, i.e. by formula (4). With this ϕ^*, the diagram (3) becomes commutative and ϕ^* is obviously continuous. If $\psi : \varprojlim G_i \longrightarrow \varprojlim H_i$ makes (3) commutative for every i, then $\rho_i\phi^* = \rho_i\psi$ for every i, thus $\phi^* = \psi$.

If $\phi^*(x) = \phi^*(y)$, then $\phi_i[f_i(x) = \phi_i f_i(y)]$. Therefore, if every ϕ_i is injective, then $f_i(x) = f_i(y)$ for all $i \in I$ and thus $x = y$. ∎

11.6. PROPOSITION. Let $X = \{G_i, f_{ij} | i,j \in I, i \leqslant j\}$, $Y = \{H_i, \sigma_{ij} | i,j \in I, i \leqslant j\}$, and $Z = \{K_i, \lambda_{ij} \mid i,j \in I, i \leqslant j\}$ be three projective systems of compact topological groups (rings) and let $\phi = (\phi_i)$ be a homomorphism from X to Y and $\psi = (\psi_i)$ a homomorphism from Y to Z. Assume that for all $i \in I$, the sequence

$$1 \longrightarrow G_i \xrightarrow{\phi_i} H_i \xrightarrow{\psi_i} K_i \longrightarrow 1$$

is exact. Then the induced sequence

$$1 \longrightarrow \varprojlim G_i \xrightarrow{\phi^*} \varprojlim H_i \xrightarrow{\psi^*} \varprojlim K_i \longrightarrow 1$$

is also exact.

Proof. By Proposition 11.5, ϕ^* is injective. A straightforward verification shows that $\mathrm{Ker}\psi^* = \mathrm{Im}\phi^*$. We are therefore left to verify that ψ^* is surjective. To this end, fix (k_i) in $\varprojlim K_i$ and consider the sets

$$A_j = \{(h_i) | (h_i) \in \prod_{i \in I} H_i, \ \psi_j(h_j) = k_j \ , \ \sigma_{ij}(h_j) = h_i, \ \text{for all } i \leqslant j\}$$

Then all the A_j are closed and nonempty. Furthermore, the collection

$$\{A_j\}, j \in I,$$

possesses the finite intersection property. Since $\prod_{i\in I} H_i$ is compact, by Tychonoff's theorem, it follows from Proposition 1.6.5, that $\bigcap_{j\in I} A_j \neq \emptyset$. Since each element of $\bigcap_{j\in I} A_j$ belongs to $\varprojlim H_i$ and is mapped by ψ^* to (k_i), the result follows. ■

For the rest of the section, we are interested in the case $\{G_i, f_{ij} | i,j \in I\}$ is a projective system of finite groups, each endowed with the discrete topology. We call the projective limit $G = \varprojlim G_i$ a profinite group. The following result gives a criterion for a compact group to be profinite.

11.7. PROPOSITION. Let G be a compact topological group and let $\{N_i | i \in I\}$ be a family of closed normal subgroups of finite index such that

(i) for every finite subset J of I, there exists $i \in I$ such that $N_i \subseteq \bigcap_{j\in J} N_j$

(ii) $\bigcap_{i\in I} N_i = 1$

Then $G \cong \varprojlim G/N_i$ topologically and algebraically.

Proof. Condition (i) guarantees that $\{G/N_i, f_{ij} | i,j \in I\}$ (where $i \leqslant j$ means $N_i \supseteq N_j$ and $f_{ij} : G/N_j \longrightarrow G/N_i$ is the natural projection) is a projective system of topological groups. Condition (ii) ensures that the map

$$\begin{cases} G \xrightarrow{\theta} \varprojlim G/N_i \\ g \longmapsto (gN_i) \end{cases}$$

is an injective homomorphism (which is obviously continuous). If

$$(g_iN_i) \in \varprojlim G/N_i,$$

then the closed sets g_iN_i have the finite intersection property. Since G is compact, there exists $g \in \bigcap_{i\in I} g_iN_i$. This element satisfies $\theta(g) = (g_iN_i)$, so θ is a continuous bijection. Since G is compact and $\varprojlim G/N_i$ is Hausdorff (Proposition 1.6.4), it follows from Proposition 1.6.6 that θ is a homeomorphism, as required. ■

We now provide some useful characterizations of profinite groups.

11.8. PROPOSITION. Let G be a topological group. Then the following conditions are equivalent:

(i) G is a profinite group

(ii) G is a Hausdorff, compact group which has a basis of open neighbourhoods of 1 consisting of normal subgroups

(iii) G is a Hausdorff, compact, totally disconnected group.

Proof. Assume that G is a profinite group. By Proposition 11.2(ii), G is compact and Hausdorff. The normal subgroups $U_S \cap G$, where

$$U_S = \prod_{i \in S} G_i \times \prod_{i \notin S} N_i,$$

S a finite subset of I, N_i a normal subgroup of G_i, form a basis of open neighbourhoods of 1. This proves that (i) implies (ii). The implication (ii) ⇒ (i) is a consequence of Propositions 11.7 and 11.3. Assume that (ii) holds and let N be the connected component of 1. If $\{U_\alpha\}$ denotes the family of all open normal subgroups of G, let $N_\alpha = U_\alpha \cap N$ for all α. The groups N_α are open normal subgroups of N; as N is connected, we must have $N_\alpha = N$ for all α. But then

$$N = \bigcap_\alpha N_\alpha = \bigcap_\alpha (N \cap U_\alpha) = N \cap (\bigcap_\alpha U_\alpha) = N \cap 1$$

the last equality being true by hypothesis (ii). Thus $N = 1$ and so (iii) holds. The implication (iii) ⇒ (ii) being a consequence of Proposition 1.6.7, the result follows. ∎

11.9. PROPOSITION. Let G be a profinite group. Then any closed subgroup (closed normal subgroup) of G is the intersection of some open subgroups (open normal subgroups).

Proof. Let H be a closed subgroup (closed normal subgroup) of G and suppose that $g \in G$ belongs to every open subgroup (open normal subgroup) that contains H. Then, for any open normal subgroup N of G, $g \in HN$ and so

$$gN \cap H \neq \emptyset.$$

Hence for any finite family $N_1, \ldots, N_s$ of open normal subgroups of G, we have

$$\bigcap_{i=1}^{s} (gN_i \cap H) = g(\bigcap_{i=1}^{s} N_i) \cap H \neq \emptyset$$

Since the sets $gN \cap H$ are closed and G is compact, it follows from Proposition

1.6.5 that there exists $h \in H$ such that $h \in gN$ for all open normal subgroups N. But then $g^{-1}h$ belongs to all open normal subgroups of G. Since G is Hausdorff and some open normal subgroups of G form a basis of open neighbourhoods of 1 (Proposition 11.8), it follows from Proposition 1.6.2 that $g^{-1}h = 1$. Thus $g \in H$ and the result follows. ■

11.10. PROPOSITION. (i) If H is a closed subgroup of a profinite group G, then H is profinite.

(ii) If N is a closed normal subgroup of a profinite group G, then G/N is profinite

(iii) The direct product $G = \prod_{i \in I} G_i$ of an arbitrary family $\{G_i | i \in I\}$ of profinite groups is profinite

(iv) Every projective limit $G = \varprojlim G_i$ of profinite groups is profinite.

Proof. (i) If N_i is an open normal subgroup of G, then $H \cap N_i$ is an open normal subgroup of H. The family of all groups $H \cap N_i$ satisfies the hypotheses of Proposition 11.7 and thus $H \cong \varprojlim H/(H \cap N_i)$ is a profinite group.

(ii) A closed normal subgroup N is an intersection of open normal subgroups of G, by Proposition 11.9. Now apply Proposition 11.7.

(iii) Let J be a finite subset of I. Given $j \in J$, let N_j denote an open normal subgroup of G_j. Then $N_J = \prod_{j \in J} N_j \times \prod_{i \in I-J} G_i$ is an open normal subgroup of G of finite index. But the intersection of all such N_J is 1. Accordingly, $G \cong \varprojlim G/N_J$ is a profinite group, by Proposition 11.7.

(iv) We know, from Proposition 11.2(i), that G is a closed subgroup of $\prod_{i \in I} G_i$. By (iii), $\prod_{i \in I} G_i$ is profinite. Now apply (i). ■

An important result for the cohomology theory of profinite groups is:

11.11. PROPOSITION. Let G be a profinite group and let H be a closed subgroup of G. Then there exists a continuous cross-section $\sigma : G/H \longrightarrow G$, with $\sigma(H) = 1$, i.e. there exists a continuous map $\sigma : G/H \longrightarrow G$ of topological spaces such that the composite map $G/H \xrightarrow{\sigma} G \longrightarrow G/H$ is the identity map.

Proof. Let $K \subseteq S$ be subgroups of G such that S/K is finite. We first show that there is a continuous cross-section $\sigma : G/H \longrightarrow G/K$ with $\sigma(H) = K$.

Apparently $S \subset H$

Indeed, let U_G be the family of all normal subgroups of G of finite index. Then $\{UK/K | U \in U_G\}$ is a basis of open neighbourhoods of $1 \in G/K$. Because S/K is finite, there exists $U \in U_G$ such that $(UK/K) \cap (S/K) = 1$. Then the natural map

$$\phi : UK/K \longrightarrow UH/H$$

is a homeomorphism. Let $g_1 = 1,\ldots,g_k$ be a transversal for U in G and put

$$\sigma(g_n uH) = g_n\phi^{-1}(uH) \qquad (n = 1,2,\ldots,k, u \in U)$$

The map σ is obviously a continuous cross-section from G/H to G/K with $\sigma(H) = K$.

Let X be the set of all pairs $\{S,\sigma\}$, where S is a subgroup of H and σ is a continuous cross-section from G/H to G/S. We may order X by setting $\{S,\sigma\} \leqslant \{S',\sigma'\}$ if $S' \subseteq S$ and σ is induced by σ'. The set X is inductive Indeed, let $\{S_i,\sigma_i\}, i \in I$ be a chain in X and let $S = \bigcap_{i\in I} S_i$. The cross-sections $\sigma_i : G/H \longrightarrow G/S_i$ induce a continuous map $\sigma' : G/H \longrightarrow \varprojlim G/S_i$, i.e. a continuous cross-section $\sigma : G/H \longrightarrow G/S$. Then $\{S,\sigma\}$ is an upper bound for the set $\{S_i,\sigma_i\}, i \in I$. By Zorn's lemma, there exists a maximal element $\{S,\sigma\}$. In this case $S = 1$. Indeed, if $S \neq 1$, then S has a proper subgroup of finite index. Then, by the foregoing, $\{S,\sigma\}$ is not maximal. This implies the required assertion. ■

We close this section by providing a number of examples.

(a) If p is a prime number, then the rings $\mathbb{Z}/p^n\mathbb{Z}, n \in \mathbb{N}$ form a projective system with respect to the canonical projections

$$\mathbb{Z}/p^n\mathbb{Z} \longrightarrow \mathbb{Z}/p^m\mathbb{Z} \qquad (n \geqslant m)$$

The corresponding projective limit

$$\mathbb{Z}_p = \varprojlim \mathbb{Z}/p^n\mathbb{Z}$$

is the *ring of p-adic integers*. Every element of $\mathbb{Z}_p$ is a sequence

$$x = (x_i + p^i\mathbb{Z}), i \in \mathbb{N},$$

where $x_i \in \mathbb{Z}$ and $x_j \equiv x_i (\text{mod } p^i\mathbb{Z})$ for $j \geqslant i$. A basic neighbourhood of x

is given by an integer $m \geqslant 0$ and it consists of all elements $y = (y_i+p^i\mathbb{Z}), i \in \mathbb{N}$ with

$$x_m \equiv y_m \pmod{p^m\mathbb{Z}}$$

The map

$$\begin{cases} \mathbb{Z} \longrightarrow \mathbb{Z}_p \\ x \longmapsto (x+p^i\mathbb{Z}) \end{cases}$$

is an embedding and we identify $\mathbb{Z}$ with its image in $\mathbb{Z}_p$. Because the series $(x_i), i \in \mathbb{N}$ converges to $x = (x_i+p^i\mathbb{Z}), i \in \mathbb{N}$ in the p-adic topology, $\mathbb{Z}$ is dense in $\mathbb{Z}_p$. The canonical projection

$$\begin{cases} \mathbb{Z}_p \longrightarrow \mathbb{Z}/p^i\mathbb{Z} \\ (x_i+p^i\mathbb{Z}) \longmapsto x_i + p^i\mathbb{Z} \end{cases}$$

has kernel $p^i\mathbb{Z}_p$. Thus $p^i\mathbb{Z}_p$ is an open subgroup of (the additive group) $\mathbb{Z}_p$ of index p^i.

(b) The rings $\mathbb{Z}/n\mathbb{Z}$, $n \in \mathbb{N}$, form a projective system with respect to the projections $\mathbb{Z}/n\mathbb{Z} \longrightarrow \mathbb{Z}/m\mathbb{Z}$ for $m|n$, where the order in $\mathbb{N}$ is now given by the divisibility $m|n$. The projective limit

$$\hat{\mathbb{Z}} = \varprojlim \mathbb{Z}/n\mathbb{Z}$$

is called the *Prüfer ring*. The groups $n\hat{\mathbb{Z}}, n \in \mathbb{N}$, are precisely the open subgroups of the profinite group $\hat{\mathbb{Z}}$, and it is easily verified that

$$\hat{\mathbb{Z}}/n\hat{\mathbb{Z}} \cong \mathbb{Z}/n\mathbb{Z}$$

Given a canonical decomposition $n = \prod_p p^{\nu_p}$ of $n \in \mathbb{N}$, we have the decomposition

$$\mathbb{Z}/n\mathbb{Z} \cong \prod_p \mathbb{Z}/p^{\nu_p}\mathbb{Z}$$

by virtue of the chinese remainder theorem. Passing to projective limits we obtain a canonical decomposition

$$\hat{\mathbb{Z}} = \prod_p \mathbb{Z}_p$$

(c) Let $\mathbb{F}_p$ be the field of p elements, p prime, and let $\bar{\mathbb{F}}_p$ be the separable closure of $\mathbb{F}_p$. For each $n \in \mathbb{N}$, we have a canonical isomorphism

$$\mathrm{Gal}(\mathbb{F}_{p^n}/\mathbb{F}_p) \cong \mathbb{Z}/n\mathbb{Z}$$

which maps the Frobenius automorphism ϕ_n of $\mathbb{F}_{p^n}$ to 1 mod $n\mathbb{Z}$. Passing to projective limits we obtain a canonical isomorphism

$$\mathrm{Gal}(\overline{\mathbb{F}}_p/\mathbb{F}_p) \cong \hat{\mathbb{Z}}$$

which sends the Frobenius automorphism ϕ of $\overline{\mathbb{F}}_p$ to $1 \in \hat{\mathbb{Z}}$ and therefore maps the group $\langle\phi\rangle$ to the dense subgroup $\mathbb{Z}$ of $\hat{\mathbb{Z}}$.

12. INFINITE GALOIS THEORY

The usual Galois correspondence between subgroups of Galois groups of finite Galois extensions and intermediate fields is not valid for infinite Galois extensions (see Example 10.5). The Krull topology introduced below restores this correspondence for closed subgroups (Theorem 12.2).

Throughout, G denotes the Galois group of the Galois extension E/F. The group G can be endowed with the *Krull topology* defined as follows. Let $\{E_i/F \mid i \in I\}$ be the set of all *finite Galois subextensions* of E/F. For each $\sigma \in G$, we take the cosets

$$\sigma\mathrm{Gal}(E/E_i) \qquad (i \in I)$$

as a basis of neighbourhoods of σ. The multiplication map

$$G \times G \longrightarrow G, \ (\sigma_1,\sigma_2) \longmapsto \sigma_1\sigma_2$$

is continuous, since the pre-image of the basic open neighbourhood $\sigma_1\sigma_2\mathrm{Gal}(E/E_i)$ of $\sigma_1\sigma_2$ contains the open neighbourhood

$$\sigma_1\mathrm{Gal}(E/E_i) \times \sigma_2\mathrm{Gal}(E/E_i)$$

of (σ_1,σ_2). Similarly, the map $G \longrightarrow G, \ \sigma \longmapsto \sigma^{-1}$ is continuous, so G becomes a topological group. Of course, if E/F is a finite extension, then the Krull topology is discrete. Note also that, by Lemma 10.2, $\mathrm{Gal}(E/E_i) \lhd G$ and the map

$$\begin{cases} G/\mathrm{Gal}(E/E_i) \longrightarrow \mathrm{Gal}(E_i/F) \\ \mathrm{Gal}(E/E_i) \longmapsto \sigma|E_i \end{cases}$$

is an isomorphism.

Partially order the set I by defining $i \leqslant j$ in case $E_i \subseteq E_j$. Given $i \leqslant j$, let

$$f_{ij} : \mathrm{Gal}(E_j/F) \longrightarrow \mathrm{Gal}(E_i/F)$$

be the restriction homomorphism. We consider the groups $\mathrm{Gal}(E_i/F)$ as discrete compact topological groups, in which case

$$\{\mathrm{Gal}(E_i/F), f_{ij} \mid i,j \in I\}$$

becomes a projective system of topological groups. The significance of the corresponding projective limit stems from the following fact

12.1. PROPOSITION. The map

$$\begin{cases} G \longrightarrow \varprojlim \mathrm{Gal}(E_i/F) \\ \sigma \longmapsto (\sigma|E_i) \end{cases}$$

is an isomorphism of topological groups. In particular, G is a profinite group.

Proof. Consider the map

$$\begin{cases} G \xrightarrow{\psi} \prod_{i\in I} \mathrm{Gal}(E_i/F) \\ \sigma \longmapsto (\sigma|E_i) \end{cases}$$

Then ψ is an algebraic homomorphism, which is injective since $\sigma|E_i = 1$ for all $i \in I$ implies $\sigma = 1$. For any given $j \in I$, the sets

$$U = \prod_{i\neq j} \mathrm{Gal}(E_i/F) \times \{\bar{\sigma}\} \qquad (\bar{\sigma} \in \mathrm{Gal}(E_j/F))$$

form a subbasis of open neighbourhoods of the product $\prod_{i\in I} \mathrm{Gal}(E_i/F)$. If $\sigma \in G$ is a lift of $\bar{\sigma}$, then $\psi^{-1}(U) = \sigma\mathrm{Gal}(E/E_j)$. Thus ψ is continuous, and

$$\psi(\sigma\mathrm{Gal}(E/E_j)) = \psi(G) \cap U$$

showing that $\psi : G \longrightarrow \psi(G)$ is open.

We now show that $\psi(G)$ is closed in $\prod_{i\in I} \mathrm{Gal}(E_i/F)$. To this end, given $s \leqslant k$ in I, consider the set

$$M_{sk} = \{(\sigma_i) \in \prod_{i\in I} \mathrm{Gal}(E_i/F) \mid \sigma_k|E_s = \sigma_s\}$$

It is clear that

$$\psi(G) = \bigcap_{s\leq k} M_{sk}$$

On the other hand, if $\mathrm{Gal}(E_s/F) = \{\sigma_1,\ldots,\sigma_n\}$ and $S_i \subseteq \mathrm{Gal}(E_k/F)$ is the set of all extensions of σ_i to E_k, then

$$M_{sk} = \bigcup_{j=1}^{n} \Big(\prod_{i\neq k,s} \mathrm{Gal}(E_i/F) \times S_j \times \{\sigma_j\}\Big),$$

showing that M_{sk} and hence $\psi(G)$ is closed. Thus ψ is a homeomorphism of G onto the closed subset $\psi(G)$ of the compact space $\prod_{i\in I} \mathrm{Gal}(E_i/F)$, i.e. G is compact.

We now use the fact that G is compact to show that

$$\psi(G) = \varprojlim \mathrm{Gal}(E_i/F)$$

which will finish the proof. If $(\sigma_i) \in \varprojlim \mathrm{Gal}(E_i/F)$, then the closed sets $\sigma_i \mathrm{Gal}(E/E_i)$ have the finite intersection property. Since G is compact, there exists $\sigma \in \bigcap_{i\in I} \sigma_i \mathrm{Gal}(E/E_i)$. This element satisfies $\psi(\sigma) = (\sigma_i)$ and the result follows. ■

12.2. THEOREM. (Main theorem of infinite Galois theory). Let E/F be a (finite or infinite) Galois extension with Galois group G.

(i) The map $K \longmapsto \mathrm{Gal}(E/K)$ is an inclusion-reversing bijection between the intermediate fields and the closed subgroups of G with the inverse map given by $H \longmapsto E^H$.

(ii) K/F is finite if and only if $\mathrm{Gal}(E/K)$ is an open subgroup of G

Proof. (i) In view of Lemma 10.1, we need only verify that $\mathrm{Gal}(E/K)$ is closed in G and that if H is a closed subgroup of G, then $H = \mathrm{Gal}(E/K)$ where $K = E^H$.

Each open subgroup of G is closed, since it is the complement of the union of its open cosets in G. If K/F is a finite subextension of E/F, then $\mathrm{Gal}(E/K)$ is open, because each $\sigma \in \mathrm{Gal}(E/K)$ has the open neighbourhood $\sigma\mathrm{Gal}(E/L) \subseteq \mathrm{Gal}(E/K)$, where L/F is the normal closure of K/F. If K/F is an arbitrary subextension of E/F, then

$$\mathrm{Gal}(E/K) = \bigcap_i \mathrm{Gal}(E/K_i)$$

where K_i/F runs through the finite subextensions of K/F. Thus $\mathrm{Gal}(E/K)$ is closed.

Let H be a closed subgroup of G and let $K = E^H$. Clearly $H \subseteq \mathrm{Gal}(E/K)$. Conversely, let $\sigma \in \mathrm{Gal}(E/K)$. If L/K is a finite Galois subextension of E/K, then $\sigma\mathrm{Gal}(E/L)$ is a basis open neighbourhood of σ in $\mathrm{Gal}(E/K)$. The map $H \longrightarrow \mathrm{Gal}(L/K)$ is surjective, since its image $\bar{H}$ has the fixed field K and is therefore equal to $\mathrm{Gal}(L/K)$ by the main theorem of finite Galois theory. Choosing a $\lambda \in H$ such that $\lambda|L = \sigma|L$, we have $\lambda \in H \cap \sigma\mathrm{Gal}(E/L)$. Thus σ belongs to the closure of H, i.e. $\sigma \in H$ since H is closed, and therefore $H = \mathrm{Gal}(E/K)$.

(ii) The "only if" part was established in (i). Assume that $H = \mathrm{Gal}(E/K)$ is an open subgroup of G. Since G is compact (Proposition 12.1), H is of finite index in G (Proposition 11.3(ii)), and this obviously implies that K/F is finite. ■

It is a consequence of Proposition 4.5 that every finite group is the Galois group of some field extension. We now generalize this result to profinite groups. The following observation will clear our path.

12.3. LEMMA. Let F be a field and let G be a profinite group of automorphisms of F such that for each $x \in F$, the stabilizer $G(x) = \{\sigma \in G | \sigma(x) = x\}$ is an open subgroup of G. Then $G = \mathrm{Gal}(F/F^G)$.

Proof. By hypothesis, for every $x_1,\ldots,x_n \in F$, the group $H = \bigcap_{i=1}^{n} G(x_i)$ is open in G. Since G is profinite, G is compact (Proposition 11.8), hence H is closed of finite index (Proposition 11.3(ii)). Therefore the intersection N of all conjugates of H is closed and of finite index, hence open (Proposition 11.3(ii)). The finite group G/N can be regarded as an automorphism group of the field $L = F^G(Gx_1,\ldots,Gx_n)$, where $Gx_i = \{\sigma(x_i) | \sigma \in G$, and it has F^G as its fixed field. Thus L/F^G is a finite Galois extension with $G/N = \mathrm{Gal}(L/F^G)$ (Proposition 4.5).

The field F is the union of all above L and 1 is the intersection of all above N. Hence

$$\mathrm{Gal}(F/F^G) = \varprojlim \mathrm{Gal}(L/F^G) = \varprojlim G/N = G$$

as required. ■

12.4. COROLLARY. (Waterhouse(1974)). Every profinite group is the Galois group of some field extension.

Proof. Let G be a profinite group and let S be the disjoint union of the sets G/H for all open subgroups H of G. Then G acts faithfully on S, and each element of S has open stabilizer. Let K be any field, take elements of S as indeterminates and form the pure transcendental extension $F = K(S)$. Then G may be regarded as an automorphism group of F. Since G satisfies the hypothesis of Lemma 12.3, the result follows. ■

We close by providing some general results pertaining to infinite Galois extensions.

12.5. PROPOSITION. Let E/F and K/F be field extensions, where E/F is Galois, and assume that E,K are subfields of some other field. Then EK/K and $E/(E \cap K)$ are Galois extensions and the map

$$\begin{cases} \mathrm{Gal}(EK/K) \longrightarrow \mathrm{Gal}(E/(E \cap K)) \\ \qquad \sigma \longmapsto \sigma|E \end{cases}$$

is an isomorphism of topological groups.

Proof. The extensions EK/K and $E/(E \cap K)$ are obviously normal and separable, hence Galois. If $\sigma \in \mathrm{Gal}(EK/K)$, then $\sigma|E$ is an F-homomorphism $E \longrightarrow EK$, hence is an element of $\mathrm{Gal}(E/F)$, since E/F is normal. The map $\sigma \longmapsto \sigma|E$ is clearly a homomorphism. If $\sigma|E$ is the identity, then σ must be the identity on EK, since $EK = K(E)$. Thus the given homomorphism is injective Let S be its image. Then S leaves $E \cap K$ fixed, and conversely, if an element $\lambda \in E$ is fixed under S, we see that λ is also fixed under $\mathrm{Gal}(EK/K)$, so $\lambda \in K$ and $\lambda \in E \cap K$. Thus $E \cap K$ is the fixed field of S. Now for the Krull topology the given map $\sigma \longmapsto \sigma|E$ is obviously continuous, whence its image S is closed since $\mathrm{Gal}(EK/K)$ is compact. Hence, by Theorem 12.2,

$$S = \mathrm{Gal}(E/(E \cap K)$$

and the result follows. ■

12.6. COROLLARY. Let E/F be a finite Galois extension and let K/F be an arbitrary extension. Then $(EK:K)$ divides $(E:F)$.

Proof. By Proposition 12.5, $(EK:K) = (E:(E \cap K))$, so our assertion follows. ∎

12.7. PROPOSITION. Let E_1/F and E_2/F be Galois extensions and assume that E_1, E_2 are subfields of some other field. Then E_1E_2/F is a Galois extension and the map

$$\begin{cases} \mathrm{Gal}(E_1E_2/F) \longrightarrow \mathrm{Gal}(E_1/F) \times \mathrm{Gal}(E_2/F) \\ \qquad\qquad \sigma \longmapsto (\sigma|E_1, \sigma|E_2) \end{cases}$$

is an injective homomorphism. Furthermore, if $E_1 \cap E_2 = F$, then it is an isomorphism.

Proof. Since E_1/F and E_2/F are normal and separable, so is E_1E_2/F and thus E_1E_2/F is a Galois extension. The given map is obviously a homomorphism. Furthermore, if σ fixes E_1 and E_2 elementwise, then σ is the identity map. Thus our homomorphism is injective.

Suppose that $E_1 \cap E_2 = F$. Owing to Proposition 12.5, given $\sigma_1 \in \mathrm{Gal}(E_1/F)$, there exists a $\sigma \in \mathrm{Gal}(E_1E_2/E_2)$ which restricts to σ_1 on E_1. Since σ induces the identity on E_2, $G_1 \times 1$ is contained in the image of the given homomorphism. A similar argument shows that $1 \times G_2$ is contained in this image. Thus our map is surjective and the result follows. ∎

12.8. COROLLARY. Let E_i/F, $1 \leqslant i \leqslant n$, be Galois extensions and let

$$E_{i+1} \cap (E_1 \ldots E_i) = F \qquad (1 \leqslant i \leqslant n-1)$$

Then

$$\mathrm{Gal}(E_1 \ldots E_n/F) \cong \mathrm{Gal}(E_1/F) \times \ldots \times \mathrm{Gal}(E_n/F)$$

Proof. The case $n = 2$ follows from Proposition 12.7. The general case is a straightforward induction on n. ∎

12.9. PROPOSITION. Let G be the Galois group of a finite Galois extension E/F. Assume that $G = G_1 \times \ldots \times G_n$ and denote by E_i the fixed field of

$$G_1 \times \dots \{1\} \times \dots \times G_n$$

where 1 occurs in the i-th place. Then E_i/F is Galois, $1 \leqslant i \leqslant n$, and

$$E_{i+1} \cap (E_1 \dots E_i) = F \ , \ E = E_1 \dots E_n \qquad (1 \leqslant i \leqslant n-1)$$

Proof. Put $H_i = G_1 \times \dots \times \{1\} \times \dots \times G_n$ so that $E_i = E^{H_i}$, $1 \leqslant i \leqslant n$. Since $H_i \lhd G$, it follows that each E_i/F is Galois (Lemma 10.3(ii)). Now $H_i = \mathrm{Gal}(E/E^{H_i}) = \mathrm{Gal}(E/E_i)$, by Proposition 4.5. Therefore, by Lemma 10.1(iii),

$$\mathrm{Gal}(E/E_1 \dots E_n) = \bigcap_{i=1}^{n} \mathrm{Gal}(E/E_i) = \bigcap_{i=1}^{n} H_i = 1$$

and thus $E = E_1 \dots E_n$. Finally, if $S_i = \mathrm{Gal}(E/E_1 \dots E_i)$, then $E_1 \dots E_i = E^{S_i}$. Furthermore,

$$S_i = \bigcap_{j=1}^{i} H_i \supseteq G_{i+1}$$

and so $G = H_{i+1} S_i$. Thus, by Lemma 10.1(iii),

$$E_{i+1} \cap (E_1 \dots E_i) = E^{H_{i+1}} \cap E^{S_i} = E^{H_{i+1} S_i} = E^G = F,$$

as required. ■

12.10. PROPOSITION. Assume that all fields below contained in some common field.

(i) If E_1/F and E_2/F are abelian extensions, then so is $E_1 E_2/F$

(ii) If E/F is abelian and K/F is any extension, then EK/K is abelian

(iii) If E/F is abelian and K is any intermediate field, then K/F and E/K are abelian.

Proof. (i) Direct consequence of Proposition 12.7.

(ii) Since E/F is abelian, so is the subextension $E/(E \cap K)$. Now apply Proposition 12.5.

(iii) Direct consequence of the fact that subgroups and factor groups of abelian groups are abelian. ■

A field extension E/F is said to *abelian*(*solvable*, *cyclic*) if E/F is Galois and $\mathrm{Gal}(E/F)$ is abelian (solvable, cyclic)

12.11. PROPOSITION. Any finite abelian extension is a composite of cyclic extensions of prime power degrees.

Proof. Apply Proposition 12.9. ■

If F is a field, the composite of all abelian extensions of F in a given algebraic closure $\bar{F}$ of F is called the *maximal abelian extension* of F, and is denoted by F^{ab}. This terminology is justified by Proposition 12.10(i) which guarantees that F^{ab}/F is an abelian extension containing any abelian subextension E/F of $\bar{F}/F$.

12.12. PROPOSITION. Let F be a field and let α,β be two elements of finite degree over F such that $F(\alpha)/F$ is Galois. Then

$$(F(\alpha,\beta):F) = (F(\beta):F)(F(\alpha):(F(\alpha)\cap F(\beta))$$

Proof. Put $E = F(\alpha)$ and $K = F(\beta)$. Then $EK = F(\alpha,\beta)$ and so, by Proposition 12.5,

$$(F(\alpha,\beta) : F(\beta)) = (F(\alpha) : (F(\alpha)\cap F(\beta)))$$

whence

$$(F(\alpha,\beta):F) = (F(\beta):F)(F(\alpha,\beta):F(\beta)) = (F(\beta):F)(F(\alpha):(F(\alpha)\cap F(\beta))$$

as required. ∎

Remark. The assumption that $F(\alpha)/F$ is Galois is indispensable. Indeed, let $F = \mathbb{Q}$, let $\alpha = \sqrt[3]{2}$ be the real cube root of 2 and let ε be a primitive cube root of 1, say

$$\varepsilon = \frac{-1+i\sqrt{3}}{2}$$

Put $\beta = \varepsilon\alpha$ and note that $\mathbb{Q}(\beta) \neq \mathbb{Q}(\alpha)$ since β is complex and α is real. Since $X^3 - 2$ is irreducible over $\mathbb{Q}$, we have $(\mathbb{Q}(\alpha):\mathbb{Q}) = (\mathbb{Q}(\beta):\mathbb{Q}) = 3$. Now $\mathbb{Q}(\alpha)\cap\mathbb{Q}(\beta)$ is a subfield of $\mathbb{Q}(\beta)$ whose degree over $\mathbb{Q}$ divides 3. Hence this degree is 3 or 1, and must be 1 since $\mathbb{Q}(\alpha) \neq \mathbb{Q}(\beta)$. In other words,

$$\mathbb{Q}(\alpha)\cap\mathbb{Q}(\beta) = \mathbb{Q}.$$

On the other hand,

$$\mathbb{Q}(\alpha,\beta) = \mathbb{Q}(\alpha,\varepsilon) = \mathbb{Q}(\alpha,i\sqrt{3})$$

Hence $(\mathbb{Q}(\alpha,\beta):\mathbb{Q}(\alpha)) = 2$ and therefore

$$6 = (\mathbb{Q}(\alpha,\beta):\mathbb{Q}) \neq (\mathbb{Q}(\alpha):\mathbb{Q})(\mathbb{Q}(\beta):\mathbb{Q}) = 9$$

as asserted. ■

12.13. PROPOSITION. Let F be a field and let α,β be two separable elements over F of coprime degrees. Then $(F(\alpha,\beta):F) = (F(\alpha):F)(F(\beta):F)$

Proof. Let E be obtained from F by adjoining all conjugates of α and β. Then E/F is a finite Galois extension. Put $m = (F(\alpha):F)$, $n = (F(\beta):F)$ and $G = \mathrm{Gal}(E/F)$. Then G transitively permutes the elements of each of the sets $A = \{\alpha_1,\ldots,\alpha_m\}$, $B = \{\beta_1,\ldots,\beta_m\}$, where α_i and β_j are the conjugates of α and β, respectively. Let $H = G_\alpha$ and $K = G_\beta$ be the stabilizers in G of α and β, respectively. Then $(G:H) = m$, $(G:K) = n$ and since $(m,n)=1$, a standard argument yields $(G: (H \cap K)) = mn$. But $H \cap K = \mathrm{Gal}(E/F(\alpha,\beta))$, hence $(E:F(\alpha,\beta)) = |H \cap K|$. It follows that

$$(F(\alpha,\beta):F) = \frac{(E:F)}{(E:F(\alpha,\beta))} = \frac{|G|}{|H\cap K|} = mn$$

as required. ■

13. WITT VECTORS

Our aim here is to consider an ingenious ring of vectors discovered by Witt(1936). Since this ring will have important applications, we study it now in its general form. Throughout, p denotes a fixed rational prime.

We begin with the polynomial ring

$$A = \mathbb{Q}[X_i, Y_j, Z_k] \qquad (i,j,k = 0,1,2,\ldots)$$

and denote by $A^{\mathbb{N}}$ the direct product of $|\mathbb{N}|$ copies of A. Thus $A^{\mathbb{N}}$ is a ring consisting of all infinite "vectors"

$$x = (x_0, x_1, \ldots, x_{n-1}, \ldots) \qquad (x_i \in A)$$

with addition and multiplication performed componentwise. We wish to define another ring structure on the set $A^{\mathbb{N}}$. In what follows, we write

$$\pi(x) = (x_0^p, x_1^p, \ldots, x_{n-1}^p, \ldots)$$

The elements $x_0, x_1, \ldots, x_{n-1}, \ldots$ of A define elements

$$x^{(0)}, x^{(1)}, \ldots, x^{(n-1)}, \ldots$$

of A be recursion formulas:

$$x^{(0)} = x_0 \quad \text{and} \quad x^{(i+1)} = (\pi(x))^{(i)} + p^{i+1}x_{i+1} \tag{1}$$

where $x = (x_0, x_1, \ldots, x_{n-1}, \ldots)$. Then (1) gives

$$x^{(0)} = x_0 \quad \text{and} \quad x^{(i+1)} = x_0^{p^{i+1}} + px_1^{p^i} + \ldots + p^{i+1}x_{i+1} \tag{2}$$

The elements $x^{(0)}, x^{(1)}, \ldots, x^{(n-1)}, \ldots$ of A define elements $x_0, x_1, \ldots, x_{n-1}, \ldots$ of A by the inverse recursion formulas:

$$x_0 = x^{(0)} \quad \text{and} \quad x_{i+1} = (1/p^{i+1})(x^{(i+1)} - x_0^{p^{i+1}} - px_1^{p^i} - \ldots - p^i x_i^p) \tag{3}$$

Let us write

$$[x^{(0)}, x^{(1)}, \ldots, x^{(n-1)}, \ldots] = (x_0, x_1, \ldots, x_{n-1}, \ldots)$$

where $x_0, x_1, \ldots, x_{n-1}, \ldots$ are defined in terms of $x^{(0)}, x^{(1)}, \ldots, x^{(n)}, \ldots$ by (3). We are now in a position to define a new ring structure on the set $A^{\mathbb{N}}$. Namely, given

$$x = (x_0, x_1, \ldots, x_{n-1}, \ldots), \; y = (y_0, y_1, \ldots, y_{n-1}, \ldots) \in A^{\mathbb{N}}$$

we define

$$x + y = [x^{(0)} + y^{(0)}, x^{(1)} + y^{(1)}, \ldots, x^{(n-1)} + y^{(n-1)}, \ldots]$$

$$xy = [x^{(0)}y^{(0)}, x^{(1)}y^{(1)}, \ldots, x^{(n-1)}y^{(n-1)}, \ldots]$$

and denote by $W(A)$ the set $A^{\mathbb{N}}$ endowed with the addition and multiplication above. It is clear that $W(A)$ is a commutative ring of characteristic 0 which is isomorphic to $A^{\mathbb{N}}$ under the mapping

$$[x^{(0)}, x^{(1)}, \ldots, x^{(n-1)}, \ldots] \longmapsto (x^{(0)}, x^{(1)}, \ldots, x^{(n-1)}, \ldots)$$

If $a \in A$, then by (2)

$$(a, 0, \ldots, 0, \ldots) = [a, a^p, \ldots, a^{p^{n-1}}, \ldots]$$

Putting $a = 0$ and $a = 1$, we conclude that the zero and identity elements of $W(A)$ are

$$(0, 0, \ldots, 0, \ldots) \quad \text{and} \quad (1, 0, \ldots, 0, \ldots)$$

respectively.

We now examine the formulas for $X + Y$ and XY where

$$X = (X_0, X_1, \ldots, X_{n-1}, \ldots) \quad \text{and} \quad Y = (Y_0, Y_1, \ldots, Y_{n-1}, \ldots)$$

Applying (2) and (3), we have

$$\begin{aligned}(X+Y)_0 &= X_0+Y_0, (XY)_0 = X_0Y_0 \\ (X+Y)_1 &= X_1 + Y_1 - (1/p)(X_0^p+Y_0^p - (X_0+Y_0)^p) \\ (XY)_1 &= X_0^pY_1 + X_1Y_0^p + pX_1Y_1\end{aligned} \tag{4}$$

In general, it is clear from the definitions that $(X+Y)_i$ and $(XY)_i$ are polynomials with rational coefficients in $X_0, Y_0, X_1, Y_1, \ldots, X_i, Y_i$. Also one sees easily that

$$(X+Y)_i = X_i + Y_i + f_i(X_0, Y_0, \ldots, X_{i-1}, Y_{i-1}) \tag{5}$$

where f_i is a polynomial in the indicated variables. The basic result which we now establish is that both $(X+Y)_i$ and $(XY)_i$ are polynomials in

$$X_0, Y_0, X_1, Y_1, \ldots, X_i Y_i$$

with *integer* coefficients. The following preliminary observation will clear our path.

13.1. LEMMA. Let $j \geqslant 1$, $0 \leqslant k \leqslant m-1$ and let $x, y \in W(A)$ be such that

$$x_i, y_i \in \mathbb{Z}[X_0, Y_0, \ldots, X_{m-1}, Y_{m-1}] \qquad (0 \leqslant i \leqslant m-1)$$

Then the following conditions are equivalent:

(i) $x_i \equiv y_i (\text{mod } p^j\mathbb{Z}[X_0, Y_0, \ldots, X_{m-1}, Y_{m-1}])$ for all $0 \leqslant i \leqslant k$

(ii) $x^{(i)} \equiv y^{(i)} (\text{mod } p^{i+j}\mathbb{Z}[X_0, Y_0, \ldots, X_{m-1}, Y_{m-1}])$ for all $0 \leqslant i \leqslant k$

Proof. Abbreviate $\mathbb{Z}[X_0, Y_0, \ldots, Y_{m-1}, Y_{m-1}]$ by $\mathbb{Z}[X_r, Y_s]$. We have $x^{(0)} = x_0$ and $y^{(0)} = y_0$, so the result is clear for $k = 0$. To prove the result by induction on k, we may assume that (i) and (ii) hold for all $i \in \{0, \ldots, k-1\}$. We have to show that

$$x_k \equiv y_k (\text{mod } p^j\mathbb{Z}[X_r, Y_s]$$

if and only if

$$x^{(k)} \equiv y^{(k)} (\text{mod } p^{k+j}\mathbb{Z}[X_r, Y_s])$$

It is clear that $x_k \equiv y_k \pmod{p^j\mathbb{Z}[X_r,Y_s]}$ if and only if

$$p^k x \equiv p^k y \pmod{p^{k+j}\mathbb{Z}[X_r,Y_s]}$$

Hence, using (1), it is enough to show that

$$(\pi(x))^{(k-1)} \equiv (\pi(y))^{(k-1)} \pmod{p^{k+j}\mathbb{Z}[X_r,Y_s]}$$

holds under the induction hypothesis. We have

$$x_i \equiv y_i \pmod{p^j\mathbb{Z}[X_r,Y_s]}, \qquad 0 \leqslant i \leqslant k-1.$$

Since $\binom{p}{t} \equiv 0 \pmod p$, $1 \leqslant t \leqslant p-1$, this gives $x_i^p \equiv y_i^p \pmod{p^{j+1}}$, $0 \leqslant i \leqslant k-1$. Hence the induction on k applies to $\pi(x)$ and $\pi(y)$ to give

$$\pi(x)^{(k-1)} \equiv \pi(y)^{(k-1)} \pmod{p^{k-1+j+1}\mathbb{Z}[X_r,Y_s]}$$

which is what was needed. ■

It is now possible to establish the following important result.

13.2. THEOREM. For any $i \in \{0,1,2,\dots\}$, $(X+Y)_i$ and $(XY)_i$ are in

$$\mathbb{Z}[X_0,Y_0,\dots,X_i,Y_i].$$

Proof. Let $X \circ Y$ denote XY or $X+Y$. Then

$$(X \circ Y)_0 \in \mathbb{Z}[X_r,Y_s], \qquad 0 \leqslant r,s \leqslant i,$$

by virtue of (4). Assume that $(X \circ Y)_k \in \mathbb{Z}[X_r,Y_s]$ for all $0 \leqslant k \leqslant i-1$. Since, by (1),

$$p^i(X \circ Y)_i = (X \circ Y)^{(i)} - (\pi(X \circ Y))^{(i-1)} \tag{6}$$

it suffices to verify that the right-hand side is in $p^i\mathbb{Z}[X_r,Y_s]$. Again, by (1),

$$X^{(i)} \equiv \pi(X)^{(i-1)} \pmod{p^i\mathbb{Z}[X_r,Y_s]}$$

and

$$Y^{(i)} \equiv \pi(Y)^{(i-1)} \pmod{p^i\mathbb{Z}[X_r,Y_s]}$$

which implies that

$$(X \circ Y)^{(i)} \equiv (\pi(X) \circ \pi(Y))^{(i-1)} \pmod{p^i\mathbb{Z}[X_r,Y_s]} \tag{7}$$

Now, for any $f \in \mathbb{Z}[X_r,Y_s]$, one has

$$f(X_0,Y_0,\dots,X_i,Y_i)^p \equiv f(X_0^p,Y_0^p,\dots,X_i^p,Y_i^p)\,(\text{mod } p\mathbb{Z}[X_r,Y_s]) \tag{8}$$

Applying (8) to $f = (X\circ Y)_k$, $0 \leqslant k \leqslant i-1$, we drive

$$(\pi(X\circ Y))_k \equiv (\pi(X)\circ \pi(Y))_k \text{ (mod } pZ[X_r,Y_s]) \qquad (0 \leqslant k \leqslant i-1) \tag{9}$$

Hence (9) and Lemma 13.1 imply

$$(\pi(X\circ Y))^{(i-1)} \equiv (\pi(X)\circ \pi(Y))^{(i-1)} \text{ (mod } p^i\mathbb{Z}[X_r,Y_s]) \tag{10}$$

Applying (7) and (10), we see that the right-hand side of (6) is in $p^i\mathbb{Z}[X_r,Y_s]$, as required. ■

For any given $i \geqslant 0$, we write

$$(X+Y)_i = a_i(X_r,Y_s), \quad (XY)_i = m_i(X_r,Y_s) \qquad (0 \leqslant r,s \leqslant i)$$

so that, by Theorem 13.2, both $a_i(X_r,Y_s)$ and $m_i(X_r,Y_s)$ are in

$$\mathbb{Z}[X_0,Y_0,\dots,X_i,Y_i] .$$

Then, for any $x,y \in W(A)$,

$$(x+y)_i = a_i(x_r,y_s) \text{ and } (xy)_i = m_i(x_r,y_s) \tag{11}$$

Let R be any commutative ring and let

$$\overline{a_i}(X_r,Y_s),\ \overline{m}_i(X_r,Y_s) \qquad (0 \leqslant r,s \leqslant i)$$

be the polynomials obtained from $a_i(X_r,Y_s)$, $m_i(X_r,Y_s)$ by replacing each coefficient $c \in \mathbb{Z}$ by the corresponding $\bar{c} = c\cdot 1_R$ in the prime subring of R. We are now ready to define the *Witt ring* $W(R)$ of R and the *n-th Witt ring* $W_n(R)$ of R, respectively.

The elements of $W(R)$ are infinite sequences

$$a = (a_i) = (a_0,a_1,\dots,a_{n-1},\dots\,) \qquad (a_i \in R)$$

with equality defined as usual. If $a = (a_i), b = (b_i)$ are in $W(R)$, then we define addition and multiplication in $W(R)$ by

$$(a+b)_i = \bar{a}_i(a_r,b_s), \qquad (ab)_i = \overline{m_i}(a_r,b_s) \qquad (0 \leqslant r,s \leqslant i) \tag{12}$$

The elements of $W_n(R)$ are finite sequences

$$a = (a_i) = (a_0,a_1,\dots,a_{n-1}) \qquad (a_i \in R)$$

with equality defined as usual, and addition and multiplication defined by (12).

We now define the following maps:

$$\begin{cases} W(R) \xrightarrow{\sigma_n} W_n(R) \\ (a_i) \longmapsto (a_0, a_1, \ldots, a_{n-1}) \end{cases}$$

$$\begin{cases} W_n(R) \xrightarrow{f_{mn}} W_m(R) \\ (a_0, \ldots, a_{n-1}) \longmapsto (a_0, \ldots, a_{m-1}) \end{cases} \qquad (m \leq n)$$

13.3. THEOREM. (i) $W(R)$ and $W_n(R)$ are commutative rings and σ_n, f_{mn} are surjective ring homomorphisms. The elements $(0,\ldots,0,\ldots)$ and $(1,0,\ldots,0,\ldots)$ are zero and identity elements of $W(R)$, while $(0,\ldots,0)$ and $(1,0,\ldots,0)$ are zero and identity elements of $W_n(R)$.

(ii) $\{W_n(R), f_{mn} | n,m \in \mathbb{N}, m \leq n\}$ is a projective system of rings and the map

$$W(R) \longrightarrow \varprojlim W_n(R), \; x \longmapsto (\sigma_n(x))$$

is a ring isomorphism.

(iii) For any subring S of R, $W(S)$ and $W_n(S)$ are identifiable with the subrings of $W(R)$ and $W_n(R)$, respectively, consisting of vectors whose coordinates are in S.

(iv) $\mathrm{Ker}\sigma_1 \supset \mathrm{Ker}\sigma_2 \supset \ldots \supset \mathrm{Ker}\sigma_n \supset \ldots$ and $\bigcap_{i=1}^{\infty} \mathrm{Ker}\sigma_i = 0$.

Proof. (i) Consider the subring $B = \mathbb{Z}[X_i, Y_j, Z_k]$ of A. If

$$x = (x_i), \; y = (y_i), \; z = (z_i)$$

are any three elements of $W(R)$, then we have a homomorphism $\omega : B \longrightarrow R$ such that $n \longmapsto n \cdot 1_R$ $n \in \mathbb{Z}, X_i \longmapsto x_i, Y_i \longrightarrow y_i, Z_i \longrightarrow z_i$. If $\lambda = (\lambda_i), \mu = (\mu_i) \in W(B)$ then by (10), $(\lambda+\mu)_i = a_i(\lambda_r, \mu_s)$ and $(\lambda\mu)_i = m_i(\lambda_r, \mu_s)$. It follows that the mapping $W(B) \longrightarrow W(R)$ $(a_i) \longmapsto (\omega(a_i))$ preserves addition and multiplication, and satisfies $X \longmapsto x$, $Y \longmapsto y$, $Z \longrightarrow z$, where $X = (X_i)$, $Y = (Y_i)$ and $Z = (Z_i)$. Then the associative, commutative, and distributive laws of addition and multiplication in $W(B)$ give the same rules for the elements x,y,z of $W(R)$. This shows that $W(R)$ is a commutative ring. The image of $0 = (0,0,\ldots,0,\ldots)$ and $1 = (1,0,\ldots,0,\ldots)$ under our homomorphism are $(0,0,\ldots,0,\ldots)$ and

$$(1,0,\ldots,0,\ldots),$$

respectively. A similar argument shows that $W_n(R)$ is a commutative ring with specified zero and identity elements. The maps σ_n, f_{mn} are obviously surjective, and are homomorphisms by virtue of (12).

(ii) The first assertion follows by the definition of f_{mn}. Furthermore, we also have $f_{mn}\sigma_n = \sigma_m$ for all $n,m \in \mathbb{N}$, $m \leqslant n$. Hence, by Proposition 11.3, the given map is a homomorphism which is obviously injective. If $(y_n) \in \varprojlim W_n(R)$, then $f_{mn}(y_n) = y_m$ for $m \leqslant n$ forces $y_n = \sigma_n(x)$ for all n, where $x_i = (y_n)_i$, as required.

(iii) Straightforward

(iv) By definition, $\mathrm{Ker}\sigma_n = \{(a_i) \in W(R) | a_0 = a_1 = \ldots = a_{n-1} = 0\}$ which clearly implies the required assertion. ■

If $x = (x_i) \in W(R)$, define

$$s(x) = (0,x_0,\ldots,x_{n-1},\ldots)$$

We shall refer to the map $s : W(R) \longrightarrow W(R)$ as the *shift*. If $a = (a_0,\ldots a_{n-1}) \in W_n(R)$, then we define

$$s_n : W_n(R) \longrightarrow W_n(R)$$

by

$$s_n(x_0,x_1,\ldots,x_{n-1}) = (0,x_0,\ldots,x_{n-2})$$

and again refer to s_n as the *shift*.

If $a \in R$, set

$$r(a) = (a,0,\ldots,0,\ldots), \; r_n(a) = (a,0,\ldots,0) \in W_n(R)$$

This defines maps

$$r : R \longrightarrow W(R), \; r_n : R \longrightarrow W_n(R)$$

Finally, define $v_n : W_n(R) \longrightarrow W_{n+1}(R)$ by

$$v_n(x_0,x_1,\ldots,x_{n-1}) = (0,x_0,\ldots,x_{n-1})$$

and, given $a = (a_i) \in W(R)$, set

$$\sum_{i=0}^{\infty} a_i = (a_0, a_0+a_1, \ldots, a_0+a_1+\ \ldots\ + a_{n-1}, \ \ldots\ )$$

13.4. THEOREM (i) The maps s, s_n and v_n preserve addition

(ii) For $x = (x_0,\ldots,x_{n-1}) \in W_n(R)$, $y = (x_0,x_1,\ldots,x_{n-1},\ldots\ ) \in W(R)$, $a \in R$,

(a) $x = \sum_{i=0}^{n-1} s_n^i(r_n(x_i)), y = \sum_{i=0}^{\infty} s^i(r(x_i))$

(b) $r_n(a)x = (ax_0, a^p x_1, \ldots, a^{p^{n-1}} x_{n-1})$

(c) $r(a)y = (ax_0, a^p x_1, \ldots, a^{p^{n-1}} x_{n-1}, \ldots\ )$

(iii) $r(ab) = r(a)r(b)$ and $r_n(ab) = r_n(a)r_n(b)$ for all $a,b \in R$, $n \in \mathbb{N}$

(iv) If $x,y \in W(R)$ are such that $x_i = 0$ or $y_i = 0$ for all $i \geqslant 1$, then

$$x + y = (x_0+y_0, x_1+y_1, \ldots, x_{n-1} + y_{n-1}, \ldots\ )$$

(v) If $x,y \in W_n(R)$ are such that $x_i = 0$ or $y_i = 0$ for all $1 \leqslant i \leqslant n-1$, then

$$x + y = (x_0+y_0, \ldots, x_{n-1} + y_{n-1})$$

Proof. (i) We first note that

$$s_n(x_0,x_1,\ldots,x_{n-1}) = \sigma_n s(x_0,x_1,\ldots,x_{n-1},\ldots\ )$$

and

$$v_n(x_0,x_1,\ldots,x_{n-1}) = \sigma_{n+1} s(x_0,x_1,\ldots,x_{n-1},\ldots\ )$$

Since both σ_n and σ_{n+1} preserve addition (Theorem 13.3(i)), it suffices to verify that s preserves addition. To prove the latter, we may harmlessly assume that $R = A$ in which case $s(x) = [0, px^{(0)}, \ldots, px^{(n-2)}, \ldots\ ]$, and hence s preserves addition.

(ii) Since $\sigma_n(y) = x$ and $\sigma_n(r(a)y) = r_n(a)x$, it suffices to prove the formulas for x. Again, we may assume that $R = A$, in which case

$$r_n(a) = [a, a^p, \ldots, a^{p^{n-1}}]$$

and hence

$$s_n^i(r_n(a)) = [0,\ldots,0,p^i a,\ldots,p^i a^{p^{n-1-i}}] \qquad (0 \leqslant i \leqslant n-1)$$

Therefore

$$\sum_{i=0}^{n-1} s_n^i(r_n(x_i)) = \sum_{i=0}^{n-1} [0,\ldots,0,p^i x_i,\ldots,p^i x_i^{p^{n-1-i}}]$$

and the latter sum is $[x^{(0)},\ldots,x^{(n-1)}] = x$ by (2). Also

$$r_n(a)x = [a,a^p,\ldots,a^{p^{n-1}}][x^{(0)},x^{(1)},\ldots,x^{(n-1)}]$$

$$= [ax^{(0)},a^p x^{(1)},\ldots,a^{p^{n-1}}x^{(n-1)}]$$

$$= (ax_0,a^p x_1,\ldots,a^{p^{n-1}}x_{n-1}) \qquad \text{(by (2) and (3))}$$

as required.

(iii) Apply (b) and (c) for $x = r_n(b)$ and $y = r(b)$, respectively.

(iv) and (v). It suffices to prove (v), since (iv) follows from (v) by applying σ_n. By (i),

$$x + y = \sum_{i=0}^{n-1} s_n^i(r_n(x_i)) + \sum_{i=0}^{n-1} s_n^i(r_n(y_i)) = \sum_{i=0}^{n-1} s_n^i(r_n(x_i) + r_n(y_i))$$

$$= \sum_{i=0}^{n-1} s_n^i(r_n(x_i+y_i))$$

since $x_i = 0$ or $y_i = 0$ for $0 \leqslant i \leqslant n-1$. ■

Next we introduce the following two maps

$$\begin{cases} W(R) \xrightarrow{\ \pi\ } W(R) \\ (x_i) \longmapsto (x_i^p) \end{cases} \qquad \begin{cases} W_n(R) \xrightarrow{\ \pi_n\ } W_n(R) \\ (x_0,\ldots,x_{n-1}) \longmapsto (x_0^p,\ldots,x_{n-1}^p) \end{cases}$$

13.5. LEMMA. (i) If $x,y \in W(R)$, then for all $i \geqslant 0$,

$$(\pi(x+y))_i \equiv (\pi(x) + \pi(y))_i \pmod{pR}$$
$$(\pi(xy))_i \equiv (\pi(x)\pi(y))_i \pmod{pR}$$

(ii) If $x,y \in W_n(R)$, then for all $i \in \{0,1,\ldots,n-1\}$,

$$(\pi_n(x+y))_i \equiv (\pi(x) + \pi(y))_i \pmod{pR}$$
$$(\pi_n(xy))_i \equiv (\pi_n(x)\pi_n(y))_i \pmod{pR}$$

Proof. This is a direct consequence of (8). ■

From now on, we concentrate on the very important special case where char$R = p$. The prime subring of R is the field $\mathbb{F}_p$ of p elements.

13.6. THEOREM. Let R be an arbitrary commutative ring of characteristic $p > 0$

and let $n \geqslant 1$.

(i) The maps π and π_n are ring homomorphisms

(ii) $p^i x = s_n^i(\pi_n^i(x))$ and $p^i y = s^i(\pi^i(y))$ for all $x \in W_n(R)$, $y \in W(R)$, $i \geqslant 1$

(iii) $W_n(\mathbb{F}_p) = \mathbb{Z}/p^n\mathbb{Z}$ is the prime subring of $W_n(R)$, the characteristic of $W_n(R)$ is p^n and the characteristic of $W(R)$ is 0.

(iv) For any $x \in W_n(R)$ and $y \in W_{n+1}(R)$, $v_n(x)y = v_n(xf_{n,n+1}\pi_{n+1}(y))$

(v) If R is a field of characteristic p, then

$$s_n^i(x)s_n^j(y) = s_n^{i+j}(\pi_n^j(x)\pi_n^i(y)) \qquad \text{for all } x,y \in W_n(R)$$

Proof. (i) Direct consequence of Lemma 13.5.

(ii) Let $X = (X_0, X_1, \ldots, X_{n-1}, \ldots) \in W(A)$. We have

$$(pX)^{(i)} = pX^{(i)} = p(\pi(x)^{(i-1)} + p^i x_i) \qquad \text{(by (1))}$$

$$\equiv p\pi(x)^{(i-1)} = (s\pi(x))^{(i)} \pmod{p^{i+1}\mathbb{Z}[X]}$$

Pg 199

Thus

$$(pX)_i \equiv (s\pi(x))_i \pmod{p\mathbb{Z}[X]}$$

(π & s commute)

by Lemma 13.1, which clearly implies $py = s\pi(y)$. Iterating the latter, we have $p^i y = s^i(\pi^i(y))$ for all $i \geqslant 1$. The corresponding assertion for x is now obtained by applying σ_n.

(iii) The prime subring of $W_n(R)$ is additively generated by $1 = (1,0,\ldots,0)$. Since $p^n 1 = s_n^n(1) = 0$ and $p^m 1 = s_n^m(1) \neq 0$ for $m < n$, 1 has additive order p^n. This proves that $\mathbb{Z}/p^n\mathbb{Z}$ is the prime subring of $W_n(R)$ and that

$$\text{char} W_n(R) = p^n.$$

On the other hand, $W_n(\mathbb{F}_p)$ contains the prime subring of $W_n(R)$ and has p^n elements. Thus $W_n(\mathbb{F}_p) = \mathbb{Z}/p^n\mathbb{Z}$. Furthermore, the above shows that the additive order of $(1,0,\ldots,0,\ldots)$ in $W(R)$ is infinite and thus $\text{char} W(R) = 0$.

(iv) Let $X = (X_0, X_1, \ldots, X_{n-1}) \in W_n(A)$ and $Y = (Y_0, Y_1, \ldots, Y_n) \in W_{n+1}(A)$. It suffices to show that

$$(v_n(X)Y)_i \equiv (v_n(xf_{n,n+1}\pi_{n+1}(Y)))_i \pmod{p\mathbb{Z}[X,Y]}$$

To this end, set

$$v_n(X)Y = (\omega_0,\omega_1,\ldots,\omega_n),\; v_n(xf_{n,n+1}\pi_{n+1}(Y)) = (t_0,t_1,\ldots,t_n)$$

Then we have to show that $\omega_i \equiv t_i \pmod{p\mathbb{Z}[X,Y]}$, $0 \leqslant i \leqslant n$. By Lemma 13.1, this is equivalent to $\omega^{(i)} \equiv t^{(i)} \pmod{p^{i+1}\,\mathbb{Z}[X,Y]}$. This holds for $i = 0$, since $\omega^{(0)} = 0 = t^{(0)}$. Since $(v_n(x))^{(i)} = px^{(i-1)}$, $1 \leqslant i \leqslant n+1$, by (2), we have that

$$\omega^{(i)} = pX^{(i-1)}Y^{(i)} \quad \text{and} \quad t^{(i)} = pX^{(i-1)}(f_{n,n+1}\pi_{n+1}(y))^{(i-1)} \qquad (1 \leqslant i \leqslant n)$$

Since $Y^{(i)} = (\pi_{n+1}(Y))^{(i-1)} + p^iY_i$, this gives the congruences

$$\begin{aligned}\omega^{(i)} &= pX^{(i-1)}Y^{(i)} \equiv pX^{(i-1)}(\pi_{n+1}(Y))^{(i-1)} \qquad (1 \leqslant i \leqslant n)\\ &\equiv pX^{(i-1)}(f_{n,n+1}\pi_{n+1}(Y))^{(i-1)}\\ &\equiv t^i \pmod{p^{i+1}\,\mathbb{Z}[X,Y]}\end{aligned}$$

and therefore proves (iv).

(v) Assume that R is a field of characteristic p. Then

$$\begin{aligned}s_n^i(\pi_n^i(x))s_n^j(\pi_n^j(x)) &= p^{i+j}xy = s_n^{i+j}(\pi_n^{i+j}(xy)) && \text{(by (ii))}\\ &= s_n^{i+j}(\pi_n^{i+j}(x)\pi_n^{i+j}(y)) && \text{(by (i))}\end{aligned}$$

Let $\bar{R}$ be an algebraic closure of R. Then $W_n(R) \subseteq W_n(\bar{R})$ and π_n is bijective on $W_n(\bar{R})$. Hence, letting $u = \pi_n^i(x)$ and $v = \pi_n^j(y)$, we have

$$s_n^i(u)s_n^j(v) = s_n^{i+j}(\pi_n^j(u)\pi_n^i(v)) \qquad \text{for all} \qquad u,v \in W_n(\bar{R})$$

which finishes the proof. ■

13.7. COROLLARY. $W(\mathbb{F}_p)$ is isomorphic to the ring of p-adic integers.

Proof. Apply Theorems 13.3(ii) and 13.6(iii). ■

13.8. THEOREM. Let R be a commutative ring of characteristic p and let $n \geqslant 1$. Then the map

$$\begin{cases} W_n(R) \longrightarrow R \\ (x_0,x_1,\ldots,x_{n-1}) \longmapsto x_0 \end{cases}$$

is a surjective ring homomorphism whose kernel I is a nilpotent ideal (in fact, $I^n = 0$).

Proof. Since $W_1(R) = R$, our map is just f_{1n} and therefore a surjective ring homomorphism. The kernel I of our homomorphism consists of all elements of the form $(0, x_0, x_1, \ldots, x_{n-2})$ and hence $I = s_n(W_n(R))$. We claim that

$$I^k \subseteq Is_n^{k-2}(I) \subseteq s_n^{k-1}(I) \qquad \text{for all } k \geq 2 \tag{13}$$

If sustained, it will follow that $I^n \subseteq s_n^n(W_n(R)) = 0$, as required.

To this end, note that $f_{n,n+1}v_n = s_n$, so applying Theorem 13.6(iv), we obtain

$$s_n(x)f_{n,n+1}(y) = s_n(xf_{n,n+1}\pi_{n+1}(y)) \qquad (x \in W_n(R),\ y \in W_{n+1}(R))$$

Since $f_{n,n+1}(y)$ can be taken to be any element z in $W_n(R)$ and since

$$f_{n,n+1}\pi_{n+1} = \pi_n f_{n,n+1},$$

this gives the relation $s_n(x)z = s_n(x\pi_n(z))$. Then

$$s_n(x)s_n(z) = s_n(x\pi_n s_n(z)) = s_n(xs_n\pi_n(z)) \in s_n(I) \tag{14}$$

and therefore $I^2 \subseteq s_n(I)$, proving (13) for $k = 2$. If we assume, by induction, $I^t \subseteq Is_n^{t-1}(I) \subseteq s_n^{t-1}(I)$ for some $t \geq 2$, then $I^{t+1} \subseteq Is_n^{t-1}(I)$. Moreover, if $x, z \in W_n(R)$, then by (14)

$$\begin{aligned} s_n(x)s_n^{t}(z) &= s_n(s_n^{t-1}(z))s_n(x) = s_n(s_n^{t-1}(z)s_n\pi_n(x)) \\ &= s_n(s_n\pi_n(x)s_n^{t-1}(z)) \in s_n(Is_n^{t-2}(I)) \subseteq s_n(s_n^{t-1}(I)) = s_n^t(I) \end{aligned}$$

and hence $I^{t+1} \subseteq s_n^t(I)$, as required. ■

The following observation will be needed to provide an application of the result above.

13.9. LEMMA. Let $J(S)$ be the Jacobson radical of an arbitrary ring S. If I is an ideal of S with $I \subseteq J(S)$, then $u \in S$ is a unit of S if and only if $u + I$ is a unit of S/I.

Proof. It is clear that if u is a unit of S, then $u + I$ is a unit of S/I. Conversely, assume that $u \in S$ is such that $u + I$ is a unit of S/I. Then $uv = 1 + x$ and $vu = 1 + y$ for some $x, y \in I$. Then $S = I + uvS = I + vuS$, so $S = uvS = vuS$ by Nakayama's lemma. Choose $w \in S$ so that $uvw = 1$. Since $1 = (uv)w = w + xw$, we have $w = 1 - xw$ also, so w itself has a right inverse.

Hence $uv \in U(S)$ and by a similar argument $vu \in U(S)$, and thus $u \in U(S)$. ■

13.10. COROLLARY. Let R be a commutative ring of characteristic p. Then, for any $n \geq 1$, an element $(x_0, x_1, \dots, x_{n-1})$ is a unit of $W_n(R)$ if and only if x_0 is a unit of R.

Proof. Apply Theorem 13.8 and Lemma 13.9. ■

13.11. COROLLARY. Let R be a commutative ring of characteristic p. Then an element $x = (x_0, x_1, \dots)$ is a unit of $W(R)$ if and only if x_0 is a unit of R.

Proof. By Theorem 13.3(ii), x is a unit of $W(R)$ if and only if for all $n \geq 1$, $\sigma_n(x) = (x_0, x_1, \dots, x_{n-1})$ is a unit of $W_n(R)$. Now apply Corollary 13.10. ■

13.12. COROLLARY. Let R be a field of characteristic p and let

$$I = \{(0, x_1, x_2, \dots) \in W(R) \mid x_i \in R\}$$

(i) $W(R)$ is an integral domain of characteristic 0 with unique maximal ideal I and with $W(R)/I \cong R$

(ii) If R is perfect, then each nonzero element of $W(R)$ can be uniquely written in the form $p^k c$ with $k \geq 0$ and $c \notin I$

Proof. That I is a unique maximal ideal of $W(R)$ is a consequence of Corollary 13.11. Since I is the kernel of the surjective ring homomorphism $\sigma_1 : W(R) \longrightarrow W_1(R) = R$, we also have $W(R)/I \cong R$. Furthermore, $\text{char} W(R) = 0$ by Theorem 13.6(iii). To prove the remaining assertions, we may assume that R is perfect.

Let $x = (x_0, x_1, \dots)$ be a nonzero element of $W(R)$ and let $k \geq 0$ be the smallest integer with $x_k \neq 0$. Since R is perfect, we may write

$$x = (0, \dots, 0, y_0^{p^k}, y_1^{p^k}, \dots) \qquad (y_i \in R,\ y_0 \neq 0)$$

Hence, by Theorem 13.6(ii), $x = p^k c$ where $c = (y_0, y_1, \dots) \notin I$. If $y = p^t d$, $d \notin I$ is another nonzero element of $W(R)$, then $xy = p^{k+t} cd \neq 0$. Thus $W(R)$ is an integral domain. Finally, if $x = p^k c = p^t d$ with $c, d \notin I$ and $k \geq t$, then $p^t(p^{k-t}c - d) = 0$. Hence $d = p^{k-t}c$ which is possible only in the case $k = t$ and hence $d = c$. ■

14. CYCLIC EXTENSIONS

Let E/F be a field extension. Recall that E/F is said to be *cyclic* (*abelian*) if it is Galois with cyclic (abelian) Galois group.

The problem that motivates this section is to describe all cyclic extensions of a given field F. Since any such extension is a composite of cyclic extensions of prime power degrees (Proposition 12.11), it is natural to concentrate on cyclic extensions E/F of degree p^n, $n \geqslant 1$, where p is a fixed prime. For an arbitrary F, this is a very difficult problem. However, in the case F contains all p^n-th roots of unity of an algebraic closure of F, a complete solution will be presented below. The cases where $p = \mathrm{char}F$ and $p \neq \mathrm{char}F$ will be treated separately. If $p \neq \mathrm{char}F$, the corresponding result is a special case of Theorem 14.4. The case where $p = \mathrm{char}F$ and $n = 1$ is given by Theorem 14.7, while its generalization to an arbitrary n constitutes the content of Theorem 14.10.

We begin by establishing two classical results.

14.1. THEOREM. (Hilbert's Theorem 90). Let E/F be cyclic of degree n with Galois group G generated by σ. Then, for any $x \in E$,

$$N_{E/F}(x) = 1 \quad \text{if and only if} \quad x = y\sigma(y)^{-1}$$

for some nonzero y in E.

Proof. For any $y \in E$, we have

$$N_{E/F}(y) = \prod_{i=0}^{n-1}\sigma^i(y) = \prod_{i=0}^{n-1}\sigma^i(\sigma(y)) = N_{E/F}(\sigma(y))$$

Thus, if $x = y\,\sigma(y)^{-1}$ for some $0 \neq y$, then

$$N_{E/F}(x) = N_{E/F}(y)N_{E/F}(\sigma(y))^{-1} = 1$$

Conversely, assume that $N_{E/F}(x) = 1$. Define the map $f : E \longrightarrow E$ by

$$f = 1 + x\sigma + x\sigma(x)\sigma^2 + \dots + x\sigma(x) \dots \sigma^{n-2}(x)\sigma^{n-1}$$

(1 is the identity map). By Proposition 4.7, f is not identically zero. Therefore there exists $\lambda \in E$ such that the element

$$y = \lambda + x\sigma(\lambda) + x\sigma(x)\sigma^2(\lambda) + \ldots + x\sigma(x) \ldots \sigma^{n-2}(x)\sigma^{n-1}(\lambda)$$

is not equal to 0. Using the assumption that $N_{E/F}(x) = 1$, we derive $y = x\sigma(y)$. Hence $x = y\sigma(y)^{-1}$ as required. ■

14.2. THEOREM. (Hilbert's Theorem 90, Additive Form). Let E/F be a cyclic extension of degree n with Galois group G generated by σ. Then, for any $x \in E$,

$$Tr_{E/F}(x) = 0 \quad \text{if and only if} \quad x = y - \sigma(y)$$

for some $y \in E$.

Proof. For any $y \in E$, we have

$$Tr_{E/F}(y) = \sum_{i=0}^{n-1} \sigma^i(y) = \sum_{i=0}^{n-1} \sigma^i(\sigma(y)) = Tr_{E/F}(\sigma(y))$$

Hence, if $x = y - \sigma(y)$ for some $y \in E$, then

$$Tr_{E/F}(x) = Tr_{E/F}(y) - Tr_{E/F}(\sigma(y)) = 0$$

Conversely, assume that $Tr_{E/F}(x) = 0$. Since E/F is Galois, it is separable, hence $Tr_{E/F}(\lambda) \neq 0$ for some $\lambda \in E$ (Proposition 6.8). Setting

$$y = Tr_{E/F}(\lambda)^{-1}(x\sigma(\lambda) + (x+\sigma(x))\sigma^2(\lambda) + \ldots + (x+\sigma(x) + \ldots + \sigma^{n-2}(x))\sigma^{n-1}(\lambda))$$

we immediately derive $x = y - \sigma(y)$. ■

As a preliminary to the next result, we prove

14.3. LEMMA. Let F be a field and let $\sigma \in \mathrm{Aut}(F)$ be of order d.

(i) If $x \in F$ is such that $\sigma(x) = x$ and $x^d = 1$, then $\sigma(y) = xy$ for some nonzero y in F.

(ii) If $\mathrm{char}F = d$, then $\sigma(x) = x + 1$ for some $x \in F$

Proof. (i) By Proposition 4.7, we may choose $\lambda \in F$ such that

$$y = \sum_{i=1}^{d} x^{-i}\sigma^i(\lambda) \neq 0.$$

Then

$$\sigma(y) = \sum_{i=1}^{d} x^{-i}\sigma^{i+1}(\lambda) = xy$$

as required.

(ii) Now choose $\lambda \in F$ such that $y = \sum_{i=1}^{d} \sigma^i(\lambda) \neq 0$. Then $\sigma(y) = y$ and

$1 = \sum_{i=1}^{d} \sigma^i(\lambda y^{-1})$. Setting $x = -\sum_{i=1}^{d} i\sigma^i(\lambda y^{-1})$, we obtain

$$\sigma(x) = -\sum_{i=1}^{d} i\sigma^{i+1}(\lambda y^{-1}) = \sum_{i=1}^{d} \sigma^{i+1}(\lambda y^{-1}) - \sum_{i=1}^{d} (i+1)\sigma^{i+1}(\lambda y^{-1})$$
$$= 1 + x$$

as we wished to show. ■

We are now ready to prove

14.4. THEOREM. Let F be a field containing a primitive n-th root of unity and let E/F be a field extension.

(i) E/F is cyclic of degree dividing n if and only if $E = F(\lambda)$ for some $\lambda \in E$ such that $\lambda^n \in F$

(ii) The minimal polynomial over F of any $\lambda \in E$ with $\lambda^n \in F$ is $X^d - \lambda^d$ where d divides n.

Proof. (i) Suppose that E/F is cyclic of degree d dividing n. Then $\mathrm{Gal}(E/F)$ is generated by an element, say σ, of order d. Let δ be a primitive d-th root of unity in F. Then, by Lemma 14.3(i), $\sigma(\lambda) = \delta\lambda$ for some nonzero λ in E. Then the minimal polynomial of λ over F has d distinct roots, namely $\lambda, \sigma(\lambda) = \delta\lambda, \ldots, \sigma^{d-1}(\lambda) = \delta^{d-1}\lambda$ and, thus, $E = F(\lambda)$. Furthermore,

$$\sigma(\lambda^d) = \sigma(\lambda)^d = (\delta\lambda)^d = \lambda^d$$

so that $\lambda^d \in F$ and $\lambda^n \in F$.

Conversely, assume that $E = F(\lambda)$ for some $\lambda \in E$ such that $\lambda^n \in F$. Let ε be a primitive n-th root of unity. Then $X^n - \lambda^n = \prod_{i=1}^{n}(X-\lambda\varepsilon^i)$ and so E is the splitting field of $X^n - \lambda^n \in F[X]$. Because $X^n - \lambda^n$ has n distinct roots, we conclude that E/F is Galois. We have an injective homomorphism

$$\mathrm{Gal}(E/F) \longrightarrow \langle\varepsilon\rangle$$

sending each $\sigma \in \mathrm{Gal}(E/F)$ to $a_\sigma \in \langle\varepsilon\rangle$ where $\sigma(\lambda) = \lambda a_\sigma$. Thus $\mathrm{Gal}(E/F)$ is cyclic of order dividing n.

(ii) Let $\lambda \in E$ be such that $\lambda^n \in F$ and put $K = F(\lambda)$. Then, by (i), K/F is cyclic of degree d dividing n and $\lambda^d \in F$. Hence $X^d - \lambda^d$ is the minimal polynomial of λ over F. ■

14.5. COROLLARY. Let E/F be a field extension of degree n and let F contain a primitive n-th root of unity. Then E/F is cyclic if and only if $E = F(\lambda)$ for some $\lambda \in E$ such that $X^n - \lambda^n$ is the minimal polynomial of λ over F.

Proof. Direct consequence of Theorem 14.4. ■

14.6. COROLLARY. Let n be a positive integer and let F be a field containing all n-th roots of unity of an algebraic closure of $\bar{F}$ of F. If $\lambda \in \bar{F}$ is separable over F and $\lambda^n \in F$, then $F(\lambda)/F$ is cyclic of degree dividing n.

Proof. If $\text{char}F \nmid n$, then F contains a primitive n-th root of unity and the result follows by virtue of Theorem 14.4(i). We may therefore assume that $\text{char}F = p > 0$ and $n = mp^t$, where $(m,p) = 1$. Setting $\mu = \lambda^{p^t}$, we have $\mu^m \in F$. Since F contains a primitive m-th root of unity, $F(\mu)/F$ is cyclic of degree dividing m (hence n), by Theorem 14.4(i). But $F(\lambda) = F(\lambda^{p^t})$, by Proposition 3.8, hence the result. ■

We now turn to the case of cyclic extensions E/F of degree p^n where $p = \text{char}F$. Our point of departure is the case where $n = 1$.

14.7. THEOREM. (Artin-Schreir). Let F be a field of characteristic $p > 0$ and let E/F be a field extension.

(i) E/F is cyclic of degree p if and only if $E = F(\lambda)$ for some $\lambda \in E$ such that $\lambda^p - \lambda \in F$ and $\lambda \notin F$.

(ii) The minimal polynomial over F of any $\lambda \in E$ with $\lambda^p - \lambda \in F$ and $\lambda \notin F$ is

$$X^p - X - (\lambda^p - \lambda)$$

(iii) For any given $a \in F$, the polynomial $X^p - X - a$ is either irreducible over F or splits into p distinct linear factors over F.

Proof. We first note that if $a \in F$ and α is a root of $X^p - X - a$, then $\alpha + i$ is also a root for $i = 0,1,\ldots,p-1$. This is so since

$$(\alpha+i)^p = \alpha^p + i^p = \alpha^p + i.$$

Thus $X^p - X - \alpha$ splits into p distinct linear factors over $F(\alpha)$ and therefore $F(\alpha)/F$ is Galois.

Now assume that $E = F(\lambda)$ for some $\lambda \in E$ such that $\lambda^p - \lambda \in F$ and $\lambda \notin F$. Since λ is a root of $X^p - X - (\lambda^p-\lambda)$, E/F is Galois. Given $\sigma \in \mathrm{Gal}(E/F)$, there exists a unique $i_\sigma \in \{0,1,\ldots,p-1\}$, such that $\sigma(\lambda^p-\lambda) = \lambda^p - \lambda + i_\sigma$. Consider the mapping

$$\begin{cases} \mathrm{Gal}(E/F) \longrightarrow \mathbb{Z}/p\mathbb{Z} \\ \qquad \sigma \longmapsto i_\sigma + p\mathbb{Z} \end{cases}$$

This is obviously an injective homomorphism. Since $\mathbb{Z}/p\mathbb{Z}$ is cyclic of order p and $E \neq F$, we conclude that $\mathrm{Gal}(E/F)$ is cyclic of order p. Thus E/F is cyclic of degree p and $X^p - X - (\lambda^p-\lambda)$ is the minimal polynomial of λ over F. This proves (ii) and the "if" part of (i).

Assume that E/F is cyclic of degree p. Let σ be a generator of $\mathrm{Gal}(E/F)$, so that σ has order p. Owing to Lemma 14.3(ii), $\sigma(\lambda) = \lambda + 1$ for some $\lambda \in F$. Then $\sigma(\lambda^p) = \sigma(\lambda)^p = \lambda^p + 1$ and $\sigma(\lambda^p-\lambda) = (\lambda^p+1) - (\lambda+1) = \lambda^p-\lambda$, so that $\lambda^p-\lambda \in F$. But $\lambda \notin F$ since $\sigma(\lambda) = \lambda + 1 \neq \lambda$. Hence $E = F(\lambda)$ and (i) is established.

To prove (iii), we need only verify that if no root of $X^p - X - \alpha$ lies in F, then $f(X) = X^p - X - \alpha$ is irreducible over F. Assume by way of contradiction that

$$f(X) = g(X)h(X)$$

with $g,h \in F[X]$ and $1 \leq \deg g < p$. Since

$$f(X) = \prod_{i=0}^{p-1}(X-\alpha-i)$$

(α is a root of $f(X)$), we see that $g(X)$ is a product of $(X-\alpha-i)$ over certain integers $i \in \{0,1,\ldots,p-1\}$. If $d = \deg g$, then the coefficient of X^{d-1} is a sum of terms $-(\alpha+i)$ taken over precisely d integers i. Hence it is equal to $-d\alpha+j$ for some integer j. But $d \neq 0$ in F, hence α lies in F, because the coefficients of $g(X)$ lie in F, a contradiction. ■

Our next aim is to characterize the cyclic extensions E/F of degree p^n

where $p = \text{char}F$. This will be achieved by analyzing conditions for the existence of an injective homomorphism

$$\text{Gal}(E/F) \longrightarrow \mathbb{Z}/p^n\mathbb{Z}$$

Precisely this was done in Theorem 14.7 in the case $n = 1$. Since $\mathbb{Z}/p^n\mathbb{Z}$ is isomorphic to the prime subring of any ring R of characteristic p^n, all homomorphisms from $\text{Gal}(E/F)$ to $\mathbb{Z}/p^n\mathbb{Z}$ are identifiable with the homomorphisms $\text{Gal}(E/F) \longrightarrow P_n$, where P_n is the additive group of the prime subring of a suitable ring R of characteristic p^n. Our ring R will be the n-th Witt ring $W_n(E)$ of E defined in the previous section. Since our n will be fixed, we shall use a simplified notation pertaining to $W_n(E)$. Namely, we shall write π (instead of π_n) for the homomorphism

$$W_n(E) \longrightarrow W_n(E),\ (x_0,x_1,\ldots,x_{n-1}) \longmapsto (x_0^p,x_1^p,\ldots,x_{n-1}^p)$$

and s (instead of s_n) for the shift homomorphism

$$W_n(E) \longrightarrow W_n(E),\ (x_0,x_1,\ldots,x_{n-1}) \longmapsto (0,x_0,\ldots,x_{n-2})$$

For any homomorphism $\sigma : E \longrightarrow E$, we let

$$\sigma(x) = (\sigma(x_0),\sigma(x_1),\ldots,\sigma(x_{n-1}))$$

for $x = (x_0,x_1,\ldots,x_{n-1}) \in W_n(E)$. Since addition and multiplication are defined by polynomials with coefficients in $\mathbb{F}_p$ and since $\sigma(a) = a$ for all $a \in \mathbb{F}_p$, σ is a homomorphism from $W_n(E)$ to $W_n(E)$. Finally, if $\sigma \in \text{Gal}(E/F)$, the induced σ on $W_n(E)$ is an automorphism leaving fixed the elements of the subring $W_n(F)$.

14.8. LEMMA. Let E be a field of characteristic p and let $\sigma \in \text{Aut}(E)$ have finite order d. Then

$$\sum_{i=1}^{d} \sigma^i(x) = -1 \qquad \text{for some } x \in W_n(E)$$

Proof. Let $\lambda_0 \in E$ be such that $y_0 = \sum_{i=1}^{d} \sigma^i(\lambda_0) \neq 0$ and let

$$\lambda = (\lambda_0,0,\ldots,0) \in W_n(E).$$

Setting $y = \sum_{i=1}^{n} \sigma^i(\lambda)$, we see that the zeroth coordinate of y is y_0. Hence, by Corollary 13.10, y is a unit of $W_n(E)$. Because $\sigma(y) = y$, we therefore have

$$1 = \sum_{i=1}^{d} \sigma^i(\lambda y^{-1})$$

and we may take $x = -\lambda y^{-1}$. ∎

Given $m \leqslant n$, note that the additive order of $s^{n-m}(1) = (0,\ldots,0,1,0,\ldots,0)$ is p^m and that

$$p^m s^{n-m}(x) = 0 \qquad \text{for all } x \in W_n(E)$$

14.9. LEMMA. Let E be a field of characteristic p and let $\sigma \in \mathrm{Aut}(E)$ have finite order p^m dividing p^n. Then

$$\sigma(x) = x + s^{n-m}(1) \qquad \text{for some } x \in W_n(E)$$

Proof. Owing to Lemma 14.8, we may choose $\lambda \in W_n(E)$ such that

$$\sum_{i=1}^{p^m} \sigma^i(\lambda) = -1$$

Applying s^{n-m}, we have

$$s^{n-m}(-1) = s^{n-m}\Big(\sum_{i=1}^{p^m} \sigma^i(\lambda)\Big) = \sum_{i=1}^{p^m} s^{n-m}\sigma^i(\lambda)$$
$$= \sum_{i=1}^{p^m} \sigma^i(s^{n-m}(\lambda))$$

Letting $x = \sum_{i=1}^{p^m} i\sigma^i(s^{n-m}(\lambda))$, we then have

$$\sigma(x) = \sum_{i=1}^{p^m} i\sigma^{i+1}(s^{n-m}(\lambda)) = \sum_{i=1}^{p^m} (i+1)\sigma^{i+1}(s^{n-m}(\lambda)) - \sum_{i=1}^{p^m} \sigma^{i+1}(s^{n-m}(\lambda))$$
$$= \sum_{j=1}^{p^m} j\sigma^j(s^{n-m}(\lambda)) - \sum_{j=1}^{p^m} \sigma^j(s^{n-m}(\lambda))$$
$$= x + s^{n-m}(1)$$

since $p^m s^{n-m}(\lambda) = 0$. ∎

We are now ready to describe cyclic extensions of degree dividing p^n.

14.10. THEOREM. Let F be a field of characteristic $p > 0$, let $n \geqslant 1$ be an integer and let E/F be an algebraic field extension. Then E/F is cyclic of degree p^m dividing p^n if and only if there exist $\lambda_0,\lambda_1,\ldots,\lambda_{n-1}$ in E such that $E = F(\lambda_0,\lambda_1,\dots,\lambda_{n-1})$ and the element $\lambda = (\lambda_0,\lambda_1,\dots,\lambda_{n-1})$ of $W_n(E)$ satisfies

$$\pi(\lambda) - \lambda \in W_n(F)$$

Furthermore, for each $a \in W_n(F)$, there exists a cyclic extension K/F of degree dividing p^n and an element $\lambda = (\lambda_0,\lambda_1,\dots,\lambda_{n-1})$ in $W_n(K)$ such that

$$\pi(\lambda) - \lambda = a$$

Proof. For the sake of clarity, we divide the proof into three steps.

Step 1. Here we prove the first assertion under the assumption that E/F is separable. Assume that $E = F(\lambda_0,\lambda_1,\dots,\lambda_{n-1})$, where $\lambda = (\lambda_0,\lambda_1,\dots,\lambda_{n-1}) \in W_n(E)$ is such that $\pi(\lambda) - \lambda \in W_n(F)$. Denote by $\bar{E}$ an algebraic closure of E. Note that, for $x \in W_n(E)$, $(\pi(x) - x) - (\pi(\lambda) - \lambda) = 0$ if and only if $\pi(x-\lambda) = x - \lambda$ that is, if and only if $x - \lambda \in W_n(\mathbb{F}_p)$. Hence $\{\lambda + i \mid i \in \mathbb{Z}/p^n\mathbb{Z}\}$ is the set of all roots of $\pi(X) - X - (\pi(\lambda) - \lambda)$ in $W_n(\bar{E})$, since $W_n(\mathbb{F}_p) = \mathbb{Z}/p^n\mathbb{Z}$ by Theorem 13.6(iii). Because $\pi(\lambda) - \lambda \in W_n(F)$, these roots are mapped into themselves under $\mathrm{Gal}(\bar{E}/F)$. Hence E is mapped into itself under $\mathrm{Gal}(\bar{E}/F)$, since E is generated over F by the coordinates of $\lambda + i, i \in \mathbb{Z}/p^n\mathbb{Z}$. Since E/F is separable, we conclude that E/F is Galois. We have an injective homomorphism

$$\begin{cases} \mathrm{Gal}(E/F) \longrightarrow \mathbb{Z}/p^n\mathbb{Z} \\ \qquad \sigma \longmapsto i_\sigma \end{cases}$$

where $\sigma(\lambda) = \lambda + i_\sigma$. Since $\mathbb{Z}/p^n\mathbb{Z}$ is cyclic of order p^n, $\mathrm{Gal}(E/F)$ is therefore cyclic of order p^m dividing p^n.

Conversely, assume that E/F is cyclic of degree p^m dividing p^n. Let σ be a generator for $\mathrm{Gal}(E/F)$, so that σ has order p^m. Owing to Lemma 14.9, we have

$$\sigma(\lambda) = \lambda + s^{n-m}(1) \qquad \text{for some } \lambda \in W_n(E)$$

Then

$$\sigma(\pi(\lambda)) = \pi\sigma(\lambda) = \pi(\lambda+s^{n-m}(1)) = \pi(\lambda) + s^{n-m}(1)$$

and therefore

$$\sigma(\pi(\lambda)-\lambda) = (\pi(\lambda) + s^{n-m}(1)) - (\lambda + s^{n-m}(1)) = \pi(\lambda)-\lambda$$

whence $\pi(\lambda) - \lambda \in W_n(F)$. Thus $F(\lambda_0,\lambda_1,\ldots,\lambda_{n-1})/F$ is cyclic of degree p^k dividing p^n, by the first part of the proof. Since $s^{n-m}(1)$ has additive order p^m, the restriction of σ to $F(\lambda_0,\lambda_1,\ldots,\lambda_{n-1})$ has order p^m. Thus $p^k = p^m$ and $E = F(\lambda_0,\lambda_1,\ldots,\lambda_{n-1})$.

Step 2. Here we show that, for each $a \in W_n(F)$, *there exists a cyclic extension* K/F *of degree dividing* p^n *and an element* $\lambda = (\lambda_0,\lambda_1,\ldots,\lambda_{n-1})$ *in* $W_n(K)$ *such that* $\pi(\lambda) - \lambda = a$. We use the notation $(\pi-1)\lambda = \pi(\lambda) - \lambda$ and denote by S the separable closure of F. We prove by induction on n that there exists

$$\lambda = (\lambda_0,\lambda_1,\ldots,\lambda_{n-1})$$

such that $(\pi-1)\lambda = a$ and $\lambda_i \in S$ for $0 \leqslant i \leqslant n-1$. The required assertion will then follow by Step 1 by taking $K = F(\lambda_0,\lambda_1,\ldots,\lambda_{n-1})$.

If $n = 1$, we take $\lambda_0 \in S$ to be a root of $X^p - X - a_0$ and $K = F(\lambda_0)$ (Theorem 14.7(i)). Next, let $n > 1$ and again choose $\lambda_0 \in S$ such that

$$\lambda_0^p - \lambda_0 = a_0.$$

Then

$$(a_0,a_1,\ldots,a_{n-1}) - (\pi-1)(\lambda_0,0,\ldots,0) = (0,b_1,\ldots,b_{n-1})$$

for suitable $b_k \in S$. If $K' = F(b_1,\ldots,b_{n-1})$, then S is the separable closure of K' and we can find, by induction $\mu_1,\ldots,\mu_{n-1} \in S$ such that

$$(b_1,\ldots,b_{n-1}) = (\pi-1)(\mu_1,\ldots,\mu_{n-1}) = (\mu_1^p,\ldots,\mu_{n-1}^p) - (\mu_1,\ldots,\mu_{n-1})$$

But then

$$(0,b_1,\ldots,b_{n-1}) = (0,\mu_1^p,\ldots,\mu_{n-1}^p) - (0,\mu_1,\ldots,\mu_{n-1}) \qquad \text{(by Theorem 13.4(i))}$$
$$= (\pi-1)(0,\mu_1,\ldots,\mu_{n-1})$$

We now have

$$(a_0,a_1,\ldots,a_{n-1}) = (\pi-1)((\lambda_0,0,\ldots,0) + (0,\mu_1,\ldots,\mu_{n-1}))$$
$$= (\pi-1)(\lambda_0,\lambda_1,\ldots,\lambda_{n-1})$$

for suitable $\lambda_i \in S$, $1 \leqslant i \leqslant n-1$. Thus $(\pi-1)\lambda = a$ for $\lambda = (\lambda_0,\lambda_1,\ldots,\lambda_{n-1})$

as required.

Step 3. Completion of the proof. In view of Steps 1 and 2, we are left to verify that if $E = F(\lambda_0,\lambda_1,\dots,\lambda_{n-1})$, where $\lambda = (\lambda_0,\lambda_1,\dots,\lambda_{n-1}) \in W_n(E)$ is such that $\pi(\lambda) - \lambda \in W_n(F)$, then E/F is separable. To this end, we apply Step 2, to find $\mu = (\mu_0,\mu_1,\dots,\mu_{n-1})$ such that $\pi(\mu) - \mu = \pi(\lambda) - \lambda$ and such that each μ_i, $0 \leq i \leq n-1$, is separable over F. But then $\pi(\mu-\lambda) = \mu - \lambda$ and $\mu - \lambda \in W_n(\mathbb{F}_p)$. Thus $\lambda = \mu - (\mu-\lambda) \in W_n(S)$, where S is the separable closure of F, and therefore E/F is separable. ∎

15. KUMMER THEORY

Let F be a field and n a positive integer. A Galois extension E/F is said to be of *exponent* n if the exponent of $\mathrm{Gal}(E/F)$ divides n, i.e. if $\sigma^n = 1$ for all $\sigma \in \mathrm{Gal}(E/F)$. By a *Kummer extension* of exponent n one understands an abelian extension E/F of finite exponent n, where F contains a primitive n-th root of unity. It follows that for any such extension the characteristic of F does not divide n. Throughout the discussion, the ground field F will be fixed, and all extensions lie in a given algebraic closure $\bar{F}$ of F. Moreover, we assume that F contains a primitive n-th root of unity, which will be denoted by ε_n. Our aim is to survey all Kummer extensions E/F. If $G = \mathrm{Gal}(E/F)$, then G endowed with the Krull topology is a compact group. We write

$$\mathrm{Hom}_c(G,\langle\varepsilon_n\rangle) = \{f : G \longrightarrow \langle\varepsilon_n\rangle \mid f \text{ is a continuous homomorphism}\}$$

Then $\mathrm{Hom}_c(G,\langle\varepsilon_n\rangle)$ is a group under multiplication of values:

$$(\alpha\beta)(g) = \alpha(g)\beta(g) \qquad (\alpha,\beta \in \mathrm{Hom}_c(G,\langle\varepsilon_n\rangle), g \in G)$$

We denote by $\mathrm{Hom}(G,\langle\varepsilon_n\rangle)$ the group of all (not necessarily continuous) homomorphisms from G to $\langle\varepsilon_n\rangle$. Then $\mathrm{Hom}_c(G,\langle\varepsilon_n\rangle)$ is a subgroup of

$$\mathrm{Hom}(G,\langle\varepsilon_n\rangle)$$

which can be characterized by

$$\mathrm{Hom}_c(G,\langle\varepsilon_n\rangle) = \{\chi \in \mathrm{Hom}(G,\langle\varepsilon_n\rangle) \mid \mathrm{Ker}\chi \text{ is an open subgroup of } G\} \qquad (1)$$

The latter is true, since a homomorphism from a topological group into a discrete group is continuous if and only if its kernel is an open subgroup. In case G is finite (hence discrete), we have $\mathrm{Hom}_c(G,<\varepsilon_n>) = \mathrm{Hom}(G,<\varepsilon_n>)$. Some further information is provided below.

Let G be a finite abelian group. Then the set $G^* = \mathrm{Hom}(G,\mathbb{C}^*)$ of all homomorphisms from G to $\mathbb{C}^*$ forms a group under multiplication of values. The elements of the group G^* are called *characters* of G. If the exponent of G divides n, then the image of any $\chi \in G^*$ is in the cyclic group of n-th roots of unity of $\mathbb{C}^*$. Thus

$$\mathrm{Hom}(G,\mathbb{C}^*) = \mathrm{Hom}(G,<\delta_n>)$$

where δ_n is a primitive n-th root of unity in $\mathbb{C}$. Given a subgroup H of G, we put

$$H^{\perp} = \{\chi \in G^* | \chi(h) = 1 \text{ for all } h \in H\}$$

We clearly have a natural isomorphism

$$H^{\perp} \cong (G/H)^*$$

15.1. LEMMA. If G is a finite abelian group, then

$$G \cong G^* \text{ (noncanonically) and } G^{**} \cong G \text{ (canonically)}$$

Proof. One immediately verifies that $(G_1 \times G_2)^* \cong G_1^* \times G_2^*$. Hence, to prove $G \cong G^*$, we may assume that G is cyclic, say of order n. Let g be a generator of G and let $\chi : G \longrightarrow \mathbb{C}^*$ be a homomorphism. Then $\chi(g)^n = 1$, i.e. $\chi(g)$ is an n-th root of unity. Let δ_n be a primitive n-th root of unity and let

$$\psi : G^* \longrightarrow <\delta_n>$$

be defined by $\psi(\chi) = \chi(g)$. It is obvious that ψ is an injective homomorphism. Since for any $k \geqslant 1$, $\chi(g) = \delta_n^k$ is an element of G^*, ψ is also surjective. Thus $G \cong G^*$.

For each $g \in G$, $\chi \in G^*$, define $\psi_g(\chi) = \chi(g)$. Then $\psi_g \in G^{**}$ and the map $g \longmapsto \psi_g$ is a homomorphism from G into G^{**}. If $\chi(g) = 1$ for all $\chi \in G^*$ and $H = <g>$, then

$$H^{\perp} = G^{\star} \cong (G/H)^{\star}$$

Hence, by the foregoing, $G \cong G/H$ which implies $H = 1$. Thus $g = 1$ and the map $g \longmapsto \psi_g$ is injective. Since $|G| = |G^{\star}| = |G^{\star\star}|$, the result follows. ■

We now return to our discussion of Kummer extensions E/F with respect to a fixed positive integer n. Throughout ε_n denotes a primitive n-th root of unity in F.

We shall use the symbol $\sqrt[n]{a}$, $a \in F$, to denote any element $\lambda \in \bar{F}$ such that $\lambda^n = a$. There are precisely n such elements, namely $\varepsilon_n^i\lambda$, $0 \leqslant i \leqslant n-1$. We observe that the field $F(\lambda)$ is the same no matter which n-th root λ of a we select. We denote this field by $F(\sqrt[n]{a})$.

We denote by $(F^{\star})^n$ the subgroup of $F^{\star}$ consisting of all n-th powers of elements of $F^{\star}$. If H is a subgroup of $F^{\star}$ containing $(F^{\star})^n$, we write F_H for the composite of all fields $F(\sqrt[n]{a})$ with $a \in H$. It is uniquely determined by H as a subfield of $\bar{F}$. Note also that F_H is the splitting field of the family of polynomials

$$X^n - a \qquad (a \in H)$$

15.2. LEMMA. (i) For any subgroup H of $F^{\star}$ containing $(F^{\star})^n$, F_H/F is a Kummer extension of exponent n.

(ii) If E/F is a Kummer extension of exponent n, then

$$E = F_H \quad \text{with} \quad H = (E^{\star})^n \cap F^{\star}$$

(iii) E/F is a Kummer extension of exponent m if and only if F contains a primitive m-th root of unity and E is the splitting field of a family of polynomials of the form $X^m - a$, $a \in F$.

Proof. (i) Owing to Theorem 14.4(i), $F(\sqrt[n]{a})/F$ is cyclic of degree dividing n. Hence, for any $\sigma \in \mathrm{Gal}(F_H/F)$, the restriction of σ^n to $F(\sqrt[n]{a})$ is 1. Therefore $\sigma^n = 1$ since F_H is generated by all $\sqrt[n]{a}$, $a \in H$. Furthermore, F_H/F is abelian since it is the composite of cyclic extensions.

(ii) Clearly, $F_H \subseteq E$. Note, further, that E/F is the composite of its finite subextensions E'/F. Because the finite abelian group $\mathrm{Gal}(E'/F)$ is the direct product of cyclic groups, each of which can be understood as the Galois group of a cyclic subextension of E'/F, we conclude that E'/F (and hence E/F) is the composite of its cyclic subextensions.

Assume that K/F is a cyclic subextension of E/F. Then $\mathrm{Gal}(K/F)$ is of order dividing n, so by Theorem 14.4(i), $K = F(\sqrt[n]{a})$ for some $a \in (E^*)^n \cap F^* = H$. Therefore $K \subseteq F_H$ and thus $E \subseteq F_H$, as required.

(iii) Assume that F contains a primitive m-th root of unity. If E is a splitting field of a family of polynomials of the form $X^m - a$, $a \in F$, then E is the composite of fields $F(\sqrt[m]{a})$. Hence, by the argument of (i), E/F is a Kummer extension of exponent m. The converse is true by virtue of (ii). ■

15.3. LEMMA. Let E/F be a Kummer extension of exponent n and let

$$H = (E^*)^n \cap F^*$$

Then, for any $a \in H$, the map

$$\begin{cases} \mathrm{Gal}(E/F) \xrightarrow{\chi_a} \langle \varepsilon_n \rangle \\ \sigma \longmapsto \dfrac{\sigma(\sqrt[n]{a})}{\sqrt[n]{a}} \end{cases}$$

is continuous homomorphism, which is independent of the choice of $\sqrt[n]{a}$.

Proof. Let λ be a fixed n-th root of a in $\overline{F}$ so that $\lambda\varepsilon_n^i$, $0 \leqslant i \leqslant n-1$, are all distinct such roots. Then, for $\sqrt[n]{a} = \lambda\varepsilon_n^i$, we have

$$\frac{\sigma(\sqrt[n]{a})}{\sqrt[n]{a}} = \frac{\sigma(\lambda)\varepsilon_n^i}{\lambda\varepsilon_n^i} = \frac{\sigma(\lambda)}{\lambda}$$

proving that χ_a is independent of the choice of $\sqrt[n]{a}$. Furthermore, given $\sigma_1, \sigma_2 \in \mathrm{Gal}(E/F)$, we have $\sigma_i(\lambda)/\lambda = \varepsilon_n^{k_i}$ for some $k_i \geqslant 1$, so that

$$\chi_a(\sigma_1\sigma_2) = \frac{(\sigma_1\sigma_2)(\lambda)}{\lambda} = \frac{\sigma_1(\lambda\varepsilon_n^{k_2})}{\lambda} = \varepsilon_n^{k_1+k_2} = \chi_a(\sigma_1)\chi_a(\sigma_2)$$

Finally, $\mathrm{Ker}\chi_a = \mathrm{Gal}(E/F(\sqrt[n]{a}))$ and $F(\sqrt[n]{a})/F$ is finite. Hence, by Theorem 12.2 (ii), $\mathrm{Ker}\chi_a$ is an open subgroup of $\mathrm{Gal}(E/F)$. Invoking (1), we deduce that χ_a is continuous, as required. ■

We shall refer to χ_a in Lemma 15.3 as the *Kummer character* of $\mathrm{Gal}(E/F)$ corresponding to a. What one commonly understands under the name Kummer theory is the content of the following

15.4. THEOREM. Let F be a field containing a primitive n-th root of unity ε_n.

(i) The map $H \longrightarrow F_H$ gives a bijection of the set of subgroups of F^* containing $(F^*)^n$ and Kummer extensions of F of exponent n. The inverse of this map is given by $E \longmapsto (E^*)^n \cap F^*$.

(ii) If E/F is a Kummer extension of exponent n and $H = (E^*)^n \cap F^*$, then the map

$$\begin{cases} H/(F^*)^n \longrightarrow \mathrm{Hom}_c(\mathrm{Gal}(E/F), <\varepsilon_n>) \\ a(F^*)^n \longmapsto \chi_a \end{cases}$$

is an isomorphism. In particular, E/F is finite if and only if $H/(F^*)^n$ is finite and if that is the case, then

$$\mathrm{Gal}(E/F) \cong H/(F^*)^n \quad \text{and} \quad (E:F) = |H/(F^*)^n|$$

Proof. Put $G = \mathrm{Gal}(E/F)$. If $a,b \in H$ and $\alpha^n = a, \beta^n = b$, then

$$(\alpha\beta)^n = ab$$

and hence

$$\sigma(\alpha\beta)/\alpha\beta = (\sigma(\alpha)/\alpha)(\sigma(\beta)/\beta) \qquad \text{for all } \sigma \in G$$

If $a = \lambda^n$ with $\lambda \in F^*$, then $\chi_a(\sigma) = \sigma(\lambda)/\lambda = 1$ for all $\sigma \in G$. Therefore the map in (ii) is a homomorphism. Assume that $a \in H$ is such that $\chi_a = 1$, i.e. $\sigma(\alpha) = \alpha$ for all $\sigma \in G$, where $\alpha^n = a$. By Lemma 15.2(ii), $E = F_H$ and so $F(\alpha)$ is a subfield of E. If α is not in F, there exists an automorphism of $F(\alpha)$ over F which is not identity. Extend this automorphism to E and call this extension σ_0. Then $\sigma_0(\alpha) \neq \alpha$, a contradiction. Thus we have an exact sequence

$$1 \longrightarrow H/(F^*)^n \longrightarrow \mathrm{Hom}_c(G, <\varepsilon_n>) \tag{2}$$

Note that the map $G \longrightarrow \mathrm{Hom}(H/(F^*)^n, <\varepsilon_n>)$, $\sigma \longmapsto f_\sigma$, where $f_\sigma(a(F^*)^n) = \chi_a(\sigma)$, is a homomorphism. If $\chi_a(\sigma) = 1$ for all $a \in H$, then for every generator α of E with $\alpha^n = a \in H$, we have $\sigma(\alpha) = \alpha$ and hence $\sigma = 1$. Thus we have an exact sequence

$$1 \longrightarrow G \longrightarrow \mathrm{Hom}(H/(F^*)^n, <\varepsilon_n>) \tag{3}$$

Applying the first isomorphism of Lemma 15.1, it follows from (2) and (3) that E/F is finite if and only if $H/(F^*)^n$ is finite and if that is the case then $G \cong H/(F^*)^n$.

To complete the proof of (ii), we need only verify that the given map is surjective. By the foregoing, this is true if E/F is a finite extension. If E/F is of infinite degree, we let $H_i/(F^*)^n$ run through the finite subgroups of $H/(F^*)^n$ and we let $E_i = F_{H_i}$. Then $H/(F^*)^n = \bigcup_i H_i/(F^*)^n$ and $E = \bigcup_i E_i$. Therefore the groups $\mathrm{Gal}(E/E_i)$ form a basis of open neighbourhoods of $1 \in G$. Because the kernel of a continuous homomorphism $\chi : G \longrightarrow <\varepsilon_n>$ is open, it must contain a subgroup $\mathrm{Gal}(E/E_i)$. Now χ yields a homomorphism

$$\bar{\chi} : \mathrm{Gal}(E_i/F) \longrightarrow <\varepsilon_n>$$

such that $\chi(\sigma) = \bar{\chi}(\sigma|E_i)$. By the surjectivity of the finite case, $\bar{\chi}$ is of the form $\bar{\chi}_a : \mathrm{Gal}(E_i/F) \to <\varepsilon_n>$, $a \in H_i$. But then

$$\chi(\sigma) = \bar{\chi}_a(\sigma|E_i) = \sigma(\sqrt[n]{a})/\sqrt[n]{a} = \chi_a(\sigma)$$

whence $\chi = \chi_a$, proving surjectivity.

To prove (i), it suffices, by Lemma 15.2(i),(ii), to show that $F_{H_1} \subseteq F_{H_2}$ implies $H_1 \subseteq H_2$, for any subgroups H_1, H_2 of F^* containing $(F^*)^n$. If $b \in H_1$, then $F(\sqrt[n]{b}) \subseteq F_{H_2}$ and $F(\sqrt[n]{b})$ is contained in a finitely generated subextension of F_{H_2}. Thus we may assume that $H_2/(F^*)^n$ is finitely generated, hence finite. Let H_3 be the subgroup of F^* generated by H_2 and b. Then $F_{H_2} = F_{H_3}$ and from what we saw above, the degree of this field over F is precisely

$$|H_2/(F^*)^n| \quad \text{or} \quad |H_3/(F^*)^n|$$

Hence $H_2 = H_3$, so $b \in H_2$ and therefore $H_1 \subseteq H_2$, as asserted. ∎

We shall now derive some consequences of interest.

15.5. COROLLARY. Let E/F be a finite Kummer extension of exponent n, let

$G = \mathrm{Gal}(E/F)$ and let $\sqrt[n]{F^*}$ denote the subgroup of E consisting of all n-th roots of elements in F^*. Then

$$G \cong \sqrt[n]{F^*}/F^* \cong (\sqrt[n]{F^*})^n/(F^*)$$

Proof. Put $H = (E^*)^n \cap F^*$. Then $H = (\sqrt[n]{F^*})^n$ and so the map

$$\sqrt[n]{F^*} \longrightarrow H/(F^*)^n, \; x \longmapsto x^n(F^*)^n$$

is a surjective group homomorphism; its kernel is F^* since F contains a primitive n-th root of unity. Thus, by Theorem 15.4(ii),

$$\sqrt[n]{F^*}/F^* \cong H/(F^*)^n \cong G$$

and the result follows. ■

15.6. COROLLARY. Let p be a prime and let F be a field containing a primitive p-th root of unity. Suppose α is an element of a field extension of F such that $\alpha^p \in F$. If $\beta \in F(\alpha)$ is such that $\beta^p \in F$, then $\beta = \alpha^k a$ for some $k \in \mathbb{Z}$ and some $a \in F$.

Proof. The assertion is trivial in case $\alpha \in F$. Assume that $\alpha \notin F$. Thus $\alpha^p \notin F^p$ since F contains all p-th roots of unity. This immediately implies (see Lemma 16.3 of the next section) that $(F(\alpha):F) = p$. Thus $F(\alpha)/F$ is a Kummer extension of exponent p and $F(\alpha) = F_H$ where $H = \langle\alpha^p\rangle(F^*)^p$. If $\beta \in F$, then $\beta = \alpha^p(\alpha^{-p}\beta)$ and $\alpha^{-p}\beta \in F$. We may therefore assume that $\beta \notin F$. Thus $F(\alpha) = F(\beta)$ and $F(\beta) = F_{H_1}$ where $H_1 = \langle\beta^p\rangle(F^*)^p$. By Theorem 15.4(i), $H = H_1$ and hence

$$\beta^p = \alpha^{pk} c^p \qquad \text{for some } k \in \mathbb{Z},\ c \in F^*$$

Taking p-th roots, we get $\beta = \alpha^k a$ for some $a \in F$. ■

It is useful sometimes to know that the condition on F containing a primitive p-th root of unity can be deleted. Thus we have more generally:

15.7. COROLLARY. Let p be a prime and let F be a field with $\mathrm{char}F \neq p$. Suppose α is an element of a field extension of F such that $\alpha^p \in F - F^p$. If $\beta \in F(\alpha)$ is such that $\beta^p \in F$, then $\beta = \alpha^k a$ for some $k \in \mathbb{Z}$ and some $a \in F$.

Proof. Since $\alpha^p \notin F^p$, we have $(F(\alpha):F) = p$. Let ε denote a primitive

p-th root of unity over F. The degree $(F(\varepsilon):F) \leqslant p-1$, hence the degrees $(F(\alpha):F)$ and $(F(\varepsilon):F)$ being relatively prime implies $F(\alpha) \cap F(\varepsilon) = F$. Consider $F(\varepsilon)(\alpha)$ over $F(\varepsilon)$. Then $\alpha^p \in F(\varepsilon)$, $\beta^p \in F(\varepsilon)$, and $F(\varepsilon)$ contains ε. Applying Corollary 15.6, we conclude that $\beta = \alpha^k a$ for some $k \in \mathbb{Z}$ and some $a \in F(\varepsilon)$. But

$$a = \beta\alpha^{-k} \in F(\varepsilon) \cap F(\alpha) = F,$$

hence the result. ∎

By the *maximal Kummer extension of* F *of exponent* n we understand the composite of all Kummer extensions of F of exponent n. It is clear that such an extension is a unique largest Kummer extension of F of exponent n.

15.8. COROLLARY. Let E be the maximal Kummer extension of F of exponent n. Then

$$\mathrm{Hom}_c(\mathrm{Gal}(E/F), \langle \varepsilon_n \rangle) \cong F^*/(F^*)^n$$

Proof. Owing to Theorem 15.4, it suffices to verify that $(E^*)^n \cap F^* = F^*$. So fix any $\lambda \in F^*$ and put $a = \sqrt[n]{\lambda}$. Then $a \in E$ since $F(a)/F$ is a Kummer extension of exponent n. Hence $\lambda = a^n \in (E^*)^n \cap F^*$ as required. ∎

16. RADICAL EXTENSIONS AND RELATED RESULTS

Let E/F be a field extension. We write $t(E^*/F^*)$ for the torsion subgroup of the factor group E^*/F^*. Following Kaplansky(1972), we say that E/F is a *radical extension* if E is of the form

$$E = F(\lambda_1, \ldots, \lambda_m) \tag{1}$$

with

$$\lambda_i^{n_i} \in F(\lambda_1, \ldots, \lambda_{i-1}) \qquad (1 \leqslant i \leqslant m) \tag{2}$$

for some $n_i \in \mathbb{N}$. It is clear that any radical extension is a finite extension. The case where $m = 1$, i.e. E is of the form $E = F(\lambda)$ with $\lambda^n \in F$ for some $n \geqslant 1$ is of particular importance. We shall refer to any such E/F as a *simple radical extension*. Thus, if E/F is a radical extension satisfying (1) and (2),

then

$$F \subseteq F(\lambda_1) \subseteq F(\lambda_1,\lambda_2) \subseteq \ldots \subseteq F(\lambda_1,\ldots,\lambda_m) = E$$

and each $F(\lambda_1,\ldots,\lambda_i)/F(\lambda_1,\ldots,\lambda_{i-1})$ is a simple radical extension.

By an *irreducible radical extension*, we understand a simple radical extension of the form $F(\lambda)/F$ such that $\lambda^n \in F$ where $n = (F(\lambda):F)$. Thus E/F is an irreducible radical extension if and only if E is of the form $E = F(\lambda)$, where λ is a root of an irreducible over F binomial of the form

$$X^n - a \qquad (a \in F)$$

Expressed otherwise, a finite extension E/F is an irreducible radical extension if and only if $E = F(\lambda)$ for some $\lambda \in E^*$ such that the order of λF^* in E^*/F^* is equal to the degree of λ over F.

16.1. LEMMA. Let E/F be a field extension of degree n and let F contain a primitive n-th root of unity. Then E/F is irreducible radical if and only if E/F is cyclic.

Proof. Direct consequence of Corollary 14.5. ■

In order to survey all irreducible radical extensions of a given field F, we must investigate conditions under which $X^n - a$ is irreducible over F, for any given $a \in F$. The following observation shows that we may harmlessly assume that n is a prime power.

16.2. LEMMA. Let F be any field, a an element in F, and m,n relatively prime positive integers. Then $X^{mn} - a$ is irreducible over F if and only if $X^m - a$ and $X^n - a$ are irreducible over F.

Proof. Since $X^{mn} - a = (X^n)^m - a = (X^m)^n - a$, if $X^{mn} - a$ is irreducible, then so are $X^m - a$ and $X^n - a$. Conversely, assume that both $X^m - a$ and $X^n - a$ are irreducible over F. Let E be the splitting field of $X^{mn} - a$ and let $\lambda \in E$ be a root of $X^{mn} - a$. Then λ^m is a root of $X^n - a$ and λ^n is a root of $X^m - a$, so

$$(F(\lambda^m):F) = n \quad \text{and} \quad (F(\lambda^n):F) = m$$

Hence both n,m divide $(F(\lambda):F)$ and therefore nm divides $(F(\lambda):F)$, since

$(n,m) = 1$. But λ is a root of $X^{mn} - a$, so $(F(\lambda):F) \leq mn$. This shows that $(F(\lambda):F) = mn$ and hence that $X^{mn} - a$ is irreducible over F. ■

We proceed to attack the prime power case. As a prelude, we investigate the prime case.

16.3. LEMMA. Let p be a prime and a an element of a field F. Then $X^p - a$ is irreducible over F if and only if $a \notin F^p$.

Proof. If $a = \mu^p$ for some $\mu \in F$, then μ is a root of $X^p - a$ and hence $X^p - a$ is not irreducible. Conversely, assume that $a \notin F^p$. We argue by contradiction and denote by f an irreducible factor of $X^p - a$ of degree k, $1 \leq k < p$. Let c be the constant term of f. The roots of $X^p - a$ (in some splitting field) all have the form εu, where u is one fixed root and ε is a p-th root of unity. Because $\pm c$ is a product of k of these roots, we have $\pm c = \delta u^k$ with $\delta^p = 1$. Since $(k,p) = 1$, there exist integers r and s such that $rk + sp = 1$. We have

$$u - u^{rk}u^{sp} = (\pm c/\delta)^r a^s$$

Hence $u\delta^r$ lies in F. But $a = (u\delta^r)^p$, a contradiction. ■

16.4. LEMMA. Let p be a prime, let a be an element of a field F and let $X^p - a$ be irreducible over F. If λ is a root of $X^p - a$, then

(i) If p is odd, or if $p = 2$ and char$F = 2$, then $\lambda \notin F(\lambda)^p$

(ii) If $p = 2$ and char$F \neq 2$, then $\lambda \in F(\lambda)^2$ if and only if $a \in -4F^4$.

Proof. (i) Assume by way of contradiction that $\lambda = \omega^p$ for some $\omega \in F(\lambda)$. The case where char$F = p$ is straightforward: since ω is a polynomial in λ and p-th powers are taken termwise we have $\omega^p \in F$, which is impossible, by Lemma 15.3. Now assume that char$F \neq p$ and adjoin to $F(\lambda)$ a primitive p-th root of unity, say ε. The resulting field E is a splitting field of $X^p - a$ over F and therefore E/F is a normal extension. Any automorphism of E/F sends λ into some $\varepsilon^i\lambda$, $0 \leq i \leq p-1$, and for every $i \in \{0,1,\dots,p-1\}$ there is an automorphism f_i sending λ into $\varepsilon^i\lambda$. Put $\omega_i = f_i(\omega)$. Then $\varepsilon^i\lambda = \omega_i^p$. The element ω is in $F(\lambda)$ but not in F, hence its minimal polynomial (say f) over F has degree p and has p distinct roots in E. If ω' is any of these roots, then there is an automorphism ψ of E/F sending ω into ω'. If $\psi(\lambda) = \varepsilon^j\lambda$ we must have $\psi(\omega) = \omega_j$. Hence the elements $\omega_0, \omega_1, \ldots, \omega_{p-1}$ are all roots of f which implies that $z = \omega_0\omega_1\ldots\omega_{p-1} \in F$. We now multiply together the equations $\varepsilon^i\lambda = \omega_i^p$, to obtain

$$\mu\lambda^p = \mu a = z^p$$

where $\mu = 1\cdot\varepsilon\cdot\varepsilon^2\ldots\varepsilon^{p-1}$. If p is odd, $\mu = 1$, and we have a contradiction $a = z^p$, proving (1).

(ii) Assume that $\lambda = \omega^2$ with $\omega = \alpha+\beta\lambda(\alpha, \beta \in F)$. From $\lambda = (\alpha+\beta\lambda)^2$ we obtain the equations $\alpha^2 + \beta^2\alpha = 0$, $2\alpha\beta = 1$. Eliminating β, we find $a = -4\alpha^4 \in -4F^4$. Conversely, if $a = -4\alpha^4$, $\alpha \in F$, we take $\beta = (1/2)\bar{\alpha}$ and verify $\lambda = (\alpha+\beta\lambda)^2$. ■

We are now ready to tackle the prime power case (the prime case will be excluded because of Lemma 16.3).

16.5. LEMMA. Let F be an arbitrary field, let $n \geqslant 2$ be a positive integer and let p be a prime. Denote by a an arbitrary element in F.

(i) If p is odd, or $p = 2$ and char$F = 2$, then $X^{p^n} - a$ is irreducible over F if and only if $a \notin F^p$.

(ii) If $p = 2$ and char$F \neq 2$, then $X^{2^n} - a$ is irreducible over F if and only if $a \notin F^2$ and $a \notin -4F^4$.

Proof. (i) If $a = \mu^p$ for some $\mu \in F$, then $X^{p^n} - a = X^{p^n} - \mu^p$ is dividible by $X^{p^{n-1}} - \mu$. Conversely assume that $a \notin F^p$. Let λ be a root of $X^{p^n} - a$ and let $\mu = \lambda^{p^{n-1}}$. Then μ is a root of $X^p - a$, hence by Lemma 16.3,

$$(F(\mu):F) = p$$

We claim that λ has degree p^{n-1} over $F(\mu)$; if sustained, it will follow that λ has degree p^n over F and $X^{p^n} - a$ is irreducible over F. That λ has degree p^{n-1} over $F(\mu)$ will be true by induction on n, provided $\mu \notin F(\mu)^p$. The desired conclusion is therefore a consequence of Lemma 16.4(i).

(ii) If $a \in F^2$, then obviously $X^{2^n} - a$ is reducible. Suppose next that $a \in -4F^4$ and write $a = -4\alpha^4$, $\alpha \in F$, $Y = X^{2^{n-2}}$. Then

$$X^{2^n} - a = Y^4 + 4\alpha^4 = (Y^2+2\alpha Y+2\alpha^2)(Y^2-2\alpha Y+2\alpha^2)$$

Conversely, assume that $a \notin F^2$ and $a \notin -4F^4$. Because $a \notin -4F^4$, it follows that $-4a$ is not a fourth power in F. Again, let λ be a root of $X^{2^n} - a$ and $\mu = \lambda^{2^{n-1}}$. Since $a \notin F^2$ and μ is a root of $X^2 - a$, we have $(F(\mu):F) = 2$ by Lemma 16.3. We must show that $(F(\lambda):F(\mu)) = 2^{n-1}$. For $n = 2$ this will be true if μ is not a square in $F(\mu)$, and for $n > 2$ this will be true by induction on n, provided μ is not a square in $F(\mu)$ and -4μ is not a fourth power in $F(\mu)$. In the latter case, $-\mu$ is a square in $F(\mu)$. So it suffices to rule out the possibility that either μ or $-\mu$ is a square in $F(\mu)$. Now these two statements are equivalent since $\mu \mapsto -\mu$ induces an automorphism of $F(\mu)$ over F. Hence, by applying Lemma 16.4(ii), the result follows. ■

Applying the foregoing, we now deduce the following result.

16.6. THEOREM. Let F be an arbitrary field, let $n \geqslant 1$ and let $a \in F$. Then

$$X^n - a \quad \text{is irreducible over } F$$

if and only if $a \notin F^p$ for all primes p dividing n and $a \notin -4F^4$ whenever $4|n$.

Proof. If p^s is the highest power of a prime p dividing n and if $a \in F^p$, then $X^{p^s} - a$ is reducible by Lemmas 16.3 and 16.5. Hence $X^n - a$ is reducible, by Lemma 16.2. If $4|n$ and $a = -4\lambda^4$, $\lambda \in F$, then

$$X^n - a = X^n + 4\lambda^4 = (X^{n/2} - 2\lambda X^{n/4} + 2\lambda^2)(X^{n/2} + 2\lambda X^{n/4} + 2\lambda^2)$$

Conversely, assume that $a \notin F^p$ for all primes p dividing n and $a \notin -4F^4$ whenever $4|n$. Let p^s be the highest power of a prime p dividing n. By Lemma 16.2, it suffices to show that $X^{p^s} - a$ is irreducible. The desired conclusion is therefore a consequence of Lemmas 16.5 and 16.3. ■

Our next aim is to prove that if E/F is a radical extension, then for any intermediate field K, $\mathrm{Gal}(K/F)$ is solvable. The following preliminary observations will clear our path.

16.7. LEMMA. Let E/F be a field extension and let $E_1/F, \ldots, E_n/F$ be radical subextensions. Then $(E_1E_2\ldots E_n)/F$ is a radical extension.

Proof. It suffices to treat the case $n = 2$. Let $E_1 = F(\lambda_1, \ldots, \lambda_m)$ and $E_2 = F(\mu_1, \ldots, \mu_k)$ exhibit the assumption that E_1/F and E_2/F are radical extensions. Then

$$E_1E_2 = F(\lambda_1,\dots,\lambda_m,\mu_1,\dots,\mu_k)$$

shows that E_1E_2/F is a radical extension. ∎

16.8. LEMMA. Let $F \subseteq K \subseteq E$ be a chain of fields.

(i) If E/F is radical, then so is E/K.

(ii) If K/F is radical and E is the normal closure of K over F, then E/F is radical.

Proof. (i) If $E = F(\lambda_1,\dots,\lambda_m)$ with $\lambda_i^{n_i} \in F(\lambda_1,\dots,\lambda_{i-1})$, $1 \leqslant i \leqslant m$, then $E = K(\lambda_1,\dots,\lambda_m)$ with $\lambda_i^{n_i} \in K(\lambda_1,\dots,\lambda_{i-1})$, as required.

(ii) Since K/F is finite, we may write $K = F(a_1,\dots,a_s)$ where a_i is a root of an irreducible polynomial $f_i(X)$ over F, $1 \leqslant i \leqslant s$. If λ_i is any root of $f_i(X)$, then $F(\lambda_1,\dots,\lambda_s)$ is F-isomorphic to K and hence $F(\lambda_1,\dots,\lambda_s)/F$ is radical. Since K is the composite of a finite number of fields of the form $F(\lambda_1,\dots,\lambda_s)$, the required assertion follows by virtue of Lemma 16.7. ∎

16.9. LEMMA. (i) Let p be a prime and E a splitting field of $X^p - 1$ over F. Then $\mathrm{Gal}(E/F)$ is cyclic.

(ii) If F is a field in which $X^n - 1$ splits into linear factors, a an arbitrary element of F and E the splitting field of $X^n - a$ over F, then $\mathrm{Gal}(E/F)$ is cyclic.

Proof. (i) If $\mathrm{char}F = p$, then $E = F$ and there is nothing to prove. If $\mathrm{char}F \neq p$, then $E = F(\varepsilon)$, where ε is a primitive p-th root of unity. Hence $\mathrm{Gal}(E/F)$ is cyclic, by Corollary 5.12.

(ii) If u is one root of $X^n - a$, then the general root has the form εu where $\varepsilon^n = 1$ and so ε lies in F. It follows that $E = F(u)$, and that an automorphism of E/F is determined by its value on u. Now all elements ε with $\varepsilon^n = 1$ form a cyclic group, say generated by δ. If $\sigma \in \mathrm{Gal}(E/F)$, then $\sigma(u) = \delta^{i_\sigma} u$ for some $i_\sigma \geqslant 1$. The map $\mathrm{Gal}(E/F) \longrightarrow \langle\delta\rangle$, $\sigma \longmapsto \delta^{i_\sigma}$ is obviously an injective homomorphism, hence the result. ∎

We are now ready to prove

16.10. THEOREM. If E/F is a radical extension, then for any intermediate field K, Gal(K/F) is solvable.

Proof. Let K be an intermediate field and let K_0 be the fixed subfield of Gal(K/F). Then K/K_0 is Galois and Gal(K/F) = Gal(K/K_0). Since, by Lemma 16.8(i), E/K_0 is radical, we may assume that $K = K_0$, i.e. K/F is Galois If L denotes a normal closure of E over F, then by Lemma 16.8(ii), L/F is radical. Thus we may also assume that E/F is normal. Hence Gal(K/F) is a homomorphic image of Gal(E/F) (Lemma 10.2), and we are left to verify that Gal(E/F) is solvable.

Let $E = F(\lambda_1,\ldots,\lambda_n)$ exhibit the assumption that E/F is a radical extension. By inserting further λ's, if necessary, we can arrange that in each case a prime power of λ_i lies in $F(\lambda_1,\ldots,\lambda_{i-1})$. We shall now argue by induction on n. By hypothesis, $\lambda_1^p \in F$ for some prime p. Let E_0 be a splitting field of $X^p - 1$ over E and let E_1 be the subfield of E_0 generated by F and the roots of $X^p - 1$. Since E_0/F is obviously normal, Gal(E/F) is a homomorphic image of Gal(E_0/F), so we need only show that Gal(E_0/F) is solvable.

The extension E_1/F is obviously Galois and hence cyclic, by Lemma 16.9(i). Since

$$\text{Gal}(E_0/F)/\text{Gal}(E_0/E_1) \cong \text{Gal}(E_1/F)$$

it suffices to show that Gal(E_0/E_1) is solvable. Now $E_0 = E_1(\lambda_1,\ldots,\lambda_n)$, for E_0 is generated over F by the λ's and the roots of $X^p - 1$, and the latter are already in E_1. Let G = Gal(E_0/E_1) and let H = Gal$(E_0/E_1(\lambda_1))$. Since $X^p - 1$ factors completely in E_1, $E_1(\lambda_1)$ is a splitting field of $X^p - \lambda_1^p$ over E_1 and hence $E_1(\lambda_1)/E_1$ is normal. Furthermore, by Lemma 16.9(ii), Gal$(E_1(\lambda)/E_1)$ is cyclic. But E/F is normal and radical, hence E_0/E_1 must be normal. It follows, from Lemma 10.2, that $H \lhd G$ and G/H is cyclic. To prove that G is solvable, it remains to show that H is solvable. This follows from our inductive assumption, since $E_0/E_1(\lambda_1)$ is a radical extension generated by $n - 1$ elements, namely $\lambda_2,\ldots,\lambda_n$. So the theorem is true. ■

Let F be a field and let $f(X) \in F[X]$. We define the *Galois group* of $f(X)$

to be the Galois group of a splitting field of $f(X)$ over F.

16.11. COROLLARY. Let F be an arbitrary field and let $f(X)$ be an irreducible polynomial over F. If there exists a radical extension E/F which contains a root of $f(X)$, then the Galois group of $f(X)$ over F is solvable.

Proof. Let E_0 be the normal closure of E over F. By Lemma 16.7, E_0/F is radical and hence we may assume that E/F is normal. Therefore E contains a splitting field K of $f(X)$ over F and, by Theorem 16.10, $\mathrm{Gal}(K/F)$ is solvable. ∎

Our next result provides a partial converse to Theorem 16.10.

16.12. THEOREM. Let F be a field of characteristic 0 and let E/F be a finite normal extension with a solvable Galois group G. Then E can be embedded in a radical extension of F.

Proof. We argue by induction on $(E:F)$. Since G is solvable, it contains a normal subgroup H of prime index p. Let K be a splitting field of $X^p - 1$ over E. Then K/F is normal and, since $\mathrm{Gal}(K/E)$ is cyclic (Lemma 16.9(i)), the group $\mathrm{Gal}(K/F)$ is solvable. Let L be the subfield of K obtained by adjoining to F the roots of $X^p - 1$. Then K/L is also normal. Since L/F is a radical extension, we are left to verify that K can be embedded in a radical extension of L.

To this end, we first show that $\mathrm{Gal}(K/L)$ is isomorphic to a subgroup of G. Indeed, consider the map

$$\begin{cases} \mathrm{Gal}(K/L) \longrightarrow \mathrm{Gal}(E/F) \\ \qquad \sigma \longmapsto \sigma|E \end{cases}$$

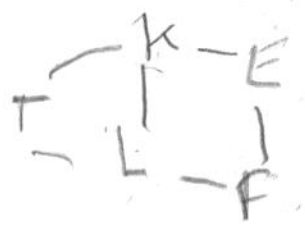

Then this map is a homomorphism, which is injective since $\sigma|E$ = identity implies that σ leaves both L and E elementwise fixed and hence $\sigma = 1$.

If $\mathrm{Gal}(K/L)$ is isomorphic to a proper subgroup of G, then by our inductive assumption K can be embedded in a radical extension of L. We may therefore assume that $\mathrm{Gal}(K/L) \cong G$. Let T be the intermediate field corresponding to H. Then $(T:L) = p$, T is normal over L, and L contains a primitive p-th root of 1. Since T/L is cyclic of degree p, it follows from Lemma 16.1, that T/L is a radical extension. Now K/T is normal with solvable Galois group H. By induction, K can be embedded in a radical extension of T. Hence K can be embedded in a radical extension of L, as required. ■

Let F be a field and let $f(X)$ be a polynomial of positive degree over F. Then the equation

$$f(X) = 0$$

is said to be *solvable by radicals over* F if each root of $f(X)$ lies in a suitable radical extension of F. In view of Lemma 16.7, this is equivalent to the condition that a splitting field of $f(X)$ over F can be embedded in a radical extension of F.

16.13. COROLLARY. (Galois' criterion). An equation $f(X) = 0$ is solvable by radicals over a field F of characteristic 0 if and only if the Galois group of $f(X)$ is solvable.

Proof. Apply Theorems 16.12 and 16.10. ■

Our next aim is to provide some applications of Theorem 16.6 which ensure the existence of extensions of large degree. As a preliminary result, we prove

16.14. LEMMA. Let F be a field of characteristic $p > 0$, let $a \in F$ and let $X^p - X - a$ be irreducible over F. If λ is a root of $X^p - X - a$, then $X^p - X - a\lambda^{p-1}$ is irreducible over $F(\lambda)$.

Proof. Assume by way of contradiction that $X^p - X - a\lambda^{p-1}$ is reducible. Then, by Theorem 14.7(iii), it has a root μ in $F(\lambda)$. We may write

$$\mu = a_0 + a_1\lambda + \dots + a_{p-1}\lambda^{p-1}$$

with $a_i \in F$. Then, since $\lambda^p = \lambda + a$, we have

$$\begin{aligned}\mu^p &= a_0^p + a_1^p\lambda^p + \dots + a_{p-1}^p\lambda^{p(p-1)} \\ &= a_0^p + a_1^p(\lambda+a) + \dots + a_{p-1}^p(\lambda+a)^{p-1} \qquad (3)\end{aligned}$$

On the other hand,

$$\mu^p = \mu + a\lambda^{p-1} = a_0 + a_1\lambda + \dots + (a_{p-1}+a)\lambda^{p-1} \qquad (4)$$

Equating the coefficients of λ^{p-1} in (3) and (4), we derive $a_{p-1}^p = a_{p-1} + a$, which contradicts the irreducibility of $X^p - X - a$ over F. ∎

We are now ready to prove the following result.

16.15. THEOREM. Let F be a field and let p be either an odd prime or $p = 2$ and $\mathrm{char}F = 2$. If F has an extension whose degree is divisible by p, then for any $n \geq 1$, F has an extension whose degree is divisible by p^n.

Proof. If $\mathrm{char}F = p$ and F is not perfect, then $\lambda \notin F^p$ for some $\lambda \in F$, hence by Theorem 16.6, $X^{p^n} - \lambda$ is irreducible for all $n \geq 1$. Therefore, if $\mathrm{char}F = p$, we may assume that F is perfect.

Let E/F be a field extension with $p|(E:F)$ and let K be the separable closure of F in E. If $\mathrm{char}F = p$, then $K = E$ since we assume that F is perfect. If $\mathrm{char}F = q = 0$, then $K = E$ by Proposition 3.7. Finally, if $\mathrm{char}F = q > 0$ and $q \neq p$, then by Propositions 3.21 and 3.24(i), $(E:K)$ is a power of q, so p divides $(K:F)$. Thus we may harmlessly assume that E/F is separable. By passing to a normal closure of E, we may further assume that E/F is Galois. Invoking Galois theory and the existence of an element of order p in $\mathrm{Gal}(E/F)$, we may assume that E/F is a Galois extension of degree p.

If $\mathrm{char}F = p$, then by Theorem 14.7, $E = F(\lambda)$ with λ a root of an irreducible polynomial over F of the form $X^p - X - a$. Hence, by Lemma 16.14, there exists an extension of F of degree p^2 and the process can be iterated to get degree p^n for any n.

Assume that $\mathrm{char}F \neq p$ and let K be a splitting field of $X^p - 1$ over E. If T is obtained from F with a primitive p-th root of unity adjoined, then K/T is Galois with Galois group cyclic of order p. By Lemma 16.1, $K = T(\lambda)$ with λ a root of an irreducible polynomial $X^p - a$, for some $a \in F$. Hence $a \notin F^p$ and, by Theorem 16.6, $X^{p^n} - a$ is irreducible over F, for any $n \geq 1$. This completes the proof of the theorem. ∎

16.16. THEOREM. Let F be a field having an extension of degree divisible by 4. Then, for any $n \geq 2$, F has an extension of degree divisible by 2^n.

Proof. If $\mathrm{char}F = 2$, then the assertion is true by appealing to Theorem 16.15. So assume that $\mathrm{char}F \neq 2$. By the argument in Theorem 16.15, we may assume that the given extension E of F is such that E/F is Galois with $(E:F) = 4$. Consider the field $E(i)$ with $i^2 = -1$. Then $E(i)/F(i)$ is Galois of degree 2 or 4. In either case, $E(i)$ contains a quadratic extension of $F(i)$ and therefore $F(i)$ contains an element a with no square root in $F(i)$. Furthermore, $-4a$ cannot have a fourth root in $F(i)$, for otherwise $-a$ is a square and so is a. Invoking Lemma 16.5(ii), we conclude that $F(i)$ has extensions of degree 2^n for all $n \geqslant 2$. So the theorem is true. ■

To present our further result, we need to introduce the notion of an ordered field. Let R be a commutative ring. We say that R is an *ordered ring* if there is a total ordering $>$ of R such that

(a) $x > x', y > y' \Rightarrow x + y > x' + y'$

(b) $x > 0, y > 0 \Rightarrow xy > 0$

In discussing the ordering on R, we shall write $x \geqslant y$ to mean $x > y$ or $x = y$, and use $<, \leqslant$ for the opposite ordering. An element $x \in R$ is said to be *positive* or *negative* according as $x > 0$ or $x < 0$.

In any ring R, a *cone* is a subset P containing 1 but not 0 and closed under addition and multiplication.

16.17. LEMMA. Let R be a commutative ordered ring

(i) The set P of positive elements of R is a cone, called a *positive cone* of R, and $R = P \cup -P \cup \{0\}$ (disjoint union), where $-P = \{-x | x \in P\}$.

(ii) Every cone P of a commutative ring S such that $S = P \cup -P \cup \{0\}$ defines an ordering on S by putting $x > y$ if and only if $x - y \in P$.

(iii) R is an integral domain of characteristic 0, and the square of any nonzero element of R is positive.

Proof. The proof of (i) and (ii) is straightforward. Given $x, y \in R$, $x, y \neq 0$, if $x, y > 0$, then $xy > 0$. Similarly, if $x, y < 0$, then $-x, -y > 0$ and thus $xy = (-x)(-y) > 0$. If x and y have opposite signs, say $x > 0 > y$, then $x, -y > 0$, hence $-xy = x(-y) > 0$ and so $xy < 0$. Similarly if $x < 0 < y$, then $xy < 0$. In each case $xy \neq 0$, so R is an integral domain. In particular, if $x = y$, then the last two cases cannot occur and $x^2 > 0$. Finally, note that $1^2 = 1 > 0$ and that $1, 1+1, \ldots$ are all > 0, hence $\mathrm{char} R = 0$. ■

With this information at our disposal, we now prove

16.18. THEOREM. Let F be a field which has a quadratic extension but no extension of degree 4. Then F is an ordered field in which every positive element has a square root.

Proof. Let P be the subset of F consisting of all nonzero squares. Then P is obviously closed under multiplication. Thus, in view of Lemma 16.17 (ii), we are left to verify that P is closed under addition and that for any $0 \neq a \in F$, either a or $-a$ is a square, but not both.

Owing to Theorem 16.15, we must have $\mathrm{char} F \neq 2$. Therefore, by Lemma 16.5 (iii), for every a in F either a is a square or $-4a$ is a fourth power; otherwise $X^4 - a$ would be irreducible over F and yield an extension of degree 4. In particular, either a or $-a$ is a square. If -1 is a square, then every element in F would be a square, and no quadratic extension of F could exist. Thus, for any $0 \neq a \in F$, either a or $-a$ is a square, but not both.

By the foregoing, we are left to verify that the sum of two squares is a square. To this end, we form the field $F(i)$ with $i^2 = -1$. In $F(i)$ every element must be a square, since otherwise there would be a quadratic extension of $F(i)$ and therefore an extension of F of degree 4. Invoking the fact that $a + bi$ is a square in $F(i)$, we find that $a^2 + b^2$ is a square in F, as required. ■

16.19. COROLLARY. (Artin-Schreir). Let E/F be a finite field extension with E algebraically closed but F not algebraically closed. Then F is an ordered field and $E = F(i)$, $i^2 = -1$.

Proof. By hypothesis, E is an algebraic closure of F. Hence the degree of any finite extension of F is bounded by $(E:F)$. It follows, from Theorems 16.15 and 16.16, that $(E:F) = 2$. Therefore, by Theorem 16.18, F is an ordered field in which every positive element has a square root, and E must be $F(i)$. ■

As a preliminary to our next result, we prove

16.20. LEMMA. Let F be a field such that for some prime p every nontrivial finite extension of F has degree divisible by p. Then every finite extension of F has degree a power of p.

Proof. Let E/F be a finite field extension. If $\text{char}F = q > 0$ and $q \neq p$, then F is perfect. Indeed, otherwise by Lemma 16.3, F has an extension of degree q. If $\text{char}F = p$ and K is the separable closure of F in E, then $(E:K)$ is a power of p (Propositions 3.21 and 3.24). Thus we may harmlessly assume that E/F is separable. Furthermore, by passing to a normal closure of E over F, we may assume that E/F is Galois. Let P be a Sylow p-subgroup of $G = \text{Gal}(E/F)$, and let K be the corresponding subfield. Then $(K:F) = (G:P)$ is prime to p. By our hypothesis, this implies $K = F$. Thus $(E:F)$ is a power of p, as required. ■

16.21. THEOREM. Let F be an ordered field in which every positive element has a square root. Assume further that every polynomial of odd degree over F has a root in F. Then $F(i)$, $i^2 = -1$, is algebraically closed.

Proof. By hypothesis, F has no finite nontrivial extensions of odd degree. Invoking Lemma 16.20, we infer that the degree of any finite extension of F is a power of 2. We claim that the only nontrivial finite extension of F is $F(i)$, which will clearly yield the result. Let E/F be a finite extension and let $G = \text{Gal}(E/F)$. By Lemma 16.17, $\text{char}F = 0$ and so E/F is separable. By passing to the normal closure of E, we may assume that E/F is Galois. We know that the order of G is a power of 2. If $(E:F) > 2$, then G has a subgroup of index 4 which in turn is contained in a subgroup of index 2. This yields a chain $F \subset F_1 \subset F_2$ of fields with $(F_1:F) = (F_2:F_1) = 2$. Since $F(i)$ is the only quadratic extension of F, we must have $F_1 = F(i)$. The latter easily implies that every element in F_1 is a square, hence F_2 cannot exist, a contradiction. ■

Let E/F be a field extension and let M be a subgroup of E^* such that F^*M/F^* is finite. When is it the case that $|F^*M/F^*| = (F(M):F)$? An answer to this question is given by the following result.

16.22. THEOREM. (Kneser(1975)). Let E/F be a separable field extension and let

M be a subgroup of $E^\star$ such that $F^\star M/F^\star$ is finite. Assume that for all odd primes p, each p-th root of 1 which lies in $F^\star M$ also lies in F and that $i = \sqrt{-1} \in F$ if $1 \pm i \in F^\star$. Then

$$|F^\star M/F^\star| = (F(M):F)$$

Proof. We first observe that $F(M)$ consists of all F-linear combinations of $F^\star M$. Therefore we may choose finitely many $x_1,\ldots,x_n$ in $F^\star M$ which is an F-basis for $F(M)$. Then $x_1F^\star,\ldots,x_nF^\star$ are distinct elements of $F^\star M/F^\star$ and so

$$(F(M):F) \leqslant |F^\star M/F^\star)| \tag{5}$$

Assume that the result is true for all Sylow subgroups of $F^\star M/F^\star$. If $N/F^\star$ is a Sylow p-subgroup of $F^\star M/F^\star$ of order p^t, then

$$p^t = (F(N):F)|(F(M):F)$$

and thus $|F^\star M/F^\star| \leqslant (F(M):F)$. Invoking (5), we may therefore assume that $F^\star M/F^\star$ is a p-group.

Choose a chain of subgroups

$$F^\star = N_0 \subset N_1 \subset \ldots \subset N_t = F^\star M$$

with $[N_s:N_{s-1}] = p$, $1 \leqslant s \leqslant t$. We show, by induction on s, that

$$(F(N_s) : F(N_{s-1})) = p \tag{6}$$

and that an element of $F(N_s)$ (respectively, $F(N_s) \cap F^\star M$ if $p = 2$ and $i \in F(N_s)$) whose p-th power is in N_s is itself in N_s. Assume that the statement is true for $s-1$. If $a \in N_s$ is such that aN_{s-1} generates N_s/N_{s-1}, then a is a root of the polynomial $f(X) = X^p - a^p$ with coefficients in $F(N_{s-1})$. Hence, if f is irreducible, then (6) holds. If f is reducible, then f has a root, say b, in $F(N_{s-1})$, by Lemma 16.3. Hence $b = a\varepsilon$ with $\varepsilon^p = 1$ and $b^p = a^p \in N_{s-1}$. By induction, $b \in N_{s-1}$ and so $\varepsilon \in N_s \subseteq F^\star M$, which implies, by hypothesis, that $\varepsilon \in F$. It follows that $a \in N_{s-1}$, contrary to the assumption that aN_{s-1} is of order $p > 1$. This establishes (6).

Now assume that $c \in F(N_s)$ satisfies $c^p \in N_s$ (and $c \in F^\star M$, when $p = 2$ and $i \in F(N_s)$). Then $c^p = a^q d$ with $0 \leqslant q < p$ and $d \in N_{s-1}$. We wish to show that $c \in N_s$. This will be achieved by proving the case $q = 0$ and by showing that the case $q > 0$ cannot occur.

Assume that $q = 0$ so that $c^p \in N_{s-1}$. Because E/F is separable, so is $F(N_s)/F(N_{s-1})$. Invoking (6), we infer that there exists an $F(N_{s-1})$-homomorphism f of $F(N_s)$ into the normal closure of $F(N_s)/F(N_{s-1})$ such that $f(a) \neq a$. Since $f(a^p) = a^p = f(a)^p$, we have $f(a) = a\varepsilon$ with ε being a primitive p-th root of 1. Similarly, $f(c^p) = c^p = f(c)^p$ and so $f(c) = c\varepsilon^r$ for some

$$0 \leqslant r \leqslant p-1$$

The latter implies that $f(a^{-r}c) = a^{-r}c$ and hence that $a^{-r}c = b \in F(N_{s-1})$. Moreover, $b^p = (a^p)^{-r}c^p \in N_{s-1}$ (and $b \in F^*M$ in case $c \in F^*M$) which shows, by induction, that $b \in N_{s-1}$ and hence that $c \in N_s$.

Assume by way of contradiction that $q > 0$ and denote by N the norm of $F(N_s)$ over $F(N_{s-1})$. Because $N(a) = (-1)^{p-1}a^p$, we have

$$((-1)^{p-1}a^p)^q = N(c)^p d^{-p}$$

For odd p, a^p is then a p-th power of an element in $F(N_{s-1})$, which contradicts (6). In case $p = 2$, we have $-a^2 = \lambda^2$ with $\lambda \in F(N_{s-1})$ which shows that $i \in F(N_s)$, $i \notin F(N_{s-1})$ and $c^2 = ad = \pm i\lambda d$. Let us write

$$c = g + ih \qquad \text{with } g,h \in F(N_{s-1})$$

Then $c^2 = (g^2-h^2) + 2ghi = \pm i\lambda d$ which implies $g^2 = h^2$ and hence $c = (1\pm i)g$. We conclude that

$$g^4 = -c^4/4 \in N_{s-1}$$

which ensures, by applying twice the induction hypothesis, that $g \in N_{s-1}$. But then $1 \pm i \in F^*M$ and so, by hypothesis, $i \in F \subseteq F(N_{s-1})$. This contradiction completes the proof of the theorem. ∎

Returning to radical extensions, we next investigate conditions under which the Galois group of a binomial $X^n - a$ is abelian.

16.23. THEOREM. (Schinzel(1977)). Let F be a field, n a positive integer not divisible by char$F$ and let m be the number of n-th roots of unity contained in

F. Then, for any given $a \in F$, the Galois group of $X^n - a$ is abelian if and only if $a^m = \lambda^n$ for some $\lambda \in F$.

Proof.(Wojcik(1982)). For any integer t not divisible by char$F$, let ε_t denote a primitive t-th root of unity over F. Let F^{ab} be the maximal abelian extension of F. If $a^m = \lambda^n$ for some $\lambda \in F$, then $\sqrt[n]{a} = \varepsilon_n^j \sqrt[m]{\lambda}$, $0 \leqslant j \leqslant n-1$. Since F contains a primitive m-th root of unity, $\sqrt[m]{\lambda} \in F^{ab}$ by virtue of Lemma 15.9(iii). Thus $\sqrt[n]{a} \in F^{ab}$ and $F(\sqrt[n]{a}, \varepsilon_n)/F$ is abelian.

Conversely, assume that the Galois group of $X^n - a$ is abelian. Let k be the maximal divisor of n for which $a^m = \lambda^k$ for some $\lambda \in F$. We need only verify that $k = n$. Obviously $k \leqslant n$ and $m|k$. Assume by way of contradiction that $k < n$. Fix a prime p dividing n/k and denote by p^s the highest power of p dividing m. Put

$$t = p^{s+1}k/m,\ L = F(\varepsilon_{p^{s+1}}) \text{ and } E = F(\sqrt[p^{s+1}]{\lambda})$$

Then

$$E \subseteq F(\sqrt[t]{a}, \varepsilon_{tm}) \subseteq F(\sqrt[n]{a}, \varepsilon_{tm}) \subseteq F^{ab}$$

since, by hypothesis, $\sqrt[n]{a} \in F^{ab}$. By the definition of k, $(E:F) = p^{s+1}$. Thus $\varepsilon_{p^{s+1}} \in E$ and $F \subseteq L \subseteq E$. Moreover, since $p^{s+1}|n$ and p^s is the highest power of p dividing m, we have $\varepsilon_{p^s} \in F$ but $\varepsilon_{p^{s+1}} \notin F$. We now distinguish two cases. Suppose first that $s = 0$. Then $(L:F)$ divides $p-1$. Hence $L = F$ and $\varepsilon_p \in F$, a contradiction. Now assume that $s \geqslant 1$. We have $(E:L) < p^{s+1}$ and $E = L(\sqrt[p^{s+1}]{\lambda})$. Hence $\lambda = \lambda_1^p$ for some $\lambda_1 \in L$, by virtue of Lemma 16.5. Thus $F(\sqrt[p]{\lambda}) \subseteq L = F(\sqrt[p]{\varepsilon_{p^s}}$ and so, by Corollary 15.6, we have $\lambda = \varepsilon_{p^s}^j \lambda^p$ for some $j \geqslant 0$ and some $\lambda_2 \in F$. But then $a^m = \lambda_2^{pk}$ with $pk|n$, contrary to the definition of k. ■

We next provide a formula for the degree of the splitting field of an irreducible binomial. In what follows for any $n \geqslant 1$ not divisible by the characteristic of a field F, we write ε_n for a primitive n-th root of unity over F. As a preparatory step, we first record the following observation.

16.24. LEMMA. Let F be a field, let $n \geqslant 1$ be an integer not divisible by char$F$ and let $X^n - a \in F[X]$ be irreducible over F with root λ. Put

$$r = \max\{k \mid k \mid n \text{ and } \varepsilon_k \in F(\lambda)\}$$

If E is any field such that $F(\varepsilon_r) \subseteq E \subseteq F(\lambda)$, then

$$E = F(\lambda^s) \quad \text{where} \quad s = (F(\lambda):E)$$

Proof. Let $f(X)$ be the minimal polynomial for λ over E. Since λ is a root of $X^n - a$, we have $f(X) \mid X^n - a$. Thus, every root of $f(X)$ is of the form $\varepsilon_n^i \lambda$ for some i, and therefore

$$f(X) = \prod_{j=1}^{s} (X - \varepsilon_n^{i_j} \lambda)$$

If $d = \sum_{j=1}^{s} i_j$, then $\prod_{j=1}^{s} \varepsilon_n^{i_j} \lambda = \varepsilon_n^d \lambda^s \in E \subseteq F(\lambda)$. Also $\lambda^s \in F(\lambda)$ and thus $\varepsilon_n^d \in F(\lambda)$. Hence, by the definition of r, $\varepsilon_n^d \in F(\varepsilon_r) \subseteq E$ and so $\lambda^s \in E$. Now $s = (F(\lambda):E)$ and $(F(\lambda):F(\lambda^s)) \leq s$, since λ is a root of $X^s - \lambda^s$ over $F(\lambda^s)$. Hence we must have $E = F(\lambda^s)$. ■

We are now ready to prove

16.25. THEOREM. (Darbi(1926), Gay and Vélez(1978)). Let F be an arbitrary field of characteristic $p \geq 0$, let n be a positive integer and let m be defined by: $m = n$ if char$F = 0$ and $n = mp^k$ with $(m,p) = 1$ if char$F = p > 0$. Assume that $X^n - a \in F[X]$ is irreducible over F with root λ and define the integer s by

$$s = \max\{t \mid t \mid m \text{ and } \lambda^{m/t} \in F(\varepsilon_m)\}$$

If E is the splitting field of $X^n - a$ over F, then

$$(E:F) = \frac{n(F(\varepsilon_m):F)}{s}$$

Proof. Our first step is to reduce the general case to the case where $m = n$, i.e. char$F \nmid n$. So assume that char$F = p > 0$. Then

$$E = F(\lambda, \varepsilon_m) = F(\lambda^{p^k}, \lambda^m, \varepsilon_m)$$

which in turn is the composite of the separable extension $F(\lambda^{p^k}, \varepsilon_m)/F$ and a purely inseparable extension $F(\lambda^m)/F$. By Lemma 16.2, both $X^m - a$ and $X^{p^k} - a$ are irreducible over F. Since λ^m is a root of $X^{p^k} - a$, we have

$(F(\lambda^m):F) = p^k$. Hence, by Proposition 3.28,

$$(E:F) = p^k(F(\lambda^{p^k}, \varepsilon_m):F)$$

Thus, if the result holds for $X^m - a$ (whose splitting field is $F(\lambda^{p^k}, \varepsilon_m)$, then

$$(E:F) = p^k m(F(\varepsilon_m):F)/s = n(F(\varepsilon_m):F)/s$$

We may therefore assume that $m = n$ and hence that char$F \nmid n$.

Put $L = F(\varepsilon_n) \cap F(\lambda)$ and $s' = (L:F)$. Since $X^n - a$ is irreducible,

$$(F(\lambda):F) = n$$

On the other hand, by Proposition 12.5,

$$(E:F(\lambda)) = (F(\varepsilon_n, \lambda):F(\lambda)) = (F(\varepsilon_n):L)$$

It therefore follows that

$$(E:F) = n(F(\varepsilon_n):F)/s'$$

and we must show that $s = s'$.

Now by the definition of r in Lemma 16.24,

$$F(\varepsilon_r) \subseteq F(\varepsilon_n) \cap F(\lambda) = L \subseteq F(\lambda)$$

and hence, by Lemma 16.24, $L = F(\lambda^q)$ for $q = (F(\lambda):L)$. Since $(F(\lambda):F) = n$, we have

$$(L:F) = n/q = s'$$

This shows that $\lambda^q \in F(\varepsilon_n)$ and $q = n/s'$. Hence, by the definition of s,

$$s \geqslant s'.$$

On the other hand, since $\lambda^{n/s} \in F(\varepsilon_n) \cap F(\lambda) = L$, we have

$$(F(\lambda):L) = q \leqslant n/s$$

whence $s' \geqslant s$, as required. ■

Let E/F be a simple radical extension, say $E = F(\lambda)$ with $\lambda \in E^*$ and a power of λ in F^*. Then we obviously have

$$(<\lambda>t(E^*)F^*)/F^* \subseteq t(E^*/F^*)$$

It is therefore natural to investigate circumstances under which the equality holds.

16.26. THEOREM. (May(1979)). Let p be a prime and let E/F be a field extension with $\text{char}F \neq p$. Assume that $E = F(\lambda)$ where $\lambda^p \in F - F^p$. If $p = 2$, further assume that $E \neq F(i)$ $(i^2 = -1)$. Then

$$t(E^*/F^*) = (<\lambda>t(E^*)F^*)/F^*$$

Proof. Owing to Lemma 16.3, $X^p - \lambda^p$ is irreducible over F and hence $(E:F) = p$. Let $\mu \in E^*$ be an element of prime-power order modulo F^*. First assume that the order is q^r for some prime $q \neq p$, and let $\mu^{q^r} = \gamma \in F^*$. If N is the norm map from E to F, then $N(\mu)^{q^r} = \gamma^p$, hence $\gamma = \gamma_1^{q^r}$ for some $\gamma_1 \in F^*$. It follows that $\mu = \gamma_1\varepsilon$ for some root of unity $\varepsilon \in F^*$ which shows that $\mu F^* \in t(E^*)F^*$.

Now assume that μ has order p^r modulo F^*, and let $\mu^{p^r} = \gamma \in F^*$. We claim that $\gamma \notin F^p$. Assume the contrary and write $\gamma = \gamma_1^p$ for some $\gamma_1 \in F^*$. Let ε be a primitive p-th root of unity. Then

$$\mu^{p^{r-1}} = \gamma_1\varepsilon^m \qquad \text{for some } m$$

We must have $p \nmid m$, for otherwise the order of μ modulo F^* would be less than p^r. For the same reason, we must have $\varepsilon \notin F^*$. It follows that

$$E \supseteq F(\varepsilon) \supseteq F$$

with $(F(\varepsilon):F) > 1$. But $(E:F) = p$ and $(F(\varepsilon):F)$ divides $p-1$. This contradiction substantiates our claim.

Except possibly when $p = 2, r > 1$ and $i \notin F$, we have $(F(\mu):F) = p^r$ by virtue of Lemmas 16.5 and 16.3, and thus $r = 1$. Consider the exceptional case, i.e. assume that $p = 2, r > 1, i \notin F$ and $(F(\mu):F) < 2^r$. Under these circumstances we know, from Lemma 16.5, that $\gamma = -4\delta^4$ for some $\delta \in F$. From

$$\mu^{2^r} = -4\delta^4,$$

we see that $i \in E$ and hence $E = F(i)$. Since this case is excluded in the hypothesis, we may therefore return to the situation where

$$\mu^p = \gamma \quad \text{and} \quad (F(\mu):F) = p$$

Then we have

$$F(\mu) = E = F(\lambda)$$

Hence, by Corollary 15.7, $\mu = \lambda^k a$ for some $k \in \mathbb{Z}$ and some $a \in F$. Thus $\mu F^* \in \langle\lambda\rangle F^*$ and the result follows. ■

We return to the study of the field $F(\alpha)$ where α is a root of an irreducible polynomial $X^p - a \in F[X]$. As an easy application of Corollary 15.7, we first show that the adjunction of two "genuinely different" p-th roots result in an extension of degree p^2.

16.27. PROPOSITION. Let p be a prime and let F be a field with char$F \neq p$. If α,β are roots of irreducible polynomials $X^p - a$, $X^p - b$ over F, then

$$(F(\alpha,\beta):F) = p^2$$

unless $b = c^p a^k$ for some $k \in \mathbb{Z}$ and some $c \in F$.

Proof. If the polynomial $X^p - b$ remains irreducible over $F(\alpha)$, then $(F(\alpha,\beta):F) = p^2$. If it is reducible over $F(\alpha)$, then it has a root γ in $F(\alpha)$. (Lemma 16.3). Since $\gamma^p = b \in F$, it follows from Corollary 15.7 that $\gamma = c\alpha^k$ for some $k \in \mathbb{Z}$, $c \in F$. Raising this equation to the p-th power, we obtain $b = c^p a^k$. ■

Assume that α,β are roots of irreducible polynomials $X^p - a$, $X^p - b$ over F. What can be said about the degree of $\alpha + \beta$, provided $(F(\alpha,\beta):F) = p^2$? To answer this question, we first make the following preliminary observation.

16.28. LEMMA. Let E/F be a finite Galois extension and let α and β be elements of E of degree m and n, respectively, over F such that

$$(F(\alpha,\beta):F) = mn.$$

(i) For any conjugate α_i of α and β_j of β, there exists $\sigma \in \mathrm{Gal}(E/F)$ such that

$$\sigma(\alpha) = \alpha_i \quad \text{and} \quad \sigma(\beta) = \beta_j$$

(ii) If no difference of two conjugates of α equals a difference of two conjugates of β, then

$$F(\alpha,\beta) = F(\alpha+\beta)$$

Proof. (i) Since E/F is a finite Galois extension, we may, by virtue of Corollary 6.5(i), choose $f \in \mathrm{Gal}(E/F)$ such that $f(\alpha_i) = \alpha$. Write $f(\beta_j) = \beta_k$ for some conjugate β_k of β. We claim that there exists $f_1 \in \mathrm{Gal}(E/F)$ such that $f_1(\alpha) = \alpha$ and $f_1(\beta_k) = \beta$; if sustained, the assertion will follow by taking $\sigma = (f_1 f)^{-1}$.

Let $\beta_1,\dots,\beta_n$ be all conjugates of β over F. Our hypothesis implies that the degree of β over $F(\alpha)$ is still n, so that the roots of its minimal polynomial over $F(\alpha)$ are still $\beta_1,\dots,\beta_n$. Since $E/F(\alpha)$ is Galois, the required automorphism exists.

(ii) Let $\alpha_1,\dots,\alpha_m$ be all conjugates of α over F. Then, by (i),

$$\alpha_i + \beta_j \qquad (1 \leqslant i \leqslant m,\ 1 \leqslant j \leqslant n)$$

are all conjugates of $\alpha + \beta$. If a coincidence occurs, then the difference of two conjugates of α will equal the difference of two conjugates of β. Thus, by hypothesis, all $\alpha_i + \beta_j$ are distinct, proving that the degree of $\alpha + \beta$ is mn. Hence $F(\alpha,\beta) = F(\alpha+\beta)$ as asserted. ■

16.29. THEOREM. Let F be a field and let p be a prime distinct from char$F$. Let α and β be roots of irreducible polynomials of the form $X^p - a$, $X^p - b$ over F. If $(F(\alpha,\beta):F) = p^2$, then $F(\alpha+\beta) = F(\alpha,\beta)$, i.e. $\alpha + \beta$ has degree p^2 over F.

Proof. Since $p \neq$ char$F$, α and β are separable over F. Hence there is a finite Galois extension E/F with $\alpha,\beta \in E$. Owing to Lemma 16.28, it suffices to verify that no difference of two conjugates of α equals a difference of two conjugates of β. Now the difference of two conjugates of α has the form $(\varepsilon^i - \varepsilon^j)\alpha$, where ε is a primitive p-th root of unity. Therefore, if a difference of two conjugates of α equals a difference of two conjugates of β, then $\alpha/\beta \in F(\varepsilon)$. But $(F(\varepsilon):F) \leqslant p-1$, while the degree of α/β over F must divide $p^2 = (F(\alpha,\beta):F)$. Thus $\alpha/\beta \in F$, contrary to $(F(\alpha,\beta):F) = p^2$. ■

17. DEGREES OF SUMS IN A SEPARABLE FIELD EXTENSION

Let F be a field and let γ be an algebraic element in a field extension of F. In what follows, we write $\deg\gamma$ for the degree of γ over F.

Let α,β be two separable elements in a field extension of F. Put $m = \deg\alpha$, $n = \deg\beta$ and assume that $(m,n) = 1$. Then, by Proposition 12.13,

$$(F(\alpha,\beta):F) = mn$$

and, by Corollary 3.32,

$$F(\alpha,\beta) = F(\gamma) \qquad \text{for some } \gamma \in F(\alpha,\beta)$$

In fact, the proof of Proposition 3.30 produces (for infinite F) an element γ of the form

$$\gamma = \alpha + \lambda\beta \qquad \text{for some } \lambda \in F$$

Thus, if F is infinite, we do have

$$\deg(\alpha+\lambda\beta) = mn \qquad \text{for some } \lambda \in F$$

Our aim is to exhibit circumstances under which the choice of $\lambda \in F$ is completely arbitrary, as long as $\lambda \neq 0$. The following is a reformulated version of a result due to Isaacs(1970).

17.1. THEOREM. Let F be an arbitrary field and let α,β be two separable elements in a field extension of F of coprime degrees m and n, respectively. Assume that either $\mathrm{char}F = 0$ or $\mathrm{char}F = p > 0$ and at least one of the following holds:

(a) $p \nmid mn$ and $p > \min(m,n)$

(b) $p \nmid mn$ and m or n is a prime power

(c) $p \nmid n$ and any prime divisor of n is $> m$

Then

$$\deg(\alpha+\lambda\beta) = mn \qquad \text{for all } 0 \neq \lambda \in F$$

Proof. For the sake of clarity, we divide the proof into a number of steps,

Step 1. Here we fix notation and make some preliminary observations.

Let the field E be obtained from F by adjoining all conjugates $\alpha_1,\dots,\alpha_m$ and $\beta_1,\dots,\beta_n$ of α and β, respectively. Then E/F is a finite Galois extension and we put $G = \mathrm{Gal}(E/F)$. The group G acts transitively on the sets

$$A = \{\alpha_1,\dots,\alpha_m\} \quad \text{and} \quad B = \{\beta_1,\dots,\beta_n\}$$

Let H and K be the stabilizers in G of α and β, respectively. Then

$$(G:H) = m, \quad (G:K) = n \tag{1}$$

and since $(m,n) = 1$, a standard argument yields

$$(G:(H \cap K)) = mn \tag{2}$$

Hence the size of the orbit of $(\alpha,\beta) \in A \times B$ under the induced action of G on $A \times B$ is mn. It follows that G is transitive on $A \times B$ and hence

$$\text{each } \alpha_i + \beta_j \text{ is a conjugate of } \alpha + \beta \tag{3}$$

$(1 \leq i \leq m,\ 1 \leq j \leq n)$. Since $|HK| = |H||K|/|H \cap K|$, it follows from (1) and (2) that

$$|HK| = \frac{|G|}{m} \cdot \frac{|G|}{n} \quad \frac{mn}{|G|} = |G|$$

Hence $(K:(H \cap K)) = m$ and therefore

$$K \text{ acts transitively on } A \tag{4}$$

Let W_1 and W_2 be the F-linear spans of A and B, respectively, let

$$V = W_1 + W_2$$

and let $U = W_1 \cap W_2$. Then G acts as a group of linear transformations of V and W_1, W_2, U are G-invariant subspaces. Replacing β by $\lambda\beta$, if necessary, it suffices to show that $\deg(\alpha+\beta) = mn$. We use induction on $|G|$, i.e. on the F-dimension of the normal closure of $F(\alpha,\beta)$. If $|G| = 1$, then $\alpha,\beta \in F$ and there is nothing to prove. By induction, we may therefore assume that if α,β are of coprime degrees m_0 and n_0, respectively, over an intermediate field $L \neq F$, and (a), (b) or (c) hold with respect to m_0, n_0, then $m_0 n_0$ is the degree of $\alpha_0 + \beta_0$ over L. *From now on, we assume that* $\deg(\alpha+\beta) < mn$ *and derive a contradiction.*

Owing to (3), not all $\alpha_i + \beta_j$ are distinct and so

$$\alpha + \beta = \alpha_k + \beta_s \qquad \text{for some } k \in \{1,\ldots,m\},\ s \in \{1,\ldots,n\} \tag{5}$$

where $\alpha \neq \alpha_k$ or $\beta \neq \beta_s$. Then

$$\alpha - \alpha_k = \beta_s - \beta \neq 0$$

and so $U \neq 0$.

Let M be the kernel of the action of G on V and let L be the fixed field of M. Then $F \subseteq F(\alpha,\beta) \subseteq L$ and L/F is Galois, so $E = L$ and hence $M = 1$. Thus

$$G \text{ acts faithfully on } V \tag{6}$$

Step 2. Here we prove that there is no proper subgroup G_0 *of* G *which acts so that the orbits* A_0 *and* B_0 *of* α *and* β *under* G_0 *satisfy*

$$m_0 | m,\ n_0 | n \quad \alpha_k \in A_0 \quad \text{and} \quad \beta_s \in B_0$$

where $m_0 = |A_0|$ and $n_0 = |B_0|$.

Assume the contrary. Then $(m_0,n_0) = 1$ and since, by (5), $\alpha + \beta = \alpha_k + \beta_s$, the number of elements in the G_0-orbit of $\alpha + \beta$ is $< m_0n_0$. Now $G_0 = \mathrm{Gal}(E/L)$ where L is the fixed field of G_0, and m_0,n_0 are the degrees of α and β, respectively, over L. Since $|G_0| < |G|$, it follows that none of (a), (b) or (c) holds with respect to m_0,n_0. But then none of (a), (b) or (c) holds with respect to m,n, a contradiction.

Step 3. Let $N \triangleleft G$ *be the kernel of the action of* G *on* A. *We now show that*

$$n\beta = u + \gamma \qquad (u \in U,\ \gamma \in W_2 \text{ is fixed by } G) \tag{7}$$

$$N \text{ fixes } n\beta \tag{8}$$

Note that G permutes the sets of cosets of U in W_1 and is transitive on the set of those cosets which contain elements of A. All of these, therefore, contain equal number of elements of A. Now $\alpha,\alpha_k \in \alpha + U$ and if $A_0 = A \cap (\alpha+U)$, then $|A_0| \,|\, m$. Let G_0 be the stabilizer of the coset $\alpha + U$ in G. Clearly, $H \subseteq G_0$ and hence G_0 is transitive on B. We claim that G_0 is transitive on A_0. Indeed, if $\alpha_i \in A_0$, then $\alpha^g = \alpha_i$ for some $g \in G$. Hence

$$(\alpha+U)^g = \alpha_i + U = \alpha + U$$

and so $g \in G_0$. This establishes transitivity and, by Step 2, we must have $G_0 = G$. Thus G stabilizes $\alpha + U$ and so $A \subseteq \alpha + U$. By a similar argument, $B \subseteq \beta + U$ and so $\beta_j = \beta + u_j$ for some $u_j \in U$. Summing over $\beta_j \in B$, we derive

$$\Sigma\beta_j = \Sigma u_j + n\beta$$

Hence $n\beta = u + \gamma$, where $u = -\Sigma u_j \in U$ and $\gamma = \Sigma\beta_j \in W_2$ is fixed by G. This proves (7) and (8), since N fixes all elements of $W_1 \supseteq U$.

Step 4. Here we show that if $\mathrm{char}F \nmid mn$, *then there is no subgroup* $L \subseteq K$ *which is transitive on* A *and such that* $\mathrm{char}F \nmid |L|$.

Assume the contrary. By (7), we have $n\beta = u + \gamma$ for some $u \in U$ and some $\gamma \in W_2$ fixed by G. As $U \subseteq W_1$, we have $u = \sum_{i=1}^{m} \lambda_i\alpha_i$ for some $\lambda_i \in F$. Now if $x \in L \subseteq K$, we have

$$\beta = \beta^x = \frac{1}{n}\sum_{i=1}^{m} \lambda_i\alpha_i^x + \frac{1}{n}\gamma \tag{9}$$

Put $\sigma = \sum_{i=1}^{m} \alpha_i$ and note that since L is transitive on A, we have

$$\sum_{x\in L} \alpha_i^x = \frac{|L|\delta}{m}$$

Summing (9) over L, it therefore follows that

$$|L|\beta = \frac{|L|}{mn}\sum_{i=1}^{m} \lambda_i\delta + \frac{|L|}{n}\gamma$$

Because γ and δ are fixed by G and $\mathrm{char}F \nmid |L|$, we conclude that β is fixed by G. This is a contradiction since $\beta \neq \beta_s$.

Step 5. Completion of the proof. Here we derive a final contradiction.

Let N be the kernel of the action of G on A. If $\mathrm{char}F \nmid n$, then by (8), N fixes β and hence $N = 1$, by (6). Thus G is isomorphic to a subgroup of S_m, so $|G|\,|\,m!$ Therefore all prime divisors of $|G|$ (in particular, those of n) are $\leqslant m$. Since $n > 1$ (recall that $\beta \neq \beta_s$), this certainly excludes (c).

Since (c) is excluded, we have $\mathrm{char}F \nmid mn$, by hypothesis. By interchanging A and B in the above argument we obtain $|G|\,|\,n!$ and hence all prime divisors of $|G|$ are $\leqslant \min(m,n)$. Hence, if $\mathrm{char}F = p > 0$ and (a) holds or if char$F$ = 0, then char$F \nmid |G|$. But this contradicts (4) and Step 4. Hence we must have char$F = p > 0$ and (b). We may assume that $m = q^e$, q prime, and let Q be a Sylow q-subgroup of K. Then $(K:(K \cap H)) = q^e$ so $K = (K \cap H)Q$ and it follows from (4) that Q is transitive on A. Since char$F \nmid q$, by hypothesis, we derive a final contradiction by applying Step 4. ∎

The following simple observation illustrates that to establish the best possible improvement of the above theorem, with conditions given in terms of m,n and char$F$, it suffices to consider only group representations

17.2. LEMMA. Let V be a finite-dimensional vector space over a field K and let a finite group G act faithfully on V as linear transformations of V. Assume that $u,v \in V$ are permuted by G into orbits of sizes m and n respectively and that $u + v$ lies in an orbit of size k. Then there exists a finite Galois extension E/F (with $K \subseteq F$) and elements $\alpha,\beta \in E$ such that

$$\deg\alpha = m, \deg\beta = n \text{ and } \deg(\alpha+\beta) = k$$

Proof. Let $t = \dim_K V$ and let $X_1,X_2,\ldots,X_t$ be indeterminates. Put $R = K[X_1,\ldots,X_t]$ and let E be the quotient field of R. We may identify V with the K-linear span of $X_1,X_2,\ldots,X_t$ in R. Then G acts faithfully as a group of K-automorphisms of R and hence of E. Let F be the fixed field of G in E (so $K \subseteq F$) and let α and β be elements of E corresponding to u and v. By Proposition 4.5, E/F is a finite Galois extension and G is the Galois group of E/F. Since for any $\gamma \in E$, $\deg\gamma$ is the size of the orbit of γ under G, the result follows. ∎

We close by giving some limitations on possible improvements of Theorem 17.1. The following two examples in conjunction with Lemma 17.2 illustrate that the conclusion of Theorem 17.1 fails in each of the cases below:

(i) char$F = 2$, $m = 3$ and $n = 4$

(ii) char$F = 3$, $m = 3$ and $n = 4$

17.3. EXAMPLE.(Isaacs(1970)). Let G be the alternating group of degree 4. Let V^* be a four dimensional vector space over $\mathbb{Z}/2\mathbb{Z}$ and let G permute a basis, $\{w,x,y,z\}$, in the natural manner. Put

$$V_0 = \{0, w + x + y + z\} \quad \text{and} \quad V = V^*/V_0$$

Then G acts faithfully on V and we put $u = w + x + V_0$ and $v = w + V_0$. Then u and v are permuted by G into orbits of sizes 3 and 4, respectively. However, the size of the orbit of $u + v$ is $4 \neq 12$. ■

17.4. EXAMPLE.(Isaacs(1970)). Let V be a four dimensional vector space over $\mathbb{Z}/3\mathbb{Z}$, with basis $\{w,x,y,z\}$. Let G be the group generated by the elements $\sigma_1,\sigma_2,\sigma_3 \in GL(V)$ whose matrices are

$$\sigma_1 = \begin{bmatrix} 1 & 1 & 0 & 0 \\ 0 & 1 & 0 & 0 \\ 0 & 0 & 1 & 0 \\ 0 & 0 & 0 & 1 \end{bmatrix}, \quad \sigma_2 = \begin{bmatrix} 1 & 0 & 0 & 0 \\ 0 & -1 & 0 & 0 \\ 0 & 1 & 1 & 0 \\ 0 & 1 & 0 & 1 \end{bmatrix}, \quad \sigma_3 = \begin{bmatrix} 1 & 0 & 0 & 0 \\ 0 & 1 & 0 & 0 \\ 0 & 0 & 0 & 1 \\ 0 & 0 & 1 & 0 \end{bmatrix}$$

Then G is the direct product of the subgroups $<\sigma_1,\sigma_2>$ of order 6 and $<\sigma_3>$ of order 2. Put $u = w$ and $v = y$. Then the orbit of u under G is $\{w,w + x,w - x\}$, and the orbit of v under G is $\{y,y + x,z,z + x\}$. However, the orbit of $u + v$ is

$$\{w + y,\ w + y + x,\ w + y - x,\ w + z,\ w + z + x,\ w + z - x\}$$

which has six elements. ■

18. GALOIS COHOMOLOGY

Let G be a group. A *G-module* is an abelian group A on which G operates such that

(i) $1a = a$ for all $a \in A$

(ii) $g(a+b) = ga + gb$ for all $g \in G$, $a,b \in A$

(iii) $(gg_1)a = g(g_1a)$ for all $g,g_1 \in G$, $a \in A$

Note that if A is a G-module, then A is a $\mathbb{Z}G$-module via

$$(\Sigma x_g g)a = \Sigma x_g(ga) \qquad (x_g \in \mathbb{Z},\ g \in G, a \in A)$$

Conversely, any given $\mathbb{Z}G$-module A can be regarded in the obvious way as a G-

module. The *fixed module* of a G-module A is defined to be the subgroup

$$A^G = \{a \in A | ga = a \text{ for all } g \in G\}$$

It is clear that A^G is the largest submodule of A on which G acts trivially.

If A,B are G-modules, the group of all abelian group homomorphisms $A \longrightarrow B$ is denoted by $\mathrm{Hom}(A,B)$. The group $\mathrm{Hom}(A,B)$ has a G-module structure defined as follows: if $\phi \in \mathrm{Hom}(A,B)$, $g\phi$ is the mapping $a \longmapsto g\phi(g^{-1}a)$, $a \in A$, $g \in G$. In particular, if X is any abelian group, we can form the G-module $\mathrm{Hom}(\mathbb{Z}G,X)$. A G-module of this type is said to be *co-induced*.

Let A be any G-module. A function f from the direct product of n copies of G to A is called an *n-cochain* of G in A. An n-cochain f is said to be *normalized* if $f(g_1,\ldots,g_n) = 0$ whenever any of the g_i is the identity element of G. The set of all n-cochains, written $C^n(G,A)$, becomes an abelian group under the multiplication of values. Extending the definition of $C^n(G,A)$ to the case $n = 0$, we put

$$C^0(G,A) = A$$

The formula

$$d_nf(g_1,\ldots,g_{n+1}) = g_1f(g_2,\ldots,g_{n+1}) + \\ + \sum_{i=1}^{n} (-1)^i f(g_1,\ldots,g_ig_{i+1},\ldots,g_{n+1}) + (-1)^{n+1}f(g_1,\ldots,g_n) \tag{1}$$

determines a homomorphism

$$d_n : C^n(G,A) \longrightarrow C^{n+1}(G,A)$$

It is a standard fact that $d_nd_{n-1} = 0$ for all $n \geqslant 1$. We set

$$Z^n(G,A) = \mathrm{Ker}d_n \quad \text{and} \quad B^n(G,A) = \mathrm{Im}d_{n-1} \qquad (n \geqslant 1)$$

and refer to the elements of $Z^n(G,A)$ and $B^n(G,A)$ as *n-cocycles* and *n-coboundaries*, respectively. The *n-th cohomology group* $H^n(G,A)$ with coefficients in A is defined by

$$H^n(G,A) = Z^n(G,A)/B^n(G,A) \qquad (n \geqslant 1)$$

$$H^0(G,A) = A^G$$

It can be easily shown that the cohomology group $H_n(G,A)$ is unaffected if we restrict ourselves to normalized n-cochains. Let us recall the standard fact which says that

$$H^n(G,A) = 0 \quad \text{for all} \quad n \geqslant 1 \quad \text{if} \quad A \quad \text{is co-induced} \tag{2}$$

Formula (1) shows that a 1-cocycle is a *crossed homomorphism*, i.e. a map

$$f : G \longrightarrow A$$

satisfying

$$f(gg') = gf(g') + f(g)$$

Similarly a 1-coboundary is a *principal crossed homomorphism*, i.e. a map

$$f : G \longrightarrow A$$

for which there exists $a \in A$ such that

$$f(g) = ga - a$$

In particular, if G acts trivially on A, then

$$H^1(G,A) = \text{Hom}(G,A)$$

From (1) we see also that a 2-cocycle is a function $f : G \times G \longrightarrow A$ such that

$$g_1f(g_2,g_3) - f(g_1g_2,g_3) + f(g_1,g_2g_3) - f(g_1,g_2) = 0 \tag{3}$$

while a 2-coboundary is a function $f : G \times G \longrightarrow A$ for which there exists a map $t : G \longrightarrow A$ such that

$$f(g_1,g_2) = g_1t(g_2) - t(g_1g_2) + t(g_1)$$

Let E/F be a finite Galois extension with $G = \text{Gal}(E/F)$. Then the additive and multiplicative groups E^+ and E^* of E are G-modules and we may consider the groups

$$H^n(G,E^+) \quad \text{and} \quad H^n(G,E^*) \qquad\qquad \text{for all} \quad n \geqslant 1$$

Our first aim is to show that the group $H^n(G,E^+)$ is always trivial. This will be derived as a consequence of a general result of independent interest, namely the normal basis theorem.

Given a finite Galois extension E/F, any F-basis of E of the form

$$\{\sigma(a)\,|\,\sigma \in \mathrm{Gal}(E/F)\}$$

for some $a \in E$, is called a *normal basis*.

18.1. THEOREM. (Normal basis theorem). Any finite Galois extension E/F has a normal basis.

Proof. Put $G = \mathrm{Gal}(E/F)$ and $n = |G|$. We first consider the case where F is finite. Then G is cyclic, say generated by σ. We may regard σ as an F-linear transformation of the F-space E. Let $f(X)$ be its minimal polynomial. Then $\deg f(X) \leqslant n = (E:F)$. But, by Corollary 4.8, $1,\sigma,\ldots,\sigma^{n-1}$ are linearly independent over E, hence also over F, and thus $\deg f(X) = n$. Since $\sigma^n = 1$, we have $f(X) = X^n - 1$. Thus the minimal and characteristic polynomial for σ are the same. By linear algebra, there is a "cyclic vector". That is there exists an element $\alpha \in E$ whose images under powers of σ span E over F, as required.

Assume that F is infinite and let $\sigma_1,\sigma_2,\ldots,\sigma_n$ be all distinct elements of G. Owing to Corollary 3.32, we may write $E = F(\lambda)$ for some $\lambda \in E$. Let $f(X)$ be the minimal polynomial of λ and let $\lambda_i = \sigma_i(\lambda)$, $1 \leqslant i \leqslant n$. Set

$$g(X) = \frac{f(X)}{(X-\lambda)f'(\lambda)}$$

and

$$g_i(X) = \sigma_i(g(X)) = \frac{f(X)}{(X-\lambda_i)f'(\lambda_i)}$$

Then $g_i(X)$ is a polynomial over E with λ_k as a root for $k \neq i$. Hence

$$g_i(X)g_k(X) \equiv 0 \pmod{f(X)} \qquad \text{for } i \neq k \tag{4}$$

In the equation

$$g_1(X) + g_2(X) + \ldots + g_n(X) - 1 = 0 \tag{5}$$

the left side is of degree at most $n - 1$. Furthermore, (5) is true for n distinct values of X, namely $\lambda_1,\lambda_2,\ldots,\lambda_n$ since $g_i(\lambda_i) = 1$ and $g_k(\lambda_i) = 0$ for $k \neq i$. Thus the left side of (5) is identically zero.

Multiplying (5) by $g_i(X)$ and applying (4), we derive

$$g_i(X)^2 \equiv g_i(X) \pmod{f(X)} \tag{6}$$

Now consider the determinant

$$D(X) = |\sigma_i\sigma_k(g(X))| \qquad (1 \leqslant i,k \leqslant n)$$

If we square it by multiplying column by column and compute its value $\pmod{f(X)}$, we obtain from (4), (5) and (6) a determinant that has 1 in the diagonal and 0 elsewhere. Thus

$$D(X)^2 \equiv 1 \pmod{f(X)}$$

Since F is infinite, we may always choose a value μ for X such that

$$D(\mu) \neq 0, \ \mu \in F$$

Now put $\alpha = g(\mu)$. Then the determinant

$$|\sigma_i\sigma_k(\alpha)| \neq 0 \tag{7}$$

Finally, consider any linear relation

$$a_1\sigma_1(\alpha) + a_2\sigma_2(\alpha) + \dots + a_n\sigma_n(\alpha) = 0 \qquad (a_i \in F)$$

Applying the automorphism σ_i to it would lead to n homogeneous equations for the n unknowns $a_i \in F$. Invoking (7), we deduce that each $a_i = 0$ and the result follows. ■

18.2. COROLLARY. Let E/F be any finite Galois extension with Galois group G. Then

$$H^n(G,E^+) = 1 \qquad \text{for all } n \geqslant 1$$

Proof. We first show that the G-module E^+ is isomorphic to the G-module FG, where FG is the group algebra of G over F. Indeed, by Theorem 18.1, the required isomorphism is given by

$$FG \longrightarrow E^+$$

$$\sum_{g\in G} \lambda_g g \longmapsto \sum_{g\in G} \lambda_g g(a)$$

where $a \in E$ is such that $\{\sigma(a)|\sigma \in G\}$ is a normal basis. Invoking (2), we are therefore left to verify that

$$\mathrm{Hom}(\mathbb{Z}G,F) \cong FG \qquad \text{as } G\text{-modules}$$

For each $\lambda \in \mathrm{Hom}(\mathbb{Z}G,F)$ and each $g \in G$, put $\lambda_g = \lambda(g)$. Then the map

$$\begin{cases} \mathrm{Hom}(\mathbb{Z}G,F) \longrightarrow FG \\ \qquad\qquad \lambda \longmapsto \Sigma\lambda_g g \end{cases}$$

is a required isomorphism. ∎

We now turn our attention to the study of $H^n(G,E^*)$. Here the situation is more complicated: the first cohomology is always trivial, but in contrast to Corollary 18.2, the groups $H^n(G,E^*)$ for $n > 1$ are not generally trivial.

18.3. THEOREM. Let E/F be a finite Galois extension with Galois group G. Then $H^1(G,E^*) = 1$

Proof. Let $f : G \longrightarrow E^*$ be a crossed homomorphism. For $\lambda \in E^*$ we put

$$\alpha = \sum_{\sigma\in G} f(\sigma)\sigma\lambda$$

Because of the linear independence of the automorphisms σ(Corollary 4.8), we may choose $\lambda \in E^*$ such that $\alpha \neq 0$. We then obtain for $\tau \in G$

$$\tau\alpha = \sum_{\sigma\in G} \tau f(\sigma)(\tau\sigma\lambda) = \sum_{\sigma\in G} f(\tau)^{-1}f(\tau\sigma)(\tau\sigma\lambda) = f(\tau)^{-1}\alpha$$

which implies that

$$f(\tau) = \frac{\tau\alpha^{-1}}{\alpha^{-1}}$$

Therefore $f \in B^1(G,E^*)$ and so $H^1(G,E^*) = 1$. ∎

In what follows, we shall fix the following notation:

E/F a finite Galois extension

$G = \mathrm{Gal}(E/F)$, N a normal subgroup of G

$\delta f \in B^2(G,E^*)$ is a coboundary corresponding to $f : G \longrightarrow E^*$, i.e.

$$(\delta f)(\sigma_1,\sigma_2) = \sigma_1(f(\sigma_2))f(\sigma_1\sigma_2)^{-1}f(\sigma_1) \qquad (\sigma_1,\sigma_2 \in G)$$

K the fixed field of N.

The above assumptions guarantee that K/F is a Galois extension whose Galois group is identifiable with G/N. Each $\alpha \in Z^2(G/N,K^*)$ determines a unique

$\alpha^* \in Z^2(G,E^*)$ given by

$$\alpha^*(\sigma_1,\sigma_2) = \alpha(\bar{\sigma}_1,\bar{\sigma}_2)$$

where $\bar{\sigma}_i = \sigma_i N$, $i = 1,2$. The map $\alpha \longmapsto \alpha^*$ induces a homomorphism

$$\text{Inf} : H^2(G/N,K^*) \longrightarrow H^2(G,E^*)$$

called *inflation map*. By restricting the cochains to $N \times N$, we also have a homomorphism

$$\text{Res} : H^2(G,E^*) \longrightarrow H^2(N,E^*)$$

called the *restriction map*. These two homomorphisms are connected by the following theorem.

18.4. THEOREM. (Hochschild(1950), Eilenberg and MacLane(1948)). The sequence

$$1 \longrightarrow H^2(G/N,K^*) \xrightarrow{\text{Inf}} H^2(G,E^*) \xrightarrow{\text{Res}} H^2(N,E^*)$$

is exact.

Proof. For the sake of clarity, we divide the proof into three steps.

Step 1. Here we show that Inf is injective.

Assume that $\alpha \in Z^2(G/N,K^*)$ is such that $\alpha^* = \delta t$ where $t : G \longrightarrow E^*$, i.e.

$$\alpha^*(\sigma_1,\sigma_2) = \sigma_1(t(\sigma_2))t(\sigma_1\sigma_2)^{-1}t(\sigma_1) \qquad (\sigma_1,\sigma_2 \in G)$$

Then $\alpha(1,\bar{\sigma}) = t(1)$ so that $t(1) \in K^*$. Now set $t_1(\sigma) = \sigma(t(1))$. Then $t_1(\sigma) \in K^*$ for all $\sigma \in G$, and $t_1(\sigma n) = t_1(\sigma)$ for all $\sigma \in G$, $n \in N$. This implies that $t_1 = \beta^*$ for some $\beta : G/N \longrightarrow K^*$, where $\beta^*(\sigma) = \beta(\bar{\sigma})$. Setting $\alpha_1 = \alpha\delta\beta^{-1}$, it suffices to verify that α_1 is a coboundary.

We have $\alpha_1^* = \delta s$, where $s(\sigma) = t(\sigma)\sigma(t(1))^{-1}$. If $n_1,n_2 \in N$ then

$$(\delta s)(n_1,n_2) = 1$$

By Theorem 18.3, we can find $\lambda \in E^*$ such that $s(n) = n(\lambda)\lambda^{-1}$ for all $n \in N$. Let us set $\mu(\sigma) = s(\sigma)\lambda\sigma(\lambda)^{-1}$. Then $\delta\mu = \delta s$, so that $\alpha_1^* = \delta\mu$. Furthermore, $\mu(n) = 1$ for all $n \in N$. This gives

$$1 = \alpha_1^*(\sigma,n) = \sigma(\mu(n))\mu(\sigma n)^{-1}\mu(\sigma) = \mu(\sigma n)^{-1}\mu(\sigma) \qquad (\sigma \in G,\ n \in N)$$

i.e. $\mu(\sigma n) = \mu(\sigma)$ for all $\sigma \in G$, $n \in N$. Also

$$1 = \alpha_1^*(n,\sigma) = n(\mu(\sigma))\mu(n\sigma)^{-1}\mu(n) = n(\mu(\sigma))\mu(n\sigma)^{-1} = n(\mu(\sigma))\mu(\sigma)^{-1} (\sigma \in G,\ n \in N)$$

Therefore $n(\mu(\sigma)) = \mu(\sigma)$ for all $\sigma \in G$, $n \in N$ which means that $\mu(\sigma) \in K^*$. Thus we have $\mu = u^*$ for some $u : G/N \longrightarrow K^*$, which shows that

$$\alpha = \alpha_1 \delta\beta = \delta u \delta\beta = \delta(u\beta),$$

as required.

Step 2. We now prove that $\mathrm{Im}\,\mathrm{Inf} \subseteq \mathrm{Ker}\,\mathrm{Res}$.

Assume that $\alpha \in Z^2(G/N,K^*)$. It suffices to show that α^* is equal to a coboundary on $N \times N$. To this end, set $\beta(n) = \alpha^*(1,1)$ for $n \in N$. Then

$$(\delta\beta)(n_1,n_2) = \alpha^*(1,1) = \alpha^*(n_1,n_2) \qquad (n_1,n_2 \in N)$$

i.e. $\alpha^* = \delta\beta$ on $N \times N$, as required.

Step 3. Here we complete the proof by showing that $\mathrm{Ker}\mathrm{Res} \subseteq \mathrm{Im}\mathrm{Inf}$.

Let $\alpha \in Z^2(G,E^*)$ and assume that, on $N \times N$, $\alpha = \delta t$ for some $t : N \longrightarrow E^*$. We may write $G = N \cup \tau_1 N \cup \ldots \cup \tau_k N$, where $1, \tau_1,\ldots,\tau_k$ is a transversal for N in G. We extend t to a cochain (still denoted by t) for E/F by setting

$$t(\tau_i n) = \tau_i(t(n))\alpha(\tau_i,n)^{-1}\alpha(1,n) \qquad (1 \leqslant i \leqslant k,\ n \in N)$$

If $\alpha_1 = \alpha\delta t^{-1}$, we have $\alpha_1(n_1,n_2) = 1$ for all $n_1,n_2 \in N$. Moreover, an easy verification shows that $\alpha_1(\tau_i n_1,n_2) = 1$ for each i. Thus $\alpha_1(\sigma,n) = 1$ for all $\sigma \in G$, $n \in N$. Now we have

$$1 = \sigma_1(\alpha_1(\sigma_2,n)) = \alpha_1(\sigma_1\sigma_2,n)\alpha_1(\sigma_1,\sigma_2 n)^{-1}\alpha_1(\sigma_1,\sigma_2)$$

so that $\alpha_1(\sigma_1,\sigma_2 n) = \alpha_1(\sigma_1,\sigma_2)$, i.e. $\alpha_1(\sigma_1,\sigma_2)$ depends only on σ_1 and the coset $\bar{\sigma}_2$ of σ_2 mod N.

Now define $s_\sigma(n) = \alpha_1(n,\sigma)$. Then we have

$$\begin{aligned} n_1(s_\sigma(n_2))s_\sigma(n_1n_2)^{-1}s_\sigma(n_1) &= n_1(\alpha_1(n_2,\sigma))\alpha_1(n_1n_2,\sigma)^{-1}\alpha_1(n_1,\sigma) \\ &= \alpha_1(n_1,n_2\sigma)^{-1}\alpha_1(n_1,n_2)\alpha_1(n_1,\sigma) \\ &= 1 \end{aligned}$$

i.e. $s_\sigma \in Z^1(N,E^*)$. Owing to Theorem 18.3, we can find $\lambda_\sigma \in E^*$ such that $s_\sigma(n) = n(\lambda_\sigma)\lambda_\sigma^{-1}$. Because s_σ depends only on $\bar{\sigma}$ and $s_1 = 1$, we may take λ_σ

to depend only on $\bar{\sigma}$ and $\lambda_1 = 1$. If we define

$$\alpha(\sigma_1,\sigma_2) = \alpha_1(\sigma_1,\sigma_2)\sigma_1(\lambda_{\sigma_2})^{-1}\lambda_{\sigma_1\sigma_2}\lambda_{\sigma_1}^{-1}$$

then α_2 is obviously in the same cohomology class as α_1 and hence represents the same element of $H^2(G,E^*)$ as do α_1 and α. Clearly, $\alpha_2(\sigma_1,\sigma_2)$ depends only on σ_1 and the coset $\bar{\sigma}_2$ of σ_2 mod N. Moreover,

$$\begin{aligned}\alpha_2(\sigma_1 n,\sigma_2) &= \alpha_1(\sigma_1 n,\sigma_2)\sigma_1 n(\lambda_{\sigma_2})^{-1}\lambda_{\sigma_1\sigma_2}\lambda_{\sigma_1}^{-1}\\ &= \alpha_1(\sigma_1 n,\sigma_2)\sigma_1(\lambda_{\sigma_2}\alpha_1(n,\sigma_2))^{-1}\lambda_{\sigma_1\sigma_2}\lambda_{\sigma_1}^{-1}\\ &= \alpha_1(\sigma_1,n\sigma_2)\sigma_1(\lambda_{\sigma_2})^{-1}\lambda_{\sigma_1\sigma_2}\lambda_{\sigma_1}^{-1}\\ &= \alpha_2(\sigma_1,\sigma_2)\end{aligned}$$

proving that $\alpha_2(\sigma_1,\sigma_2)$ depends only on $\bar{\sigma}_1$ and $\bar{\sigma}_2$. Finally, by our construction, $\alpha_2(n,\sigma) = 1$ which implies

$$n(\alpha_2(\sigma_1,\sigma_2)) = \alpha_2(n\sigma_1,\sigma_2)\alpha_2(n,\sigma_1\sigma_2)^{-1}\alpha_2(n,\sigma_1) = \alpha_2(\sigma_1,\sigma_2)$$

Thus $\alpha_2(\sigma_1,\sigma_2) \in K^*$ and therefore $\alpha_2 = t^*$ for some $t \in Z^2(G/N,K^*)$. This means that the cohomology class of α belongs to Im Inf, as required. ■

18.5. COROLLARY. If N is of order m and if c is an arbitrary element of $H^2(G,E^*)$, then $c^m \in \text{Im}(\text{Inf})$ where $\text{Inf} : H^2(G/N,K^*) \longrightarrow H^2(G,E^*)$ is the inflation map.

Proof. Let $\alpha \in Z^2(G,E^*)$. Define $t : G \longrightarrow E^*$ by

$$t(\sigma) = \prod_{n\in N}\alpha(\sigma,n)$$

Then we have

$$\begin{aligned}(\delta t)(\sigma_1,\sigma_2) &= \prod_{n\in N}(\sigma_1(\alpha(\sigma_2,n))\alpha(\sigma_1\sigma_2,n)^{-1}\alpha(\sigma_1,n))\\ &= \prod_{n\in N}(\alpha(\sigma_1,\sigma_2 n)^{-1}\alpha(\sigma_1,\sigma_2)\alpha(\sigma_1,n))\\ &= \alpha(\sigma_1,\sigma_2)^m\prod_{n\in N}(\alpha(\sigma_1,\sigma_2 n)^{-1}\alpha(\sigma_1,n))\end{aligned}$$

In particular, if $\sigma_2 \in N$ this yields

$$(\delta t)(\sigma_1,\sigma_2) = \alpha(\sigma_1,\sigma_2)^m$$

whence c^m is in Ker Res , where $\text{Res} : H^2(G,E^*) \longrightarrow H^2(N,E^*)$ is the restriction map. The desired conclusion now follows by virtue of Theorem 18.4. ■

Two Galois extensions E_1/F and E_2/F are said to be *isomorphic* if E_1 is F-isomorphic to E_2.

18.6. THEOREM. Let E_1/F and E_2/F be isomorphic finite Galois extensions and let $G_i = \mathrm{Gal}(E_i/F)$, $i = 1,2$. Then there is an isomorphism

$$f : H^2(G_1,E_1^*) \longrightarrow H^2(G_2,E_2^*)$$

such that for any F-isomorphism $g : E_1 \longrightarrow E_2$ the mapping

$$\begin{cases} Z^2(G_1,E_1^*) \longrightarrow Z^2(G_2,E_2^*) \\ \alpha \longmapsto \alpha_g \end{cases}$$

where $\alpha_g(\sigma_1,\sigma_2) = g(\alpha(g^{-1}\sigma_1 g, g^{-1}\sigma_2 g))$, induces the isomorphism f.

Proof. It is clear that the mapping $\alpha \longmapsto \alpha_g$ is a homomorphism which maps coboundaries to coboundaries. Because it has an inverse, it induces an isomorphism of $H^2(G_1,E_1^*)$ onto $H^2(G_2,E_2^*)$. If λ is another F-isomorphism $E_1 \longrightarrow E_2$, then $\lambda g^{-1} = \mu$, say, is an automorphism of E_2/F, and we have $\alpha_\lambda = (\alpha_g)_\mu$.

We now claim that, for any $\beta \in Z^2(G_2,E_2^*)$, $\beta\beta_\mu^{-1}$ is a coboundary. In fact, if we put $t(\sigma) = \beta(\sigma,\mu)\beta(\mu,\mu^{-1}\sigma\mu)^{-1}$, then we have

$$(\sigma t)(\sigma_1,\sigma_2) = \beta(\sigma_1,\sigma_2)\beta_\mu(\sigma_1,\sigma_2)^{-1}$$

as can easily be verified. Since $\alpha_\lambda = (\alpha_g)_\mu$, we conclude that α_λ is cohomologous to α_g, as required. ∎

The unique isomorphism f of Theorem 18.6 is called the *natural isomorphism*. Next we introduce the following important homomorphism.

18.7. LEMMA. Let E/F be a finite Galois extension with Galois group G and, for any $\alpha \in Z^2(G,E^*)$, let $\bar{\alpha}$ denote the cohomology class of α. Given $c \in H^2(G,E^*)$, say $c = \bar{\alpha}$, define

$$\begin{cases} G \xrightarrow{c'} F^*/N_{E/F}(E^*) \\ \sigma \longmapsto (\prod_{\tau\in G}\alpha(\tau,\sigma))N_{E/F}(E^*) \end{cases}$$

Then c' is a well-defined homomorphism and the map

$$\begin{cases} H^2(G,E^*) \longrightarrow \mathrm{Hom}(G,F^*/N_{E/F}(E^*)) \\ c \longmapsto c' \end{cases}$$

is a homomorphism.

Proof. Given $\alpha \in Z^2(G,E^*)$, put

$$\bar{\alpha}(\sigma) = \prod_{\tau\in G}\alpha(\tau,\sigma)$$

It is easy to verify that, for all $\sigma_1,\sigma_2 \in G$, we have

$$\sigma_1(\bar{\alpha}(\sigma_2)) = \bar{\alpha}(\sigma_2)$$

which means that $\bar{\alpha}$ maps G into F^*. If $\beta \in Z^2(G,E^*)$ is such that $\bar{\beta} = \bar{\alpha}$, say $\alpha\beta^{-1} = \delta g$, then

$$\bar{\alpha}(\sigma)\bar{\beta}(\sigma)^{-1} = \prod_{\tau\in G}\tau(g(\sigma))g(\tau\sigma)^{-1}g(\tau) = N_{E/F}(g(\sigma))$$

This shows that the map c' is well-defined.

By the multiplicative version of (3), we have

$$\tau\alpha(\sigma_1,\sigma_2) = \alpha(\tau\sigma_1,\sigma_2)\alpha(\tau,\sigma_1\sigma_2)^{-1}\alpha(\tau,\sigma_1)$$

whence

$$N_{E/F}(\alpha(\sigma_1,\sigma_2)) = \prod_{\tau\in G}\tau(\alpha(\sigma_1,\sigma_2)) = \bar{\alpha}(\sigma_2)\bar{\alpha}(\sigma_1\sigma_2)^{-1}\bar{\alpha}(\sigma_1)$$

which shows that

$$\bar{\alpha}(\sigma_1\sigma_2) = \bar{\alpha}(\sigma_1)\bar{\alpha}(\sigma_2)N_{E/F}(\alpha(\sigma_1,\sigma_2)^{-1})$$

This proves that c' is a homomorphism.

Finally, fix $c = \bar{\alpha}$, $d = \bar{\beta}$ in $H^2(G,E^*)$. Then, for any $\sigma \in G$,

$$\begin{aligned}(c'd')(\sigma) = c'(\sigma)d'(\sigma) &= \Big(\prod_{\tau\in G}\alpha(\tau,\sigma)\Big)\Big(\prod_{\tau\in G}\beta(\tau,\sigma)\Big)N_{E/F}(E^*)\\ &= \prod_{\tau\in G}(\alpha\beta)(\tau,\sigma)N_{E/F}(E^*)\\ &= (cd)'(\sigma)\end{aligned}$$

proving that the map $c \longmapsto c'$ is a homomorphism. ■

As an application, we prove

18.8. THEOREM. Let E/F be a finite cyclic extension with Galois group G. Then the map

$$\begin{cases} H^2(G,E^*) \longrightarrow \mathrm{Hom}(G,F^*/N_{E/F}(E^*))\\ c \longmapsto c'\end{cases}$$

is an isomorphism.

Proof. By Lemma 18.7, the given map is a homomorphism. Assume that $c' = 1$. Then, if $c = \bar{\alpha}$, we can find an element $g \in C^1(G,E^*)$ such that

$$\bar{\alpha}(\sigma) = N_{E/F}(g(\sigma)) \qquad \text{for all } \sigma \in G$$

Since α is cohomologous to a normalized cocycle, we may assume that α is normalized, in which case $\alpha(1) = 1$. Then $\alpha\delta g^{-1}$ is still normalized and, furthermore, we have $\overline{\alpha\delta g^{-1}}(\sigma) = 1$ for all $\sigma \in G$. Hence we may assume that α is normalized and that $\bar{\alpha}(\sigma) = 1$ for all $\sigma \in G$.

Let σ be a fixed generator of G. We define an element $t \in C^1(G,E^*)$ by setting $t(1) = 1 = t(\sigma)$ and, for $1 < a < n = (E:F)$,

$$t(\sigma^a) = \prod_{i=1}^{a-1} \alpha(\sigma^i,\sigma)$$

Then a direct computation shows that $\delta t = \alpha^{-1}$, proving that the map $c \longmapsto c'$ is injective.

Finally, if λ is an arbitrary element in F^*, define

$$\alpha(\sigma^a,\sigma^b) = 1 \qquad \text{for } a + b < n$$

and

$$\alpha(\sigma^a,\sigma^b) = \lambda \qquad \text{for } a + b \geqslant n$$

Then it is easy to verify that α is a cocycle and that $\bar{\alpha}(\sigma) = \lambda$. Since a homomorphism of G into $F^*/N_{E/F}(E^*)$ is determined by the image of σ, this shows that the map $c \longmapsto c'$ is surjective, as required. ■

Let E/F be a finite field extension and let K be an intermediate field. Then, by Proposition 6.6, $N_{E/F} = N_{K/F}N_{E/K}$. Hence the map

$$N_{K/F} : K \longrightarrow F$$

induces a homomorphism

$$\bar{N}_{K/F} : K^*/N_{E/K}(E^*) \longrightarrow F^*/N_{E/F}(E^*)$$

18.9. THEOREM. (Nakayama(1936)). Let E/F be a finite Galois extension with Galois group G, let K be an intermediate field and let $H = \mathrm{Gal}(E/K)$. For any given $c \in H^2(G,E^*)$ let $c' \in \mathrm{Hom}(G,F^*/N_{E/F}(E^*))$ be defined as in Lemma 18.7 and let

$$\mathrm{Res} : H^2(G,E^*) \longrightarrow H^2(H,E^*)$$

be the restriction map. Then

$$c' = \bar{N}_{K/F} \circ (\mathrm{Res}(c)') \quad \text{on } H$$

Proof. Let $\tau_1, \tau_2, \ldots, \tau_k$ be a left transversal for H in G. Then

$$\tau_1, \tau_2, \ldots, \tau_k$$

induce all distinct F-homomorphisms $K \longrightarrow E$ and if $a \in K$, then by Corollary 6.5(ii),

$$N_{K/F}(a) = \prod_{i=1}^{k} \tau_i(a) \tag{8}$$

Let $\alpha \in Z^2(G,E^*)$ be such that $c = \bar{\alpha}$. Then for $h_1 \in H$, $\mathrm{Res}(c)'(h_1)$ is the coset mod $N_{E/K}(E^*)$ of the element $\prod_{h\in H} \alpha(h,h_1)$. Therefore $\bar{N}_{K/F}(\mathrm{Res}(c)'(h_1))$ is the coset mod $N_{E/F}(E^*)$ of

$$\begin{aligned} N_{K/F}\Big(\prod_{h\in H}\alpha(h,h_1)\Big) &= \prod_{i=1}^{k}\prod_{h\in H}\tau_i(\alpha(h,h_1)) && \text{(by (8))} \\ &= \prod_{i=1}^{k}\prod_{h\in H}\alpha(\tau_i h,h_1)\alpha(\tau_i,hh_1)^{-1}\alpha(\tau_i,h) && \text{(by (3))} \\ &= \prod_{i=1}^{k}\prod_{h\in H}\alpha(\tau_i h,h_1) \\ &= \bar{\alpha}(h_1) \end{aligned}$$

whence the result. ■

Let E/F be a finite Galois extension with Galois group G, let N be a normal subgroup of G and let K be the fixed field of N. Then K/F is a Galois extension with Galois group G/N. Let

$$\mathrm{Inf} : H^2(G/N,K^*) \longrightarrow H^2(G,E^*)$$

be the inflation map. By Theorem 18.4, Inf is injective and so, by Corollary 18.5, for any given $c \in H^2(G,E^*)$ there exists a unique $d_c \in H^2(G/N,K^*)$ such that

$$c^m = \mathrm{Inf}(d_c) \qquad (m = |N|)$$

The map

$$\begin{cases} H^2(G,E^\star) \longrightarrow H^2(G/N,K^\star) \\ \qquad c \longmapsto d_c \end{cases}$$

is obviously a homomorphism.

Since $N_{E/F}(E^\star) \subseteq N_{K/F}(K^\star)$, there is a natural homomorphism

$$\pi_{E,K}: F^\star/N_{E/F}(E^\star) \longrightarrow F^\star/N_{K/F}(K^\star)$$

Finally, we denote by π the natural homomorphism $G \longrightarrow G/N$.

18.10. THEOREM. (Akizuki(1936)). Let E/F be a finite Galois extension with Galois group G, let N be a normal subgroup of G and let K be the fixed field of N. For each $c \in H^2(G,E^\star)$, let $c' \in \mathrm{Hom}(G,F^\star/N_{E/F}(E^\star))$ be defined as in Lemma 18.7. Then, for each $c \in H^2(G,E^\star)$, the diagram

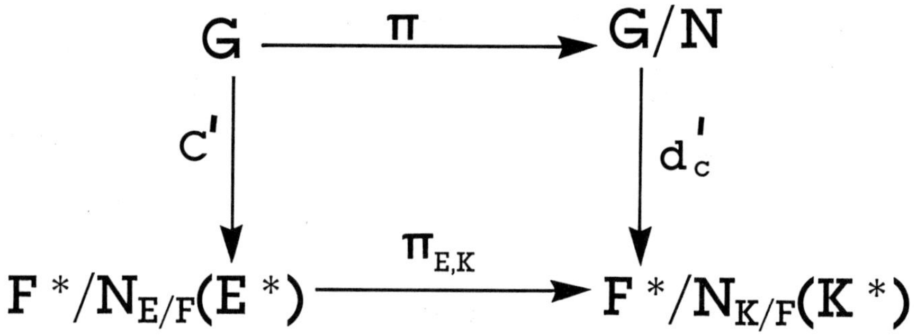

is commutative.

Proof. Let $\alpha \in Z^2(G,E^\star)$ and $\beta \in Z^2(G/N,K^\star)$ be such that $c = \bar{\alpha}$ and $d_c = \bar{\beta}$. We must show that for all $\sigma \in G$.

$$\bar{\alpha}(\sigma)N_{K/F}(K^\star) = \bar{\beta}(\sigma N)N_{K/F}(K^\star) \tag{9}$$

where

$$\bar{\alpha}(\sigma) = \prod_{\tau\in G}\alpha(\tau,\sigma) \quad \text{and} \quad \bar{\beta}(\sigma N) = \sum_{\lambda\in G/N}\beta(\lambda,\sigma N)$$

As in the proof of Corollary 18.5, set

$$t(\sigma) = \prod_{n\in N}\alpha(\sigma,n)$$

That proof shows that if

$$\ell(\sigma_1,\sigma_2) = \alpha(\sigma_1,\sigma_2)^m\ (\delta t)(\sigma_1,\sigma_2)^{-1}$$

then

$$\ell(\sigma_1,\sigma_2) = \prod_{n\in N}\alpha(\sigma_1,\sigma_2 n)\alpha(\sigma_1,n)^{-1}$$

For $\sigma_1 \in G$ and $n_1 \in N$, we have therefore $\ell(\sigma_1,n_1) = 1$. Moreover, for $\sigma_1,\sigma_2 \in G$, $\ell(\sigma_1,\sigma_2)$ depends only on σ_1 and the coset $\bar{\sigma}_2 = \sigma_2 N$. Now we retrace the computation we have made in the last part of the proof of Theorem 18.4. There we found that if $s_\sigma(n) = \ell(n,\sigma)$ for $n \in N$, then there exist elements $\lambda_\sigma \in E^*$ such that

$$s_\sigma(n_1) = n_1(\lambda_\sigma)\lambda_\sigma^{-1}$$

i.e. we must have

$$n_1(\lambda_\sigma)\lambda_\sigma^{-1} = \prod_{n\in N}\alpha(n_1,\sigma n)\alpha(n_1,n)^{-1}$$

Moreover, we may take λ_σ to depend only on the coset $\bar{\sigma} = \sigma N$ and $\lambda_1 = 1$. Now we know from our earlier computation (see Step 3 in the proof of Theorem 18.4) that if

$$s(\sigma_1,\sigma_2) = \ell(\sigma_1,\sigma_2)\sigma_1(\lambda_{\sigma_2})^{-1}\lambda_{\sigma_1\sigma_2}\lambda_{\sigma_1}^{-1}$$

then

$$s(\sigma_1,\sigma_2) \in K^*$$

and depends only on the cosets $\bar{\sigma}_1$ and $\bar{\sigma}_1$. Clearly, s is the natural extension to $Z^2(G,E^*)$ of β in $Z^2(G/N,K^*)$. Because N is a normal subgroup of G, we may rewrite $\ell(\sigma_1,\sigma_2)$ as

$$\ell(\sigma_1,\sigma_2) = \prod_{n\in N}\alpha(\sigma_1,n\sigma_2)\alpha(\sigma_1,n)^{-1}$$

By the coboundary formula, this gives

$$\ell(\sigma_1,\sigma_2) = \prod_{n\in N}\alpha(\sigma_1 n,\sigma_2)\sigma_1(\alpha(n,\sigma_2))^{-1}$$

Now let $\tau_1 N,\ldots,\tau_r N$ be all the distinct cosets of N in G. Then, by what we have seen above, $d'_c\pi(\sigma)$ is the coset mod $N_{K/F}(K^*)$ of the element

$$\prod_{i=1}^{r} s(\tau_i,\sigma) = \bar{\beta}(\bar{\sigma})$$

We have

$$\prod_{i=1}^{r} s(\tau_i,\sigma) = \prod_{i=1}^{r} \left(\prod_{n\in N}\alpha(\tau_i n,\sigma)\tau_i(\alpha(n,\sigma))^{-1}\right)\tau_i(\lambda_\sigma)^{-1}\lambda_{\tau_i\sigma}\lambda_{\tau_i}^{-1}$$
$$= \left(\prod_{\tau\in G}\alpha(\tau,\sigma)\right)\left(\prod_{i=1}^{r}\tau_i(\lambda_\sigma\prod_{n\in N}\alpha(n,\sigma))\right)^{-1}$$

Set $\gamma_\sigma = \lambda_\sigma \prod_{n\in N}\alpha(n,\sigma)$. Then, for $n_1 \in N$,

$$n_1(\gamma_\sigma) = n_1(\lambda_\sigma)\prod_{n\in N} n_1(\alpha(n,\sigma))$$
$$= \lambda_\sigma \prod_{n\in N}\alpha(n_1,n\sigma)\alpha(n_1,n)^{-1}n_1(\alpha(n,\sigma))$$
$$= \lambda_\sigma \prod_{n\in N}\alpha(n_1 n,\sigma)$$
$$= \gamma_\sigma$$

Hence $\gamma_\sigma \in K^*$ and we have

$$\prod_{i=1}^{r} s(\tau_i,\sigma) = \left(\prod_{\tau\in G}\alpha(\tau,\sigma)\right)N_{K/F}(\gamma_\sigma)^{-1}$$

which shows that

$$\bar{\alpha}(\sigma) = N_{K/F}(\gamma_\sigma)\bar{\beta}(\bar{\sigma})$$

This proves (9) and hence the result. ■

We close by providing the isomorphism class of $H^2(G,E^*)$ in case G is cyclic. In what follows we use the following standard fact of cohomology theory.

18.11. PROPOSITION. If A is a G-module, where G is a finite cyclic group, then

$$H^2(G,A) \cong A^G/(\sum_{g\in G} g)A$$

Proof. See Weiss(1969)). ■

18.12. THEOREM. Let E/F be a finite cyclic extension with Galois group G. Then

$$H^2(G,E^*) \cong F^*/N_{E/F}(E^*)$$

Proof. If $A = E^*$, then $A^G = F^*$ and (in the additive notation)

$$N_{E/F}(E^*) = (\sum_{g\in G} g)A$$

The desired conclusion is therefore a consequence of Proposition 18.11. ■

Remark. This result can also be obtained as an easy consequence of Theorem 18.8.

19. THE BRAUER GROUP OF A FIELD

Throughout this section, F denotes a field and all F-algebras are assumed to be *finite-dimensional.*

Let E/F be a finite Galois extension and let $G = \mathrm{Gal}(E/F)$. Our aim is to provide an interpretation of the group $H^2(G,E^*)$ in terms of the Brauer group $Br(F)$ of F. It will be shown that if $Br(E/F)$ is the subgroup of $Br(F)$ consisting of those $[A]$ which are split by E, then

$$H^2(G,E^*) \cong Br(E/F)$$

As a consequence, we shall deduce that every Brauer class contains a crossed product. We begin by providing all relevant definitions and establishing some results pertaining to the theory of central simple algebras.

Let A be an algebra over F. Then the centre $Z(A)$ of A contains an isomorphic copy $F\cdot 1 = \{\lambda\cdot 1 \mid \lambda \in F\}$ of F. For convenience, we shall identify F and $F\cdot 1$. We say that A is a *central simple* F-algebra if A is simple and if $Z(A) = F$.

Assume that A is a central simple F-algebra. Then there is a division algebra D with $Z(D) = F$ and a natural number n such that

$$A \cong M_n(D) \cong D \otimes_F M_n(F)$$

The algebra D is unique up to F-isomorphism, while n is unique. We shall refer to D as the *skewfield component* of A. Two central simple F-algebras A,B are said to be *similar*, written $A \sim B$, if their skewfield components are F-isomorphic. This is clearly an equivalence relation and we denote by $[A]$ the equivalence class of A. We shall refer to A as the Brauer class of A. Our first aim is to show that the Brauer classes of central simple F-algebras form an abelian group with respect to the multiplication induced by the tensor product. This will be achieved with the aid of some preliminary results presented below.

Let V be an R-module, where R is an arbitrary ring, and let $S = \mathrm{End}_R(V)$.

Then V is also an S-module, the operation of S on V being given by

$$\phi v = \phi(v) \qquad \text{for all } \phi \in S,\ v \in V$$

Each $r \in R$ induces an S-isomorphism $f_r : V \longrightarrow V$ given by

$$f_r(v) = rv \qquad \text{for all } v \in V$$

Thus we get a ring homomorphism

$$\begin{cases} R \longrightarrow \operatorname{End}_S(V) \\ r \longmapsto f_r \end{cases}$$

We say that R *acts densely* on V if for each $\theta \in \operatorname{End}_S(V)$ and each finite family $v_1, v_2, \ldots, v_n$ in V there exists $r \in R$ such that

$$\theta(v_i) = f_r(v_i) \qquad (1 \leqslant i \leqslant n)$$

19.1. LEMMA. Let V be an R-module, let $S = \operatorname{End}_R(V)$ and let V be finitely generated as S-module. Then R acts densely on V if and only if the homomorphism

$$\begin{cases} R \longrightarrow \operatorname{End}_S(V) \\ r \longmapsto f_r \end{cases}$$

is surjective.

Proof. If the given homomorphism is surjective, then obviously R acts densely on V. Conversely, assume that R acts densely on V and let $v_1, \ldots, v_n$ be a generating set for the S-module V. If $\theta \in \operatorname{End}_S(V)$, then by hypothesis there exists $r \in R$ with $\theta(v_i) = f_r(v_i)$. Thus, for any $\lambda_1, \lambda_2, \ldots, \lambda_n$ in S,

$$\theta\left(\sum_{i=1}^{n} \lambda_i v_i\right) = f_r\left(\sum_{i=1}^{n} \lambda_i v_i\right),$$

proving that $f_r = \theta$. ■

19.2. LEMMA. Assume that an R-module V is a finite direct sum of irreducible submodules. Then R acts densely on V.

Proof. Put $S = \operatorname{End}_R(V)$ and fix $v \in V$, $f \in \operatorname{End}_S(V)$. We claim that

$$rv = f(v)$$

for some $r \in R$. Because V is completely reducible, $V = Rv \oplus W$ for some R-

submodule W. Let $\pi : V \longrightarrow Rv$ be the projection. Then $\pi \in S$ and so

$$f(v) = f(\pi v) = \pi f(v)$$

This shows that $f(v) \in Rv$, as claimed.

Let $\theta \in \text{End}_S(V)$ and let $v_1,\ldots,v_n \in V$. Define $\theta_n : V^n \longrightarrow V^n$ by

$$\theta_n(x_1,\ldots,x_n) = (\theta(x_1),\ldots,\theta(x_n))$$

and put $S' = \text{End}_R(V^n)$. Then S' is a full matrix ring over S. Using this and the assumption on V, one immediately verifies that θ_n lies in $\text{End}_S(V^n)$. Hence, by the first paragraph, there exists an element $r \in R$ such that

$$(rv_1,\ldots,rv_n) = (\theta(v_1),\ldots,\theta(v_n))$$

which is what we wanted to prove. ■

For any F-algebra A we can form the so-called enveloping algebra $A \otimes_F A^0$, where A^0 denotes the F-algebra opposite to A. Given $a \in A$, let a_ℓ denote left multiplication by a on A, and a_r right multiplication. Then the maps

$$\begin{cases} A \longrightarrow A_\ell \subseteq \text{End}_F(A) \\ a \longmapsto a_\ell \end{cases}$$

and

$$\begin{cases} A^0 \longrightarrow A_r \subseteq \text{End}_F(A) \\ a \longmapsto a_r \end{cases}$$

where

$$A_\ell = \{a_\ell | a \in A\} \quad \text{and} \quad A_r = \{a_r | a \in A\}$$

are injective homomorphisms of F-algebras. By the associativity of A, we have

$$a_\ell b_r = b_r a_\ell \qquad \text{for all } a,b \in A$$

Hence A becomes a left $(A \otimes_F A^0)$-module upon defining

$$(a \otimes b)x = (a_\ell b_r)(x) = axb \qquad (a,b,x \in A)$$

19.3. LEMMA. Let A be an F-algebra and let $R = A \otimes_F A^0$

(i) $\text{End}_R(A) = Z(A)_\ell \cong Z(A)$

(ii) A is simple if and only if A is an irreducible R-module

(iii) If A is simple, then $Z(A)$ is a field

Proof. (i) The inclusion one way is clear. Conversely, assume that $f \in \text{End}_R(A)$. Then, by an easy computation, $f = f(1)_\ell$, and one then checks that $f(1) \in Z(A)$ since $f \in \text{End}_R(A)$.

(ii) The two-sided ideals in A are precisely the R-submodules of A

(iii) If A is simple, then for any $0 \neq a \in Z(A)$, $Aa = Aa = A$. Hence

$$a \in U(A) \cap Z(A) = U(Z(A)),$$

as required. ■

19.4. LEMMA. Let A be a central simple F-algebra

(i) $A \otimes_F A^\circ$ acts densely on A

(ii) $A \otimes_F A^\circ \cong M_n(F)$ where $n = \dim_F A$.

Proof. (i) Put $R = A \otimes_F A^\circ$ and note that A is an irreducible R-module by Lemma 19.3(ii). By Lemma 19.2, it follows that R acts densely on A.

(ii) By Lemma 19.3(i), $\text{End}_R(A) \cong F$ and so, by (i) and Lemma 19.1, the map $R \longrightarrow \text{End}_F(A)$, $r \longmapsto f_r$, is a surjective homomorphism of F-algebras. Since $\text{End}_F A \cong M_n(F)$, we have

$$\dim_F R = n^2 = \dim_F (\text{End}_F A)$$

and the result follows. ■

19.5. LEMMA. Let U,V be vector spaces over F and let U',V' be subspaces of U and V, respectively. Then

$$(U' \otimes_F V) \cap (U \otimes_F V') = U' \otimes_F V'$$

Proof. We can write

$$U = U' \oplus U'' \quad \text{and} \quad V = V' \oplus V''$$

for some subspaces U'',V'' of U and V, respectively. Accordingly,

$$U \otimes_F V = (U' \otimes_F V') \oplus (U' \otimes_F V'') \oplus (U'' \otimes_F V') \oplus (U'' \otimes_F V'')$$

Now the left-hand side of the required equality is

$$[(U' \underset{F}{\otimes} V') \oplus (U' \underset{F}{\otimes} V'')] \cap [(U' \underset{F}{\otimes} V') \oplus (U'' \underset{F}{\otimes} V')]$$

and is clearly $U' \underset{F}{\otimes} V'$. ■

The next observation describes a useful lattice isomorphism.

19.6. LEMMA. Let A be a central simple F-algebra and let B be any F-algebra. Then the map $I \longrightarrow A \underset{F}{\otimes} I$ is a lattice isomorphism between the ideals of B and those of $A \underset{F}{\otimes} B$. The inverse of this isomorphism is given by

$$J \longmapsto J \cap B$$

where B is identified with $1 \otimes B$

Proof. We claim that the given maps are inverse to each other. This will prove the result, for they establish a bijection which is clearly order-preserving and hence a lattice isomorphism.

Let I be an ideal in B. Then $A \underset{F}{\otimes} I$ is obviously an ideal in $A \underset{F}{\otimes} B$ and

$$(A \underset{F}{\otimes} I) \cap B = I$$

by virtue of Lemma 19.5.

Conversely, let J be an ideal in $A \underset{F}{\otimes} B$ and put $I = J \cap B$. Then clearly $A \underset{F}{\otimes} I \subseteq J$, so we need only verify that $J \subseteq A \underset{F}{\otimes} I$. To this end, note that each nonzero $c \in J$ can be written as $c = a_1 \otimes b_1 + \ldots + a_r \otimes b_r$ with $0 \neq b_i \in B$ and with $\{a_1, \ldots, a_r\}$ being a subset of an F-basis of A. By Lemma 19.4(i), $A \underset{F}{\otimes} A^\circ$ acts densely on A. Hence there exist $x_j, y_j \in A$ for $j \in \{1,2,\ldots,r\}$ such that $\sum_j x_j a_i y_j = \delta_i$, $1 \leq i \leq r$. But then

$$\sum x_j c y_j = \sum_{i,j} x_j a_i y_j \otimes b_i = 1 \otimes b_1 \in J$$

Therefore $b_1 \in I$ and similarly $b_i \in I$ for $i \in \{2,\ldots,r\}$, proving that $J \subseteq A \underset{F}{\otimes} I$. ■

19.7. LEMMA. Let A_1 and A_2 be F-algebras and let B_i be a subalgebra of A_i, $i = 1,2$. Then

$$C_{A_1 \underset{F}{\otimes} A_2}(B_1 \underset{F}{\otimes} B_2) = C_{A_1}(B_1) \otimes C_{A_2}(B_2)$$

Proof. It is obvious that

$$C_{A_1}(B_1) \otimes_F C_{A_2}(B_2) \subseteq C_{A_1 \otimes_F A_2}(B_1 \otimes_F B_2)$$

To prove the opposite inclusion, fix an F-basis $\{v_i\}$ of A_2. Then every element of $A_1 \otimes_F A_2$ can be uniquely written in the form $\Sigma a_i \otimes v_i$ with $a_i \in A_1$. Given any $b \in B_1$, we have

$$(\Sigma a_i \otimes v_i)(b \otimes 1) - (b \otimes 1)(\Sigma a_i \otimes v_i) = \Sigma(a_i b - b a_i) \otimes v_i$$

Therefore, if $\Sigma a_i \otimes v_i \in C_{A_1 \otimes A_2}(B_1 \otimes B_2)$, then $a_i b = b a_i$ for all $b \in B_1$. Hence $a_i \in C_{A_1}(B_1)$ and so

$$\Sigma a_i \otimes v_i \in C_{A_1}(B_1) \otimes_F A_2$$

Applying Lemma 19.5, we infer that

$$\begin{aligned} C_{A_1 \otimes_F A_2}(B_1 \otimes_F B_2) &\subseteq (A_1 \otimes_F C_{A_2}(B_2)) \cap (C_{A_1}(B_1) \otimes_F A_2) \\ &= C_{A_1}(B_1) \otimes_F C_{A_2}(B_2), \end{aligned}$$

as asserted. ■

19.8. COROLLARY. Let A_1 and A_2 be F-algebras. Then

$$Z(A_1 \otimes_F A_2) = Z(A_1) \otimes_F Z(A_2)$$

Proof. Apply Lemma 19.7 for the case $B_1 = A_1$ and $B_2 = A_2$. ■

19.9. COROLLARY. Let A be a central simple F-algebra and let B be any F-algebra.

(i) $A \otimes_F B$ is simple if and only if B is simple

(ii) $Z(A \otimes_F B) \cong Z(B)$

In particular, the tensor product of central simple F-algebras is again central simple.

Proof. Apply Lemma 19.6 and Corollary 19.8. ■

We are now ready to prove

19.10. PROPOSITION. The Brauer classes of central simple F-algebras form an abelian group with respect to the multiplication induced by the tensor product.

Proof. Let A and B be central simple F-algebras. Then $A \otimes_F B \cong B \otimes_F A$ is again a central simple F-algebra by virtue of Corollary 19.9. Suppose that

$$A \cong D_1 \otimes M_n(F) \quad \text{and} \quad B \cong D_2 \underset{F}{\otimes} M_m(F)$$

Then

$$A \underset{F}{\otimes} B \cong (D_1 \underset{F}{\otimes} M_n(F)) \otimes (D_2 \underset{F}{\otimes} M_m(F)) \cong (D_1 \underset{F}{\otimes} D_2) \otimes M_{nm}(F)$$

and so $A \underset{F}{\otimes} B \sim D_1 \underset{F}{\otimes} D_2$. Hence, if $A \sim A'$, $B \sim B'$, then

$$A' \underset{F}{\otimes} B' \sim D_1 \underset{F}{\otimes} D_2 \sim A \underset{F}{\otimes} B,$$

proving that the multiplication $[A]\,[B] = [A \underset{F}{\otimes} B]$ is well defined.

The multiplication of Brauer classes is associative by the corresponding law for tensor products. Taking into account that $A \underset{F}{\otimes} F \cong A$ and that, by Lemma 19.4 (ii), $A \underset{F}{\otimes} A^\circ \cong M_n(F)$, $n = \dim_F A$, we also have

$$[A]\,[F] = [A]\,, \quad [A]\,[A^\circ] = [F]$$

This shows that $[F]$ is the identity element and that $[A^\circ]$ is the inverse of $[A]$. ∎

The group of Brauer classes of central simple F-algebras is called the *Brauer group* of F and is denoted by $Br(F)$.

Let A be an F-algebra. A field extension E/F is said to be a *splitting field* for A if

$$A \underset{F}{\otimes} E \cong M_r(E)$$

for some positive integer r. Our next aim is to show that A always has a splitting field E which is a finite extension of F. The following basic result is known as the Skolem-Noether theorem

19.11. THEOREM. Let A be a central simple F-algebra and let B be a simple subalgebra of A. Then every isomorphism of F-algebras

$$\phi : B \longrightarrow B' \subseteq A$$

can be extended to an inner automorphism of A, that is, there exists a unit a in A such that

$$\phi(b) = aba^{-1} \qquad \text{for all } b \in B$$

In particular, isomorphic simple subalgebras of A are conjugate and hence have isomorphic centralizers.

Proof. Let V be an irreducible (left) A-module, and let $D = \mathrm{End}_A(V)$. Then D is a division algebra with centre $F \cdot 1_V$ and $B \otimes_F D$ is a simple F-algebra by Corollary 19.9(i). We may view V as a $B \otimes_F D$-module via

$$(b \otimes d)v = b(dv) \qquad (b \in B,\ d \in D,\ v \in V)$$

Using the automorphism ϕ, we can define a second left $B \otimes_F D$-module V', whose underlying vector space over F is V, and with the action of $B \otimes_F D$ given by

$$(b \otimes d)v' = \phi(b)(dv') \qquad (b \in B,\ d \in D,\ v' \in V\ )$$

Then V and V' are of the same F-dimension. Because $B \otimes_F D$ is simple artinian, we infer that there exists an isomorphism of $B \otimes_F D$-modules $\theta : V \longrightarrow V'$. Since $\theta \in \mathrm{End}_D(V)$, it follows from Lemmas 19.1 and 19.2 that θ is a left multiplication a_ℓ for some $a \in A$. Furthermore, $a \in U(A)$ since θ is an isomorphism. Finally, since $\theta(=a_\ell)$ is a $B \otimes_F D$-isomorphism, we have

$$a(b(dv)) = \phi(b)d(av) \quad \text{for all } b \in B,\ d \in D,\ v \in V$$

Taking $d = 1$, and applying the fact that A acts faithfully on V, it follows that $ab = \phi(b)a$, so $\phi(b) = aba^{-1}$ for all $b \in B$. ■

19.12. COROLLARY. Let A be a central simple F-algebra and let B be a simple subalgebra with centre E. If $\dim_F B = k$, then

$$A \otimes_F B^\circ \cong C_A(B) \otimes_F M_k(F) \cong M_k(C_A(B))$$

In particular, if E is a subfield of A containing F, then

$$A \otimes_F E \cong C_A(E) \otimes_F M_k(F) , \quad \text{where } k = (E:F)$$

Proof. We may identify $M_k(F)$ with the algebra of all F-linear transfromations of B. As such it contains the subalgebras B_ℓ of left multiplications and B_r of right multiplications.

Clearly $B_\ell \cong B$ and $B_r \cong B^\circ$ and, by Corollary 19.9, $A \otimes_F M_k(F)$ is central simple. Now $B \otimes_F F$ and $F \otimes_F B_\ell$ are isomorphic simple subalgebras of $A \otimes_F M_k(F)$. Hence, by Theorem 19.11 and Lemma 19.7,

$$C_A(B) \otimes_F M_k(F) \cong A \otimes_F B_r \cong A \otimes_F B^\circ$$

Since $C_A(B) \otimes M_k(F) \cong M_k(C_A(B))$, the result follows. ■

19.13. COROLLARY. Let A be a central simple F-algebra and let B be a simple subalgebra with centre E.

(i) $C_A(B)$ is simple with centre E

(ii) $C_A(C_A(B)) = B$

(iii) $\dim_F A = (\dim_F B)(\dim_F C_A(B))$

Proof. Comparing dimensions in Corollary 19.12, we have

$$(\dim_F A)(\dim_F B) = (\dim_F B)^2 \dim_F C_A(B),$$

proving (iii). Since $A \otimes_F B^\circ$ is simple, so is $C_A(B)$ by Corollaries 9.12 and 9.9. Replacing B by $C_A(B)$ in (iii), we therefore find

$$\dim_F A = (\dim_F C_A(B))(\dim_F C_A(C_A(B)))$$

whence

$$\dim_F C_A(C_A(B)) = \dim_F B$$

Since we have clearly $C_A(C_A(B)) \supseteq B$, it follows that $C_A(C_A(B)) = B$, proving (ii).

Finally, if K is the centre of $C_A(B)$, then $K \supseteq E$, and since

$$C_A(C_A(B)) = B$$

we also have $E \supseteq K$, i.e. $E = K$. ■

19.14. PROPOSITION. Let A be a central simple F-algebra, say $A = M_n(D)$ where D is a division ring with centre F, and let E be a maximal subfield of D. Then $A \otimes_F E \cong M_{nk}(E)$ where $k = (E:F)$. In particular, $\dim_F A = n^2k^2$ is a perfect square.

Proof. Applying Lemma 19.12 for the case $A = D$ and taking into account that $C_D(E) = E$, we deduce that

$$D \otimes_F E \cong M_k(E)$$

Hence

$$A \otimes_F E \cong M_n(D \otimes_F E) \cong M_n(M_k(E)) \cong M_{nk}(E)$$

as required. ∎

Let A be a central simple F-algebra. It is a consequence of Proposition 19.13 that

$$\dim_F A = r^2$$

for some integer r and that there is a finite extension E of F such that

$$A \otimes_F E = M_r(E)$$

Thus A always has a splitting field E which is a finite extension of F. We say that $[A]$ is *split* by E, or that E is a *splitting field for the class* $[A]$, if E is a splitting field for A. That the notion of a splitting field for $[A]$ is well defined is a consequence of the following simple observation.

19.15. LEMMA. Let E be a field extension of F. Then the map

$$\begin{cases} Br(F) \longrightarrow Br(E) \\ [A] \longmapsto [A \otimes_F E] \end{cases}$$

is a homomorphism whose kernel consists of those $[A]$ which are split by E.

Proof. Let A be a central simple F-algebra and write $A \cong D \otimes_F M_n(F)$. Then

$$A \otimes_F E \cong D \otimes_F M_n(F) \otimes_F E \cong D \otimes_F M_n(F)$$

and hence $[A \otimes_F E] = [D \otimes_F E]$. Thus the given map is well defined. It is also a homomorphism, since

$$(A \otimes_F B) \otimes_F E \cong (A \otimes_F E) \otimes (B \otimes_F E)$$

The final assertion is a consequence of the definition of the splitting field for $[A]$. ∎

We denote by $Br(E/F)$ the kernel of the homomorphism

$$\begin{cases} Br(F) \longrightarrow Br(E) \\ [A] \longmapsto [A \otimes_F E] \end{cases}$$

Hence, by Lemma 19.15, $Br(E/F)$ is the subgroup of $Br(F)$ consisting of those $[A]$ which are split by E.

We next present a criterion for $[A] \in Br(F)$ to be an element of $Br(E/F)$.

19.16. LEMMA. Let A be a central simple F-algebra and let E/F be a finite field extension. Then $[A] \in Br(E/F)$ if and only if there exists $B \in [A]$ such that B contains E as a self-centralizing subfield.

Proof. Let D be the division algebra in the class $[A]$, and let n be the least integer such that $E \subseteq M_n(D)$. Put $B = M_n(D)$, $V = D^n$ and observe that V is an E-module. By the minimality of n, V is an irreducible E-module. Now any $0 \neq c \in C_B(E)$ determines a nonzero endomorphism of V, hence c is invertible. Thus $C_B(E)$ is a division algebra. Furthermore, by Corollary 19.13(i), E is the centre of $C_B(E)$. Invoking Corollary 19.12, we also have

$$C_B(E) \sim B \underset{F}{\otimes} E \sim D \underset{F}{\otimes} E$$

By Lemma 19.15, this shows that E splits D if and only if $C_B(E) = E$, i.e. if and only if E is a self-centralizing subfield of B. ■

Our final observation on central simple F-algebras is given by

19.17. LEMMA. Let A be a central simple F-algebra and let e be a nonzero idempotent of A. Then

$$[A] = [eAe] \qquad \text{in } Br(F)$$

Proof. By Wedderburn's theorem, $A \cong M_n(D)$ and $[A] = [D]$ for some division algebra D and some $n \geqslant 1$. By changing the basis of an n-dimensional vector space over D on which A acts irreducibly and faithfully, i.e. applying an inner automorphism of A, we may assume that

$$e = \begin{bmatrix} I_r & 0 \\ 0 & 0 \end{bmatrix}$$

where I_r is the identity $r \times r$-matrix. Then

$$eAe \cong \begin{bmatrix} I_r & 0 \\ 0 & 0 \end{bmatrix} M_n(D) \begin{bmatrix} I_r & 0 \\ 0 & 0 \end{bmatrix} = \begin{bmatrix} M_r(D) & 0 \\ 0 & 0 \end{bmatrix}$$

and so $eAe \cong M_r(D)$. Thus $[eAe] = [D] = [A]$ as required. ■

Let E/F be a finite Galois extension and let $G = \mathrm{Gal}(E/F)$. Consider the group $Z^2(G,E^*)$ with respect to the natural action of G on E^*. As has been remarked earlier, $H^2(G,E^*)$ is unaffected if we restrict ourselves to normalized 2-cocycles. For this reason, we shall assume that all 2-cocycles are *normalized*. For any given $\alpha \in Z^2(G,E^*)$, we shall construct a central simple F-algebra $E^{\alpha}G$, called the *crossed product of* G *over* E. This crossed product $E^{\alpha}G$ was introduced by Noether and played a significant role in the classical theory of central simple algebras. That the crossed products $E^{\alpha}G$ occur in an unavoidable way in the study of central simple algebras will be illustrated by the fact that any central simple F-algebra A such that E is a self-centralizing subfield of A is isomorphic to $E^{\alpha}G$ for some $\alpha \in Z^2(G,E^*)$. In what follows, we write $\bar{\alpha}$ for the cohomology class of $\alpha \in Z^2(G,E^*)$.

For the rest of this section, E/F *denotes a finite Galois extension with Galois group* G. Given $\alpha \in Z^2(G,E^*)$, we denote by $E^{\alpha}G$ a free left E-module with basis $\{\bar{g} | g \in G\}$. Thus each element of $E^{\alpha}G$ can be uniquely written in the form

$$\Sigma x_g \bar{g} \qquad (x_g \in E,\ g \in G)$$

We define the multiplication on $E^{\alpha}G$ distributively by using

$$(r_1\bar{x})(r_2\bar{y}) = r_1{}^{x}r_2\,\alpha(x,y)\overline{xy} \qquad (r_1,r_2 \in E,\ x,y \in G) \quad (1)$$

where

$${}^{g}\lambda = g(\lambda) \qquad \text{for all} \quad \lambda \in E,\ g \in G$$

Recall that for all $x,y,z \in G$

$$\alpha(x,y)\alpha(xy,z) = {}^{x}\alpha(y,z)\alpha(x,yz) \quad (2)$$

$$\alpha(x,1) = \alpha(1,x) = 1 \quad (3)$$

Note also that, by (1),

$$\bar{g}\lambda\bar{g}^{-1} = {}^{g}\lambda\ ,\ \bar{x}\bar{y} = \alpha(x,y)\overline{xy} \qquad (g,x,y \in G,\ \lambda \in E) \quad (4)$$

19.18. LEMMA. For any $\alpha \in Z^2(G,E^*)$, $E^{\alpha}G$ is an F-algebra with identity element $\bar{1}$.

Proof. It suffices to verify that the multiplication given by (1) is associative. Indeed, in this case $E^{\alpha}G$ is obviously a ring with identity $\bar{1}$ (the latter can be seen by applying (1) and (3)) and, by (4), $\{\lambda\cdot\bar{1}|\lambda \in F\}$ is an isomorphic copy of F contained in $Z(E^{\alpha}G)$. Thus $E^{\alpha}G$ is an F-algebra.

Fix $r_1, r_2, r_3 \in E$ and $x, y, z \in G$. We first show that

$$^{x}(^{y}\lambda) = \alpha(x,y)^{xy}\lambda\, \alpha(x,y)^{-1} \qquad \text{for all } \lambda \in E \quad (5)$$

Indeed, applying (4) we have

$$^{x}(^{y}\lambda) = \bar{x}(\bar{y}\lambda\bar{y}^{-1})\bar{x}^{-1} = \alpha(x,y)\overline{xy}\,\lambda\,\overline{xy}^{-1}\alpha(x,y)^{-1} = \alpha(x,y)^{xy}\lambda\,\alpha(x,y)^{-1}$$

proving (5). We now have

$$\begin{aligned}(r_1\bar{x})[(r_2\bar{y})(r_3\bar{z})] &= r_1{}^{x}[r_2{}^{y}r_3\,\alpha(y,z)]\,\alpha(x,yz)\overline{xyz}\\ &= r_1{}^{x}r_2{}^{x}(^{y}r_3)^{x}\alpha(y,z)\alpha(x,yz)\overline{xyz}\\ &= r_1{}^{x}r_2\,\alpha(x,y)^{xy}r_3\,\alpha(x,y)^{-1\,x}\alpha(y,z)\alpha(x,yz)\overline{xyz} && \text{(by (5))}\\ &= r_1{}^{x}r_2\,\alpha(x,y)^{xy}r_3\,\alpha(x,y)^{-1}\alpha(x,y)\alpha(xy,z)\overline{xyz} && \text{(by (2))}\\ &= [(r_1\bar{x})(r_2\bar{y})](r_3\bar{z}),\end{aligned}$$

as required. ∎

In what follows, we identify E with its copy $\{\lambda\cdot 1|\lambda \in E\}$ in $E^{\alpha}G$.

19.19. LEMMA. For any $\alpha \in Z^2(G,E^*)$, the following properties hold:

(i) $E^{\alpha}G$ is a central simple F-algebra and $C_{E^{\alpha}G}(E) = E$. In particular, by Lemma 19.16, $[E^{\alpha}G] \in Br(E/F)$.

(ii) $E^{\alpha}G \cong M_n(F)$ for some $n \geqslant 1$ if and only if α is a coboundary.

Proof. (i) Given $x = \Sigma x_g\bar{g} \in E^{\alpha}G$ and $\lambda \in E$, it follows from (1) and (4) that

$$\lambda x = \Sigma\lambda x_g\bar{g} \quad \text{and} \quad x\lambda = \Sigma x_g{}^{g}\lambda\bar{g}$$

Hence $\lambda x = x\lambda$ if and only if $\lambda x_g = {}^{g}\lambda x_g$ for all $g \in G$. Therefore, if $x \neq 0$ and $x_g \neq 0$ for some $g \in G$, then $x \in C_{E^{\alpha}G}(E)$ implies $\lambda = {}^{g}\lambda$ for all $\lambda \in E$, i.e. $g = 1$. This shows that $C_{E^{\alpha}G}(E) = E$. It is clear that $F \subseteq Z(E^{\alpha}G)$. On the other hand, if $\mu \in Z(E^{\alpha}G)$ then $\mu \in E = C_{E^{\alpha}G}(E)$. Furthermore, by (4) ${}^{g}\mu = \mu$ for all $g \in G$, hence $\mu \in F$. Thus $Z(E^{\alpha}G) = F$.

Given $x = \Sigma x_g \bar{g} \in E^{\alpha}G$, let $\ell(x)$ denote the number of nonzero x_g. Assume that I is a nonzero ideal of $E^{\alpha}G$. Among the nonzero elements $x = \Sigma x_g \bar{g}$ of I choose such that has the smallest $\ell(x)$. Multiplying x by an appropriate $\overline{g^{-1}}$, if necessary, we may assume that $x_1 \neq 0$. Now for any $\lambda \in E$

$$\lambda x - x\lambda = \Sigma y_g \bar{g} \in I \ , \quad \text{where} \quad y_g = \lambda x_g - x_g{}^g\lambda$$

Since $y_g \neq 0$ implies $x_g \neq 0$ and since $y_1 = 0 \neq x_1$, we have $\ell(\lambda x - x\lambda) < \ell(x)$ Thus, by hypothesis, $\lambda x - x\lambda = 0$ i.e. $x \in C_{E^{\alpha}G}(E) = E$. Hence $I = E^{\alpha}G$ and so $E^{\alpha}G$ is simple.

(ii) Assume that α is a coboundary, say $\alpha = \delta t$ for some $t : G \longrightarrow E^*$ with $t(g) = 1$. For each $g \in G$, put $\tilde{g} = t(g)^{-1}\bar{g}$. Then $\{\tilde{g} | g \in G\}$ is an E-basis of $E^{\alpha}G$ and, for $x, y \in G$,

$$\begin{aligned} \tilde{x}\tilde{y} &= t(x)^{-1}\bar{x}t(y)^{-1}\bar{y} = t(x)^{-1x}t(y)^{-1}\alpha(x,y)\overline{xy} \\ &= t(xy)^{-1}\overline{xy} = \widetilde{xy} \end{aligned} \tag{6}$$

Given $\lambda \in E$, let $\phi(\lambda) \in \text{End}_F(E)$ be defined by $\phi(\lambda)(\mu) = \lambda\mu$ for all $\mu \in E$. Then applying (6), we deduce that the map

$$\begin{cases} E^{\alpha}G \longrightarrow \text{End}_F(E) \\ \sum_{g\in G} x_g\tilde{g} \longmapsto \sum_{g\in G} \phi(x_g)g \end{cases}$$

is a homomorphism of F-algebras. Since $E^{\alpha}G$ is simple and both $E^{\alpha}G$ and $\text{End}_F(E)$ are of the same F-dimension equal to $|G|^2$, we conclude that

$$E^{\alpha}G \cong \text{End}_F(E) \cong M_n(F) \tag{7}$$

where $n = |G|$.

Conversely, assume that $E^{\alpha}G \cong M_n(F)$ for some $n \geqslant 1$. By counting F-dimensions, we must have $n = |G|$. Hence, by (7), $E^{\alpha}G \cong E^{\beta}G$ where $\beta(x,y) = 1$ for all $x, y \in G$. Let $\phi : E^{\alpha}G \longrightarrow E^{\beta}G$ be an isomorphism of F-algebras. By Theorem 19.11, the isomorphism $E \longrightarrow \phi(E)$ can be extended to an inner automorphism of $E^{\beta}G$. Thus there exists a unit x of $E^{\beta}G$ such that $\phi(\lambda) = x^{-1}\lambda x$ for all $\lambda \in E$. Hence, if $\theta \in \text{Aut}(E^{\beta}G)$ is defined by $\theta(y) = xyx^{-1}$ for all $y \in E^{\beta}G$, then

$$(\theta\lambda)(\lambda) = \lambda \qquad \text{for all } \lambda \in E$$

Choose an E-basis $\{\tilde{g} | g \in G\}$ with $\tilde{x}\tilde{y} = \widetilde{xy}$ for all $x,y \in G$. Since

$$\bar{g}\lambda\bar{g}^{-1} = \tilde{g}\lambda\tilde{g}^{-1} = {}^{g}\lambda \qquad \text{for all } g \in G,\ \lambda \in E$$

we have

$$(\theta\phi)(\bar{g}\lambda\bar{g}^{-1}) = (\theta\phi)(\bar{g})\lambda(\theta\phi)(\bar{g})^{-1} = \tilde{g}\lambda\tilde{g}^{-1}$$

Hence $\tilde{g}^{-1}(\theta\phi)(\bar{g})$ centralizes E and therefore lies in E. It follows that

$$(\theta\phi)(\bar{g}) = t(g)\tilde{g} \qquad (g \in G,\ t(g) \in E)$$

But then

$$\begin{aligned} \alpha(x,y) &= (\theta\phi)(\alpha(x,y)) = (\theta\phi)(\bar{x}\,\bar{y}\,\overline{xy}^{-1}) \\ &= t(x)\tilde{x}\,t(y)\tilde{y}\,\tilde{y}^{-1}\,\tilde{x}^{-1}\,t(xy)^{-1} \\ &= t(x)\,{}^{x}t(y)t(xy)^{-1}, \end{aligned}$$

proving that α is a coboundary. ■

19.20. LEMMA. Let A be a central simple F-algebra such that E is a self-centralizing subfield of A. Then $A \cong E^{\alpha}G$ for some $\alpha \in Z^2(G,E^*)$.

Proof. For each $g \in G$, the map $g : E \longrightarrow E$ is an F-isomorphism of simple subalgebras of A. Hence, by Theorem 19.11, there exists $\bar{g} \in U(A)$ such that

$$\bar{g}\lambda\bar{g}^{-1} = g(\lambda) \qquad \text{for all } \lambda \in E,\ g \in G \qquad (8)$$

By Corollary 19.13(iii), we also have $\dim_F A = (\dim_F E)^2 = |G|^2$. In order to prove that

$$A = \bigoplus_{g \in G} E\bar{g} \qquad (9)$$

it suffices to show that $\sum_{g\in G} E\bar{g} = \bigoplus_{g\in G} E\bar{g}$, for then both sides will have F-dimension $|G|^2$.

If the sum $\sum_{g\in G} E\bar{g}$ is not direct, let

$$s = \lambda_1\bar{g}_1 + \ldots + \lambda_k\bar{g}_k = 0 \qquad (\lambda_i \in E^*)$$

be a relation with minimal k. It is clear that $k > 1$ since each $\bar{g} \in U(A)$. Choosing $\lambda \in E^*$ such that $g_1(\lambda) \neq g_2(\lambda)$, one easily verifies that

$$g_1(\lambda)s - s\lambda = 0$$

gives a shorter nontrivial relation connecting $\bar{g}_2,\dots,\bar{g}_k$. This is a contradiction and thus (9) is established.

Suppose now that $x,y \in G$. Then the equation (8) shows that $\bar{x}\,\bar{y}\,\overline{xy}^{-1}$ commutes with each $\lambda \in E$. Since E is its own centralizer in A, we obtain

$$\bar{x}\,\bar{y} = \alpha(x,y)\overline{xy} \qquad\qquad \text{with} \quad \alpha(x,y) \in E^*$$

The associativity of mulitplication in A implies that $\alpha : G \times G \longrightarrow E^*$ is a cocycle and thus we have shown that $A \cong E^{\alpha}G$ for some $\alpha \in Z^2(G,E^*)$. ■

The next property is crucial for the proof of the main result.

19.21. LEMMA. For all $\alpha,\beta \in Z^2(G,E^*)$, there is a similarity of F-algebras

$$E^{\alpha}G \otimes_F E^{\beta}G \sim E^{\alpha\beta}G$$

Proof. Let $C = A \otimes_F B$, where

$$A = E^{\alpha}G = \bigoplus_{g\in G} E\bar{g} \quad \text{and} \quad B = E^{\beta}G = \bigoplus_{g\in G} E\tilde{g}$$

Then C is a central simple F-algebra by Lemma 19.20 and Corollary 19.9. Our aim is to find an idempotent $e \in E \otimes_F E \subseteq C$ such that $eCe \cong E^{\alpha\beta}G$. This will yield the desired result by appealing to Lemma 19.17.

We first observe that the subfields $E \otimes 1$ and $1 \otimes E$ of $E \otimes_F E$ commute elementwise. Since E is separable over F, there exists $\lambda \in E$ such that $E = F(\lambda)$. Let $f(X) \in F[X]$ be the minimal polynomial of λ. Then $\deg f(X) = n$, where $n = (E:F)$. Now define

$$e = \prod(\lambda \otimes 1 - 1 \otimes g(\lambda))/\prod(\lambda - g(\lambda)) \otimes 1$$

where these products are taken over all $g \in G - \{1\}$. Observe that the denominator is not zero since $\lambda \neq g(\lambda)$ for each $g \in G - \{1\}$. The numerator is also distinct from zero in $E \otimes_F E$, since the elements $\{\lambda^i \otimes 1 | 0 \leqslant i \leqslant n-1\}$ are linearly independent over $1 \otimes E$. Hence $e \neq 0$ in $E \otimes_F E$.

Next we note that

$$f(X) = (X-\lambda)\prod(X-g(\lambda))$$

where g ranges over $G - \{1\}$. Therefore, we must have

$$(\lambda \otimes 1 - 1 \otimes \lambda)\prod(\lambda \otimes 1 - 1 \otimes g(\lambda)) = f(\lambda \otimes 1) = 0$$

which shows that $(1 \otimes \lambda)e = (\lambda \otimes 1)e$ in $E \otimes_F E$. Thus, by induction,

$$(\lambda^i \otimes 1)e = (1 \otimes \lambda^i)e \qquad \text{for all } i \geqslant 0$$

Because $E = \bigoplus_{i=0}^{n-1} F\lambda^i$ and multiplication in $E \otimes_F E$ is commutative, we derive

$$(\mu \otimes 1)e = e(\mu \otimes 1) = e(1 \otimes \mu) = (1 \otimes \mu)e \text{ for all } \mu \in E \quad (10)$$

Hence

$$\begin{aligned} e^2 &= e\prod(\lambda \otimes 1 - 1 \otimes g(\lambda))/\prod(\lambda - g(\lambda)) \otimes 1 \\ &= e\prod((\lambda - g(\lambda)) \otimes 1)/\prod(\lambda - g(\lambda)) \otimes 1 \\ &= e, \end{aligned}$$

proving that e is an idempotent in $E \otimes_F E$.

We are left to verify that

$$eCe \cong E^{\alpha\beta}G \quad (11)$$

To this end, note that

$$\begin{aligned} eCe &= \sum_{x,y \in G} e(E \otimes E)(\bar{x} \otimes \tilde{y})e \\ &= \sum_{x,y \in G} e(E \otimes 1)e \cdot e(1 \otimes E)e(\bar{x} \otimes \tilde{y})e \end{aligned}$$

But $e(1 \otimes E)e = e(E \otimes 1)e$ by (10), and $e(E \otimes 1)e = E'$ is a field F-isomorphic to E. We now compute $e(\bar{x} \otimes \tilde{y})e$, by using the formulas

$$\bar{x}\mu = x(\mu)\bar{x}, \quad \tilde{y}\mu = y(\mu)\tilde{y} \qquad (\mu \in E,\ x,y \in G)$$

We have

$$e(\bar{x} \otimes \tilde{y})e = (\bar{x} \otimes \tilde{y})\{\prod(x(\lambda) \otimes 1 - 1 \otimes yz(\lambda))/\prod(x(\lambda) - xz(\lambda) \otimes 1\}e$$

where z ranges over all elements of $G - \{1\}$. From (10), we obtain

$$e(\bar{x} \otimes \tilde{y})e = (\bar{x} \otimes \tilde{y})e\prod(x(\lambda) - yz(\lambda)) \otimes 1/\prod(x(\lambda) - xz(\lambda)) \otimes 1$$

If $x \neq y$, then $x(\lambda) - yz(\lambda) = 0$ for $z = y^{-1}x \in G - \{1\}$, and so

$$e(\bar{x} \otimes \tilde{y})e = 0$$

On the other hand, when $x = y$ we obtain

$$e(\bar{x} \otimes \tilde{y})e = (\bar{x} \otimes \tilde{y})e$$

A similar argument proves that

$$e(\bar{x} \otimes \tilde{y})e = e(\bar{x} \otimes \tilde{y})$$

The foregoing shows that

$$eCe = \bigoplus_{g \in G} E' \hat{g}$$

where

$$E' = e(E \otimes 1)e \quad \text{and} \quad \hat{g} = e(\bar{g} \otimes \tilde{g})e \qquad (g \in G)$$

and also shows that

$$\hat{g} = (\bar{g} \otimes \tilde{g})e = e(\bar{g} \otimes \tilde{g}) \qquad (g \in G)$$

Clearly, $\mathrm{Gal}(E'/F) \cong G$ and, for all $\mu \in E$, $g \in G$,

$$\hat{g}\, e(\mu \otimes 1)e = e(\bar{g} \otimes \tilde{g})(\mu \otimes 1)e = e(g(\mu) \otimes 1)e\, \hat{g}$$

Hence, conjugation by $\hat{g}$ acts as g on E'. Moreover, for all $x,y \in G$,

$$\begin{aligned}\hat{x}\hat{y} &= e(\bar{x}\bar{y} \otimes \tilde{x}\tilde{y})e = e(\alpha(x,y)\overline{xy} \otimes \beta(x,y)\widetilde{xy})e \\ &= e(\alpha(x,y)\beta(x,y) \otimes 1)e\, \widehat{xy}\end{aligned}$$

Thus the g's multiply according to the cocycle $\alpha\beta$. This proves (11) and the result follows. ■

We have now come to the demonstration for which this section has been developed.

19.22. THEOREM. Let E/F be a finite Galois extension, let $G = \mathrm{Gal}(E/F)$ and for any $\alpha \in Z^2(G,E^*)$, let $E^{\alpha}G$ be the crossed product of G over E with respect to the natural action of G on E. Then the map

$$\begin{cases} H^2(G,E^*) \longrightarrow Br(E/F) \\ \quad \bar{\alpha} \longmapsto [E^{\alpha}G] \end{cases}$$

is an isomorphism.

Proof. Owing to Lemma 19.19(i), $[E^{\alpha}G] \in Br(E/F)$. Consider the map

$$\begin{cases} Z^2(G,E^*) \xrightarrow{f} Br(E/F) \\ \quad \alpha \longmapsto [E^{\alpha}G] \end{cases}$$

Then, by Lemma 19.21, f is a homomorphism. Furthermore, by Lemma 19.19(ii),

$$\mathrm{Ker} f = B^2(G,E^*)$$

and thus f induces an injective homomorphism $\bar{f} : H^2(G,E^*) \longrightarrow Br(E/F)$ given by $\bar{f}(\bar{\alpha}) = [E^{\alpha}G]$. Now assume that $[A] \in Br(E/F)$. Then, by Lemma 19.16, there exists $B \in [A]$ such that B contains E as a self-centralizing subfield. Invoking Lemma 19.20, we deduce that $B \cong E^{\alpha}G$ for some $\alpha \in Z^2(G,E^*)$. Thus

$$\bar{f}(\bar{\alpha}) = [E^{\alpha}G] = [B] = [A]$$

and the result follows. ∎

Our next aim is to show that every Brauer class contains a crossed product. This will be achieved with the aid of the following result.

19.23. PROPOSITION. Every Brauer class of a field F has a splitting field which is a finite Galois extension of F.

Proof. Let D be a division algebra in a given Brauer class of F. We claim that D contains a maximal subfield E such that E/F is a separable extension; if sustained it will follow, by Lemma 19.16, that E splits D. Since E is contained in some finite Galois extension K/F, K will also split D, as required.

To substantiate our claim, we may harmlessly assume that $D \neq F$ and that $\mathrm{char}F = p > 0$. Our first task is to show that D contains a proper separable field extension of F. To this end, note that each element of D is algebraic over F. Hence, if there exists $\lambda \notin F$ such that λ is separable over F, then $F(\lambda)$ is the required extension. Assume by way of contradiction, that there are no separable extensions over F. Then, each element $x \in D$ is purely inseparable, i.e. $x^{p^r} \in F$ for some $r = r(x)$. We may therefore choose $\lambda \notin F$ such that $\lambda^p \in F$. Let f be the mapping $x \longmapsto x\lambda - \lambda x$. Then $f \neq 0$ since $\lambda \notin F$ and $Z(D) = F$. On the other hand, we have $f^p(x) = x\lambda^p - \lambda^p x = 0$ for all $x \in D$. Choose $b \in D$ such that $f(b) \neq 0$ and let i be such that $f^{i-1}(b) \neq 0$ but $f^i(b) = 0$. Then $c = f^{i-1}(b)$ has the form $c = u\lambda - \lambda u$ and satisfies $f(c) = 0$, i.e. $c\lambda = \lambda c$. Setting $v = \lambda^{-1}c$, we see that v commutes with λ and

$$c = \lambda v = u\lambda - \lambda u$$

Hence, putting $w = uv^{-1}$, we derive

$$\lambda = (u\lambda - \lambda u)v^{-1} = uv^{-1}\lambda - \lambda uv^{-1} = w\lambda - \lambda w$$

Thus $w = 1 + \lambda w\lambda^{-1}$, but $w^q \in F$ for some $q = p^n$. But then

$$\begin{aligned} w^q &= 1 + (\lambda w\lambda^{-1})^q = 1 + \lambda w^q\lambda^{-1} \\ &= 1 + w^q \quad (\text{since } w^q \in F) \end{aligned}$$

and so $1 = 0$, a contradiction. Consequently, D contains a proper separable field extension of F.

Let E/F be a separable field extension in D of maximal degree. Then E is a maximal separable extension of F in D. By Corollary 19.13(i), $C_D(E)$ is simple with centre E. But D is a division algebra, hence so is $C_D(E)$. Therefore, if $C_D(E) \neq E$, then E has a proper separable extension E', by the first part. This would imply that E' is separable over F, contrary to the maximality of E. Thus $C_D(E) = E$, i.e. E is a maximal subfield of D. ■

19.24. COROLLARY. Let A be a central simple F-algebra. Then there exists a finite Galois extension E/F and $\alpha \in Z^2(G,E^*)$, where $G = \text{Gal}(E/F)$, such that A is similar to the crossed product $E^{\alpha}G$. In particular, every Brauer class in $Br(F)$ contains a crossed product.

Proof. By Proposition 19.23, there exists a finite Galois extension E/F such that $[A] \in Br(E/F)$. Now apply Theorem 19.22. ■

20. AN INTERPRETATION OF $H_0^3(G,E^*)$

Throughout this section, E/F is a finite Galois extension with Galois group G. As usual, we define the n-th cohomology group $H^n(G,E^*)$ with respect to *normalized* n-cochains. This does not affect the isomorphism class of $H^n(G,E^*)$ and is very useful in computations. To motivate our discussion, we first introduce a distinguished subgroup $H_0^n(G,E^*)$ of $H^n(G,E^*)$.

Let K/F be a finite Galois extension with $K \supseteq E$ and put

$$H = \text{Gal}(K/F), \quad S = \text{Gal}(K/E)$$

Given $\sigma \in H$, we put $\sigma' = \sigma|E$ so that the map

$$\begin{cases} H \longrightarrow G \\ \sigma \longmapsto \sigma' \end{cases} \tag{1}$$

is a surjective homomorphism with kernel S. Each $\alpha \in C^n(G,E^*)$ determines $\alpha_K \in C^n(H,K^*)$ given by

$$\alpha_K(\sigma_1,\ldots,\sigma_n) = \alpha(\sigma'_1,\ldots,\sigma'_n) \tag{2}$$

It is clear that if $\alpha \in Z^n(G,E^*)$, then $\alpha_K \in Z^n(H,K^*)$. We say that $\alpha \in Z^n(G,E^*)$ is *split* by K if α_K is a coboundary. The map

$$\begin{cases} Z^n(G,E^*) \longrightarrow Z^n(H,K^*) \\ \alpha \longmapsto \alpha_K \end{cases}$$

is obviously a homomorphism (which carries coboundaries to coboundaries); hence it induces a homomorphism

$$\phi_K : Z^n(G,E^*) \longrightarrow H^n(H,K^*)$$

whose kernel consists of all $\alpha \in Z^n(G,E^*)$ which are split by K. We put

$$Z_0^n(G,E^*) = \bigcup_K \mathrm{Ker}\phi_K$$

where K ranges over all finite Galois extensions of F containing E. Expressed otherwise, $Z_0^n(G,E^*)$ consists of those $\alpha \in Z^n(G,E^*)$ which are split by a suitable finite Galois extension of F containing E. It is clear that $Z_0^n(G,E^*)$ is a subgroup of $Z^n(G,E^*)$ containing $B^n(G,E^*)$. We put

$$H_0^n(G,E^*) = Z_0^n(G,E^*)/B^n(G,E^*)$$

Our aim is to provide an interpretation of $H_0^3(G,E^*)$ in terms of the Brauer group of E (the group $H_0^2(G,E^*)$ turns out to be trivial). A precise description of the main result is given below.

For convenience, we first recall that the elements α of $Z^3(G,E^*)$ are precisely those maps $G \times G \times G \longrightarrow E^*$ satisfying

$$\alpha(\sigma_1,\sigma_2,\sigma_3) = 1 \qquad \text{if any } \sigma_i = 1 \tag{3}$$

$$\sigma_1\alpha(\sigma_2,\sigma_3,\sigma_4)\alpha(\sigma_1,\sigma_2\sigma_3,\sigma_4)\alpha(\sigma_1,\sigma_2,\sigma_3) = \alpha(\sigma_1\sigma_2,\sigma_3,\sigma_4)\alpha(\sigma_1,\sigma_2,\sigma_3\sigma_4) \tag{4}$$

Also, for any normalized 2-cochain $\beta : G \times G \longrightarrow E^*$, the corresponding coboundary $\delta\beta : G \times G \times G \longrightarrow E^*$ is defined by

$$(\delta\beta)(\sigma_1,\sigma_2,\sigma_3) = \sigma_1\beta(\sigma_2,\sigma_3)\beta(\sigma_1\sigma_2,\sigma_3)^{-1}\beta(\sigma_1,\sigma_2\sigma_3)\beta(\sigma_1,\sigma_2)^{-1} \tag{5}$$

A *G-normal algebra* is defined to be any (finite-dimensional) central simple E-algebra A for which every F-automorphism of E can be extended to a (ring) automorphism of A. It will be shown that the Brauer classes of G-normal algebras form a subgroup $Br_G(E)$ of $Br(E)$ containing the image $Br_F(E)$ of the homomorphism

$$\begin{cases} Br(F) \longrightarrow Br(E) \\ [A] \longmapsto [A \otimes_F E] \end{cases}$$

The main result asserts that

$$H_0^3(G,E^*) \cong Br_G(E)/Br_F(E)$$

20.1. LEMMA. Let A and B be E-algebras such that the given automorphism σ of E can be extended to ring automorphisms f and g of A and B, respectively. Then there is a unique ring automorphism $f \otimes g$ of $A \otimes_E B$ such that

$$(f \otimes g)(a \otimes b) = f(a) \otimes g(b)$$

Proof. Since any ring automorphism of $A \otimes_E B$ is uniquely determined by its values on $a \otimes b$, $a \in A$, $b \in B$, it suffices to show that $f \otimes g$ is a ring automorphism of $A \otimes_E B$. Consider the map $A \times B \xrightarrow{\psi} A \otimes_E B$, $(a,b) \longmapsto f(a) \otimes g(b)$. Then ψ is obviously biadditive and, for any $\lambda \in E$,

$$\begin{aligned}\psi(a,\lambda b) &= f(a) \otimes g(\lambda b) = f(a) \otimes \sigma(\lambda)g(b) = \sigma(\lambda)f(a) \otimes g(b)\\ &= f(\lambda a) \otimes g(b) = \psi(\lambda a,b)\end{aligned}$$

Hence $f \otimes g : A \otimes_E B \longrightarrow A \otimes_E B$ is a $\mathbb{Z}$-module homomorphism with

$$(f \otimes g)^{-1} = f^{-1} \otimes g^{-1}$$

Since $f \otimes g$ obviously preserves multiplication, the result follows. ∎

20.2. COROLLARY. If A and B are G-normal algebras, then so is $A \otimes_E B$.

Proof. Let σ be any F-automorphism of E. Then there are ring automorphisms f and g of A and B, respectively, which extend σ. Applying Lemma 20.1, we see that $f \otimes g$ is a ring automorphism of $A \underset{E}{\otimes} B$ such that

$$(f \otimes g)(\lambda \otimes 1) = \sigma(\lambda) \otimes 1 = \sigma(\lambda)(1 \otimes 1) \qquad \text{for all } \lambda \in E$$

as required. ■

20.3. LEMMA. Let A and B be E-algebras and let f_i, g_i be ring automorphisms of A,B respectively, extending the automorphism λ_i of E for $i = 1,2$. Then

$$(f_1 \otimes g_1)(f_2 \otimes g_2) = f_1f_2 \otimes g_1g_2$$

Proof. By Lemma 20.1, both sides are ring automorphisms of $A \underset{E}{\otimes} B$. Since

$$\begin{aligned}(f_1 \otimes g_1)(f_2 \otimes g_2)(a \otimes b) &= (f_1f_2)(a) \otimes (g_1g_2)(b) \qquad (a \in A,\ b \in B)\\ &= (f_1f_2 \otimes g_1g_2)(a \otimes b),\end{aligned}$$

the result follows. ■

As a preliminary to the next result, let us recall the following standard fact. Assume that $R = M_n(D_1) = M_r(D_2)$ where D_1 and D_2 are division rings and let $\{e_{ij}\}$ and $\{v_{ij}\}$ be the matrix units of $M_n(D_1)$ and $M_r(D_2)$, respectively. Then $r = n$ and there exists a unit u of R such that

(i) $D_2 = u^{-1}D_1u$

(ii) $v_{ij} = u^{-1}e_{ij}u$ for all $i,j \in \{1,2,\ldots,n\}$

(see Jacobson (1968,p.26)).

Now assume that $R = M_n(D)$ where D is a division ring and let ϕ be an automorphism of R. Then $R = M_n(D')$ where $D' = \phi(D)$. Hence, by the above, $\phi(D) = u^{-1}Du$ for some unit u of R. It follows that $\psi = i_u\phi$, where

$$i_u(x) = uxu^{-1},$$

is an automorphism of D such that

$$\psi(D) = D \quad \text{and} \quad \psi|Z(D) = \phi|Z(D) \tag{6}$$

20.4. LEMMA. (i) If an algebra A is G-normal, then so are all similar algebras

(ii) The Brauer classes of G-normal algebras form a subgroup $Br_G(E)$ containing the image $Br_F(E)$ of the homomorphism

$$\begin{cases} Br(F) \longrightarrow Br(E) \\ [A] \longmapsto [A \underset{F}{\otimes} E] \end{cases}$$

Proof. (i) Let D be a division algebra over E with centre E. It suffices to show that $M_n(D)$ is G-normal if and only D is G-normal. If D is G-normal, then obviously so is $M_n(D)$. Conversely, assume that $M_n(D)$ is G-normal and let $\lambda \in G$. By hypothesis, λ extends to a ring automorphism, say ϕ, of $M_n(D)$. Replacing ϕ by ψ as in (5), the assertion follows.
(ii) It is clear that if A is G-normal, then so is $A°$. Hence, by Corollary 20.2, $Br_G(E)$ is a subgroup of $Br(E)$. Let A be a central simple F-algebra and let λ be an F-automorphism of E. Then, the map $\psi : A \underset{F}{\otimes} E \longrightarrow A \underset{F}{\otimes} E$ given by $\psi(a \otimes e) = a \otimes \lambda(e)$ is a ring automorphism which restricts to λ on $1 \otimes E$, as required. ■

The following observation will allow us to tie together G-normal algebras and certain E^*-valued 3-cocycles.

20.5. LEMMA. Let A be a G-normal algebra and, for each $x \in G$, let $\sigma(x)$ be a (ring) automorphism of A extending x, with $\sigma(1) = 1$. For any unit u of A, define $i(u) \in \text{Aut}A$ by $i(u)(a) = uau^{-1}$, $a \in A$. Then there exist maps

$$f : G \times G \longrightarrow U(A) \quad \text{and} \quad t : G \times G \times G \longrightarrow E^* \quad \text{such that}$$

$$\sigma(x)\sigma(y) = i[f(x,y)]\sigma(xy) \quad \text{for all } x,y \in G \tag{7}$$

$$f(1,y) = f(x,1) = 1 \quad \text{for all } x,y \in G \tag{8}$$

$$\sigma(x)[f(y,z)]f(x,yz) = t(x,y,z)f(x,y)f(xy,z) \tag{9}$$

$$t(x,y,z) = 1 \quad \text{if } x,y, \text{ or } z = 1 \tag{10}$$

Proof. The product $\sigma(x)\sigma(y)\sigma(xy)^{-1}$ is an automorphism of A which induces on the centre E the identity $1 = xy(xy)^{-1}$; it is therefore an inner automorphism (by Theorem 19.11), say $i[f(x,y)]$ for some $f(x,y) \in U(A)$. Since $\sigma(1) = 1$, we may and do choose $f(1,y) = f(x,1)$, which proves (7) and (8).

Using the equation (7) and the identities

$$i(uv) = i(u)i(v), \; \lambda i(u) = i(\lambda(u))\lambda \qquad (\lambda \in G, \; u,v \in U(A))$$

we may calculate the triple product $\sigma(x)\sigma(y)\sigma(z)$ in two ways as

$$[\sigma(x)\sigma(y)]\sigma(z) = i[f(x,y)f(xy,z)]\sigma(xyz)$$
$$\sigma(x)[\sigma(y)\sigma(z)] = i\{\sigma(x)[f(y,z)]\ f(x,yz)\}\sigma(xyz)$$

These results must be identical; hence if we write b_1, b_2 for the two arguments of i on the right, we have $i(b_1) = i(b_2)$ or $i(b_1 b_2^{-1}) = 1$, which means that $b_2 b_1^{-1} = t$ is an element in the centre E of A. This proves (9). Finally, the special choices made above also prove (10). ■

The following result will be derived at the end of the section as a consequence of some general facts of cohomology theory.

20.6. LEMMA. Let $t : G \times G \times G \longrightarrow E^*$ be as in Lemma 20.5. Then

(i) $t \in Z^3(G,E^*)$

(ii) For fixed choices $\sigma(x)$ a different choice of $f(x,y) \in U(A)$ replaces t by a cohomologous cocycle , and a suitable different choice of $f(x,y)$ replaces t by any specified cohomologous cocycle

(iii) If the choice of the extensions $\sigma(x)$ of x to automorphisms of A is altered, and if suitable new values of $f(x,y)$ are chosen, the cocycle t is unaltered.

In what follows we call t a *Teichmüller cocycle* of the G-normal algebra A. Our next aim is to provide a complete description of the normality of a crossed product and the form of its Teichmüller cocycle. This will be achieved with the aid of the following three preliminary results.

Let K/F be a finite Galois extension with $K \supseteq E$ and put $H = \mathrm{Gal}(K/F)$, $S = \mathrm{Gal}(K/E)$. For any $\alpha \in Z^2(S,K^*)$, we denote by $K^\alpha S$ the corresponding crossed product of S over K. Recall, from Lemma 19.19, that $K^\alpha S$ is a central simple E-algebra with $C_{K^\alpha S}(K) = K$.

20.7. LEMMA. With the notation above, assume that K_1/E is a field extension and that there is an isomorphism $f : K \longrightarrow K_1$ such that $f|E$ is an F-automorphism of E. Then K and K_1 are E-isomorphic.

Proof. Put $\lambda = f|E$ and note that, since K/F is normal, λ^{-1} can be extended to an F-automorphism, say ψ, of K. Then $f\psi : K \longrightarrow K_1$ is obviously an E-isomorphism. ■

20.8. LEMMA. The algebra $K^{\alpha}S$ is G-normal if and only if every F-automorphism of K can be extended to a ring automorphism of $K^{\alpha}S$.

Proof. Since each $g \in G = \mathrm{Gal}(E/F)$ can be extended to an F-automorphism of K, the given condition is obviously sufficient. Suppose conversely that $K^{\alpha}S$ is G-normal. For each $\lambda \in H$, the extension $\sigma(\lambda')$ of $\lambda' = \lambda|E$ maps the subfield K of $K^{\alpha}S$ onto an isomorphic subfield K_1 containing $\lambda'(E) = E$. Hence by Lemma 20.7, there exists an E-isomorphism $K_1 \longrightarrow K$, which by Theorem 19.11, can be extended to an automorphism ϕ of $K^{\alpha}S$. The product $\phi\sigma(\lambda')$ is an automorphism of $K^{\alpha}S$ whose restriction to K is an element, say λ_1 of H, such that $\lambda_1|E = \lambda'$. Thus $\lambda\lambda_1^{-1} = \mu$ is an E-automorphism of K. By Theorem 19.11, μ can be extended to an E-automorphism ψ of $K^{\alpha}S$. Finally, $\psi\phi\sigma(\lambda')$ is a ring automorphism of $K^{\alpha}S$ carrying K into K by the automorphism

$$\mu\lambda_1 = \lambda\lambda_1^{-1}\lambda_1 = \lambda;$$

it is the required extension of λ to $K^{\alpha}S$. ■

In view of (1), we have an exact sequence

$$1 \longrightarrow S \longrightarrow H \longrightarrow G \longrightarrow 1$$

where $H \longrightarrow G$ is the restriction map. Hence H may be described as a group extension of S by G by choosing for each $g \in G$ an extension $v(g) \in H$, with $v(1) = 1$, and defining the automorphism $s \longmapsto g{*}s$ of the normal subgroup S by

$$g{*}s = v(g)s\,v(g)^{-1}$$

Then there is a "factor set" $\theta(x,y)$ of elements of S such that

$$v(x)v(y) = \theta(x,y)v(xy) \qquad (x,y \in G) \qquad (11)$$

$$v(g)s = (g{*}s)v(g) \qquad (g \in G,\ s \in S) \qquad (12)$$

Any elements $h_1, h_2 \in H$ can be expressed uniquely as

$$h_1 = s_1v(g_1)\ ,\quad h_2 = s_2v(g_2) \qquad (s_1,s_2 \in S,\ g_1,g_2 \in G) \qquad (13)$$

Their product, in view of (11) and (12), is given by

$$[s_1v(g_1)]\,[s_2v(g_2)] = s_1'v(g_1g_2) \qquad (s_1' = s_1(g_1{*}s_2)\theta(g_1,g_2)) \qquad (14)$$

The factor set Θ satisfies the associativity condition

$$[x\star\Theta(y,z)]\,\Theta(x,yz) = \Theta(x,y)\Theta(xy,z) \qquad (x,y,z \in G) \qquad (15)$$

and is related to the $\star$ operation by

$$[x\star(y\star s)]\,\Theta(x,y) = \Theta(x,y)\,[(xy)\star s] \qquad (x,y \in G,\ s \in S) \qquad (16)$$

20.9. LEMMA. The crossed product $K^\alpha S$ is G-normal if and only if for any $g \in G$, $s \in S$ there exist $m(g,s) \in K^\star$ such that

$$m(1,s) = m(g,1) = 1$$

and

$$m(g,ss_1)\,[v(g)\alpha(s,s_1)] = m(g,s)\,[(g\star s)m(g,s_1)]\,\alpha(g\star s,g\star s_1) \qquad (17)$$

Proof. Suppose first that $K^\alpha S$ is G-normal. Then, by Lemma 20.8, $v(g)$ can be extended to a ring automorphism, say $\sigma(g)$, of $K^\alpha S$, with $\sigma(1) = 1$. We know that multiplication in $K^\alpha S$ is determined by

$$\bar{s}\,\bar{s}_1 = \alpha(s,s_1)\overline{ss_1} \qquad (s,s_1 \in S) \qquad (18)$$

$$\bar{s}\,\lambda = s(\lambda)\bar{s} \qquad (\lambda \in K) \qquad (19)$$

In (19) set $\lambda = v(g)^{-1}\mu$ for any $\mu \in K$ and apply the automorphism $\sigma(g)$ to obtain

$$[\sigma(g)\bar{s}]\,\mu = [(g\star s)\mu]\,[\sigma(g)\bar{s}]$$

On the other hand, (19) implies that

$$\overline{g\star s}\,\mu = [(g\star s)\mu]\,\overline{g\star s}$$

These two equations assert that $[\sigma(g)\bar{s}]\,\overline{g\star s}^{\,-1}$ commutes with every $\mu \in K$, hence lies in K, so we may write

$$\sigma(g)\bar{s} = m(g,s)\overline{g\star s} \qquad (20)$$

for some $m(g,s) \in K^\star$ with $m(g,1) = m(1,s) = 1$. Applying $\sigma(g)$ to both sides of (18), we obtain

$$m(g,s)\,\overline{g\star s}\,m(g,s_1)\,\overline{g\star s_1} = [v(g)\alpha(s,s_1)]\,m(g,ss_1)\,\overline{g\star ss_1}$$

The left side, by applying (18) and (19), is

$$m(g,s)[(g*s)m(g,s_1)]\alpha(g*s,g*s_1)\,\overline{g*ss_1}$$

Comparison with the right side gives the required equation (17).

Conversely, given such a function m, we may use (20) to define the effect of $\sigma(g)$ on $\bar{s}$, and to define $\sigma(g) : K^{\alpha}S \longrightarrow K^{\alpha}S$ additively by using

$$\sigma(g)(\lambda\bar{s}) = [v(g)\lambda]\,[\sigma(g)\bar{s}] \qquad (\lambda \in K)$$

By reversing the above computation, one immediately verifies that $\sigma(g)$ is a ring automorphism of $K^{\alpha}S$ extending the given F-automorphism g of E. ■

We are now ready to provide a complete description of the normality of a crossed product and the form of its Teichmüller cocycle.

20.10.THEOREM. (Eilenberg and MacLane(1948)). Let E/F and K/F be finite Galois extensions with $K \supseteq E$, let $G = \mathrm{Gal}(E/F)$, $H = \mathrm{Gal}(K/F)$, $S = \mathrm{Gal}(K/E)$, and for a given $\alpha \in Z^2(S,K^*)$, let $K^{\alpha}S$ be the corresponding crossed product of S over K. Given $t \in Z^3(G,E^*)$, let $t_K \in Z^3(H,K^*)$ be defined by (2). Then $K^{\alpha}S$ is G-normal with t as one of its Teichmüller cocycles if and only if there exists a cochain $\beta \in C^2(H,K^*)$ such that

$$\delta\beta = t_K \tag{21}$$

$$\beta(s_1,s_2) = \alpha(s_1,s_2) \qquad (\text{for all } s_1,s_2 \in S) \tag{22}$$

Proof. Assume that $K^{\alpha}S$ is G-normal with t as one of its Teichmüller cocycles. We may extend, by Lemma 20.8, $v(g)$, $g \in G$, to an automorphism $\sigma(g)$ of $K^{\alpha}S$, with $\sigma(1) = 1$. Owing to Lemmas 20.5 and 20.6, t can be obtained from a suitable unit $f(x,y)$ of $K^{\alpha}S$ satisfying (7). Consider the map

$$\Omega : H \longrightarrow \mathrm{Aut}(K^{\alpha}S)$$

given by

$$\Omega(sv(g)) = i(\bar{s})\sigma(g) \qquad (s \in S,\ g \in G)$$

where $i(\bar{s})$ denotes the inner automorphism of $K^{\alpha}S$ obtained by conjugation with $\bar{s}$. If $h_1,h_2 \in H$, say $h_1 = s_1v(g_1)$, $h_2 = s_2v(g_2)$, then

$$\begin{aligned}\Omega(h_1)\Omega(h_2) &= i(\bar{s}_1)\sigma(g_1)i(\bar{s}_2)\sigma(g_2)\\ &= i[\bar{s}_1(\sigma(g_1)\bar{s}_2)]\sigma(g_1)\sigma(g_2)\end{aligned}$$

$$= \hat{\iota}[\bar{s}_1(\sigma(g_1)\bar{s}_2)f(g_1,g_2)]\,\sigma(g_1g_2) \qquad \text{(by (7))}$$

Define $\beta(h_1,h_2) \in U(K^{\alpha}S)$, using the notation of (14) for the product h_1h_2, by

$$\beta(h_1,h_2)s_1' = \bar{s}_1(\sigma(g_1)\bar{s}_2)f(g_1,g_2) \tag{23}$$

Then the previous equation becomes

$$\Omega(h_1)\Omega(h_2) = \hat{\iota}[\beta(h_1,h_2)]\,\Omega(h_1h_2) \tag{24}$$

This shows that conjugation by $\beta = \beta(h_1,h_2)$ induces on $K^{\alpha}S$ an automorphism $\Omega(h_1)\Omega(h_2)\Omega(h_1h_2)^{-1}$ which is on K the identity $h_1h_2(h_1h_2)^{-1}$; in other words, $\beta \in K^{\alpha}S$ commutes with all elements of K, hence $\beta(h_1,h_2) \in K^*$. Clearly $\beta(1,h_2) = \beta(h_1,1) = 1$, so that β is a 2-cochain of H over K^*.

In the definition (23) of β choose $h_1 = s_1$, $h_2 = s_2$; then $s_1' = s_1s_2$, $g_1 = g_2 = 1$, and $\beta(s_1,s_2)\overline{s_1s_2} = \bar{s}_1\bar{s}_2$. It follows from (18) that β, restricted to $S \times S$ coincides with α, as asserted in (22).

Now we calculate the coboundary $\delta\beta$, for arguments h_1,h_2, and $h_3 = s_3v(g_3)$, and with the notation (13), (14), and

$$h_2h_3 = s_2'v(g_2g_3)\ , \qquad s_2' = s_2(g_2*s_3)\theta(g_2,g_3)$$

Because multiplication in H is associative,

$$(h_1h_2)h_3 = h_1(h_2h_3) = \varepsilon v(g_1g_2g_3)$$

where $\varepsilon \in S$ is determined from both products as

$$\varepsilon = s_1'[(g_1g_2)*s_3]\,\theta(g_1g_2,g_3) = s_1(g_1*s_2')\theta(g_1,g_2g_3)$$

Using the definition (23) repeatedly, we calculate that

$$\begin{aligned}
\beta(h_1,h_2)\beta(h_1h_2,h_3)\bar{\varepsilon} &= \beta(h_1,h_2)\beta(h_1h_2,h_3)\overline{s_1'[(g_1g_2)*s_3]\,\theta(g_1g_2,g_3)}\\
&= \beta(h_1,h_2)\bar{s}_1'\,(\sigma(g_1g_2)\bar{s}_3)f(g_1g_2,g_3)\\
&= \bar{s}_1(\sigma(g_1)\bar{s}_2)f(g_1,g_2)\,[\sigma(g_1g_2)\bar{s}_3]f(g_1g_2,g_3)\\
&= xf(g_1,g_2)f(g_1g_2,g_3)
\end{aligned} \tag{25}$$

where

$$\begin{aligned}
x &= \bar{s}_1(\sigma(g_1)\bar{s}_2)f(g_1,g_2)(\sigma(g_1g_2)\bar{s}_3)f(g_1,g_2)^{-1}\\
&= \bar{s}_1(\sigma(g_1)\bar{s}_2)(\sigma(g_1)\sigma(g_2)\bar{s}_3)
\end{aligned}$$

On the other hand, using the second expression for ε, we have

$$\begin{aligned}[][h_1\beta(h_2,h_3)]\beta(h_1,h_2h_3)\bar{\varepsilon} &= [h_1\beta(h_2,h_3)]\bar{s}_1[\sigma(g_1)\overline{s_2'}]f(g_1,g_2g_3) \\ &= \bar{s}_1[s_1^{-1}h_1\beta(h_2,h_3)][\sigma(g_1)\overline{s_2'}]f(g_1,g_2g_3) \\ &= \bar{s}_1\sigma(g_1)[\beta(h_2,h_3)\overline{s_2'}]f(g_1,g_2g_3) \\ &= \bar{s}_1\sigma(g_1)[\bar{s}_2(\sigma(g_2)\bar{s}_3)f(g_2,g_3)]f(g_1,g_2g_3) \\ &= x[\sigma(g_1)f(g_2,g_3)]f(g_1,g_2g_3)\end{aligned}$$

By the relation (9), this becomes

$$xt(g_1,g_2,g_3)f(g_1,g_2)f(g_1g_2,g_3)$$

Applying (25) and the fact that $t(g_1,g_2,g_3)$ lies in the centre E of $K^\alpha S$, we deduce that

$$t(g_1,g_2,g_3)\beta(h_1,h_2)\beta(h_1h_2,h_3) = [h_1\beta(h_2,h_3)]\beta(h_1,h_2h_3)$$

But $h_1' = g_1$, $h_2' = g_2$, $h_3' = g_3$, so by (5) and (2) this equation proves that $\delta\beta = t_K$, which is (21) in the theorem.

Conversely, assume that there exists a cochain $\beta \in C^2(H,K^*)$ satisfying (21) and (22), for a given $t \in Z^3(G,E^*)$. Define $m(g,s) \in K^*$ for $g \in G$, $s \in S$ by

$$m(g,s)\beta(g{\star}s,v(g)) = \beta(v(g),s) \tag{26}$$

To show that $K^\alpha S$ is G-normal, we must demonstrate that m satisfies (17), which by definition of m reduces to the identity

$$\begin{aligned}&\beta(v(g),ss_1)[v(g)\beta(s,s_1)]\beta(g{\star}s,v(g))[(g{\star}s)\beta(g{\star}s_1,v(g))] \\ &\qquad = \beta(v(g),s)[(g{\star}s)\beta(v(g),s_1)]\beta(g{\star}s,g{\star}s_1)\beta(g{\star}(ss_1),v(g))\end{aligned}$$

Since $\delta\beta = t_K$, it follows that $\delta\beta(h_1,h_2,h_3)$ vanishes when any one argument lies in the subgroup S. Hence

$$1 = \delta\beta(v(g),s,s_1)[\delta\beta(g{\star}s,v(g),s_1)]^{-1}\delta\beta(g{\star}s,\ g{\star}s,\ v(g))$$

If these terms are expanded out by the definition of $\delta\beta$, four of the resulting twelve terms cancel, and the remaining eight terms give precisely the desired identity (17).

We are therefore left to verify that t is the Teichmüller cocycle of $K^\alpha S$.

Define the unit $f(x,y)$ of $K^\alpha S$, $x,y \in G$, by

$$f(x,y) = \beta(v(x),\ v(y))\beta(\theta(x,y),\ v(xy)^{-1})\ \overline{\theta(x,y)} \tag{27}$$

To prove that f satisfies (7), we must demonstrate that for all $a \in K^\alpha S$,

$$[\sigma(x)\sigma(y)a]\,f(x,y) = f(x,y)\,[\sigma(xy)a] \tag{28}$$

where $\sigma(x)$ is the automorphism of $K^\alpha S$ defined from the function m above as in (20). It suffices to prove (28) for $a \in K$ and $a = \bar{s}$, since these elements generate the crossed product $K^\alpha S$. For $a \in K$, the commutativity of K reduces (28) to

$$[v(x)v(y)a]\ \overline{\theta(x,y)} = \overline{\theta(x,y)}\,[v(xy)a]$$

which follows from (19) and by the definition (11) of the factor set f.

If $a = \bar{s}$, the desired identity becomes, by the definition (20) of $\sigma(x)$,

$$[v(x)m(y,s)]\,m(x,y{\star}s)\ \overline{x{\star}y{\star}s}\ f(x,y) = f(x,y)m(xy,s)\ \overline{xy{\star}s}$$

Using the definitions of m and f together with multiplication rules (18) and (19), the bar terms reduce to $\overline{(x{\star}y{\star}s)\theta(x,y)}$ on the left and $\overline{\theta(x,y)[(xy){\star}s]}$ on the right, so they are equal by virtue of (16). Thus these bar terms may be dropped. There remains an identity involving some twelve factors in the function β with various arguments and operators. A straightforward calculation shows that this identity follows from the identity

$$\begin{aligned}1 = {} & \delta\beta(v(x),v(y),s)\delta\beta^{-1}(v(x),\ y{\star}s,v(y))\delta\beta(x{\star}y{\star}s,v(x),v(y)) \\ & \delta\beta^{-1}(x{\star}y{\star}s,\ \theta(x,y),\ v(xy))\delta\beta(\theta(x,y),\ (xy){\star}s,\ v(xy))\delta\beta^{-1}(\theta(x,y),v(xy),s)\end{aligned}$$

in which each term is 1 since $\delta\beta = t_K$, and at least one of the arguments in each term lies in S. Indeed, the latter identity involves 24 terms, four from each coboundary involved. Each coboundary involves two factors in which products of the arguments occur, and each of these factors cancels with corresponding factors from the next or the preceding coboundary. The remaining twelve factors give precisely the desired identity, and thus complete the proof of (28).

The elements $f(x,y)$, as defined in (27), can now be used to calculate a Teichmüller cocycle t_1 of $K^\alpha S$ as

$$[\sigma(x)f(y,z)]\,f(x,yz) = t_1(x,y,z)f(x,y)f(xy,z)$$

We are left to verify that this identity holds with $t_1 = t$. In this identity we may substitute the definition of $f(x,y)$ given by (27), use the rule

$$\sigma(x)\bar{s} = m(x,s)\,\overline{x\star s}$$

to apply $\sigma(x)$, replace m by its definition (26) in terms of β, and use the multiplication rules of $K^{\alpha}S$ to move all the bar terms to the right end on each side. The left side gives $\overline{x\star\theta(y,z)\theta(x,yz)}$, the right side $\overline{\theta(x,y)\theta(xy,z)}$ and these are equal, by virtue of (15). The bar terms may then be dropped, leaving an identity in t and some twelve factors β, with various arguments. This identity can be obtained from the identity

$$\begin{aligned} 1 = {} & [t(x,y,z)\delta\beta^{-1}(v(x),v(y),v(z))]\,\delta\beta(v(x),\theta(y,z),v(yz)) \\ & \delta\beta^{-1}(x\star\theta(y,z),v(x),v(yz))\delta\beta(x\star\theta(y,z),\theta(x,yz),v(xyz)) \\ & \delta\beta^{-1}(\theta(x,y),\theta(xy,z),v(xyz))\delta\beta(\theta(x,y),v(xy),v(z)), \end{aligned}$$

by expanding out each $\delta\beta$ by the definition of the coboundary. In this expansion each term involving a product of two arguments of $\delta\beta$ cancels with a similar term from a neighbouring $\delta\beta$; the remaining terms give exactly the twelve factors β mentioned above. The last displayed identity again holds since $\delta\beta = t_K$, for this shows at once that the first term in brackets above is 1, and that the remaining terms are 1 since in each case at least one argument is in S. This proves that t is the Teichmüller cocycle of $K^{\alpha}S$, as required. ∎

The following property will enable us to take full advantage of the results so far obtained.

20.11. LEMMA. Let A be a G-normal algebra with Teichmüller cocycle t, let K/F be a finite Galois extension with $K \supseteq E$ and let $H = \mathrm{Gal}(K/F)$. Then $A \otimes_E K$ is an H-normal algebra over K with a Teichmüller cocycle t_K given by (2).

Proof. Select $\sigma(x)$, $f(x,y)$ and $t(x,y,z)$, $x,y,z \in G$, to be as in Lemma 20.5. Each automorphism $\lambda \in H$ restricts to an automorphism $\lambda' \in G$, and the extension $\sigma(\lambda')$ agrees with λ on the field E of scalars in $A \otimes_E K$. Invoking

Lemma 20.1, there is an automorphism $\sigma^*(\lambda) = \sigma(\lambda') \otimes \lambda$ of $A \underset{F}{\otimes} K$ which agrees on the respective factors with $\sigma(\lambda')$ and λ and which has $\sigma^*(1) = 1$. Thus λ is extendible to an automorphism of $A \underset{E}{\otimes} K$, proving that $A \underset{E}{\otimes} K$ is H-normal.

Applying Lemma 20.3 twice, we have

$$\begin{aligned} \sigma^*(\lambda)\sigma^*(\mu) &= [\sigma(\lambda') \otimes \lambda][\sigma(\mu') \otimes \mu] = \sigma(\lambda')\sigma(\mu') \otimes \lambda\mu \\ &= i[f(\lambda',\mu')]\sigma(\lambda'\mu') \otimes \lambda\mu \qquad \text{(by (7))} \\ &= (i[f(\lambda',\mu')] \otimes 1)(\sigma((\lambda\mu)') \otimes \lambda\mu) \\ &= (i[f(\lambda',\mu')] \otimes 1)(\sigma^*(\lambda\mu)) \end{aligned}$$

The automorphism $i[f(\lambda',\mu')] \otimes 1$ is conjugation in $A \underset{E}{\otimes} K$ by the element $f^*(\lambda,\mu) = f(\lambda',\mu') \otimes 1$ and thus

$$\sigma^*(\lambda)\sigma^*(\mu) = i[f^*(\lambda,\mu)]\sigma^*(\lambda\mu)$$

i.e. f^* satisfies condition (7) with respect to H. The corresponding Teichmüller cocycle t^* of $A \underset{E}{\otimes} K$ can therefore be derived from f^* by formula (9) as

$$t^*(\lambda,\mu,\nu) = t(\lambda',\mu',\nu') \otimes 1$$

Thus $t^* = t_K$ and the result follows. ∎

We have now come to the demonstration for which this section has been developed.

20.12. THEOREM. (Eilenberg and Maclane(1948), Teichmüller(1940)). Let E/F be a finite Galois extension with Galois group G, let $Br_G(E)$ be the subgroup of $Br(E)$ consisting of Brauer classes of G-normal algebras, and let $\overline{t(A)}$ be the cohomology class of a Teichmüller cocycle $t(A)$ of a G-normal algebra A. Then the map

$$\begin{cases} Br_G(E) \xrightarrow{T} H^3(G,E^*) \\ \quad [A] \longmapsto \overline{t(A)} \end{cases}$$

is a homomorphism such that

$$\text{Ker}T = Br_F(E) \quad \text{and} \quad \text{Im}T = H_0^3(G,E^*)$$

In particular

$$H_0^3(G,E^*) \cong Br_G(E)/Br_F(E)$$

Here $Br_F(E)$ is the subgroup of $Br(E)$ consisting of all Brauer classes of the form $[B \otimes_F E]$, where $[B] \in Br(F)$.

Proof. Let A_1, A_2 be two G-normal algebras with Teichmüller cocycles t_1, t_2, respectively. We claim that $A_1 \otimes_E A_2$ has $t_1 t_2$ as its Teichmüller cocycle. Indeed, choose extensions $\sigma_i(x)$ and elements $f_i(x,y)$ in A_i, $x,y \in G$, to satisfy (7) and (9) for $i = 1,2$. Then, by Lemma 20.1, $\sigma(x) = \sigma_1(x) \otimes \sigma_2(x)$ is an extension of the automorphism x to $A_1 \otimes_E A_2$. Moreover,

$$f(x,y) = f_1(x,y) \otimes f_2(x,y)$$

is a unit of $A_1 \otimes_E A_2$ and

$$i[f(x,y)] = i[f_1(x,y)] \otimes i[f_2(x,y)]$$

Applying Lemma 20.3, we deduce that σ and f are related by (7), while calculating (9) one deduces that $t = t_1 t_2$ is a Teichmüller cocycle of $A_1 \otimes_E A_2$. In particular, for any $n \geq 1$, $M_n(E)$ has $t = 1$ as a Teichmüller cocycle. Therefore, any two similar G-normal algebras have the same Teichmüller cocycles. Thus T is indeed a homomorphism.

To prove that $\operatorname{Ker} T = Br_F(E)$, we must show that

(a) For any central simple F-algebra B, $B \otimes_F E$ has a Teichmuller cocycle 1

(b) If A is a G-normal algebra with Teichmüller cocycle 1, then $B \otimes_F E \cong M_k(A)$ for $k = (E:F)$ and for some central simple F-algebra B. In particular,

$$[A] = [B \otimes_F E]$$

Property (a) is the special case $K = E$, $E = F$ of Lemma 20.10. To prove (b), consider A with $t(A) = 1$. Using (9), we can construct an algebra B with elements uniquely represented as sums $\Sigma a_g \bar{g}$ with coefficients $a_g \in A$ for each $g \in G$ with multiplication table determined by the distributive law and the rules

$$\bar{g}\,a = [\sigma(g)a]\,\bar{g} \qquad (g \in G,\ a \in A) \tag{29}$$
$$\bar{x}\,\bar{y} = f(x,y)\overline{xy} \qquad (x,y \in G)$$

This construction is exactly analogous to the construction of a crossed product algebra over a field. Since $t(x,y,z) = 1$ for all $x,y,z \in G$, it follows from (9) that B is an associative algebra with $\bar{1}$ as the identity element. In particular, A is identifiable with the subalgebra $\{a\cdot\bar{1} | a \in A\}$. A repetition of the argument in Lemma 19.19 shows that B is simple. Furthermore, a straightforward computation shows that $C_B(E) = A$; hence by (29), $Z(B) = F$. Thus B is a central simple F-algebra. Applying Corollary 19.12, we have $B \otimes_F E \cong M_k(A)$ where $k = (E:F)$, which clearly implies (b).

We now claim that the assertion $\mathrm{Im}T = H_0^3(G,E^*)$ is a consequence of the following two statements (K is a finite Galois extension of F)

(c) If the G-normal algebra A is split by $K \supseteq E$, then for any Teichmüller cocycle t of A, t_K is a coboundary.

(d) If $t \in Z^3(G,E^*)$ is such that t_K is a coboundary, then t is the Teichmüller cocycle of some G-normal algebra split by K.

Indeed, if (c) is true, then choosing as K a splitting field for A (see Proposition 19.23), we see that $H_0^3(G,E^*) \supseteq \mathrm{Im}T$. On the other hand, (d) ensures that $\mathrm{Im}T \supseteq H_0^3(G,E^*)$, as claimed.

Property (c) is a consequence of Lemma 20.11. Indeed, if A is split by K, then $A \otimes_E K \cong M_n(E)$ for some $n \geq 1$ and every Teichmüller cocycle for $A \otimes_E K$ (with arguments in H) is a coboundary, and Lemma 20.11 shows that t_K is a coboundary.

To prove (d), suppose that $t \in Z^3(G,E^*)$ is such that t_K is a coboundary. Then $t_K = \delta\beta$ for some 2-cochain β of H in K^*. In particular,

$$\delta\beta(x,y,z) = 1$$

for all $x,y,z \in S$, so that the definition $\alpha(x,y) = \beta(x,y)$ gives an element of $Z^2(S,K^*)$. We may therefore construct the crossed product $K^\alpha S$ of S over K. Applying Theorem 20.9, we conclude that $K^\alpha S$ is G-normal with t as one of its Teichmüller cocycles. This proves (d) and hence the result. ■

20.13. COROLLARY. Let $T : Br_G(E) \longrightarrow H^3(G,E^*)$ be the homomorphism of Theorem 20.12. If K/F is a finite Galois extension of F with $K \supseteq E$, then T carries the elements of $Br_G(E)$ which are split by K onto those cohomology classes $\bar{t} \in H^3(G,E^*)$ for which $t \in Z^3(G,E^*)$ is split by K.

Proof. This is a direct consequence of properties (c) and (d) established in the course of the proof of Theorem 20.12. ■

20.14. COROLLARY. Further to the assumptions and notations of Theorem 20.12, assume that G is cyclic. Then, for any G-normal algebra A, there exists a central simple F-algebra B such that

$$B \otimes_F E \cong M_k(A) \qquad \text{where } k = (E:F)$$

Proof. The general reduction theory for cohomology groups (see Eilenberg and MacLane(1947a,§16)) proves that $H^3(G,E^*) \cong H^1(G,E^*)$. Hence, by Theorem 18.3, $H^3(G,E^*) = 1$. Thus the Teichmüller cocycle of A is 1. The desired conclusion now follows by virtue of property (b) established in the course of the proof of Theorem 20.12. ■

Turning our attention to $H_0^2(G,E^*)$, we now prove

20.15. THEOREM. (Eilenberg and MacLane(1948)). Let E/F be a finite Galois extension with Galois group G. Then $H_0^2(G,E^*) = 1$.

Proof. Let $\alpha \in Z_0^2(G,E^*)$, i.e. $\alpha_K = \delta\beta$ for some $\beta \in C^1(H,K^*)$, where K/F is a suitable finite Galois extension with Galois group H and with $K \supseteq E$. Define $\beta_0(s) = \beta(s)$ for all $s \in S$. Then, since α is normalized, $\delta\beta_0 = 1$ and $\beta_0 \in Z^1(S,K^*)$. Hence, by Theorem 18.3, there is a constant $b \in K^*$ such that $\beta_0(s)b = s(b)$ for all $s \in S$. Define β_1 as $\beta_1(h) = \beta(h)b/h(b)$ for all $h \in H$. Then β_1 is cohomologous to β, and $\delta\beta_1 = \alpha_K$, while β_1 is 1 on the subgroup S. In particular, because α is normalized,

$$\alpha_K(h,s) = \alpha_K(s,h) = 1 \qquad \text{for all } s \in S$$

and so

$$\beta_1(hs) = [h\beta_1(s)]\beta_1(h), \quad \beta_1(sh) = [s\beta_1(h)]\beta_1(s)$$

Because $\beta_1(s) = 1$, $\beta_1(hs) = \beta_1(h) = \beta_1(sh)$ is an element invariant under s. Hence each value $\beta_1(h)$ lies in E^*. Now each automorphism $g \in G$ has an extension $v(g) \in H$ and $v(x)v(y) = \eta v(xy)$ for some $\eta = \eta(x,y) \in S$. Then, for all $x,y \in G$,

$$\begin{aligned}\alpha(x,y) &= \alpha_K(v(x),v(y)) = \delta\beta_1(v(x),v(y))\\ &= v(x)\beta_1(v(y))[\beta_1(\eta v(xy))]^{-1}\beta_1(v(x))\\ &= x\beta_1(v(y))[\beta_1(v(xy))]^{-1}\beta_1(v(x))\end{aligned}$$

and therefore $\alpha = \delta\beta_2$, where $\beta_2(x) = \beta_1(v(x))$ is a cochain of G over E^*. So the theorem is true. ■

We close by showing how Lemma 20.6 can be derived as a consequence of a general cohomology theory. The following simple observation will clear our path.

20.16. LEMMA. Let A be a central simple algebra over a field E. Then

$$E^* = Z(U(A))$$

Proof. It suffices to show that $Z(U(A)) \subseteq E^*$. The algebra A can be represented as the algebra of all $n \times n$ matrices, with matrix units e_{ij}, over a division algebra D with centre E. Then $e = \Sigma e_{ii}$ is the identity element of A and, for any $r \neq s$,

$$\begin{aligned}(e+e_{rs})(e-e_{rs}) &= e + e_{rs} - e_{rs} = e\\ (e-e_{rs})(e+e_{rs}) &= e - e_{rs} + e_{rs} = e\end{aligned}$$

proving that $e + e_{rs} \in U(A)$. If $a = \Sigma d_{ij}e_{ij}$, $d_{ij} \in D$, is any element of $Z(U(A))$, then $a(e+e_{rs}) = (e+e_{rs})a$ or $ae_{rs} = e_{rs}a$. Multiplication gives

$$\sum_i d_{ir}e_{is} = ae_{rs} = e_{rs}a = \sum_j d_{sj}e_{rj}$$

Comparing the coefficients, it follows that $d_{rr} = d_{ss}$ for $r \neq s$ and $d_{ir} = 0$ for $i \neq r$. Hence, if $d = d_{rr}$, then $a = \Sigma de_{ii} = de$. Because $Z(D) = E$, we have $0 \neq d \in E$, as required. ■

For the rest of this section, G denotes an arbitrary group. A *G-kernel* (N,θ) is defined to be a group N together with a homomorphism

$$\theta : G \longrightarrow \mathrm{Out}(N) = \mathrm{Aut}(N)/\mathrm{In}(N)$$

where $\mathrm{In}(N)$ is the normal subgroup of the automorphism group $\mathrm{Aut}(N)$ of N consisting of inner automorphisms. Each inner automorphism of N leaves the elements of $Z(N)$ fixed, so that each coset $\theta(x)$ determines a unique automorphism $\theta_0(x)$ of $Z(N)$. Thus we have a homomorphism

$$\theta_0 : G \longrightarrow \mathrm{Aut}(Z(N))$$

and therefore $Z(N)$ is a G-module via $g \cdot n = \theta_0(g)n$ for all $n \in Z(N)$, $g \in G$. The pair $(Z(N), \theta_0)$ is called the *centre* of the G-kernel (N, θ).

20.17. EXAMPLE. Let E/F be a finite Galois extension with Galois group G, let A be a G-normal algebra and, for each automorphism $g \in G$, let $\sigma(g)$ be an extension of g to a ring automorphism of A. Then $\sigma(g)$ is also an automorphism of $U(A)$ and, by (7), $\sigma(x)\sigma(y) = \sigma(xy)$ modulo the inner automorphsims of $U(A)$. Setting $N = U(A)$, it follows from Lemma 20.16 that $Z(N) = E^*$. Thus, if $\theta(x)$ is the coset of $\sigma(x)$ modulo the inner automorphisms of $U(A)$, then

$$(U(A), \theta) \text{ is a } G\text{-kernel with centre } (E^*, \theta_0)$$

Let (N, θ) be a G-kernel with centre $(Z(N), \theta_0)$. For each $\theta(x)$, choose an automorphism $\sigma(x)$ of N in $\theta(x)$ with $\sigma(1) = 1$. Since $\theta : G \longrightarrow \mathrm{Out}(N)$, is a homomorphism,

$$\sigma(x)\sigma(y)\sigma(xy)^{-1} \in \mathrm{In}(N) \qquad \text{for all } x, y \in G$$

Hence, we may choose elements $f(x,y) \in N$ with $f(1,y) = f(x,1) = 1$ such that

$$\sigma(x)\sigma(y) = i[f(x,y)]\sigma(xy) \qquad \text{for all } x, y \in G \tag{30}$$

where

$$i[f(x,y)](n) = f(x,y)n\, f(x,y)^{-1} \qquad \text{for all } n \in N$$

Since $\sigma(x)[\sigma(y)\sigma(z)] = [\sigma(x)\sigma(y)]\sigma(z)$, an application of (30) yields

$$i[f(x,y)f(xy,z)] = i[(\sigma(x)f(y,z))f(x,yz)]$$

Thus there exist elements $t(x,y,z)$ in $Z(N)$ such that

$$[\sigma(x)f(y,z)]f(x,yz) = t(x,y,z)f(x,y)f(xy,z) \tag{31}$$

Thus t is normalized 3-cochain of G with values in $Z(N)$. Hence, in the situation of Example 20.17, we have

20.18. LEMMA. Each Teichmüller 3-cochain t of a G-normal algebra A is a 3-cochain of the corresponding G-kernel $(U(A), \theta)$ with centre (E^*, θ_0).

Proof. Compare (30, (31) with (7), (9). ∎

We can now derive Lemma 20.6 as a consequence of Lemma 20.18 and the following general result.

20.19. THEOREM. (Eilenberg and MacLane(1947b)). Let t be a 3-cochain of G with values in $Z(N)$ corresponding to the G-kernel (N,θ).

(i) The cochain t is a cocycle

(ii) If the choice of f in (30) is changed, then t is changed to a cohomologous cocycle. By suitable changing the choice of f, the cocycle t may be changed to any cohomologous cocycle.

(iii) If the choice of the automorphisms σ is changed, then a suitable new selection of f leaves the cocycle t unaltered.

Proof. (i) Let us calculate the expression

$$A = \sigma(x)[\{\sigma(y)f(z,u)\}f(y,zu)]f(x,yzu)$$

in two ways. First using (31) three times, and recalling that t lies in $Z(N)$, we have

$$\begin{aligned} A &= \sigma(x)[t(y,z,u)f(y,z)f(yz,u)]f(x,yzu) \\ &= \sigma(x)[t(y,z,u)f(y,z)]t(x,yz,u)f(x,yz)f(xyz,u) \\ &= [xt(y,z,u)]t(x,yz,u)t(x,y,z)B \end{aligned}$$

where B is given by

$$B = f(x,y)f(xy,z)f(xyz,u)$$

Alternatively, by first applying (30), we have

$$\begin{aligned} A &= f(x,y)[\sigma(xy)f(z,u)]f(x,y)^{-1}[\sigma(x)f(y,zu)]f(x,yzu) \\ &= f(x,y)[\sigma(xy)f(z,u)]t(x,y,zu)f(xy,zu) \\ &= t(x,y,zu)t(xy,z,u)B \end{aligned}$$

Comparison gives

$$[xt(y,z,u)]t(x,yz,u)t(x,y,z) = t(x,y,zu)t(xy,z,u)$$

which, by (4), shows that $t \in Z^3(G,Z(N))$.

(ii) Any other choice of $f(x,y)$ in (30) has the form

$$f'(x,y) = g(x,y)f(x,y) \qquad (g(x,y) \in Z(N))$$

where g is any normalized 2-cochain of G in $Z(N)$. Using (31) to calculate t' from f', we have

$$\begin{aligned} t'(x,y,z) &= t(x,y,z)\{[xg(y,z)]g(xy,z)^{-1}g(x,yz)g(x,y)^{-1}\}^{-1} \\ &= t(x,y,z)\delta g(x,y,z)^{-1}, \end{aligned}$$

proving (ii).

(iii) If $\sigma'(x)$ is another selection of automorphisms in $\theta(x)$, then

$$\sigma'(x) = i[n(x)]\sigma(x) \qquad (n(x) \in N)$$

However,

$$\begin{aligned} \sigma'(x)\sigma'(y) &= i[n(x)]\sigma(x)i[n(y)]\sigma(y) \\ &= i[n(x)(\sigma(x)n(y))f(x,y)]\ \sigma(xy) \\ &= i[n(x)(\sigma(x)n(y))f(x,y)n(xy)^{-1}]\sigma'(xy) \end{aligned}$$

Thus we may select $f'(x,y)$ as the expression in brackets, so that

$$\begin{aligned} f'(x,y)n(xy) &= n(x)[\sigma(x)n(y)]f(x,y) \\ &= [\sigma'(x)n(y)]n(x)f(x,y) \end{aligned}$$

It therefore follows that

$$\begin{aligned} [\sigma'(x)f'(y,z)]f'(x,yz)n(xyz) &= \sigma'(x)[f'(y,z)\,n\,(yz)]n(x)f(x,yz) \\ &= \sigma'(x)[(\sigma'(x)n(z))n(y)]n(x)[\sigma(x)f(y,z)]f(x,yz) \\ &= [\sigma'(x)\sigma'(y)n(z)][\sigma'(x)n(y)]n(x)t(x,y,z)f(x,y)f(xy,z) \\ &= t(x,y,z)[\sigma'(x)\sigma'(y)n(z)]f'(x,y)n(xy)f(xy,z) \\ &= t(x,y,z)f'(x,y)[\sigma'(xy)n(z)]n(xy)f(xy,z) \\ &= t(x,y,z)f'(x,y)f'(xy,z)n(xyz) \end{aligned}$$

which shows that

$$[\sigma'(x)f'(y,z)]f'(x,yz) = t(x,y,z)f'(x,y)f'(xy,z)$$

Thus the cocycle t is unaltered and the result follows. ∎

21. A COGALOIS THEORY FOR RADICAL EXTENSIONS

Throughout this section, E/F denotes a *finite field extension*. We put

$$\mathrm{Cog}(E/F) = \{\lambda F^* \in E^*/F^* \mid \lambda^m \in F^* \quad \text{for some} \quad m \geq 1\}$$

Hence $\mathrm{Cog}(E/F)$ is the torsion subgroup $t(E^*/F^*)$ of E^*/F^*. Given a subgroup H of $\mathrm{Cog}(E/F)$, we write

$$E_H = \{a \in E \mid a = b_1 + \dots + b_n \quad \text{with} \quad b_i F^* \in H,\ 1 \leq i \leq n, \quad \text{for some} \quad n \geq 1\} \tag{1}$$

Then E_H is a subfield of E containing F and such that

$$(E_H:F) \leq |H|$$

Indeed, E_H is obviously an F-subalgebra of E consisting of F-linear combinations of elements λ_i, $i \in I$, where $H = \{\lambda_i F^* \mid i \in I\}$ (in particular, the above inequality always holds). Since E/F is finite, E_H is an artinian integral domain, hence a field (Proposition 1.1.3).

The following terminology is chosen to reflect the duality. Following Greither and Harrison(1986), we call E/F *cogalois* with cogalois group $\mathrm{Cog}(E/F)$ if:

(i) (*Conormality*) $|\mathrm{Cog}(E/F)| \leq (E:F)$ (2)

(ii) (*Coseparability*) $E = E_H$ with $H = \mathrm{Cog}(E/F)$ (3)

For example, one immediately verifies that

$$\mathbb{Q}(\sqrt{2})/\mathbb{Q} \quad \text{is cogalois with} \quad \mathrm{Cog}(\mathbb{Q}\sqrt{2}/\mathbb{Q}) = \{\mathbb{Q}^*, \sqrt{2}\mathbb{Q}^*\}$$

while

$$\mathbb{Q}(\sqrt{-3})/\mathbb{Q} \quad \text{is not cogalois}$$

since $|\mathrm{Cog}(\mathbb{Q}(\sqrt{-3})/\mathbb{Q})| = 3$; more precisely

$$\mathrm{Cog}(\mathbb{Q}(\sqrt{-3})/\mathbb{Q}) = \{\mathbb{Q}^*, \sqrt{-3}\mathbb{Q}^*, ((-1 + \sqrt{-3}/2)\mathbb{Q}^*\}$$

Our aim is to present a theory which is dual to the usual Galois theory of fields. Namely, we show that

(a) If E/F is cogalois and K is an intermediate field, then E/K and K/F are cogalois.

(b) The maps $K \longmapsto \mathrm{Cog}(K/F)$ and $H \longmapsto E_H$ are inverse bijections between the lattice of intermediate fields of E/F, and of subgroups of $\mathrm{Cog}(E/F)$, respectively.

We also study the connection between $\mathrm{Gal}(E/F)$ and $\mathrm{Cog}(E/F)$ in the special case where E/F is both Galois and cogalois. All the results presented are extracted from an important work of Greither and Harrison(1986).

21.1. LEMMA. (i) If E/F is cogalois, then E/F is a radical extension such that every set of representatives for $\mathrm{Cog}(E/F)$ forms a basis of E over F. In particular,

$$|\mathrm{Cog}(E/F)| = (E:F)$$

(ii) If $F \subseteq K \subseteq E$ is a chain of fields, then the sequence

$$1 \longrightarrow \mathrm{Cog}(K/F) \longrightarrow \mathrm{Cog}(E/F) \longrightarrow \mathrm{Cog}(E/K)$$

is exact.

Proof. (i) Let $\lambda_1,\ldots,\lambda_n$ be a full set of representatives for $\mathrm{Cog}(E/F)$. Then, by (3) and (1), each element of E is an F-linear combination of $\lambda_1,\ldots,\lambda_n$. Hence

$$(E:F) \leqslant n = |\mathrm{Cog}(E/F)|$$

and so $(E:F) = n$, by virtue of (2). Since $E = F(\lambda_1,\ldots,\lambda_n)$ and $\lambda_i^{\mu_i} \in F$ for some $\mu_i \geqslant 1$, (i) follows.

(ii) Let $t(E^*/F^*)$ be the torsion subgroup of E^*/F^*. Since the sequence

$$1 \longrightarrow K^*/F^* \longrightarrow E^*/F^* \longrightarrow E^*/K^*$$

is exact, the induced sequence

$$1 \longrightarrow t(K^*/F^*) \longrightarrow t(E^*/F^*) \longrightarrow t(E^*/K^*)$$

is also exact. ∎

The proof of the next result will require the following useful observation.

21.2. LEMMA. If $0 \longrightarrow A \xrightarrow{i} B \xrightarrow{j} C \longrightarrow 0$ is an exact sequence of G-modules, then we have an exact sequence

$$0 \longrightarrow A^G \xrightarrow{i} B^G \xrightarrow{j} C^G \xrightarrow{\alpha} H^1(G,A)$$

Proof. We identify A with its image $i(A)$, so that i becomes inclusion Let $c \in C^G$ and assume that $b \in B$ is such that $j(b) = c$. Then

$$j(gb-b) = gc-c = 0,$$

i.e. $gb - b \in A$, for all $g \in G$. Now

$$f : G \longrightarrow A, \; g \longmapsto gb - b$$

is a crossed homomorphism and we associate with $c \in C^G$ the class $\alpha c = f$

$$\bmod B^1(G,A).$$

The map $c \longmapsto \alpha c$ is well defined, since if $b' = b + a$, $a \in A$, is another preimage of c, then the crossed homomorphism

$$f'(g) = gb' - b' = f(g) + ga - a$$

differs from $f(g)$ only by the crossed homomorphism $f_a(g) = ga - a$ which is contained in $B^1(G,A)$. Note that if $\alpha c = 0$, then $f(g) = gb - b = ga - a$ for some $a \in A$. Hence $g(b-a) = b - a$, i.e. $b - a = b' \in B^G$ and $j(b') - j(b) = c$. This proves the exactness of our sequence. ∎

Following Greither and Harrison(1986), we say that E/F is *pure* if the following holds:

Every p-th root of unity in E^*, where p is a prime or $p = 4$, lies in F^*. For example, $\mathbb{Q}(\sqrt{2})/\mathbb{Q}$ is pure but $\mathbb{Q}(\sqrt{-3})/\mathbb{Q}$ is not. The following lemma exhibits a family of radical extensions which are, at the same time, cogalois.

21.3. LEMMA. Assume that $E = F(\lambda)$ with $\lambda^p = a \in F$, $(E:F) = p$ prime, and E/F is pure and separable. Then

$$\mathrm{Cog}(E/F) = \langle \lambda F^* \rangle \quad \text{and} \quad E/F \text{ is cogalois}$$

Proof. It clearly suffices to verify that $\mathrm{Cog}(E/F) = \langle \lambda F^* \rangle$. We first show that

$$\mathrm{Cog}(E/F) \text{ has no elements of prime order } q \neq p \tag{4}$$

Indeed, assume that $\mu \in E - F$ is such that $\mu^q = b \in F$. Since

$$q \nmid p, \; (F(\mu):F) \neq q$$

and so $X^q - b$ is reducible over F. Hence, by Lemma 16.3, $b = c^q$ for some $c \in F$ so that c is a root of $X^q - b$. Then $(\mu/c)^q = 1$ and therefore, by pureness, $\mu/c \in F$, contrary to the assumption that $\mu \notin F$.

It will now be shown that

$$\text{Cog}(E/F) \text{ has no elements of order } p^2 \tag{5}$$

Assume by way of contradiction that $\mu^p \in E - F$ is such that $\mu^{p^2} = b \in F$. Since $(E:F) < p^2$, $X^{p^2} - b$ must be reducible over F. If p is odd, then by Lemma 16.5, $b \in F^p$, say $b = \beta^p$ for $\beta \in F$. Then $(\mu^p/\beta)^p = 1$, so $\mu^p/\beta \in F$, contrary to $\mu^p \notin F$. If $p = 2$ and b has a square root in F, the same argument works. If $p = 2$ and b has no square root in F, then $b = -4d^4$ for some $d \in F$ (Lemma 16.5). Hence $-4b = d_1^4$ for some $d_1 \in F$ and so

$$(d_1/\mu)^4 = -4$$

Setting $t = d_1/\mu$, we have

$$0 = t^4 + 4 = (t^2-2t+2)(t^2+2t+2)$$

and thus either $t - 1$ or $t + 1$ is a fourth root of 1. But E/F is pure, so $t \in F$ and therefore $\mu \in F$, a contradiction.

Let $\mu F^* \in \text{Cog}(E/F)$. Owing to (4) and (5), $\mu^p \in F$ and we may assume $\mu \notin F$. Thus $E = F(\mu)$ for any $\mu \in E - F$. Furthermore, because E/F is separable, $\text{char} F \neq p$. Let $K = F(\varepsilon)$ be a splitting field of $X^p - 1$ over F, with ε a primitive p-th root of unity. Setting

$$L = E(\varepsilon) = K(\lambda)$$

we have $(L:E) \leqslant p-1$ and $(K:F) \leqslant p-1$. Furthermore,

$$p(L:E) = (L:F) = (L:K)(K:F)$$

which shows that $p \mid (L:K) = p$ and $\lambda \notin K$. Thus L/K is a Kummer extension of exponent p, by virtue of Lemma 15.2. Put

$$\sqrt[p]{K^*} = \{x \in L^* \mid x^p \in K^*\}$$

Since $(L:K) = p$, it follows from Corollary 15.5 that $\sqrt[p]{K^*}/K^*$ is cyclic of order p. Since μ and λ are in $\sqrt[p]{K^*}$ and since $\lambda \notin K^*$, we conclude that

$$\mu = \lambda^m k \qquad \text{for some } m \in \mathbb{N},\ k \in K^*$$

With $\mu^p = d \in F$, it therefore follows that

$$d = a^m k^p$$

Let $G = \text{Gal}(K/F)$ and let $\phi : K^* \longrightarrow (K^*)^p$ be defined by $\phi(x) = x^p$. Then the exact sequence of G-modules

$$1 \longrightarrow <\varepsilon> \longrightarrow K^* \longrightarrow (K^*)^p \longrightarrow 1$$

gives an exact sequence

$$(K^*)^G \longrightarrow ((K^*)^p)^G \longrightarrow H^1(G,<\varepsilon>)$$

by virtue of Lemma 21.2. Since $|G|$ and $|<\varepsilon>|$ are coprime, $H^1(G,<\varepsilon>) = 1$, and so the homomorphism $(K^*)^G \longrightarrow ((K^*)^p)^G$ is surjective. But $(K^*)^G = F^*$ and $k^p = d/a^m$ is in $((K^*)^p)^G$, so

$$k^p = \alpha^p \qquad \text{for some } \alpha \in F^*$$

Consequently, $d = a^m\alpha^p = \mu^p$ and so $\mu/(\lambda^m\alpha)$ is a p-th root of 1. It follows that $\mu/(\lambda^m\alpha) \in F^*$ and thus $\mu \in \lambda^m F^*$, as required. ■

21.4. LEMMA. If $F \subseteq K \subseteq E$ is a chain of fields, and K/F and E/K are both conormal, then E/F is conormal.

Proof. By definition, K/F is conormal if and only if $|\text{Cog}(K/F)| \leqslant (K:F)$. The desired conclusion is therefore a consequence of Lemma 21.1(ii). ■

We are now ready to characterize cogalois extensions.

21.5. THEOREM. (Greither and Harrison(1986)). Let E/F be a finite field extension. Then E/F is cogalois if and only if E/F is coseparable, separable and pure.

Proof. First assume that E/F is cogalois. Then E/F is coseparable by definition. Assume by way of contradiction that E/F is not separable. Then F is infinite, char$F = p > 0$ and, by Corollary 3.13, $p|(E:F)$. Since, by Lemma 21.1(i), $(E:F) = |\text{Cog}(E/F)|$, there is an element $\lambda F^* \in \text{Cog}(E/F)$ of order p, i.e. $\lambda \notin F^*$ and $\lambda^p \in F^*$. Then for all $\mu \in F$, $\mu + \lambda \notin F$ by

$$(\mu+\lambda)^p = \mu^p + \lambda^p \in F.$$

One immediately verifies that $(\mu+\lambda)F^* \neq (d+\lambda)F^*$ for $\mu \neq d \in F$. Therefore $\mathrm{Cog}(E/F)$ contains infinitely many elements $(\mu+\lambda)F^*$, $\mu \in F$, which is a contradiction.

To prove purity, let $\varepsilon^p = 1$ with $\varepsilon \in E$, where p is a prime or $p = 4$. We must show that $\varepsilon \in F$. Assume first that p is a prime. If $\mathrm{char}F = p$, then $\varepsilon = 1$ and there is nothing to prove. Hence we may assume that $\mathrm{char}F \neq p$ and $\varepsilon \neq 1$. Since

$$1 + \varepsilon + \dots + \varepsilon^{p-1} = 0$$

it is not the case that $1\cdot F^*, \varepsilon F^*, \dots, \varepsilon^{p-1}F^*$ are distinct elements in $\mathrm{Cog}(E/F)$ (see Lemma 21.1(i)). Hence some $\varepsilon^i/\varepsilon^j \in F$ $(i \neq j)$, so $\varepsilon \in F$.

Now assume that $p = 4$. Again we may harmlessly assume that $\mathrm{char}F \neq 2$, $\varepsilon^2 \neq 1$. One checks that $(1+\varepsilon)^4 = -4 \in F^*$. But

$$1 + \varepsilon - (1+\varepsilon) = 0,$$

so it is not the case that the elements $1\cdot F^*$, εF^*, $(1+\varepsilon)F^*$ of $\mathrm{Cog}(E/F)$ are distinct (see Lemma 21.1(i)). Since any of the three possible equality relations between them implies $\varepsilon \in F$, purity is established.

Conversely, assume that E/F is coseparable, separable and pure. We may choose a finite subgroup G of $\mathrm{Cog}(E/F)$ such that $E_G = E$ (e.g. take $G = \mathrm{Cog}(E/F)$). Since G is abelian, there is a chain

$$1 = H_0 \subseteq H_1 \subseteq \dots \subseteq H_n = G$$

of subgroups of G such that H_i/H_{i-1} is cyclic of prime order p_i, $1 \leqslant i \leqslant n$ Hence $E_{H_i}/E_{H_{i-1}}$ is pure (since E/F is pure) and, by construction, it is also coseparable. Because H_i/H_{i-1} is cyclic of order p_i, E_{H_i} is obtained from $E_{H_{i-1}}$ by adjoining a p_i-th root. Note also that $E_{H_i}/E_{H_{i-1}}$ is separable, since so is E/F. Invoking Lemma 21.3, we conclude that $E_{H_i}/E_{H_{i-1}}$ is cogalois, and thus conormal. Hence, by inductive application of Lemma 21.4, we deduce that E/F is also conormal. Thus E/F is cogalois and the result follows. ■

With the aid of the above result, we now provide the following examples of cogalois extensions.

21.6. EXAMPLE. Let $a_1, a_2, \ldots, a_s$ be positive rational numbers, let $n_i \in \mathbb{N}$ and let

$$E = \mathbb{Q}(\sqrt[n_1]{a_1}, \ldots, \sqrt[n_s]{a_s}) \subseteq \mathbb{R}$$

Since $E \subseteq \mathbb{R}$, $E/\mathbb{Q}$ is obviously pure. It is also clear that $E/\mathbb{Q}$ is coseparable *Thus* $E/\mathbb{Q}$ *is cogalois,* by virtue of Theorem 21.5. ∎

21.7. EXAMPLE. Let p be an odd prime, let n be a positive integer and let ε_{p^n} be a primitive p^n-th root of unity over $\mathbb{Q}$. Then

$$\mathbb{Q}(\varepsilon_{p^n})/\mathbb{Q}(\varepsilon_p) \text{ is cogalois} \qquad \text{for all } n \geq 1$$

Indeed, $\mathbb{Q}(\varepsilon_{p^n})/\mathbb{Q}(\varepsilon_p)$ is certainly coseparable and separable. To show purity one can use the fact that $\mathbb{Q}(\varepsilon_{p^n})$ and $\mathbb{Q}(\varepsilon_q)$ are linearly disjoint if $q \neq p$ is a prime or $q = 4$. ∎

As a preliminary to the next result, we recall that a *lattice* is a partially ordered set in which any two elements x,y have both a least upper bound $x \vee y$ and a greatest lower bound $x \wedge y$ in the set. Hence the set of subgroups of a given group G forms a lattice, denoted by $L(G)$, with inclusion as the partial ordering and with

$$H_1 \vee H_2 = \langle H_1, H_2 \rangle, \; H_1 \wedge H_2 = H_1 \cap H_2 \qquad (H_i \in L(G))$$

Similarly, if F is a field, then the set $L(F)$ of subfields of F forms a lattice with inclusion as the partial ordering and with

$$F_1 \vee F_2 = F_1 F_2, \; F_1 \wedge F_2 = F_1 \cap F_2 \qquad (F_i \in L(F))$$

A *homomorphism* of lattices is a mapping $f : L \longrightarrow L'$ such that

$$f(x \vee y) = f(x) \vee f(y), \; f(x \wedge y) = f(x) \wedge f(y)$$

for all $x,y \in L$. It is clear that a lattice homomorphism *preserves the ordering*: $a \leq b$ implies $f(a) \leq f(b)$, but not every order-preserving mapping between lattices is a lattice homomorphism. Let L and L' be two lattices and let $f : L \longrightarrow L'$ be a bijection. Then one immediately verifies that the following conditions are equivalent:

(a) f is an isomorphism of lattices

(b) for any $x,y \in L$, $x \geqslant y$ if and only if $f(x) \geqslant f(y)$

As an application of Theorem 21.5, we now prove the main theorem.

21.8. THEOREM. (Greither and Harrison(1986)). Assume that a finite field extension E/F is cogalois and that K is an intermediate field. Then

(i) E/K and K/F are again cogalois

(ii) $E_H = K$ for $H = \text{Cog}(K/F)$

(iii) $\text{Cog}(E_S/F) = S$ for every subgroup S of $\text{Cog}(E/F)$

(iv) The maps $K \longmapsto \text{Cog}(K/F)$ and $S \longmapsto E_S$ are inverse isomorphisms of lattices

(v) $\text{Cog}(E/K)$ is canonically isomorphic to $\text{Cog}(E/F)/\text{Cog}(K/F)$.

Proof. (i) Owing to Theorem 21.5, E/F is coseparable, separable and pure. Hence E/K is coseparable, separable and pure. Invoking Theorem 21.5 again, we deduce that E/K is cogalois. By Lemma 21.1 (i), $|\text{Cog}(E/F)| = (E:F)$ and $|\text{Cog}(E/K)| = (E:K)$. Therefore, applying Lemma 21.1 (ii), we have

$$(K:F) \leqslant |\text{Cog}(K/F)| \tag{6}$$

If $\text{Cog}(K/F) = \{\lambda_1 F^*,\ldots,\lambda_r F^*\}$, then $\lambda_1 F^*,\ldots,\lambda_r F^*$ are distinct elements of $\text{Cog}(E/F)$. Hence, by Lemma 21.1 (i), the elements $\lambda_1,\ldots,\lambda_r$ are F-linearly independent. By (6), the λ_i form a basis for K over F, which clearly implies that K/F is cogalois.

(ii) It is clear that $E_H \subseteq K$. On the other hand, we have

$$(E_H:F) = |\text{Cog}(K/F)|$$

and, by (i), we also have

$$(K:F) = |\text{Cog}(K/F)|$$

Hence $E_H = K$ as required.

(iii) Clearly $S \subseteq \text{Cog}(E_S/F)$ and, by (i), $|\text{Cog}(E_S/F)| = (E_S:F)$. Since $(E_S:F) \leqslant |S|$, we have $S = \text{Cog}(E_S/F)$.

(iv) Direct consequence of (ii) and (iii)

(v) By Lemma 21.1. (ii), there is a natural injection

$$\mathrm{Cog}(E/F)/\mathrm{Cog}(K/F) \longrightarrow \mathrm{Cog}(E/K) \tag{7}$$

Since E/F, K/F and E/K are cogalois, it follows from Lemma 21.1(i) that the orders of the groups in (7) are equal, hence the result. ■

Our next aim is to examine extensions E/F which are cogalois and Galois at the same time. A duality between the involved cogalois and Galois groups will also be exhibited.

21.9. LEMMA. Let E/F be a finite field extension.

(i) E/F is coseparable and Galois if and only if E/F is separable and E is the splitting field of some polynomial

$$(X^{n_1}-a_1)(X^{n_2}-a_2)\ \ldots\ (X^{n_s}-a_s) \tag{8}$$

with $n_i \in \mathbb{N}$ and $a_i \in F$

(ii) If E/F is cogalois and Galois and char$F = 0$, then E is the splitting of the polynomial (8) and for p dividing any n_i, F contains a primitive p-th root of unity. Here p is an arbitrary prime or $p = 4$.

Proof. (i) The "if" part is obvious. Conversely, assume that E/F is coseparable and Galois. Since E/F is coseparable, there is a finite subgroup H of $\mathrm{Cog}(E/F)$ such that $E = E_H$. We may write

$$H = \langle\lambda_1 F^*\rangle \times \ldots \times \langle\lambda_s F^*\rangle \qquad (\lambda_i \in E^*)$$

with the order of $\lambda_i F^*$ equal to $q_i = p_i^{e_i}$, a prime power. We claim that E is the splitting field of

$$(X^{q_1}-a_1)(X^{q_2}-a_2)\ \ldots\ (X^{q_s}-a_s)$$

with $a_i = \lambda_i^{q_i}$, $1 \leqslant i \leqslant s$. Since $E = F(\lambda_1,\ldots,\lambda_s)$ and λ_i is a root of $X^{q_i} - a_i$, it suffices to verify that E contains a primitive q_i-th root of unity for all $i \in \{1,\ldots,s\}$. Because E/F is normal, E contains all conjugates $\lambda_i\varepsilon_1,\ldots,\lambda_i\varepsilon_r$ of λ_i over F. Here all ε_j are q_i-th roots of unity. Assume by way of contradiction that no ε_j, $1\leqslant j \leqslant r$, is primitive. Since every conjugate of $\lambda_i^{(q_i/p_i)}$ is a (q_i/p_i)-th power of a conjugate of λ_i, this implies that $\lambda_i^{(q_i/p_i)}$ is equal to all its conjugates, so $\lambda_i^{(q_i/p_i)} \in F$, which is a contradiction.

(ii) If E/F is cogalois and Galois, then E/F is coseparable and Galois, hence by (i), E is the splitting field of the polynomial (8). Let p be a prime or $p = 4$, and let $p|n_i$. Since $X^{n_i} - a_i$ splits completely over E and $\mathrm{char}F = 0$, E contains a primitive p-th root of unity, say ε_p. But E/F is also pure by Theorem 21.5. Hence $\varepsilon_p \in F$ and the result follows. ∎

21.10. LEMMA. Let E/F be a finite field extension which is coseparable and Galois, let $G = \mathrm{Gal}(E/F)$ and let C be a subgroup of $\mathrm{Cog}(E/F)$. Then the map

$$\begin{cases} G \times C \xrightarrow{f} t(E^*) \\ (g, \lambda F^*) \longmapsto g(\lambda)\lambda^{-1} \end{cases}$$

is well defined and satisfies the following properties:

(i) $f(g,c_1c_2) = f(g,c_1)f(g,c_2)$ for all $g \in G$, $c_1,c_2 \in C$

(ii) For any $c \in C$, the map $\chi_c : G \longrightarrow t(E^*)$, $g \longmapsto f(g,c)$ is a crossed homomorphism, where $t(E^*)$ is regarded as a G-module in a natural way.

(iii) If $f(g,C) = 1$ and $C = \mathrm{Cog}(E/F)$, then $g = 1$

(iv) If $f(G,c) = 1$, then $c = 1$

(v) The map $C \longrightarrow Z^1(G,t(E^*))$, $c \longmapsto \chi_c$ is an injective homomorphism.

Proof. If $\lambda F^* \in C$, say $\lambda^n = \mu \in F$, then for any $g \in G$,

$$g(\lambda^n) = g(\lambda)^n = \mu = \lambda^n$$

Hence $g(\lambda)\lambda^{-1} \in t(E^*)$. If $\mu = \lambda\beta$ with $\beta \in F^*$, then $g(\mu)\mu^{-1} = g(\lambda)\lambda^{-1}$, proving that the given map is well defined.

(i) Write $c_1 = \lambda_1 F^*$ and $c_2 = \lambda_2 F^*$. Then

$$\begin{aligned} f(g,c_1c_2) &= f(g,\lambda_1\lambda_2 F^*) = g(\lambda_1\lambda_2)(\lambda_1\lambda_2)^{-1} \\ &= [g(\lambda_1)\lambda_1^{-1}]\,[g(\lambda_2)\lambda_2^{-1}] \\ &= f(g,c_1)f(g,c_2) \end{aligned}$$

as required.

(ii) Given $g_1,g_2 \in G$ and $c = \lambda F^* \in C$, we have

$$\begin{aligned} [g_1\chi_c(g_2)]\chi_c(g_1) &= g_1(g_2(\lambda)\lambda^{-1})g_1(\lambda)\lambda^{-1} \\ &= (g_1g_2)(\lambda)g_1(\lambda)^{-1}g_1(\lambda)\lambda^{-1} \\ &= \chi_c(g_1g_2) \end{aligned}$$

as asserted.

(iii) Assume that $C = \mathrm{Cog}(E/F)$ and $f(g,c) = 1$ for $c \in C$. Then $g(\lambda) = \lambda$ for all $\lambda \in E^*$ such that $\lambda F^* \in C$. But E/F is coseparable, hence generated over F by all such λ. Therefore $g(\lambda) = \lambda$ for all $\lambda \in E$, i.e. $g = 1$.

(iv) Assume that $f(g,c) = 1$ for all $g \in G$, where $c = \lambda F^* \in C$. Then $g(\lambda) = \lambda$ for all $g \in G$. Since E/F is a Galois extension, we conclude that $\lambda \in F^*$, i.e. $c = 1$.

(v) Since $\chi_c(g) = f(g,c)$, the given map is a homomorphism, by virtue of (i). The injectivity is a consequence of (iv). ■

21.11. THEOREM. (Greither and Harrison(1986)). Let E/F be a finite field extension which is both Galois and cogalois, with $G = \mathrm{Gal}(E/F)$ and $H = \mathrm{Cog}(E/F)$, let $L(G)$ and $L(H)$ be the lattices of subgroups of G and H, respectively, and for any subgroup S of G, let E^S be the fixed field of S and

$$S^{\perp} = \{h \in H | f(S,h) = 1\}$$

where f is as in Lemma 21.10. Then

$$E^S = E_{S^{\perp}} \tag{9}$$

and the map

$$\begin{cases} L(G) \longrightarrow L(H) \\ S \longmapsto S^{\perp} \end{cases}$$

is a duality (i.e. anti-isomorphism).

Proof. By Lemma 21.10(i), $S^{\perp}$ is a subgroup of H and, by definition,

$$S^{\perp} = \{\lambda F^* \in H | s(\lambda) = \lambda \text{ for all } s \in S\} \tag{10}$$

Let $x \in E_{S^{\perp}}$ so that $x = \lambda_1 + \dots + \lambda_n$ with $\lambda_i F^* \in S^{\perp}$ and some $n \geqslant 1$. Then, by (10), $s(\lambda_i) = \lambda_i$ for all $s \in S$, $i \in \{1,2,\dots,n\}$. Hence $x \in E^S$ and therefore $E_{S^{\perp}} \subseteq E^S$. Assume by way of contradiction that $E^S \nsubseteq E_S$. Since E/F is cogalois and E^S is an intermediate field, Theorem 21.8(ii) tells us that

$$E^S = E_W$$

for $W \nsubseteq S^{\perp}$. Choose $wF^* \in W - S^{\perp}$. Since $w \in E_W = E^S$, we have $wF^* \in S^{\perp}$, a contradiction.

Let $L_F(E)$ be the lattice of subfields of E containing F. Then, by Galois theory, $L(G) \longrightarrow L_F(E)$, $S \longmapsto E^S$ is a duality. On the other hand, by Theorem 21.8(iv), the map $L_F(E) \longrightarrow L(H)$, $E^S \longmapsto \mathrm{Cog}(E^S/F)$ is an isomorphism. Hence the map $L(G) \longrightarrow L(H)$, $S \longmapsto \mathrm{Cog}(E^S/F)$ is a duality. But, by (9) and Theorem 21.8(iii), we have

$$\mathrm{Cog}(E^S/F) = \mathrm{Cog}(E_{S^\perp}/F) = S^\perp,$$

hence the result. ∎

We next present the following example of a nonabelian Galois and cogalois extension.

21.12. EXAMPLE. (Greither and Harrison(1986)). Let ε_n denote a primitive n-th root of unity over $\mathbb{Q}$. We claim that

$$\mathbb{Q}(\varepsilon_q, \sqrt[9]{5})/\mathbb{Q}(\varepsilon_3)$$

is a *nonabelian Galois and cogalois extension*.

Put $F = \mathbb{Q}(\varepsilon_3)$ and $E = F(\varepsilon_9, \sqrt[9]{5})(=\mathbb{Q}(\varepsilon_9, \sqrt[9]{5}))$. Then $(E:F) = 27$ and E/F is coseparable. To prove purity, it suffices to show that $\varepsilon_4 \notin E$ and $\varepsilon_p \notin E$ for all odd primes $p \neq 3$. This can be verified by using the fact that $F(\sqrt[9]{5})(\varepsilon_4)$ is quadratic over $F(\sqrt[9]{5})$ but E is cubic over $F(\sqrt[9]{5})$ together with the fact that $p - 1$ does not divide $54 = (E:\mathbb{Q})$ for any odd prime $p \neq 3$. Hence E/F is cogalois. On the other hand, since E is a splitting field of $X^9 - 5$ over F, E/F is also Galois. Now $\mathrm{Gal}(E/F) = \langle \alpha, \beta \rangle$ where

$$\alpha(\sqrt[9]{5}) = \varepsilon_9\sqrt[9]{5},\ \alpha(\varepsilon_9) = \varepsilon_9$$
$$\beta(\sqrt[9]{5}) = \sqrt[9]{5},\ \beta(\varepsilon_9) = \varepsilon_9^4$$

Since $\beta\alpha\beta^{-1} = \alpha^4 \neq \alpha$, $\mathrm{Gal}(E/F)$ is nonabelian, as asserted. ∎

Let $r, m \in \mathbb{N}$. Following Greither and Harrison(1986), we say that m is *related* to r, if m divides r, every prime p dividing r also divides m, and $4|m$ in case $4|r$. If m is related to r, we put

$$D_{r,m} = \mathrm{Ker}(U(\mathbb{Z}/r\mathbb{Z}) \longrightarrow U(\mathbb{Z}/m\mathbb{Z}))$$

Let

$$1 \longrightarrow H \longrightarrow G \xrightarrow{\psi} D_{r,m} \longrightarrow 1 \tag{11}$$

be an extension of groups. Following Greither and Harrison(1986), we say that this extension is *allowable* if H is a finite abelian group with $H^r = 1$ (hence H is a $\mathbb{Z}/r\mathbb{Z}$-module in a natural way) and if the action of $D_{r,m}$ on H determined by this extension is the natural one. The latter means that if

$$t : D_{r,m} \longrightarrow G$$

is any right inverse to ψ, then for any $x = \mu + r\mathbb{Z} \in D_{r,m}$

$$xh = t(x)h\ t(x)^{-1} = h^{\mu} \qquad \text{for all } h \in H \tag{12}$$

For example, if $G = H \rtimes D_{r,m}$ (semidirect product) with respect to the action of $D_{r,m}$ on H given by (12), then G is the split allowable extension of H by $D_{r,m}$.

The group $D_{r,m}$ acts naturally on the additive group $\mathbb{Z}/r\mathbb{Z}$: for any

$$\mu + r\mathbb{Z} \in D_{r,m},$$

$$(\mu + r\mathbb{Z})(k + r\mathbb{Z}) = \mu k + r\mathbb{Z}$$

21.13. LEMMA. Assume that m is related to r and let $D = D_{r,m}$. Then

(i) D is cyclic of order r/m

(ii) $(\mathbb{Z}/r\mathbb{Z})^D = (r/m)(\mathbb{Z}/r\mathbb{Z})$

(iii) $H^2(D, \mathbb{Z}/r\mathbb{Z}) = 0$

Proof. (i) Since m is related to r, the canonical decompositions of m and r are of the form $m = p_1^{\alpha_1}\cdots p_s^{\alpha_s}$, $r = p_1^{\beta_1}\cdots p_s^{\beta_s}$ with $0 < \alpha_i \leqslant \beta_i$. Since

$$U(\mathbb{Z}/r\mathbb{Z}) = \prod_{i=1}^{s} U(\mathbb{Z}/p_i^{\beta_i}\mathbb{Z}) \quad \text{and} \quad U(\mathbb{Z}/m\mathbb{Z}) = \prod_{i=1}^{s} U(\mathbb{Z}/p_i^{\alpha_i}\mathbb{Z})$$

it suffices to consider the case where $m = p^{\alpha}$, $r = p^{\beta}$ with $0 < \alpha < \beta$, p prime. Now the natural map $U(\mathbb{Z}/p^{\beta}\mathbb{Z}) \longrightarrow U(\mathbb{Z}/p^{\alpha}\mathbb{Z})$ is obviously surjective. If p is odd, then by Proposition 5.11, both groups are cyclic of orders $p^{\beta-1}(p-1)$, $p^{\alpha-1}(p-1)$, respectively. Hence D is cyclic of order $p^{\beta-\alpha} = r/m$. Assume that $m = 2^{\alpha}$, $r = 2^{\beta}$. Since $\beta \geqslant 2$, $4|r$ and so $\alpha \geqslant 2$, by hypothesis. If

$\alpha > 2$ and D is not cyclic, then by Proposition 5.11, $U(\mathbb{Z}/2^{\alpha}\mathbb{Z})$ is cyclic which is impossible by Proposition 5.11. Hence we may assume that $\alpha = 2$. By Proposition 5.11(ii), $U(\mathbb{Z}/2^{\beta}\mathbb{Z}) \cong \mathbb{Z}_2 \times \mathbb{Z}_{2^{\beta-2}}$ and the element $5 + 2^{\beta}\mathbb{Z}$ generates a subgroup of order $2^{\beta-2}$. Since this element is contained in D and

$$|U(\mathbb{Z}/4\mathbb{Z})| = 2,$$

we must have $D = \langle 5 + 2^{\beta}\mathbb{Z}\rangle$. Thus D is cyclic of order $2^{\beta-2} = r/m$, as required.

(ii) Let $a = (r/m)s + r\mathbb{Z}$ for some $s \in \mathbb{Z}$ and let $x = \mu + r\mathbb{Z}$ be an arbitrary element of D. Then $\mu = 1 + tm$ for some $t \in \mathbb{Z}$ and so

$$xa = (1+tm)(r/m)s + r\mathbb{Z} = (r/m)s + tr + r\mathbb{Z} = a,$$

proving that $(r/m)(\mathbb{Z}/r\mathbb{Z}) \subseteq (\mathbb{Z}/r\mathbb{Z})^D$. Conversely, let $a = k + r\mathbb{Z} \in (\mathbb{Z}/r\mathbb{Z})^D$. Since m is related to r, $(1+m,r) = 1$ and hence $x = (1+m) + r\mathbb{Z} \in D$. Since $xa = a$, we have

$$mk = (1+m)k - k \in r\mathbb{Z}$$

which shows that $a \in (r/m)(\mathbb{Z}/r\mathbb{Z})$, as required.

(iii) Since D is cyclic and $(\mathbb{Z}/r\mathbb{Z})^D = (r/m)(\mathbb{Z}/r\mathbb{Z})$, it suffices, by Proposition 18.11, to show that

$$r/m + r\mathbb{Z} \in (\sum_{d\in D} d)(\mathbb{Z}/r\mathbb{Z}) \tag{13}$$

Now $D = \{(1+km) + r\mathbb{Z} \mid 0 \leqslant k \leqslant (r/m) - 1\}$ and so

$$\begin{aligned}(\sum_{d\in D} d)(1+r\mathbb{Z}) &= \sum_{k=0}^{(r/m)-1} (1+km) + r\mathbb{Z} \\ &= \frac{r}{m} + \frac{r}{2}\left(\frac{r}{m} - 1\right) + r\mathbb{Z} \\ &= \frac{r}{m} + \frac{ra}{2} + r\mathbb{Z} \end{aligned} \tag{14}$$

with $a = (r/m) - 1 \in \mathbb{Z}$. If r/m is odd, then a is even and

$$\frac{r}{m} + \frac{ra}{2} \equiv \frac{r}{m} \pmod{r}$$

Hence, by (14), $r/m + r\mathbb{Z} = (\sum_{d\in D} d)(1+r\mathbb{Z})$, proving (13). Assume that r/m is even. Since m is related to r, 8 must divide r, and $4|m$. We then have

$$\left(\frac{r}{m}+\frac{ra}{2}\right)\left(1-\frac{ma}{2}\right) = \frac{r}{m}+\frac{ra}{2}-\frac{ra}{2}-\frac{ra^2m}{4}$$

$$= \frac{r}{m}-\frac{m}{4}ra^2$$

$$\equiv \frac{r}{m} \pmod{r}$$

Hence, by (14),

$$\frac{r}{m}+r\mathbb{Z} = \left(1-\frac{ma}{2}\right)\left(\frac{r}{m}+\frac{ra}{2}+r\mathbb{Z}\right) \in \left(\sum_{d\in D} d\right)(\mathbb{Z}/r\mathbb{Z}),$$

which again proves (13). ∎

Assume that

$$1 \longrightarrow H \longrightarrow G \xrightarrow{\psi} D_{r,m} \longrightarrow 1$$

is an allowable extension of H by $D_{r,m}$. Since $\mathbb{Z}/r\mathbb{Z}$ is a $D_{r,m}$-module, we can regard $\mathbb{Z}/r\mathbb{Z}$ as a G-module via

$$g(k+r\mathbb{Z}) = \psi(g)(k+r\mathbb{Z})$$

For any subgroup S of G, we put

$$X(S) = Z^1(S,\mathbb{Z}/r\mathbb{Z})$$

$$B(S) = B^1(S,\mathbb{Z}/r\mathbb{Z})$$

so that

$$X(S)/B(S) = H^1(S,\mathbb{Z}/r\mathbb{Z})$$

21.14. LEMMA. Assume that m is related to r and that

$$1 \longrightarrow H \longrightarrow G \xrightarrow{\psi} D_{r,m} \longrightarrow 1$$

is an allowable extension of H by $D = D_{r,m}$. Then

(i) $|X(G)| \geqslant |G|$

(ii) For subgroups $S \subseteq T$ of G, we have

$$|\mathrm{Ker}(X(T) \xrightarrow{\mathrm{Res}} X(S)| \leqslant (T:S)$$

In particular,

$$|X(T)|/|X(S)| \leqslant (T:S)$$

(iii) For subgroups $S \subseteq T$ of G, the restriction map Res: $X(T) \longrightarrow X(S)$ is surjective and

$$|X(S)| = |S| \qquad \text{for any subgroup } S \text{ of } G$$

(iv) For subgroups $S \subset T$ of G, there is a $\chi \in X(G)$ with $\chi(S) = 0, \chi(T) \neq 0$.

Proof. (i) For each $a \in \mathbb{Z}/r\mathbb{Z}$, let $\chi_a \in B(G)$ be defined by $\chi_a(g) = ga - a$. Then the map $a \mapsto \chi_a$ induces an exact sequence

$$0 \longrightarrow (\mathbb{Z}/r\mathbb{Z})^G \longrightarrow \mathbb{Z}/r\mathbb{Z} \longrightarrow B(G) \longrightarrow 0$$

By definition, $(\mathbb{Z}/r\mathbb{Z})^G = (\mathbb{Z}/r\mathbb{Z})^D$ and, by Lemma 21.13(ii), we conclude that $|B(G)| = r/m$. Hence

$$|X(G)| = |H^1(G, \mathbb{Z}/r\mathbb{Z})|\,(r/m)$$

Since $|G| = |H|\,|D| = |H|(r/m)$, by Lemma 21.13(i), it suffices to verify that

$$|H^1(G, \mathbb{Z}/r\mathbb{Z})| \geq |H| \tag{15}$$

The following exact sequence comes from the Lyndon-Hochschild spectral sequence (see MacLane (1963, p.354)):

$$H^1(G, \mathbb{Z}/r\mathbb{Z}) \xrightarrow{\text{Res}} H^1(H, \mathbb{Z}/r\mathbb{Z})^G \longrightarrow H^2(D, \mathbb{Z}/r\mathbb{Z}) \tag{16}$$

We observe

(a) H operates trivially on $\mathbb{Z}/r\mathbb{Z}$, and $rH = 0$, so

$$H^1(H, \mathbb{Z}/r\mathbb{Z}) = \text{Hom}(H, \mathbb{Z}/r\mathbb{Z}) \tag{17}$$

has exactly $|H|$ elements.

(b) The operation of G on $H^1(H, \mathbb{Z}/r\mathbb{Z})$ comes from conjugation and G operates on H and $\mathbb{Z}/r\mathbb{Z}$ through D by scalar multiplication. From (17) we therefore deduce that the operation of G on $H^1(H, \mathbb{Z}/r\mathbb{Z})$ is trivial.

Applying (a), (b), (16) and Lemma 21.13, we deduce that (15) holds.

(ii) By induction, one can perform a reduction to the case that there are no groups properly between S and T. Then the following holds

(c) There is an element $t \in T - S$ with $t^p \in S$, p prime

(d) $T = \langle S, T \rangle$

(e) $(T:S) \geq p$

Now let $\chi \in \mathrm{Ker}(X(T) \longrightarrow X(S))$. We have to show that there are at most p choices for χ. By (d), the latter will follow provided we show that we have at most p possible values for $\chi(t)$.

Let $\psi(t) = 1 + ma + r\mathbb{Z}$. By repeated use of the definition of a crossed homomorphism, we get:

$$\begin{aligned} 0 = \chi(t^p) &= \chi(t) + (1+ma)\chi(t) + \dots + (1+ma)^{p-1}\chi(t) \\ &= c\chi(t) \quad \text{with } c = \sum_{i=0}^{p-1} (1+ma)^i \end{aligned}$$

We claim that $c + r\mathbb{Z}$ and $p + r\mathbb{Z}$ are associated in $\mathbb{Z}/r\mathbb{Z}$; if sustained, it will follow that the equation $0 = c\chi(t)$ has at most p solutions in $\mathbb{Z}/r\mathbb{Z}$, and we are done.

To substantiate our claim, we must prove two facts:

(f) If a prime $q \neq p$ divides r, then $q \nmid c$

(g) If $p|r$ then $p|c$, and if $p^2|r$ then $p^2 \nmid c$

Assume that a prime $q \neq p$ divides r. Since m is related to r, we must also have $q|m$. Then

$$c \equiv \sum_{i=0}^{p-1} (1+0) \equiv p \pmod{q},$$

proving (f).

Assume that $p|r$ (hence $p|m$). Then

$$c \equiv \sum_{i=0}^{p-1} (1+0) \equiv 0 \pmod{p}$$

Now assume that $p^2|r$. Since $p^2|m^2$, the binomial theorem yields

$$c \equiv \sum_{i=0}^{p-1} (1+ima) \equiv p + map\,\frac{p-1}{2} \pmod{p^2}$$

If $p \neq 2$, then p^2 divides $map(p-1)/2$, so $c \equiv p \pmod{p^2}$ and $p^2 \nmid c$. If $p = 2$, then (since $4|r$) 4 already divides m, and again we obtain $c \equiv 2 \pmod 4$, as required.

(iii) By (ii), applied to $T = S$ and $S = 1$, we have

$$|X(S)| \leqslant |S| \tag{18}$$

Again, by (ii)

$$|\mathrm{Ker}(X(G) \longrightarrow X(S))| \leqslant (G:S) \tag{19}$$

Finally, (i) says that

$$|X(G)| \geqslant |G| = |S|(G:S) \tag{20}$$

Taking (18), (19) and (20) together, we conclude that all inequalities (18), (19), (20) are in fact equalities and $X(G) \longrightarrow X(S)$ is surjective. This implies that $X(T) \longrightarrow X(S)$ is also surjective, which proves (iii).

(iv) By (iii), we have $|X(T)| > |X(S)|$. Hence the restriction map

$$X(T) \longrightarrow X(S)$$

cannot be injective. The desired coclusion now follows by taking an extension χ to $X(G)$ of any $\alpha \neq 0$ in $\mathrm{Ker}(X(T) \longrightarrow X(S))$. ■

We are now ready to prove

21.15. THEOREM. (Greither and Harrison(1986)). Suppose that $m \in \mathbb{N}$ is related to $r \in \mathbb{N}$ and let G be an allowable extension of H by $D_{r,m}$. Put

$$X(G) = Z^1(G, \mathbb{Z}/r\mathbb{Z})$$

where the action of G on $\mathbb{Z}/r\mathbb{Z}$ is induced by the natural action of $D_{r,m}$ on $\mathbb{Z}/r\mathbb{Z}$. Define $f : G \times X(G) \longrightarrow \mathbb{Z}/r\mathbb{Z}$ by

$$f(g,\chi) = \chi(g)$$

and, for any $S \in L(G)$, $W \in L(X(G))$ ($L(G)$ and $L(X(G))$ are lattices of subgroups of G and $X(G)$, respectively), put

$$S^{\perp} = \{\chi \in X(G) | f(S,\chi) = 0\}$$

$$W^{\perp} = \{g \in G | f(g,W) = 0\}$$

Then the maps

$$\begin{cases} L(G) \longrightarrow L(X(G)) \\ \quad S \longmapsto S^{\perp} \end{cases} \qquad\qquad \begin{cases} L(X(G)) \longrightarrow L(G) \\ \quad W \longmapsto W^{\perp} \end{cases}$$

are mutually inverse order-inverting bijections.

Proof. One immediately verifies that $S^{\perp}$ and $W^{\perp}$ are again subgroups and that the given maps are order inverting. Also, by definition, we have

$$S^{\perp\perp} \supseteq S \quad \text{and} \quad W^{\perp\perp} \supseteq W$$

By Lemma 21.14(iv), if $S \subset T$ then $S^{\perp} \supset T^{\perp}$. In particular, if $S \subset S^{\perp\perp}$, then $S^{\perp} \supset S^{\perp\perp\perp}$. But obviously $S^{\perp\perp\perp} \supseteq S^{\perp}$, so $S^{\perp} \supset S^{\perp}$, a contradiction. Thus $S = S^{\perp\perp}$ and we are left to verify that $W = W^{\perp\perp}$.

Every $g \in G$ defines in an obvious way an element g' in $\text{Hom}(X(G),\mathbb{Z}/r\mathbb{Z})$, and $g' = 0$ implies $g = 1$ by Lemma 21.14(iv). Because $X(G)$ is abelian of exponent dividing r, and

$$|\text{Hom}(X(G),\mathbb{Z}/r\mathbb{Z})| = |X(G)| = |G|$$

(Lemma 21.14(iii)), the map $g \longmapsto g'$ is a bijection from G onto

$$\text{Hom}(X(G),\mathbb{Z}/r\mathbb{Z})$$

From the duality theory of finite abelian groups, it follows that for any

$$W \subset W_1 \subseteq X(G)$$

there exists $\lambda \in \text{Hom}(X(G),\mathbb{Z}/r\mathbb{Z})$ with $\lambda(W) = 0$, $\lambda(W_1) \neq 0$. But $\lambda = g'$ for some $g \in G$, so $f(g,W) = 0$, $f(g,W_1) \neq 0$. This implies $W^{\perp\perp} = W$, as required. ■

We are now ready to prove a useful companion to Theorem 21.11

21.16. THEOREM. (Greither and Harrison(1986)). Let E/F be a finite field extension such that E is a splitting field of the polynomial

$$(X^{n_1}-a_1)(X^{n_2}-a_2)\ \cdots\ (X^{n_s}-a_s) \qquad (n_i \in \mathbb{N}, a_i \in F^*)$$

where a_i, n_i are such that for p dividing any n_i, F contains a primitive p-th root of unity (p is an arbitrary prime or $p = 4$).

(i) If r is the least common multiple of the n_i, then E contains a primitive r-th root of unity, say ε

(ii) Pick $\lambda_i \in E$ with $\lambda_i^{n_i} = a_i$ and denote by C the subgroup of $\text{Cog}(E/F)$ generated by εF^* and all $\lambda_i F^*$. Then there is a lattice isomorphism

$$L_F(E) \longrightarrow L(C)$$

where $L_F(E)$ is the lattice of all subfields of E containing F and $L(C)$ is the lattice of all subgroups of C.

(iii) Any intermediate field K is generated over F by certain elements of the form $\varepsilon^{i}\lambda_1^{k_1}\lambda_2^{k_2}\ldots\lambda_s^{k_s}$, $0 \leqslant k_j < n_j$, $1 \leqslant j \leqslant s$, $0 \leqslant i < r$.

Proof. (i) Our assumption on roots of unity guarantees that char$F \nmid n_i$, $1 \leqslant i \leqslant s$. Hence by Proposition 3.1, each $X^{n_i} - a_i$ has only simple roots. This implies that E contains a primitive n_i-th root of unity (hence also a primitive r-th root of unity) and that E/F is Galois (Proposition 4.1)

(ii) Put $H = \mathrm{Gal}(E/F(\varepsilon))$, $G = \mathrm{Gal}(E/F)$ and $D = \mathrm{Gal}(F(\varepsilon)/F)$. Then the restriction homomorphism $G \longrightarrow D$ induces an exact sequence

$$1 \longrightarrow H \longrightarrow G \longrightarrow D \longrightarrow 1 \tag{21}$$

Note that D is canonically a subgroup of $U(\mathbb{Z}/r\mathbb{Z})$ (identify $\sigma \in D$ with $\mu + r\mathbb{Z}$ if $\sigma(\varepsilon) = \varepsilon^{\mu}$). Let us show that $D = D_{r,m}$ for some $m \geqslant 1$ related to r. To this end, define m' to be the product of all primes dividing r if $4 \nmid r$ and twice the latter product if $4|r$. Then m' is the smallest divisor of r related to r. By Lemma 21.13, $D_{r,m'}$ is cyclic, and by the hypothesis concerning roots of unity, $D \subseteq D_{r,m'}$. Hence, by Lemma 21.13(i), the order of D divides r/m', so $|D| = r/m$ for some m such that $m'|m|r$. In particular, m is related to r and $D_{r,m} \subseteq D_{r,m'}$.
Since $D_{r\ m'}$ is cyclic, it contains exactly one subgroup of order r/m, and thus $D = D_{r,m}$. It is now immediate that (21) is an allowable extension. Note also that E/F is coseparable, since $E = E_C$.

Since E/F is Galois and coseparable and C is a subgroup of Cog(E/F), we may apply Lemma 21.10. Let $f : G \times C \longrightarrow t(E^*)$ be as in Lemma 21.10 and, for any subgroup S of G, put

$$S^{\perp} = \{c \in C | f(S,c) = 1\}$$

Note that the image of f is contained in $<\varepsilon> \cong \mathbb{Z}/r\mathbb{Z}$, so the conclusion of Lemma 21.10 is applicable to $\mathbb{Z}/r\mathbb{Z}$ (instead of $t(E^*)$). Hence there is an injective homomorphism

$$C \longrightarrow Z^1(G,\mathbb{Z}/r\mathbb{Z}) = X(G) \tag{22}$$

By Lemma 21.14(iii), $|X(G)| = |G| = (E:F)$. Since $E = E_C$, we must have $|C| \geqslant (E:F)$. Thus (22) is an isomorphism and $|C| = (E:F)$, which implies that

C forms an F-basis of E. Since $C \cong X(G)$, we have $L(C) \cong L(X(G))$. But, by Theorem 21.15, $L(X(G))$ is dual to $L(G)$ and, by Galois theory, $L(G)$ is dual to $L_F(E)$. Thus $L(C) \cong L_F(E)$, as required.

(iii) Let $F \subseteq K \subseteq E$ be a chain of fields and put $s = (E:K)$. Then $K = E^S$ for a subgroup S of G of order s. By Lemma 21.14(iii),

$$S^{\perp} = \text{Ker}(X(G) \longrightarrow X(S))$$

has index s in C, since $|C| = |X(G)| = |G|$. The field $E_{S^{\perp}}$ is contained in $E^S = K$. Because C forms an F-basis of E, we have $(E_{S^{\perp}}:F) = |S|$, so

$$(E:E_{S^{\perp}}) = (H:S^{\perp}) = s = (E:K)$$

and thus $K = E_{S^{\perp}}$. This means that K is generated over F by certain elements of the form $\varepsilon^i\lambda_1^{k_1}\lambda_2^{k_2}\cdots\lambda_s^{k_s}$, as required. ∎

22. ABELIAN p-EXTENSIONS OVER FIELDS OF CHARACTERISTIC p

Throughout this section, F denotes a field of characteristic $p > 0$. A finite Galois extension E/F is said to be an *abelian p-extension* if the Galois group of E/F is an abelian p-group. Our aim is to examine abelian p-extensions of F.

Let E/F be a finite Galois extension with Galois group G. Given $n \geqslant 1$, we write $W_n(E)$ for the n-th Witt ring of E. Recall that, by Theorem 13.3(iii), $W_n(F)$ is identifiable with the subring of $W_n(E)$ consisting of vectors whose coordinates are in F. The natural action of G on E can be extended to the action of G on $W_n(E)$ by setting

$$g(a_0,a_1,\ldots,a_{n-1}) = (g(a_0),g(a_1),\ldots,g(a_{n-1})) \qquad (g \in G,\ a_i \in E) \tag{1}$$

It is clear that $a \mapsto ga$ is an automorphism of $W_n(E)$ and that the set of these automorphisms is a group isomorphic to G. In other words, G acts faithfully as a group of automorphisms of $W_n(E)$. Thus $W_n(F)$ is the fixed subring of G, i.e.

$$W_n(F) = \{a \in W_n(E) \mid ga = a \text{ for all } g \in G\} \tag{2}$$

For any $a \in W_n(E)$ we define its *trace* $Tr(a)$ by

$$Tr(a) = \sum_{g \in G} ga$$

It is clear that $Tr(a)$ is fixed by all elements of G, so by (2)

$$Tr(a) \in W_n(F) \qquad \text{for all} \quad a \in W_n(E)$$

22.1. LEMMA. Let E/F be a finite Galois extension with Galois group G. and let $n \geq 1$. Then there exists $a \in W_n(E)$ such that $Tr(a)$ is a unit of $W_n(E)$.

Proof. If $a = (a_0, a_1, \ldots, a_{n-1}) \in W_n(E)$, then the first component of $Tr(a)$ is

$$\sum_{g \in G} ga_0 = Tr_{E/F}(a_0)$$

since the first components are added in forming a sum in $W_n(E)$. By Proposition 6.8, $Tr_{E/F}(a_0) \neq 0$ for some $a_0 \in E$. If a_0 is chosen in this way and

$$a = (a_0, \ldots, a_{n-1}),$$

then

$$Tr(a) = (Tr_{E/F}(a_0), \ldots)$$

has nonzero first component. Thus, by Corollary 13.10, $Tr(a)$ is a unit of $W_n(E)$, as required. ■

As an application, we now provide the following useful observation:

22.2. PROPOSITION. Let E/F be a finite Galois extension with Galois group G, and let the action of G on the additive group of $W_n(E)$ be defined by (1). Then $H^1(G, W_n(E)) = 0$.

Proof. Let $f : G \longrightarrow W_n(E)$ be a crossed homomorphism, i.e.

$$f(gg') = gf(g') + f(g)$$

for all $g, g' \in G$. Owing to Lemma 22.1, we may choose a in $W_n(E)$ such that $Tr(a)$ is a unit of $W_n(E)$. Put

$$b = Tr(a)^{-1}\Big(\sum_{g \in G} f(g)(ga)\Big)$$

Then, for all $x \in G$,

$$\begin{aligned} b - xb &= Tr(a)^{-1}\Big(\sum_{g\in G}(f(xg)(xga) - xf(g)(xga)\Big) \\ &= Tr(a)^{-1}\Big(\sum_{g\in G} f(x)(xga)\Big) \\ &= Tr(a)^{-1}f(x)Tr(a) \\ &= f(x), \end{aligned}$$

proving that $f \in B^1(G,W_n(E))$, as required. ∎

We next recall that, by Theorem 13.6, the map

$$\pi : W_n(E) \longrightarrow W_n(E),\ (x_0,\dots,x_{n-1}) \longmapsto (x_0^p,\dots,x_{n-1}^p)$$

is a ring homomorphism.

22.3. LEMMA. The map $\rho : W_n(E) \longrightarrow W_n(E)$ defined by

$$\rho(x) = \pi(x) - x$$

is an endomorphism of the additive group of $W_n(E)$ whose kernel is the prime subring $W_n(\mathbb{F}_p)$ of $W_n(E)$ (having order p^n, by Theorem 13.6(iii))

Proof. It is clear that ρ preserves addition. An element

$$(x_0,x_1,\dots,x_{n-1}) \in W_n(E)$$

is in $\mathrm{Ker}\rho$ if and only if $x_i^p = x_i$ for all i, or equivalently, if and only if $x_i \in \mathbb{F}_p$ for all i. ∎

In what follows, given $a_1,a_2,\dots,a_r \in W_n(E)$, we write

$$F(a_1,a_2,\dots,a_r)$$

for the field obtained from F by adjoining all coordinates of a_i, $1 \leqslant i \leqslant r$. We are now ready to prove our first major result.

22.4. THEOREM. Let F be a field of characteristic $p > 0$, let E/F be a finite Galois extension with Galois group G which is an abelian p-group of exponent p^e and, for any given $n \geqslant e$, let

$$W_{n,p}(E) = \{x \in W_n(E) \mid \rho(x) \in W_n(F)\}$$

(i) $W_{n,p}(E)$ is a subgroup of the additive group of $W_n(E)$ containing $W_n(F)$, and the map

$$\begin{cases} W_{n,p}(E) \longrightarrow \mathrm{Hom}(G, W_n(\mathbb{F}_p)) \\ \qquad a \longmapsto \chi_a \end{cases}$$

defined by $\chi_a(g) = ga - a$, is a homomorphism

(ii) The sequence of additive groups

$$0 \longrightarrow W_n(F) \longrightarrow W_{n,p}(E) \longrightarrow \mathrm{Hom}(G, W_n(\mathbb{F}_p)) \longrightarrow 0$$

is exact. In particular,

$$W_{n,p}(E)/W_n(F) \cong G$$

(iii) For any given $a_1, a_2, \ldots, a_r$ in $W_{n,p}(E)$,

$$E = F(a_1, \ldots, a_r) \text{ if and only if } W_{n,p}(E)/W_n(F) = \langle a_1 + W_n(F), \ldots, a_r + W_n(F) \rangle$$

Proof. (i) By Lemma 22.3, ρ preserves addition, hence $W_{n,p}(E)$ is a subgroup of $W_n(E)$. It is obvious that $W_n(F) \subseteq W_{n,p}(E)$. Given $a \in W_{n,p}(E)$ and $g \in G$, we have $\rho(a) = \pi(a) - a \in W_n(F)$ and so $g(\pi(a) - a) = \pi(a) - a$ or $g\pi(a) - \pi(a) = ga - a$. Hence

$$\pi(\chi_a(g)) = \pi(ga) - \pi(a) = g\pi(a) - \pi(a) = ga - a = \chi_a(g)$$

proving that $\chi_a(g) \in W_n(\mathbb{F}_p)$. Also we have

$$\begin{aligned} \chi_a(xy) &= (xy)a - a = (xy)a - xa + xa - a \\ &= x(ya-a) + (xa-a) = (ya-a) + (xa-a) \\ &= \chi_a(x) + \chi_a(y), \end{aligned}$$

whence $\chi_a \in \mathrm{Hom}(G, W_n(\mathbb{F}_p))$. Let $a, b \in W_{n,p}(E)$. Then

$$\chi_{a+b}(g) = g(a+b) - (a+b) = (ga-a) + (gb-b) = (\chi_a + \chi_b)(g),$$

proving that the map $a \longmapsto \chi_a$ is a homomorphism.

(ii) Given $a \in W_{n,p}(E)$, $\chi_a(G) = 0$ if and only if a is fixed by G, hence by (2), $W_n(F)$ is the kernel of the map $a \longmapsto \chi_a$.

To prove that the map $a \longmapsto \chi_a$ is surjective, fix $\chi \in \mathrm{Hom}(G, W_n(\mathbb{F}_p))$. Then, for all $x, y \in G$, $\chi(xy) = \chi(x) + \chi(y)$ and, since $\chi(y) \in W_n(\mathbb{F}_p)$, we also have

$$\chi(xy) = x\chi(y) + \chi(x)$$

Thus $\chi \in Z^1(G, W_n(E))$ and, by Proposition 22.2, there exists $a \in W_n(E)$ such that $\chi(g) = ga - a$ for all $g \in G$. Since $\chi(g) \in W_n(\mathbb{F}_p)$, $\pi(\chi(g)) = \chi(g)$ and this gives

$$g(\pi(a) - a) = \pi(a) - a \qquad \text{for all } g \in G$$

Hence $\rho(a) = \pi(a) - a \in W_n(F)$ and so $a \in W_n(E)$, proving that the map $a \longmapsto \chi_a$ is surjective .

Finally, since $W_n(\mathbb{F}_p)$ is of exponent $p^n \geq p^e$, we have $\text{Hom}(G,W_n(\mathbb{F}_p)) \cong G$ and hence $G \cong W_{n,p}(E)/W_n(F)$, proving (ii).

(iii) First assume that $W_{n,p}(E)/W_n(F) = \langle a_1+W_n(F),\ldots,a_r + W_n(F)\rangle$. Then, by (ii), $\text{Hom}(G,W_n(\mathbb{F}_p)) = \langle\chi_{a_1},\ldots,\chi_{a_r}\rangle$. Now let $E' = F(a_1,\ldots,a_r)$ and let H be the subgroup of G corresponding to E' (i.e. the Galois group of E/E'). If $h \in H$, we have $ha_i = a_i$, $1 \leq i \leq r$, so $\chi_{a_i}(h) = 0$. This implies that

$$\chi(h) = 0 \qquad \text{for all } \chi \in \text{Hom}(G,W_n(\mathbb{F}_p))$$

and hence that $h = 1$. Thus $H = 1$ and therefore $E' = F(a_1,\ldots,a_r) = E$.

Conversely, let $E = F(a_1,\ldots,a_r)$ for some $a_1,\ldots,a_r$ in $W_{n,p}(E)$. We first claim that $\chi_{a_i}(g) = 1$ for all $i \in \{1,\ldots,r\}$ implies $g = 1$. Indeed, if $\chi_{a_i}(g) = 1$, then $ga_i = a_i$ and hence $ga = a$ for all $a \in E$. But then $g = 1$, as claimed. It therefore follows that $\text{Hom}(G,W_n(\mathbb{F}_p)) = \langle\chi_{a_1},\ldots,\chi_{a_r}\rangle$. In particular, if $a \in W_{n,p}(E)$, then χ_a can be written in the form

$$\chi_a = k_1\chi_{a_1} + \ldots + k_r\chi_{a_r}$$

Hence, for every $g \in G$, we have

$$ga - a = k_1(ga_1-a_1) + \ldots + k_r(ga_r-a_r)$$

whence

$$g(a-k_1a_1 - \ldots - k_ra_r) = a - k_1a_1 - k_2a_2 - \ldots - k_ra_r$$

Thus $a - k_1a_1 - k_2a_2 - \ldots - k_ra_r \in W_n(F)$, as we wished to show. ■

22.5. COROLLARY. Let F be a field of characteristic $p > 0$, let E/F be a finite Galois extension with Galois group G which is an abelian p-group of exponent p and let

$$E_p = \{\lambda \in E | \lambda^p - \lambda \in F\}$$

(i) E_p is a subgroup of the additive group of E containing F and the map

$$\begin{cases} E_p \longrightarrow \mathrm{Hom}(G, \mathbb{F}_p) \\ a \longmapsto \chi_a \end{cases}$$

defined by $\chi_a(g) = ga - a$, is a homomorphism.

(ii) The sequence of additive groups

$$0 \longrightarrow F \longrightarrow E_p \longrightarrow \mathrm{Hom}(G, \mathbb{F}_p) \longrightarrow 0$$

is exact. In particular,

$$E_p/F \cong G$$

(iii) For any given $a_1, a_2, \ldots, a_r$ in E_p,

$$E = F(a_1, \ldots, a_r) \text{ if and only if } E_p/F = \langle a_1 + F, \ldots, a_r + F\rangle$$

Proof. Apply Theorem 22.4 in the special case where $n = 1$. ■

For the rest of this section, we concentrate on existence of abelian p-extensions of a given exponent. In what follows the map ρ is defined as in Lemma 22.3 and $W_{n,p}(E)$ is defined in Theorem 22.4. Note that, by the definition of $W_{n,p}$,

$$\rho(W_{n,p}(E)) = W_n(F) \cap \rho(W_n(E)) \tag{3}$$

22.6. LEMMA. Under the hypotheses of Theorem 22.4, the following holds:

$$\rho(W_{n,p}(E))/\rho(W_n(F)) \cong W_{n,p}(E)/W_n(F) \cong G \tag{4}$$

Proof. Consider the homomorphism

$$\begin{cases} W_{n,p}(E) \longrightarrow \rho(W_{n,p}(E))/\rho(W_n(F)) \\ a \longmapsto \rho(a) + \rho(W_n(F)) \end{cases}$$

An element $a \in W_{n,p}(E)$ is in the kernel of this homomorphism if and only if $\rho(a) = \rho(x)$ for some $x \in W_n(F)$. This is equivalent to $\rho(a-x) = 0$ which means that $a - x \in W_n(\mathbb{F}_p)$. Hence it is clear that the kernel of this homomorphism is $W_n(F)$. Since the given homomorphism is obviously surjective, the result follows. ■

22.7. LEMMA. Let $a = (a_0, a_1, \ldots, a_{n-1}) \in W_n(F)$. Then there exists a finite separable field extension E/F such that $E = F(b_0, b_1, \ldots, b_{n-1})$ and the element $b = (b_0, b_1, \ldots, b_{n-1})$ of $W_n(E)$ satisfies $\rho(b) = a$.

Proof. The case $n = 1$ is obvious. Indeed, all we have to do is to construct a separable extension $E = F(b)$ generated by a root b of the polynomial $X^p - X - a$. Because the derivative $(X^p - X - a)' = -1$, the given polynomial has distinct roots, so any root b of $X^p - X - a$ will satisfy the condition.

By induction, we may assume that we have already constructed a separable extension $L = F(b_0, b_1, \ldots, b_{n-2})$ such that the element

$$b' = (b_0, b_1, \ldots, b_{n-2}) \in W_{n-2}(L)$$

satisfies $\rho(b') = (a_0, a_1, \ldots, a_{n-2})$. Consider the polynomial ring $L[X]$ and the Witt ring $W_n(L[X])$. We take the vector

$$Y = (b_0, b_1, \ldots, b_{n-2}, X) \in W_n(L[X])$$

and we form

$$\rho(Y) = (b_0^p, b_1^p, \ldots, b_{n-2}^p, X^p) - (b_0, b_1, \ldots, b_{n-2}, X)$$

Then, there exists $f(X) \in L[X]$ such that

$$\rho(Y) = (a_0, a_1, \ldots, a_{n-2}, f(X))$$

Therefore

$$(a_0, a_1, \ldots, a_{n-2}, f(X)) + (b_0, b_1, \ldots, b_{n-2}, X) = (b_0^p, b_1^p, \ldots, b_{n-2}^p, X^p)$$

Invoking formula (5) of Sec. 13, we deduce that

$$X^p = f(X) + X + \lambda$$

for some $\lambda \in L$. Hence $f(X) = X^p - X - \lambda$. By looking at the derivative of $f(X)$, it follows that $f(X) = a_{n-1}$ has distinct roots. If $E = L(b_{n-1})$ where $f(b_{n-1}) = a_{n-1}$, then E is separable over L and thus $E = F(b_0, b_1, \ldots, b_{n-1})$ is separable over F. Moreover, by our construction,

$$b = (b_0, b_1, \ldots, b_{n-1}) \in W_n(E)$$

satisfies $\rho(b) = a$. ■

The following result demonstrates that any finite subgroup of $W_n(F)/\rho(W_n(F))$

is a Galois group of some abelian p-extension of F.

22.8. THEOREM. Let F be a field of characteristic $p > 0$, and let S be a subgroup of the additive group of $W_n(F)$ containing $\rho(W_n(F))$ and such that $S/\rho(W_n(F))$ is finite. Then there exists an abelian p-extension E/F such that the exponent of Gal(E/F) is p^e with $e \leqslant n$ and

$$S = W_n(F) \cap \rho(W_n(E)) \tag{5}$$

$$\text{Gal}(E/F) \cong S/\rho(W_n(F)) \tag{6}$$

Proof. We first observe that (6) follows from (3), (4) and (5). Let $s^{(1)}, s^{(2)}, \ldots, s^{(r)}$ be elements of S such that the cosets $s^{(i)} + \rho(W_n(F))$ generate $S/\rho(W_n(F))$. By repeated application of Lemma 22.7, we can find nr elements $b_k^{(i)}$, $1 \leqslant i \leqslant r$, $0 \leqslant k < n-1$ together with a finite separable extension E/F generated by them such that

$$\rho(b_0^{(i)}, \ldots, b_{n-1}^{(i)}) = (s_0^{(i)}, \ldots, s_{n-1}^{(i)}) \text{ in } W_n(E) \text{ for all } i \in \{1,2,\ldots,r\} \tag{7}$$

where $s^{(i)} = (s_0^{(i)}, \ldots, s_{n-1}^{(i)})$.

Let K/F be a finite Galois extension with $K \supseteq E$. We form $W_n(K)$ and let the Galois group G of K/F act in $W_n(K)$ as before. If $g \in G$ and

$$b^{(i)} = (b_0^{(i)}, \ldots, b_{n-1}^{(i)}),$$

then $\rho(b^{(i)}) = s^{(i)}$ gives $\rho(gb^{(i)}) = g\rho(b^{(i)}) = gs^{(i)} = s^{(i)}$. Hence

$$\rho(gb^{(i)} - b^{(i)}) = 0 \qquad (1 \leqslant i \leqslant r)$$

and therefore $gb^{(i)} - b^{(i)} \in W_n(\mathbb{F}_p) \subseteq W_n(F)$. This implies that $g(E) \subseteq E$ and hence that E/F is Galois. Furthermore, by (7), each $s^{(i)}$ is in $\rho(W_n(E))$ and so

$$S \subseteq \rho(W_n(E)) \cap W_n(F) \tag{8}$$

In what follows, we put $G = \text{Gal}(E/F)$.

If $x, y \in G$, then $xb^{(i)} = b^{(i)} + c^{(i)}$ and $yb^{(i)} = b^{(i)} + d^{(i)}$ for some $c^{(i)}, d^{(i)} \in W_n(F)$. It therefore follows that

$$(yx)b^{(i)} = b^{(i)} + d^{(i)} + c^{(i)} = (xy)b^{(i)}$$

which proves that G is abelian. Also $x^k b^{(i)} = b^{(i)} + kc^{(i)}$, so

$$x^{p^n} b^{(i)} = b^{(i)}$$

since $W_n(E)$ has characteristic p^n. This shows that $x^{p^n} = 1$ for all $x \in G$, so G is a finite ableian p-group of exponent p^e with $e \leqslant n$. In view of (8), we are left to verify that

$$\rho(W_n(E)) \cap W_n(F) \subseteq S \tag{9}$$

Let $\chi_i \in \mathrm{Hom}(G, W_n(\mathbb{F}_p))$ be defined by $\chi_i(g) = gb^{(i)} - b^{(i)}$ (see Theorem 22.4(i)). Then it is clear that $\chi_i(g) = 0$ for all $i \in \{1,2,\ldots,r\}$ implies that $g = 1$. Hence

$$\mathrm{Hom}(G, W_n(\mathbb{F}_p)) = \langle \chi_1, \chi_2, \ldots, \chi_r \rangle$$

Therefore, if a is any element of $W_n(E)$ such that $\rho(a) \in W_n(F)$, then we have

$$\chi_a = n_1\chi_1 + n_2\chi_2 + \ldots + n_r\chi_r \qquad (n_i \in \mathbb{Z})$$

This implies that $a = \Sigma n_i b^{(i)} + \lambda$ for some $\lambda \in W_n(F)$, by virtue of Theorem 22.4(ii). Then

$$\rho(a) = \Sigma n_i s^{(i)} + \rho(\lambda) \in S,$$

proving (9) and hence the result. ■

In order to apply the preceding result to cyclic extensions, we need the following auxiliary result.

22.9. LEMMA. If $\lambda_0, \lambda_1, \ldots, \lambda_{n-1} \in F$, then $\lambda_0 \in \rho(F)$ if and only if

$$\lambda = (\lambda_0, \lambda_1, \ldots, \lambda_{n-1})$$

satisfies

$$p^{n-1}\lambda \in \rho(W_n(F))$$

Proof. Owing to Theorem 13.6(ii), we have

$$p^{n-1}\lambda = (0, \ldots, 0, \lambda_0^{p^{n-1}})$$

On the other hand,

$$(0,\dots,0,\lambda_0) - (0,\dots,0,\lambda_0^{p^{n-1}}) = (0,\dots,0,\lambda_0) - (0,\dots,0,\lambda_0^p)$$
$$+ (0,\dots,0,\lambda_0^p) - (0,\dots,0,\lambda_0^{p^2}) + \dots$$
$$+ (0,\dots,0,\lambda_0^{p^{n-2}}) - (0,\dots,0,\lambda_0^{p^{n-1}}) \in \rho(W_n(F))$$

Thus $p^{n-1}\lambda \in \rho(W_n(F))$ if and only if $(0,\dots,0,\lambda_0) \in \rho(W_n(F))$.

Assume that $(0,\dots,0,\lambda_0) \in \rho(W_n(F))$, say

$$(0,\dots,0,\lambda_0) = \pi(\alpha) - \alpha$$

where $\alpha = (\alpha_0,\alpha_1,\dots,\alpha_{n-1})$. Let $\sigma = f_{n-1,n}$, where $f_{n-1,n}$ is defined prior to Theorem 13.3. Then $\sigma\pi(\alpha) - \sigma(\alpha) = \sigma(0,\dots,0,\lambda_0) = 0$. Since $\sigma\pi = \pi\sigma$, we have $\pi(\sigma(\alpha)) = \sigma(\alpha)$, i.e. $(\alpha_0^p,\dots,\alpha_{n-2}^p) = (\alpha_0,\dots,\alpha_{n-2})$. Setting

$$\mu = (\alpha_0,\alpha_1,\dots,\alpha_{n-2},0),$$

we therefore have $\pi(\mu) = \mu$, so $\beta = \alpha - \mu$ satisfies $\pi(\beta) - \beta = (0,\dots,0,\lambda_0)$. Moreover,

$$\beta = (0,\dots,0,\beta_{n-1})$$

This implies that $\beta_{n-1}^p - \beta_{n-1} = \lambda_0$, so $\lambda_0 \in \rho(F)$. Conversely, if this condition holds so that $\lambda_0 = \alpha_{n-1}^p - \alpha_{n-1}$, then $\pi(\alpha) - \alpha = (0,\dots,0,\lambda_0)$ for $\alpha = (0,\dots,0,\alpha_{n-1})$. ∎

We are now ready to provide the following generalization of Theorem 14.7(i).

22.10. THEOREM. Let F be an arbitrary field of characteristic $p > 0$. Then the following conditions are equivalent:

(a) For each $n \geqslant 1$, there exists a cyclic extension E/F of degree p^n

(b) There exists an element $\mu \in F$ which cannot be written in the form $\mu = x^p - x$ for any $x \in F$.

(c) There exists an element $\mu \in F$ such that $X^p - X - \mu$ is irreducible over F

Proof. The equivalence of (b) and (c) is a consequence of Theorem 14.7(iii). Note also that (b) is equivalent to $\rho(F) \neq F$. If (a) holds, then $\rho(F) \neq F$ by Theorem 14.7(i). We are therefore left to verify that $\rho(F) \neq F$ implies (a).

Assume that $\rho(F) \neq F$ and choose $\lambda_0 \in F - \rho(F)$. Let $\lambda = (\lambda_0,\lambda_1,\dots,\lambda_{n-1})$ where the λ_i, $i > 0$, are any elements of F. By Lemma 22.9, $p^{n-1}\lambda \notin \rho(W_n(F))$

and this implies that the subgroup S of $W_n(F)$ generated by λ and $\rho(W_n(F))$ has the property that $S/\rho(W_n(F))$ is cyclic of order p^n. The desired conclusion is therefore a consequence of Theorem 22.8. ∎

23. FORMALLY REAL FIELDS

To motivate our discussion, let us recall the following piece of information. According to Lemma 16.17, every ordered commutative ring R is an integral domain of characteristic 0 in which the square of any nonzero element is positive. In particular, R has the quotient field F.

23.1. LEMMA. Let R be an ordered commutative ring and let F be the quotient field of R. Then there is a unique ordering of F extending the ordering of R, namely:

For any $x,y \in R$, $y \neq 0$, $x/y > 0$ if and only if $xy > 0$ (1)

Proof. Since $y^2 > 0$ and $xy = (x/y)y^2$, any ordering on F extending that on R must satisfy (1). We are therefore left to verify that (1) defines an ordering. If $x/y = x_1/y_1$, then $xy_1 = x_1y$ and so $xyy_1^2 = x_1y_1y^2$. Thus

$$xy > 0 \Leftrightarrow xyy_1^2 > 0 \Leftrightarrow x_1y_1y^2 > 0 \Leftrightarrow x_1y_1 > 0,$$

proving that (1) is well-defined.

In view of Lemma 16.17(ii), it suffices to verify that the set of positive elements of F is closed under addition and multiplication. If $x/y > 0$ and $x_1/y_1 > 0$, i.e. if $xy > 0$ and $x_1y_1 > 0$, then

$$(xy_1+x_1y)yy_1 = xyy_1^2 + x_1y_1y^2 > 0$$

and

$$xx_1yy_1 = (xy)(x_1y_1) > 0,$$

as required. ∎

23.2. LEMMA. There is only one way of making $\mathbb{Q}$ and $\mathbb{R}$ into an ordered field, namely the usual ordering.

Proof. In any ordering of $\mathbb{Z}$ all the elements $1, 1+1, \ldots$ must be positive. Since these numbers, their negatives and 0 exhaust $\mathbb{Z}$, there is only one way of ordering $\mathbb{Z}$, namely the "natural" ordering of integers. Hence, by Lemma 23.1, the same applies to $\mathbb{Q}$.

In any ordering of $\mathbb{R}$ squares must be positive. Since the squares, their negatives and 0 exhaust $\mathbb{R}$, no other ordering is possible apart from the usual one. ■

Note that Lemma 23.2 fails for arbitrary ordered fields. The field $\mathbb{Q}(\sqrt{2})$ is the simplest example of a field with different orderings. Indeed, there are two ways of embedding $\mathbb{Q}(\sqrt{2})$ into $\mathbb{R}$ and correspondingly two ways of ordering it. It is therefore appropriate to investigate the following problem:
What elements of a field must be positive under every ordering?
Of course, under any ordering the sums of squares of non-zero elements are always positive. In this section, among other results, we demonstrate that this condition is sufficient to make a field orderable.

A field F is called *formally real* if

$$\sum_{i=1}^{n} \lambda_i^2 = 0 \Rightarrow \lambda_i = 0,\ 1 \leqslant i \leqslant n \qquad (\lambda_i \in F)$$

23.3. LEMMA. (i) Every ordered field is formally real
(ii) Every formally real field is of characteristic 0
(iii) A field is formally real if and only if -1 cannot be written as a sum of squares.

Proof. (i) In any ordered field squares of nonzero elements are positive, hence so are their sum.
(ii) If F is a formally real field, then for any $n \in \mathbb{N}$,

$$n \cdot 1 = 1^2 + 1^2 + \dots + 1^2 \neq 0$$

and so $\operatorname{char} F = 0$.
(iii) If F is not formally real, then $\lambda_1^2 + \dots + \lambda_n^2 = 0$ with $\lambda_1 \neq 0$ and so

$$-1 = (\lambda_2/\lambda_1)^2 + \dots + (\lambda_n/\lambda_1)^2$$

Conversely, if $\lambda_1^2 + \dots + \lambda_k^2 = -1$, then $1^2 + \lambda_1^2 + \dots + \lambda_k^2 = 0$. ■

Let F be an arbitrary field. In what follows we write $P(F)$ for the set of elements of F which are sums of squares.

23.4. LEMMA. Let F be an arbitrary field. Then $P(F)$ is closed under addition and multiplication and contains λ^{-1} for every $0 \neq \lambda \in P(F)$. Furthermore, if char$F \neq 2$ and F is not formally real, then $F = P(F)$.

Proof. Evidently $P(F)$ contains 0 and is closed under addition and multiplication. If $0 \neq \lambda \in P(F)$, then $\lambda = \Sigma\lambda_i^2$ and so

$$\lambda^{-1} = \lambda(\lambda^{-1})^2 = \Sigma(\lambda_i\lambda^{-1})^2 \in P(F)$$

Assume that char$F \neq 2$ and F is not formally real. By Lemma 23.3(iii), $-1 \in P(F)$ and, if λ is any element of F, then

$$\lambda = \left(\frac{1+\lambda}{2}\right)^2 - \left(\frac{1-\lambda}{2}\right)^2 = \left(\frac{1+\lambda}{2}\right)^2 + (-1)\left(\frac{1-\lambda}{2}\right)^2 \in P(F)$$

since $P(F)$ is closed under addition and multiplication. ∎

We now turn our attention to algebraic extensions of formally real fields. A formally real field is said to be *real closed* if it has no proper algebraic extensions that are formally real. In other words, a real closed field is a maximal formally real field in its algebraic closure. For example, $\mathbb{R}$ is real closed. We wish to extend a given formally real field to a real closed field and to prove that the latter can be ordered in just one way. First, however, it will be convenient to provide a characterization of real closed fields.

23.5. THEOREM. A field F is real closed if and only if the following three conditions hold:

(a) -1 is not a sum of squares

(b) Every element of F is either a square or the negative of a square

(c) Every polynomial of odd degree over F has a root in F.

Proof. Assume that (a), (b) and (c) hold. By Lemma 23.3(iii), property (a) implies that F is formally real. An application of (b) shows, by Lemma 16.17 (ii), that F can be ordered by declaring the nonzero squares to be positive. Moreover, this is the only way of defining an ordering. Let i be a root of $X^2 + 1$ in an algebraic closure of F. Owing to Theorem 16.21, $F(i)$ is algebraically closed. By (a), $F(i) \neq F$ and so $(F(i):F) = 2$, which shows that F

is a maximal formally real subfield of $F(i)$. Thus F is real closed.

Conversely, assume that F is real closed. Then (a) holds by virtue of Lemma 23.3(iii). To prove (b), assume that $\lambda \in F$ is not a square. Then we can construct the proper algebraic extension $E = F(\sqrt{\lambda})$. By hypothesis, E is not formally real, so there exist μ_i, γ_i not all zero in F such that

$$\Sigma(\mu_i + \gamma_i\sqrt{\lambda})^2 = 0$$

This gives

$$\Sigma(\mu_i^2 + \gamma_i^2\lambda) + 2(\Sigma\mu_i\gamma_i)\sqrt{\lambda} = 0$$

Because $\sqrt{\lambda} \notin F$, we have $2\Sigma\mu_i\gamma_i = 0$ and $\Sigma\mu_i^2 + \lambda\Sigma\gamma_i^2 = 0$. Since F is formally real, $\Sigma\gamma_i^2 \neq 0$. Then $-\lambda = (\Sigma\mu_i^2)(\Sigma\gamma_i^2)^{-1}$. Using the properties of the set $P(F)$ stated in Lemma 23.4, it follows that $-\lambda \in P(F)$. Because $-1 \notin P(F)$, this implies that $\lambda \notin P(F)$. Hence we have shown that, if an element of F is not a square, then it is not a sum of squares. Expressed otherwise, if $\lambda \in P(F)$, then λ is a square. Furthermore, we have seen that, if λ is not a square, then $-\lambda \in P(F)$ and this now ensures that $-\lambda$ is a square, proving (b).

Let $f(X) \in F[X]$ be a polynomial of odd degree n. The case $n = 1$ being trivial, we use induction on n. If $f(X)$ is reducible, one of its factors is of odd degree, so it has a root in F. Therefore we may assume that $f(X)$ is irreducible. Let $E = F(\lambda)$ where λ is a root of $f(X)$. Then $E \neq F$, so E is not formally real. Hence we have a relation $\Sigma f_i(\lambda)^2 = -1$ where $f_i(X)$ is a polynomial of degree $\leqslant n - 1$. It follows that

$$\Sigma f_i(X)^2 = -1 + f(X)g(X) \tag{2}$$

for some $g(X) \in F[X]$. The leading coefficient of $\Sigma f_i(X)^2$ is positive in the ordering in F (uniquely determined by (a) and (b)), and the degree of this polynomial is even and $\leqslant 2(n-1)$. Thus $\deg g(X)$ is odd and $\leqslant 2(n-1) - n = n - 2$. Therefore there exists $\mu \in F$ such that $g(\mu) = 0$. Substituting this μ in the relation (2) gives $\Sigma f_i(\mu)^2 = -1$, contrary to the formal reality of F. ∎

23.6. COROLLARY. (i) Every real closed field can be ordered in only one way, and any automorphism necessarily preserves ordering.

(ii) If F is a real closed field, then $F(i)$ (where $i^2 = -1$) is algebraically closed.

(iii) Any polynomial over a real closed field splits into linear and quadratic factors.

Proof. Properties (i) and (ii) were established in the course of the proof of Theorem 23.5. If F is real closed, then $F(i)$ is algebraically closed and of degree two over F. Hence any irreducible polynomial over F has degree 1 or 2, proving (iii). ■

23.7. THEOREM. (i) Any formally real field F can be embedded in a real closed field which is algebraic over F.

(ii) A field can be ordered if and only if it is formally real.

Proof. (i) Let $\bar{F}$ be an algebraic closure of F. Consider the formally real subfields of $\bar{F}$ containing F. We claim that the union of any chain $\{K_\lambda\}$ of formally real fields is formally real; if sustained, it will follow that the formally real subfields of $\bar{F}$ containing F form an inductive collection and by Zorn's lemma there is a maximal element L say, which is a real closed field containing F and contained in $\bar{F}$.

To substantiate our claim, assume that we have an equation

$$-1 = \sum_{i=1}^{n} a_i^2$$

holding in the union. Then $a_i \in K_{\lambda_i}$ say and so (3) would hold in the largest of the fields $K_{\lambda_1}, \ldots, K_{\lambda_n}$, a contradiction.

(ii) If F is an ordered field, then the square of any nonzero element is positive. Hence F must be a formally real field. Conversely, assume that F is a formally real field. Then, by (i) and Corollary 23.6(i), F can be ordered. ■

An element λ of a field F is said to be *totally positive* if $\lambda > 0$ in every ordering of F. In particular, it will be understood that in a field which has no ordering every element is totally positive. Therefore every element of a field which is not formally real is totally positive.

23.8. THEOREM. Let F be a field of characteristic $\neq 2$ and let λ be a nonzero element of F. Then λ is totally positive in F if and only if λ is a sum of square of elements of F.

Proof. If $\lambda = \Sigma\mu_i^2$, $\mu_i \neq 0$, then $\mu_i^2 > 0$ in every ordering of F, hence the same is true for λ. Conversely, assume that λ is not a sum of squares in F. We denote by $\bar{F}$ an algebraic closure of F and consider the collection of subextensions E/F of $\bar{F}/F$ in which λ is not a sum of squares. By hypothesis, F is contained in this collection and the letter is now obviously inductive. Hence, by Zorn's lemma, it contains a maximal element M. If M is not formally real, then by Lemma 23.4, every element of M is a sum of squares. But λ is not a sum of squares in M, hence M is formally real and therefore can be ordered, by Theorem 23.7(ii).

We next claim that $-\lambda$ is a square in M. Indeed, otherwise $M \subset M(\sqrt{-\lambda}) \subseteq \bar{F}$ and we must have $\lambda = \Sigma(a_i+b_i\sqrt{-\lambda})^2$ with $a_i, b_i \in M$. This gives

$$\lambda = \Sigma a_i^2 - \lambda\Sigma b_i^2 + 2(\Sigma a_i b_i\sqrt{-\lambda})$$

It follows that $\Sigma a_i b_i = 0$, so $\lambda(1+\Sigma b_i^2) = \Sigma a_i^2$. Then $1 + \Sigma b_i^2 \neq 0$ by the formal reality of M. Thus $\lambda = (\Sigma a_i^2)(1+\Sigma b_i^2)^{-1}$ is a sum of squares in M, by Lemma 23.4, which contradicts the choice of M. Hence $-\lambda = \alpha^2$ with $\alpha \in M$. This implies that $-\lambda > 0$ and $\lambda < 0$ in every ordering of M. Because M can be ordered, the induced ordering in F gives an ordering of F in which $\lambda < 0$. Hence λ is not totally positive. ■

Let F be an ordered field. We say that F has the *intermediate value property for polynomials*, provided the following condition holds:
If $f(X) \in F[X]$ and $\alpha,\beta \in F$ are such that $f(\alpha)f(\beta) < 0$, then $f(\gamma) = 0$ for some γ between α and β.

We next provide some additional characterizations of real closed fields.

23.9. THEOREM. An ordered field is real closed if and only if it has the intermediate value property for polynomials.

Proof. Let F be an ordered field. If the intermediate value property holds, then F satisfies (b), (c) of Theorem 23.5, while (a) holds in any ordered fields. Hence, by Theorem 23.5, F is real closed.

Conversely, assume that F is real closed. Let $f(X) \in F[X]$ and let

$\lambda,\mu \in F$, say $\lambda < \mu$ such that $f(\lambda)f(\mu) < 0$. We have to find $\alpha \in F$, $\lambda < \alpha < \mu$, such that $f(\alpha) = 0$. By Corollary 23.6(iii), we can write $f(X)$ as a product of linear and quadratic factors, irreducible over F. But each irreducible quadratic factor has the form

$$X^2 + aX + b = (X+\frac{a}{2})^2 + (b - \frac{a^2}{4})$$

where $b^2 - (a^2/4) > 0$, by irreducibility. Therefore $X^2 + aX + b > 0$ for all values of X, so $f(X)$ can change sign only when a linear factor changes sign, and this happens only at a root of $f(X)$, lying between λ and μ. ■

23.10. THEOREM. A field F is real closed if and only if $F(i) \neq F$ $(i^2 = -1)$ and $F(i)$ is algebraically closed.

Proof. If F is real closed, then $i \notin F$ since F is formally real. Hence $F(i) \neq F$ and, by Corollary 23.6(ii), $F(i)$ is algebraically closed.

Conversely, assume that $F(i) \neq F$ and $F(i)$ is algebraically closed. Then, by Corollary 16.19, F is an ordered field. It follows that F is a maximal formally real field in its algebraic closure $F(i)$. Thus F must be real closed. ■

As an easy consequence of Theorem 23.10, we derive

23.11. COROLLARY. Let E/F be a field extension, where E is real closed. Then the algebraic closure of F in E is also real closed.

Proof. Let $K = E(i)$ where $i^2 = -1$. Then, by Theorem 23.10, K is algebraically closed. Hence the algebraic closure L of F in K is algebraically closed. If $\lambda + \mu i$, $\lambda,\mu \in E$ is in L, then so is $\lambda - \mu i$. Thus

$$\lambda = (1/2)(\lambda+\mu i+\lambda-\mu i) \in L$$

and therefore $\mu \in L$. Because $\lambda,\mu \in E$, it follows that $\lambda,\mu \in E_1$, the algebraic closure of F in E. But then $L = E_1(i)$ and, since $i \notin E_1$, we see that E_1 satisfies the conditions of Theorem 23.10. Thus E_1 is a real closed field. ■

Our next aim is to derive a classical result, namely Sturm's theorem, which will allow us to determine the precise number of roots in a real closed field of a polynomial $f(X)$. Our exposition is borrowed heavily from Weber's work Lehrbuch der Algebra(1898), Vol I, pp. 301-313.

In what follows F denotes a real closed field and $f(X) \in F[X]$ a polynomial of positive degree. Given $\alpha, \beta \in F$ with $\alpha \leqslant \beta$, we put

$$[\alpha,\beta] = \{x \in F \mid \alpha \leqslant x \leqslant \beta\}$$

Following Weber, we say that a sequence of polynomials

$$f_0(X) = f(X), f_1(X), \ldots, f_n(X) \qquad (f_i(X) \in F[X])$$

is a *Sturm sequence* for $f(X)$ in $[\alpha,\beta]$ if the following four conditions hold:

$f_n(X)$ has no roots in $[\alpha,\beta]$ (4)

$f_0(\alpha) \neq 0$, $f_0(\beta) \neq 0$ (5)

If $\gamma \in [\alpha,\beta]$ is a root of $f_i(X)$ with $0 < i < n$, then (6)

$$f_{i-1}(\gamma)f_{i+1}(\gamma) < 0$$

If $\gamma \in [\alpha,\beta]$ is a root of $f_0(X)$, then there exist $\gamma_1 \leqslant \gamma$ and $\gamma_2 \geqslant \gamma$ such that

$$\begin{aligned} f_0(x)f_1(x) &< 0 \quad \text{for all } \gamma_1 \leqslant x < \gamma \\ f_0(x)f_1(x) &> 0 \quad \text{for all } \gamma < x \leqslant \gamma_2 \end{aligned} \tag{7}$$

We now establish the existence of Sturm sequences for any polynomial with distinct roots. It will be convenient to introduce the following definition.

Let $f(X)$ be an arbitrary element of $F[X]$. By the *standard sequence* for $f(X)$, we understand the sequence:

$$\begin{aligned} f_0(X) &= f(X),\ f_1(X) = f'(X) \\ f_0(X) &= q_1(X)f_1(X) - f_2(X), \quad \deg f_2(X) < \deg f_1(X) \\ &\ \ \vdots \\ f_{i-1}(X) &= q_i(X)f_i(X) - f_{i+1}(X), \quad \deg f_{i+1}(X) < \deg f_i(X) \\ &\ \ \vdots \\ f_{n-1}(X) &= q_n(X)f_n(X) \end{aligned} \tag{8}$$

It is obvious that $f_n(X)$ is the greatest common divisor of $f(X)$ and $f'(X)$ and this is a divisor of all the $f_i(X)$.

23.12. LEMMA. Further to the notation above, define the sequence

$$g_0(X), g_1(X), \ldots, g_n(X) \tag{9}$$

of polynomials over F by

$$g_i(X) = f_i(X) f_n(X)^{-1} \qquad (0 \leqslant i \leqslant n)$$

(i) The sequence (9) is a Sturm sequence for $g_0(X)$ in any interval $[\alpha,\beta]$ such that $g_0(\alpha) \neq 0$, $g_0(\beta) \neq 0$.

(ii) If $f(X)$ has no multiple roots, then the standard sequence (8) is a Sturm sequence for $f(X)$ in any interval $[\alpha,\beta]$ such that $f_0(\alpha) \neq 0$, $f_0(\beta) \neq 0$.

Proof. (i) Condition (5) holds by hypothesis. Since $g_n(X) = 1$, (4) also holds. Dividing the polynomials in (8) by $f_n(X)$, we obtain

$$g_{i-1}(X) = q_i(X) g_i(X) - g_{i+1}(X) \qquad (0 < i < n)$$

To prove (6), assume that $\gamma \in [\alpha,\beta]$ is a root of $g_i(X)$ with $0 < i < n$. Then $g_{i-1}(\gamma) \neq 0$ and $g_{i+1}(\gamma) \neq 0$, since otherwise all the $g_k(\gamma) = 0$ from a certain point on, contrary to $g_n(X) = 1$. Therefore $g_{i-1}(\gamma) g_{i+1}(\gamma) \neq 0$ and, taking into account that

$$g_{i-1}(\gamma) = q_i(\gamma) g_i(\gamma) - g_{i+1}(\gamma) = -g_{i+1}(\gamma)$$

we derive $g_{i-1}(\gamma) g_{i+1}(\gamma) < 0$, proving (6).

To prove (7), assume that $\gamma \in [\alpha,\beta]$ is a root of $g_0(X)$. Then there exist $m \geqslant 1$ and $h(X) \in F[X]$ with $h(\gamma) \neq 0$ such that $f(X) = (X-\gamma)^m h(X)$. Hence

$$f'(X) = (X-\gamma)^m h'(X) + m(X-\gamma)^{m-1} h(X)$$

and

$$f_n(X) = (X-\gamma)^{m-1} t(X) \qquad (t(\gamma) \neq 0)$$

Thus $h(X) = t(X)u(X)$ where $u(\gamma) \neq 0$ and $h'(X) = t(X)v(X)$ for some $v(X)$. It therefore follows that

$$g_0(X) = (X-\gamma)u(X), \quad u(\gamma) \neq 0$$
$$g_1(X) = (X-\gamma)v(X) + mu(X)$$

so $g_1(\gamma) = mu(\gamma) \neq 0$. Now select an interval $[\gamma_1,\gamma_2]$ containing γ in its interior such that $u(x) \neq 0$ and $g_1(x) \neq 0$ in $[\gamma_1,\gamma_2]$. Then, by Theorem 23.9, $g_1(x)$ and $u(x)$ are either both positive or both negative in $[\gamma_1,\gamma_2]$. Thus $g_0(x)g_1(x) = (x-\gamma)g_1(x)u(x)$ has the same sign as $x-\gamma$ in $[\gamma_1,\gamma_2]$. This shows that

$$g_0(x)g_1(x) < 0 \quad \text{for all} \quad \gamma_1 \leqslant x < \gamma$$

and

$$g_0(x)g_1(x) > 0 \quad \text{for all} \quad \gamma < x \leqslant \gamma_2,$$

proving (7).

(ii) Assume that $f(X)$ has no multiple roots. Then the greatest common divisor of $f(X)$ and $f'(X)$ is 1. Hence the sequence $f_0(X), f_1(X), \ldots, f_n(X)$ differs from $g_0(X), g_1(X), \ldots, g_n(X)$ by a nonzero factor in F. Therefore the sequence of $f_i(X)$ is a Sturm sequence for $f(X)$ in $[\alpha,\beta]$. ■

Let $\lambda_1, \lambda_2, \ldots, \lambda_n$ be a sequence of elements in a real closed field F. By the *number of sign changes* in this sequence, we understand the number of times there is a change from a positive to a negative value, or vice versa, ignoring zeros. For example, the sequence 2,0,3,-1,0,0,4 has two changes. The following result contains the burden of the proof of Sturm's theorem.

23.13. LEMMA. Let F be a real closed field and let $f(X) \in F[X]$ be of positive degree. Assume that there exists a Sturm sequence

$$f_0(X) = f(X),\ f_1(X), \ldots, f_n(X)$$

for $f(X)$ in $[\alpha,\beta]$, and for any $\gamma \in F$ denote by $\delta(\gamma)$ the number of sign changes in the sequence $f_0(\gamma), f_1(\gamma), \ldots, f_n(\gamma)$. Then the number of distinct roots of $f(X)$ in $(\alpha,\beta) = \{x \in F \mid \alpha < x < \beta\}$ is $\delta(\alpha) - \delta(\beta)$.

Proof. We may split the interval $[\alpha,\beta]$ into subintervals by the roots of the polynomials $f_i(X)$ of the given Sturm sequence. In this way we obtain a sequence $\alpha = \alpha_0 < \alpha_1 < \ldots < \alpha_k = \beta$ such that none of the $f_i(X)$ has a root in (α_j, α_{j+1}). Put $\alpha'_j = (1/2)(\alpha_{j-1} + \alpha_j)$, $1 \leqslant j \leqslant k$, to ensure that

$\alpha'_j \in (\alpha_{j-1}, \alpha_j)$ and denote by $\delta(\alpha'_j)$ the number of sign changes in the sequence

$$f_0(\alpha'_j),\ f_1(\alpha'_j),\dots,f_n(\alpha'_j)$$

Then we obviously have

$$\delta(\alpha) - \delta(\beta) = \delta(\alpha) - \delta(\alpha'_1) + \sum_{j=1}^{k-1} (\delta(\alpha'_j) - \delta(\alpha'_{j+1})) + \delta(\alpha'_k) - \delta(\beta) \tag{10}$$

Note also that $f_0(\alpha) \neq 0$, $f_0(\beta) \neq 0$, $f_n(\alpha_j) \neq 0$, $f_n(\alpha'_j) \neq 0$.

We now claim that if no $f_i(\alpha) = 0$, $0 < i < n$, then $\delta(\alpha) = \delta(\alpha'_1)$. Indeed, $f_s(\alpha)f_s(\alpha'_1) > 0$ for $s = 0,1,\dots,n$, since otherwise, by Theorem 23.9, one of the $f_s(X)$ has a root in (α,α'_1), contrary to the property of the intervals

$$(\alpha_j,\alpha_{j+1})$$

This proves that $\delta(\alpha) = \delta(\alpha'_1)$, as claimed.

Assume that $f_i(\alpha) = 0$ for some i, $0 < i < n$. Then, applying (6), we have $f_{i-1}(\alpha)f_{i+1}(\alpha) < 0$. But $f_{i-1}(X)$ and $f_i(X)$ have no roots in (α,α_1), hence $f_{i-1}(\alpha)f_{i-1}(\alpha'_1) > 0$ and $f_{i+1}(\alpha)f_{i+1}(\alpha'_1) > 0$. Thus $f_{i-1}(\alpha'_1)f_{i+1}(\alpha') < 0$, so $f_{i-1}(\alpha)$, 0, $f_{i+1}(\alpha)$ and $f_{i-1}(\alpha'_1)$, $f_i(\alpha'_1)$, $f_{i+1}(\alpha'_1)$ contribute the same number of sign changes to $\delta(\alpha)$ and $\delta(\alpha'_1)$, respectively. By considering all the i, we conclude that $\delta(\alpha) - \delta(\alpha'_1) = 0$. By a similar argument, we also have

$$\delta(\alpha'_k) - \delta(\beta) = 0$$

Again, by the same argument, we see that if α_j, $1 \leqslant j \leqslant k-1$, is not a root of $f(X) = f_0(X)$, then $\delta(\alpha'_j) - \delta(\alpha'_{j+1}) = 0$. We are left to examine the case when $f(\alpha_j) = 0$ for $1 \leqslant j \leqslant k-1$.

By (7) and the choice of the α'_j, we have

$$f_0(\alpha'_j)f_1(\alpha'_j) < 0 \quad \text{and} \quad f_0(\alpha'_{j+1})f_1(\alpha'_{j+1}) > 0$$

It follows that the sequence $f_0(\alpha'_j),f_1(\alpha'_j)$ has one sign change, while the sequence $f_0(\alpha'_{j+1}),f_1(\alpha'_{j+1})$ has none. The previous argument now shows that $f_{i-1}(\alpha'_j),f_i(\alpha'_j)$, $f_{i+1}(\alpha'_j)$ and $f_{i-1}(\alpha'_{j+1})$, $f_i(\alpha'_{j+1})$, $f_{i+1}(\alpha'_{i+1})$ have the same number of sign changes if $i > 1$. Thus, if $f(\alpha_j) = 0$, then $\delta(\alpha'_j) - \delta(\alpha'_{j+1}) = 1$. This shows that $\delta(\alpha'_j) - \delta(\alpha'_{j+1}) = 0$ or 1 according as $f(\alpha_j) \neq 0$ or $f(\alpha_j) = 0$. Consequently, by (10), $\delta(\alpha) - \delta(\beta)$ is the number of α_j such that $f(\alpha_j) = 0$. ■

We are now ready to prove

23.14. THEOREM. (Sturm's theorem). Let F be a real closed field and let

$$f(X) \in F[X]$$

be of positive degree. If $f_0(X) = f(X), f_1(X),\dots,f_n(X)$ is the standard sequence (8) for $f(X)$ and if $[\alpha,\beta]$ is an interval such that $f(\alpha) \neq 0$, $f(\beta) \neq 0$, then the number of distinct roots of $f(X)$ in (α,β) is $\delta(\alpha) - \delta(\beta)$, where $\delta(\gamma)$ denotes the number of sign changes in the sequence $f_0(\gamma), f_1(\gamma),\dots,f_n(\gamma)$.

Proof. As in Lemma 23.12, we put $g_i(X) = f_i(X)f_n(X)^{-1}$, $0 \leqslant i \leqslant n$. Then one immediately verifies that, apart from multiplicities, the polynomials $f(X)$ and $g_0(X)$ have the same roots in (α,β). By Lemma 23.12(i), the sequence

$$g_0(X), g_1(X),\dots,g_n(X)$$

is a Sturm sequence for $g_0(X)$ in $[\alpha,\beta]$. Hence, by Lemma 23.13, the number of these roots is $\delta_g(\alpha) - \delta_g(\beta)$, where $\delta_g(\gamma)$ is the number of sign changes in $\{g_i(\gamma)\}$. But

$$f_i(\gamma) = g_i(\gamma)f_n(\gamma) \quad \text{and} \quad f_n(\alpha) \neq 0,\ f_n(\beta) \neq 0,$$

so $\delta_g(\alpha) = \delta(\alpha)$ and $\delta_g(\beta) = \delta(\beta)$. Thus $\delta(\alpha) - \delta(\beta)$ is equal to the number of distinct roots of $f(X)$ in (α,β). ■

Let F be an ordered field. By a *real closure* of F, we understand a real closed field E algebraic over F and such that the unique ordering of E (see Corollary 23.6(i)) extends the ordering of F. The following observation will enable us to construct the real closure.

23.15. LEMMA. Let F be an ordered field and let E be the field obtained by adjoining to F the square roots of all positive elements of F. Then E is formally real.

Proof. We may clearly assume that $E = F(\sqrt{a_1},\sqrt{a_2},\dots,\sqrt{a_n})$ for some finitely many positive elements $a_1,\dots,a_n$ of F. We prove, by induction on $(E:F)$, that if $\Sigma\lambda_i\mu_i^2 = 0$ for $\lambda_i > 0$ in F and μ_i in E, then all $\mu_i = 0$. This will clearly yield the result.

If $(E:F) = 1$, then there is nothing to prove. We may assume that

$$E \supset L = F(\sqrt{a_1},\ldots,\sqrt{a_{n-1}}),$$

so by induction the result holds for L. Now assume that $\Sigma\lambda_i\mu_i^2 = 0$, $\mu_i \in E$, $\lambda_i > 0$ in F. Write

$$\mu_i = u_i + v_i\sqrt{a_n} \qquad (u_i, v_i \in L)$$

Then

$$\Sigma\lambda_i u_i^2 + \Sigma a_n\lambda_i v_i^2 + 2(\Sigma\lambda_i u_i v_i)\sqrt{a_n} = 0$$

Because $\sqrt{a_n} \notin L$, $\Sigma\lambda_i u_i v_i = 0$, so $\Sigma\lambda_i u_i^2 + \Sigma a_n\lambda_i v_i^2 = 0$. Taking into account that $\lambda_i, a_n\lambda_i \in F$, $u_i, v_i \in L$ and $\lambda_i > 0$, $a_n\lambda_i > 0$, we conclude that every u_i and v_i is zero. But then each μ_i is zero, as required. ■

We are now ready to prove

23.16. THEOREM. (i) Any ordered field has a real closure

(ii) If F_1 and F_2 are two ordered fields and $f : F_1 \longrightarrow F_2$ is an order isomorphism between them, then there is a unique isomorphism f^* between their real closures extending f, and f^* is an order preserving isomorphism

(iii) If E_1 and E_2 are two real closures of a given ordered field F, then there is an order preserving F-isomorphism of E_1 onto E_2.

Proof. (i) Let F be an ordered field and let E be the field obtained by adjoining to F the square roots of all positive elements of F. Then, by Lemma 23.15 E is formally real. Hence, by Theorem 23.7(i), E can be embedded in a real closed field R which is algebraic over F. But every positive element of F is a square in E, so it is positive in R and thus the ordering on R induces the given ordering on F. This proves that R is a real closure of F.

(ii) Let R_1 and R_2 be real closures of F_1, F_2, respectively. Any field isomorphism $R_1 \longrightarrow R_2$ is order preserving since it maps squares to squares. Moreover, any order isomorphism $f^* : R_1 \longrightarrow R_2$ extending f is uniquely determined. Indeed, any element $\alpha \in R_1$ has a minimal polynomial $f(X) \in F_1[X]$, and by Sturm's theorem the roots $\alpha_1 < \alpha_2 < \ldots < \alpha_t = \alpha < \ldots < \alpha_n$ of $f(X)$ in R_1 are equal in number to the roots $\alpha_1^* < \alpha_2^* < \ldots < \alpha_n^*$ in R_2 of the corresponding polynomial $f^*(X) \in F_2[X]$. Because f^* maps the α's to the α^*,s and preserves their order, α must be mapped to α^*_t, which shows that f^* is uniquely determined.

We have actually defined a bijection $f^* : R_1 \longrightarrow R_2$ and we are left to verify that it is a homomorphism. To this end, we first note that any finite subset of R_1 lies in a finite, hence simple extension of F_1. Thus we need only show that for α as above the canonical isomorphism $\psi : F_1(\alpha) \longrightarrow F_2(f^*(\alpha))$ agrees with f^*.

Fix any $\beta \in F_1(\alpha)$ and let its conjugates over F_1 in R_1 be

$$\beta_1 < \dots < \beta_s = \beta < \dots < \beta_k$$

We may choose in R_1 elements $\gamma_1, \dots, \gamma_{n-1}, \delta_1, \dots, \delta_{k-1}$ such that

$$\gamma_i^2 = \alpha_{i+1} - \alpha_i, \quad \delta_j^2 = \beta_{j+1} - \beta_j$$

The F_1-subalgebra of R_1 generated by all $\alpha, \beta, \gamma, \delta$ is a simple extension, $F(\lambda)$ say, and there is an isomorphism $\sigma : F_1(\lambda) \longrightarrow F_2(f^*(\lambda))$ extending f. But $\alpha_{i+1} = \alpha_i + \gamma_i^2 > \alpha_i$, so $\sigma(\alpha_{i+1}) > \sigma(\alpha_i)$ and similarly $\sigma(\beta_{j+1}) > \sigma(\beta_j)$. Hence σ preserves the order of the α's and therefore $\sigma(\alpha) = f^*(\alpha)$. The conclusion is that σ extends ψ and so $\psi(\beta) = \sigma(\beta) = f^*(\beta)$, as required.

(iii) Take the identity mapping on F and apply (ii). ■

As an illustration, let us take the field $\mathbb{Q}$ of rational numbers. This is an ordered field (in fact, by Lemma 23.2, it admits only one ordering). Hence, by Theorem 23.16, any two real closures of $\mathbb{Q}$ are $\mathbb{Q}$-isomorphic. We can say even more, namely

23.17. LEMMA. Let $\bar{\mathbb{Q}}$ be the algebraic closure of $\mathbb{Q}$ in $\mathbb{C}$. Then $\bar{\mathbb{Q}} \cap \mathbb{R}$ is a real closure of $\mathbb{Q}$. In particular, the real closure of $\mathbb{Q}$ can be identified with the field of real algebraic numbers.

Proof. Since $\mathbb{Q}$ admits only one ordering and since $(\bar{\mathbb{Q}} \cap \mathbb{R})/\mathbb{Q}$ is an algebraic extension, it suffices to verify that $\bar{\mathbb{Q}} \cap \mathbb{R}$ is real closed. If

$$a + bi \in \bar{\mathbb{Q}}$$

with $a, b \in \mathbb{R}$, then $a - bi \in \bar{\mathbb{Q}}$. Hence $a, b \in \bar{\mathbb{Q}} \cap \mathbb{R}$ and therefore

$$\bar{\mathbb{Q}} = (\bar{\mathbb{Q}} \cap \mathbb{R})(i)$$

Since $\bar{\mathbb{Q}}$ is algebraically closed and $i \notin \bar{\mathbb{Q}} \cap \mathbb{R}$, the desired conclusion follows by virtue of Theorem 23.10. ■

It is now an easy matter to provide a survey of all orderings of algebraic number fields.

23.18. THEOREM. Let F be an algebraic number field of degree n, say $F = \mathbb{Q}(\lambda)$, and let $\bar{\mathbb{Q}}$ be the algebraic closure of $\mathbb{Q}$ in $\mathbb{C}$. Let $\lambda_1, \ldots, \lambda_r$ be all distinct real conjugates of λ and, for each $i \in \{1,2,\ldots,r\}$, let

$$f_i : F \longrightarrow \bar{\mathbb{Q}} \cap \mathbb{R}$$

be the $\mathbb{Q}$-homomorphism sending λ to λ_i (if λ has no real conjugates, then set $r = 0$). Then there exist exactly r distinct orderings of F, namely those by declaring $x > 0$ in F if and only if $f_i(x) > 0$ in $\bar{\mathbb{Q}} \cap \mathbb{R}$, $1 \leqslant i \leqslant r$. In particular, F is an ordered field if and only if the minimal polynomial of λ over $\mathbb{Q}$ has at least one real root.

Proof. It is clear that each f_i provides an ordering of F. Now suppose we have any ordering of F and let L be a real closure of F relative to this ordering. Because F is algebraic over $\mathbb{Q}$, it is clear that L is a real closure of $\mathbb{Q}$. Hence, by Lemma 23.17 and Theorem 23.16, there is an order preserving $\mathbb{Q}$-isomorphism $f : L \longrightarrow \bar{\mathbb{Q}} \cap \mathbb{R}$. The restriction of f to F coincides with one of the f_i and it is obvious that the given ordering of F is the same as the ordering determined by f_i.

Finally, assume that f_i and f_j provide the same ordering of F. Then we have an order preserving isomorphism $\mathbb{Q}(\lambda_i) \longrightarrow \mathbb{Q}(\lambda_j)$ which sends λ_i to λ_j. Since $\bar{\mathbb{Q}} \cap \mathbb{R}$ is a real closure of $\mathbb{Q}(\lambda_i)$ and $\mathbb{Q}(\lambda_j)$, it follows from Theorem 23.16 that there is a $\mathbb{Q}$-automorphism of $\bar{\mathbb{Q}} \cap \mathbb{R}$ sending λ_i to λ_j. On the other hand, since $\bar{\mathbb{Q}} \cap \mathbb{R}$ is a real closure of $\mathbb{Q}$, Theorem 23.16 shows that the identity map is the only $\mathbb{Q}$-automorphism of $\bar{\mathbb{Q}} \cap \mathbb{R}$. Thus we must have $\lambda_i = \lambda_j$, as we wished to show. ■

23.19. COROLLARY. (Hilbert-Landau). Let F be an algebraic number field and let

$f_1\ f_2\ \cdots\ f_r$ $(r \geqslant 0)$ be all distinct $\mathbb{Q}$-homomorphisms of F into $\bar{\mathbb{Q}} \cap \mathbb{R}$, where $\bar{\mathbb{Q}}$ is the algebraic closure of $\mathbb{Q}$ in $\mathbb{C}$. Then an element $\lambda \neq 0$ in F is a sum of squares in F if and only if $f_i(\lambda) > 0$ for $i = 1,2,\ldots,r$.

Proof. Apply Theorems 23.8 and 23.18. ∎

24. TRANSCENDENTAL EXTENSIONS

Let E/F be a field extension and let S be any subset of E. Recall that S (or the elements of S) is said to be *algebraically independent* over F if the elements

$$\prod_{s\in S} s^{\nu(s)} \qquad (\nu(s) \geqslant 0)$$

(finitely many $\nu(s) \neq 0$) are F-linearly independent. It is clear that S is algebraically independent over F if and only if so is every finite subset of S. In the sepcial case where $S = \{s\}$, we see that S is algebraically independent over F if and only if $\{1,s,s^2,\ldots,s^n,\ldots\}$ is an F-linearly independent set. If $s \in E$ is algebraically independent over F, then we also say that s is *transcendental* over F.

A field extension E/F is said to be *transcendental* if it is not algebraic that is, if E contains at least one element which is transcendental over F. It is clear that any extension of a transcendental extension of F is itself a transcendental extension of F.

We say that the field extension E/F is *purely transcendental* if $E = F(S)$ for some algebraically independent subset S of E over F.

It will be often convenient to deal with the families of elements of E, say a family $\{a_i\}_{i\in I}$. We say that such a family is algebraically independent over F if its elements are distinct (i.e. $a_i \neq a_j$ if $i \neq j$) and if the set consisting of the elements in this family is algebraically independent over F.

24.1. LEMMA. Let E/F be a field extension and let $S = \{s_i\}_{i\in I}$ be a family of elements of E. Then the following conditions are equivalent:

(i) S is algebraically independent over F

(ii) The map

$$\begin{cases} F[(X_i)_{i\in I}] \longrightarrow E \\ \qquad X_i \longmapsto s_i \end{cases}$$

is an injective homomorphism of F-algebras

(iii) For each $s \in S$, s is transcendental over $F(S')$, where $S' = S - \{s\}$.

Proof. The equivalence of (i) and (ii) follows from Proposition 1.2.2. If there exists $s \in S$ such that s is algebraic over $F(S')$, then s is a root of a polynomial in $F[(X_i)i \in I]$ and hence the map in (ii) has a nontrivial kernel. Conversely, assume that S is not algebraically independent over F. Then S has n distinct elements $s_1,\ldots,s_n$ such that $\Sigma a_{e_1\ldots e_n} s_1^{e_1} s_2^{e} \ldots s_n^{e_n} = 0$ is a nontrivial linear combination of the monomials $s_1^{e_1}\ldots s_n^{e_n}$. Hence s_i (for some i) is algebraic over $F(s_1,\ldots,s_{i-1},s_{i+1},\ldots,s_n)$, as required. ∎

24.2. LEMMA. (i) If E/F is a purely transcendental extension, then there exists a set I such that E is F-isomorphic to the quotient field of the polynomial ring $F[(X_i), i \in I]$.

(ii) If E_1/F and E_2/F are purely transcendental extensions, say $E_i = F(S_i)$ where S_i is algebraically idependent over F, $i = 1,2$, and if $|S_1| = |S_2|$, then E_1 is F-isomorphic to E_2.

Proof. (i) If S is a subset of E such that $E = F(S)$, then E is the quotient field of $F[S]$. If S is independent over F, then by Lemma 24.1(ii), $F[S] \cong F[(X_i)i \in I]$ where $S = \{s_i | i \in I\}$. Hence E is F-isomorphic to the quotient field of $F[(X_i)i \in I]$.

(ii) Direct consequence of (i). ∎

Let E/F be a field extension. We can introduce an ordering among algebraically independent subsets of E, by ascending inclusion. These subsets are obviously inductively ordered and hence there exist maximal elements. A subset S of E which is algebraically idependent over F and is maximal with respect to the inclusion ordering is called a *transcendency basis* of E over F. Thus a subset S of E is a transcendency basis of E over F if and only if

(a) S is algebraically independent over F

(b) $E/F(S)$ is algebraic

24.3. THEOREM. If S and S' are transcendency bases of E over F, then $|S| = |S'|$.

Proof. Suppose first that $|S|$ is finite, say $S = \{s_1, s_2, \ldots, s_n\}$. We may assume that $n \leqslant |S'|$ and hence we may choose n distinct elements, say

$$t_1, \ldots, t_n$$

in S'. It suffices to show that if we renumber $s_1, \ldots, s_n$ suitably, then E is algebraic over $F_i = F(t_1, \ldots, t_i, s_{i+1}, \ldots, s_n)$ for $i = 0,1,\ldots n$ (since the case $i = n$ implies that $S' = \{t_1, \ldots, t_n\}$). The case $i = 0$ being obvious, we may assume that $i \geqslant 1$ and that E is algebraic over F_{i-1}. Then t_i is algebraic over F_{i-1}, and so we must have a relation

$$\lambda_0 t_i^{k-1} + \ldots + \lambda_k = 0$$

where

$$\lambda_0, \ldots, \lambda_k \in F[t_1, \ldots, t_{i-1}, s_i, \ldots, s_n],\ \lambda_0 \neq 0$$

Because $t_1, \ldots, t_i$ are algebraically independent, some of $s_i, \ldots, s_n$ must appear in the relation. We may renumber $s_i, \ldots, s_n$ so that s_i appears properly. Then s_i is algebraic over F_i, and so E is algebraic over F_i. This establishes the case where $|S|$ is finite.

Now assume that $|S|$ is infinite and that $|S| \geqslant |S'|$. Each element x of S' is algebraic over $F(K_x)$, where K_x is a suitable finite subset of S. Put

$$K = \bigcup_x K_x$$

Because every element of S' is algebraic over $F(K)$, we have $K = S$. But then

$$|S| = |\cup K_x| \leqslant \aleph_0 |S'| = |S'|,$$

hence $|S| = |S'|$ as required.

By the *transcendency degree* of E over F, we understand the cardinality of any transcendency basis of E over F. Theorem 24.3 guarantees that the notion of transcendency degree is well defined.

24.4 PROPOSITION. (i) If a field E is generated by a subset S over a field F, then there is a subset S_1 of S which is a transcendency basis of E over F.

(ii) If a field E is finitely generated over its subfield F and if S is a transcendency basis of E over F, then S and $E/F(S)$ are finite.

Proof. (i) There is a maximal member S_1 in the set of algebraically independent subsets of S.

(ii) Direct consequence of (i), Theorem 24.3 and the fact that every finitely generated algebraic extension is finite. ∎

In what follows, we write tr.d.(E/F) for the transcendency degree of E over F. It is clear that E is algebraic over F if and only if tr.d. $(E/F) = 0$.

24.5. THEOREM. Let $F \subseteq E \subseteq K$ be a chain of fields and let S_1 and S_2 be transcendency bases of E/F and K/E, respectively. Then $S_1 \cup S_2$ is a transcendency basis of K/F and

$$\text{tr.d. } (K/F) = \text{tr.d. } (K/E) + \text{tr.d. } (E/F)$$

In particular, by Proposition 24.4(i), if S is a subset of E such that $E/F(S)$ is algebraic, then S contains a transcendency basis of E/F.

Proof. Clearly $S_1 \cap S_2 = \emptyset$, so it suffices to verify that $S_1 \cup S_2$ is a transcendency basis of K/F.

Let $\{a_1, a_2, \ldots, a_m, b_1, b_2, \ldots, b_n\}$ be any finite subset of $S_1 \cup S_2$, where we assume that the a_i are in S_1 and the b_j are in S_2. Assume that there exists a polynomial $f(\{X\},\{Y\}) = f(X_1,\ldots,X_m,Y_1,\ldots,Y_n)$ in $m + n$ variables over F such that $f(\{a\},\{b\}) = 0$. The polynomial $f(\{a\},Y)$ in the n indeterminates Y_j has coefficients in E and must be zero since the b_j are algebraically independent over E. Because the a_i are algebraically independent over F, it follows that $f(\{X\},\{Y\})$, considered as a polynomial in $\{Y\}$ with coefficients in $F[\{X\}]$, must be zero. Thus $f(\{X\},\{Y\}) = 0$ and therefore $S_1 \cup S_2$ is algebraically independent over F.

By hypothesis, E is an algebraic extension of $F(S_1)$. Hence $E(S_2)$ is an algebraic extension of $F(S_1)(S_2) = F(S_1 \cup S_2)$. However, K is an algebraic extension of $E(S_2)$. Thus K is an algebraic extension of $F(S_1 \cup S_2)$ and the result follows. ■

24.6. THEOREM. Let E_1/F and E_2/F be two extensions of F contained in a common field K. Then

$$\text{tr.d.}\ (E_1E_2/E_1) \leqslant \text{tr.d.}\ (E_2/F) \tag{1}$$

and

$$\text{tr.d.}\ (E_1E_2/F) \leqslant \text{tr.d.}\ (E_1/F) + \text{tr.d.}\ (E_2/F) \tag{2}$$

Proof. Let S be a transcendency basis of E_2/F. Since $E_1E_2 = E_1(E_2)$ and every element of E_2 is algebraic over $F(S)$, we see that E_1E_2 is algebraic over $E_1(S)$. Hence, by Theorem 24.5, S contains a transcendency basis of E_1E_2/E_1, which proves (1).

By Theorem 24.5, we have

$$\text{tr.d.}\ (E_1E_2/F) = \text{tr.d.}\ (E_1E_2/E_1) + \text{tr.d.}\ (E_1/F)$$

The inequality (2) is therefore a consequence of (1). ■

24.7. THEOREM. If a field E is finitely generated over its subfield F and if K is an intermediate field, then K is finitely generated over F.

Proof. Suppose first that K is algebraic over F. Let $x_1,\dots,x_n$ be a transcendency basis of E over F. If elements $y_1,\dots,y_r \in K$ are linearly independent over F, then they are linearly independent over $F(x_1,\dots,x_n)$. Therefore, by Proposition 24.4(iii),

$$(K:F) \leqslant (E:F(x_1,\dots,x_n)) < \infty$$

Now we consider the general case. Let S be a transcendency basis of K over F. Then, by the algebraic case, K is finitely generated over $F(S)$, and hence over F, too. ■

3
Valuation theory

Valuation theory plays a fundamental role in the study of structure of fields. In selecting material for this chapter we have held steadfast to a single-minded purpose: to present mainly those results deemed essential for our subsequent investigation of multiplicative structure of fields. After recording some frequently used observations, we examine in detail completions and extensions of valuations. Further topics include Dedekind domains, valuations and places, and ramification theory. The chapter culminates with a study of complete discrete valued fields, where the Witt ring of a field plays a crucial role in many constructions. Among other results, we provide a characterization of local fields by proving that a field F is local if and only if it is one of the following types:

(i) A finite extension of the rational p-adic field $\mathbb{Q}_p$

(ii) A formal power serves field in one variable over a finite field

In the final section, dealing with cyclotomic extensions of p-adic fields, we exhibit an explicit formula for the inertia field of a p-adic field of finite degree over $\mathbb{Q}_p$.

1. VALUATIONS

Our aim in this section is to record certain frequently used facts concerning valuations in general.

Let F be an arbitrary field. A *valuation* v of F is any mapping

$$v : F \longrightarrow \mathbb{R}$$

satisfying the following three properties:

(1) $v(x) \geq 0$ for all $x \in F$, and $v(x) = 0$ if and only if $x = 0$

(2) $v(xy) = v(x)v(y)$ for all $x,y \in F$

(3) $v(x+y) \leq v(x) + v(y)$ for all $x,y \in F$

If instead of (3) v satisfies the stronger condition

(4) $v(x+y) \leq \max\{v(x),v(y)\}$

then v is said to be a *non-archimedean valuation*. All the remaining valuations are called *archimedean*.

One way of obtaining archimedean valuations is as follows: the ordinary absolute value $|\ |$ on $\mathbb{C}$ is an archimedean valuation, whose restriction to any subfield is an archimedean valuation on that subfield. Now suppose that F can be embedded in $\mathbb{C}$, and let $\mu : F \longrightarrow \mathbb{C}$ be an embedding. Define

$$v_\mu : F \longrightarrow \mathbb{R}$$

by setting

$$v_\mu(x) = |\mu(x)|$$

Then v_μ is an archimedean valuation of F. This example will play an important role in the special case where F is an algebraic number field.

A classical example of a non-archimedean valuation is the *p-adic valuation* v_p of $\mathbb{Q}$ defined as follows. Let p be a fixed rational prime. Then $v_p(0) = 0$ and for $x \neq 0$ and $x = p^\nu y$, where the numerator and denominator of y are prime to p,

$$v_p(x) = p^{-\nu}$$

Let v be a valuation of F. Then (1) and (2) imply that v restricts to a homomorphism of the multiplicative group F^* into the group of positive real numbers. The image $v(F^*)$ of this homomorphism is called the *value group* of v. In case $v(F^*) = \{1\}$, the valuation v is called *trivial*.

We next collect together some elementary facts about valuations.

1.1. LEMMA. Let v be a valuation of a field F.

(i) $v(x/y) = v(x)/v(y)$ for all $x \in F, y \in F^*$

(ii) $v(-x) = v(x)$ for all $x \in F$

(iii) $v(x-y) \geqslant |v(x) - v(y)|$ for all $x,y \in F$

(iv) If F is an algebraic extension of a finite field, then F possesses only the trivial valuation.

Proof. (i) Direct consequence of (2).

(ii) Because $v(-x) = v(-1)v(x)$, it suffices to show that $v(-1) = 1$. The latter is true since

$$v(-1)^2 = v(1) = 1 \quad \text{and} \quad v(-1) > 0$$

(iii) If we apply (3) to $(x-y) + y = x$, we find that

$$v(x-y) \geqslant v(x) - v(y)$$

Applying (ii), we must also have

$$v(x-y) = v(y-x) \geqslant v(y) - v(x)$$

as required.

(iv) By hypothesis, any $x \in F$ belongs to a finite subfield of F. Therefore, if $x \in F^*$, then x is a root of 1. But then $v(x)$ is a positive root of 1 in $\mathbb{R}$ and thus $v(x) = 1$. ■

1.2. LEMMA. Let S be an integral domain with quotient field F. If

$$v : S \longrightarrow \mathbb{R}$$

is any mapping satisfying (1), (2), and (3), then v extends to a valuation of F by setting

$$v(x/y) = v(x)/v(y) \qquad \text{for all } x \in S,\ 0 \neq y \in S$$

Proof. The map $v : F \longrightarrow \mathbb{R}$ is obviously well-defined and satisfies (1) and (2). To prove (3), let $x = x_1/x_2$ and $y = y_1/y_2$ with $x_i, y_i \in S$ and $x_2 \neq 0$, $y_2 \neq 0$. Then

$$\begin{aligned} v(x+y) &= v(x_1y_2+y_1x_2)/v(x_2y_2) \\ &\leqslant (v(x_1)v(y_2) + v(y_1)v(x_2))/v(x_2)v(y_2) \\ &= v(x) + v(y) \end{aligned}$$

as required. ■

Two valuations v_1 and v_2 of a field F are called *equivalent* if there exists a positive $r \in \mathbb{R}$ such that

$$v_2(x) = v_1(x)^r \qquad \text{for all } x \in F$$

It is clear that a valuation which is equivalent to the trivial valuation is itself the trivial valuation. A *prime* of F is an equivalence class of nontrivial valuations of F.

Let v_1 be a valuation of F and let r be an arbitrary positive real number. Then the function $v(x) = v_1(x)^r$ satisfies conditions (1) and (2) but not necessarily (3). However, v satisfies the following condition: there exists a constant $c > 0$ such that

$$(3') \qquad v(x+y) \leqslant c \max\{v(x), v(y)\} \qquad \text{for all } x,y \in F$$

(for the above v we only have to put $c = 2^r$)

Any map $v : F \longrightarrow \mathbb{R}$ satisfying (1), (2) and (3') is called a *generalized valuation* of F. The name is justified by the fact that any valuation is a generalized valuation (by taking $c = 2$ in (3')).

Generalized valuations v_1, v_2 are said to be *equivalent* if there exists a positive real number r such that $v_2(x) = v_1(x)^r$ for all $x \in F$.

1.3. LEMMA. An arbitrary generalized valuation is equivalent to a valuation .

Proof. Let v be any generalized valuation of a field F. We first note that if (3') holds for $c > 2$, then, setting $r = \log_c 2$, v^r satisfies (3') for $c = 2$. Moreover, if v satisfies (3') for c, then it satisfies (3') for any $c' \geqslant c$. We may therefore assume that v satisfies (1), (2) and (3') (for $c = 2$).

By hypothesis, $v(x_1+x_2) \leqslant 2\max\{v(x_1), v(x_2)\}$. Hence, by induction

$$v\Big(\sum_{i=1}^{2^r} x_i\Big) \leqslant 2^r\{\max\, v(x_i)\}$$

and so for any $n > 0$, we have

$$v\Big(\sum_{i=1}^{n} x_i\Big) \leqslant 2^r \max\{v(x_i)\} \leqslant 2n \max\{v(x_i)\}$$

where $2^{r-1} < n \leqslant 2^r$, on inserting $2^r - n$ zero summands. In particular

$$v(n) \leqslant 2nv(1) = 2n$$

But now

$$(v(x+y))^n = v((x+y)^n) \leq 2(n+1) \sum_{i=0}^{n} v(\tbinom{n}{i}x^i y^{n-i})$$
$$\leq 4(n+1) \sum_{i=0}^{n} \tbinom{n}{i} v(x)^i v(y)^{n-i}$$
$$= 4(n+1)(v(x) + v(y))^n$$

whence

$$v(x+y) \leq \sqrt[n]{4(n+1)}(v(x) + v(y))$$

Hence, letting n tend to infinity, we obtain (3). ■

A field F is called a *topological field* if F is also a topological space such that the operations sum, product reciprocal are continuous.

1.4. LEMMA. Assumed that a field F is provided with a valuation v.

(i) F is a metric space with the distance between x and y given by $v(x-y)$. In particular, v gives rise to a Hausdorff topology on F, by taking as basis for the neighbourhoods of a point $x_0 \in F$ the sets $\{x \in F | v(x-x_0) < \varepsilon\}$, where ε ranges over all positive real numbers.

(ii) With respect to the given topology, the field F is a topological field.

Proof. (i) Direct consequence of (1) and (3).

(ii) The continuity of the addition and multiplication follows from

$$v(x'+y') - (x+y)) \leq v(x'-x) + v(y'-y)$$
$$v(x'y'-xy) \leq v(x)v(y'-y) + v(y)v(x'-x) + v(x'-x)v(y'-y)$$

To prove that the mapping $x \longmapsto x^{-1}$ is continuous at any $x_0 \neq 0$, it suffices to show that the inequality $v(x-x_0) < (1/2)\min\{v(x_0), \varepsilon v(x_0)^2\}$ implies

$$v(x^{-1}-x_0^{-1}) < \varepsilon$$

Since $v(x-x_0) < (1/2)v(x_0)$ implies

$$v(x_0) - v(x) \leq v(x_0-x) = v(x-x_0) < (1/2)v(x_0)$$

and hence that $(v(x)v(x_0))^{-1} < 2v(x_0)^{-2}$, we have

$$v(x^{-1}-x_0^{-1}) = (v(x)v(x_0))^{-1}v(x-x_0)$$
$$< 2\, v(x_0)^{-2} \cdot \tfrac{1}{2}\, \varepsilon\, v(x_0^2) = \varepsilon\ ,$$

as required. ■

A topological field whose topology is induced by a valuation is called a *valuation field*. It is clear that the trivial valuation induces the discrete topology.

Let v be a valuation of a field F. If $\{x_n\}$ is a sequence of elements of F and $x \in F$, then x is called the limit of this sequence if

$$\lim_{n\to\infty} v(x-x_n) = 0$$

Note that if $v(x) < 1$, then $\lim_{n\to\infty} x^n = 0$ since $v(x^n-0) = v(x)^n \longrightarrow 0$ as $n \longrightarrow \infty$. This implies that if v induces the discrete topology, then v is trivial. Indeed, if $v(x) < 1$ for some $x \in F$, then $\lim_{n\to\infty} x^n = 0$ which is impossible in discrete topology. Similarly, the existence of x with $v(x) > 1$ leads to a contradiction. In view of this observation, properties (i) and (ii) in the result below are equivalent for any valuations v_1 and v_2.

1.5. PROPOSITION. Let v_1 and v_2 be two valuations of a field F and let v_1 be nontrivial. Then the following conditions are equivalent:

(i) v_1 and v_2 are equivalent

(ii) v_1 and v_2 determine the same topology

(iii) $v_1(x) < v_1(y)$ implies $v_2(x) < v_2(y)$ for all $x,y \in F$

(iv) $v_1(x) < 1$ implies $v_2(x) < 1$ for all $x \in F$

Proof. We first note that (iii) is equivalent to (iv) by Lemma 1.1(i) and that the implication (i) $\Rightarrow$ (iv) is obvious. We are therefore left to verify (ii) $\Leftrightarrow$ (iv) and (iv) $\Rightarrow$ (i).

(ii) $\Leftrightarrow$ (iv): Assume that v_1 and v_2 determine the same topology. Because v_1 is nontrivial the set $\{x \in F | v_1(x) < 1\}$ is nonempty. Since this set coincides with the set of those $x \in F$ for which $\lim_{n\to\infty} x^n = 0$, property (iv) holds.

Conversely, assume that (iv) holds. Then $v_1(x) > 1$ implies $v_2(x) > 1$ since $v_1(x^{-1}) < 1$. Thus v_1 and v_2 determine the same topology.

(iv) $\Rightarrow$ (i): Since v_1 is nontrivial, there exists an element $x_0 \in F$ such that $v_1(x_0) > 1$. Put $a = v_1(x_0)$, $b = v_2(x_0)$ and $r = (\log b)/(\log a)$. Given $x \in F^*$, we have $v_1(x) = v_2(x)^{\alpha}$ for some $\alpha \in \mathbb{R}$. If m,n are integers such that

$m/n > \alpha$ and $n > 0$, we have

$$v_1(x) < v_1(x_0)^{m/n}$$

whence $v_1(x^n/x_0^m) < 1$ and thus

$$v_2(x) = v_2(x_0)^{\alpha} \qquad \text{for all } x \in F^*$$

Accordingly, $v_2(x) = v_1(x)^r$ for all $x \in F$, as required. ∎

When we say that v is a valuation of a valuation field F, we assume that v induces the topology of F. By Proposition 1.5, such a valuation is unique up to equivalence. Two valuation fields (F_i, v_i), $i = 1,2$, are said to be *isomorphic* if there is a field isomorphism $f : F_1 \longrightarrow F_2$ such that the valuation v_2' of F_2 given by $v_2'(f(x)) = v_1(x)$ is equivalent to v_2.

Our next aim is to characterize archimedean valuations. The traditional formulation of the archimedean axiom runs as follows:

(5) If x and y are nonzero elements of F, then there exists a positive integer n such that $v(nx) > v(y)$.

It will be shown below that archimedean valuations are precisely those which obey axiom (5). We need the following elementary observation.

1.6. LEMMA. Let v be a valuation of a field F. If v is non-archimedean and $v(x) \neq v(y)$, then

$$v(x+y) = \max\{v(x), v(y)\}$$

Proof. Since $v(x) \neq v(y)$, we may assume that $v(x) < v(y)$. Then, by (4), we have $v(x+y) \leqslant v(y)$ and, on the other hand,

$$v(y) = v((x+y)-x) \leqslant \max\{v(x+y), v(x)\}$$

by Lemma 1.1(ii). Because the equality $v(y) \leqslant v(x)$ is ruled out, we conclude that $v(y) \leqslant v(x+y)$. Hence

$$v(x+y) = v(y) = \max\{v(x), v(y)\}$$

and the result follows. ∎ We are now in a position to characterize the non-archimedean valuations.

1.7. PROPOSITION. Let v be a valuation of F. Then the following conditions are equivalent:

(i) v is non-archimedean

(ii) $v(n) \leqslant 1$ for any $n \in \mathbb{N}$

(iii) There exists $c \in \mathbb{R}$ such that $v(n) \leqslant c$ for all $n \in \mathbb{N}$

(iv) v does not satisfy (5)

Proof. (i) $\Rightarrow$ (ii): It follows from (4) that $v(n) \leqslant v(1) = 1$

(ii) $\Rightarrow$ (iii): Obvious

(iii) $\Rightarrow$ (iv): For any given $x \in F$, we have

$$v(nx) = v(n)v(x) \leqslant cv(x) \qquad \text{for all } n \in \mathbb{N}$$

and therefore, by choosing y such that $v(y) \geqslant cv(x)$, property (5) will fail. Such a choice is always possible if v is nontrivial, since $v(x) > 1$ implies $\lim_{n\to\infty} v(x^n) = \infty$. However, if v is trivial, then (5) clearly does not hold.

(iv) $\Rightarrow$ (i): Choose $x,y \neq 0$ so that, for all $n \in \mathbb{N}$, the inequality

$$v(nx) \leqslant v(y)$$

holds. Owing to Lemma 1.1(i), we have $v(n) \leqslant v(y/x)$ for all $n \in \mathbb{N}$. If, for some $n_0 \in \mathbb{N}$, $v(n_0) > 1$, then $v(n_0^k) > v(y/x)$ for a suitable $k \in \mathbb{N}$ which is impossible. Hence $v(n) \leqslant 1$ for all $n \in \mathbb{N}$ and therefore

$$v((x+y)^n) \leqslant \sum_{k=0}^{n} v(\tbinom{n}{k}x^k y^{n-k}) \leqslant \sum_{k=0}^{n} v(x)^k v(y)^{n-k}$$
$$\leqslant (n+1)\max\{v(x)^n, v(y)^n\}$$

The conclusion is that

$$v(x+y) \leqslant (n+1)^{1/n}\max\{v(x), v(y)\}$$

and as n tends to infinity, we obtain (4). ∎

1.8. COROLLARY. Assume that char$F \neq 0$. Then all valuations of F are non-archimedean.

Proof. Because the prime subfield of F is finite, condition (iii) of Proposition 1.7 is trivially satisfied. ∎

Let v be a valuation of a field F. It is a consequence of Propositions 1.5 and 1.7 that v is archimedean if and only if any valuation equivalent to v

is archimedean. We say that a prime of F is *archimedean* if at least one (hence all) of its elements is archimedean. Next we establish the following classical result.

1.9. THEOREM. (Approximation theorem, Artin and Whaples(1945)). Let

$$v_1, v_2, \ldots, v_n$$

be nontrivial nonequivalent valuations of a field F. Then, for any

$$\lambda_1, \ldots, \lambda_n \in F$$

and any $\varepsilon > 0$, there exists $\lambda \in F$ such that

$$v_i(\lambda - \lambda_i) < \varepsilon \qquad i \in \{1,2,\ldots,n\}$$

Proof. We first observe that for any valuation v,

$$\lim \frac{a^r}{1+a^r} = \begin{cases} 0 & \text{if} \quad v(a) < 1 \\ 1 & \text{if} \quad v(a) > 1 \end{cases} \tag{6}$$

It will next be shown that there exists $c \in F$ such that

$$v_1(c) > 1, \; v_i(c) < 1 \quad \text{for} \quad i \in \{2,\ldots,n\} \tag{7}$$

Owing to Proposition 1.5, there exists $a \in F$ such that $v_1(a) > 1$, $v_2(a) \leqslant 1$ and $b \in F$ such that $v_1(b) \leqslant 1$, $v_2(b) > 1$. Then $c = ab^{-1}$ satisfies

$$v_1(c) > 1, \; v_2(c) < 1$$

which proves (7) for $n = 2$. We may therefore assume that $n > 2$ and use induction on n. By hypothesis, there exists $a \in F$ such that $v_1(a) > 1$, $v_i(a) < 1$, $3 \leqslant i \leqslant n$, and there exists $b \in F$ such that $v_1(b) > 1$, $v_2(b) < 1$. In case $v_2(a) \leqslant 1$, we put $c_r = a^r b$; now $v_1(c_r) > 1$, $v_2(c_r) < 1$ for all r, and if r is large enough, $v_i(c_r) < 1$ for $i \in \{3,\ldots,n\}$ also. In case $v_2(a) > 1$, we put $c_r = a^r b/(1+a^r)$; by (6), as $r \longrightarrow \infty, v_1(c_r) \longrightarrow v_1(b) > 1, v_2(c_r) \longrightarrow v_2(b) < 1$, $v_i(c_r) \longrightarrow 0$, $3 \leqslant i \leqslant n$, hence (7) is satisfied by $c = c_r$ when r is large enough.

By the foregoing, we may choose $c \in F$ satisfying (7). From (6) it follows that the sequence $c^r/(1+c^r)$ tends to 1 at v_1 and to 0 at $v_i, 2 \leqslant i \leqslant n$. Hence, given $\delta > 0$, we can find $u_i \in F$ such that $v_i(u_i - 1) < \delta$, $v_j(u_i) < \delta$ for $i,j \in \{1,\dots,n\}$, $i \neq j$. We now put $\lambda = \Sigma\lambda_i u_i$. Then

$$v_i(\lambda - \lambda_i) \leqslant v_i(\lambda_i(u_i - 1)) + \sum_{j \neq i} v_i(\lambda_j u_j) < \delta M,$$

where $M = \max_{i}\{\Sigma_j v_i(\lambda_j)\}$. Hence, by choosing $\delta < \varepsilon/M$, we obtain the required element λ. ■

Let us now return to non-archimedean valuations. It is usual and sometimes more convenient to write these valuations additively or, more precisely, to replace them by exponential valuations defined below.

Consider the set $\mathbb{R}_\infty$ obtained by adjoining to $\mathbb{R}$ symbol ∞. Then the linear order in $\mathbb{R}$ can be extended to that in $\mathbb{R}_\infty$ by putting

$$\alpha < \infty \quad \text{for all } \alpha \text{ in } \mathbb{R}$$

Moreover, $\mathbb{R}_\infty$ can be endowed with a commutative monoid structure by setting

$$\infty + \infty = \infty, \quad \alpha + \infty = \infty + \alpha = \infty \qquad (\alpha \in \mathbb{R})$$

We now define an *exponential valuation* ν of F as any mapping

$$\nu : F \longrightarrow \mathbb{R}_\infty$$

satisfying the following properties

(8) $\nu(x) = \infty$ if and only if $x = 0$

(9) $\nu(xy) = \nu(x) + \nu(y)$ for all $x,y \in F$

(10) $\nu(x+y) \geqslant \min\{\nu(x),\nu(y)\}$ for all $x,y \in F$

The image $\nu(F^*)$ of F^* is called the *value group* of ν. Two exponential valuations ν_1 and ν_2 of F are called *equivalent* if there exists $r > 0$ in $\mathbb{R}$ such that $\nu_1(x) = r\nu_2(x)$ for all $x \in F$. If the value group of ν is $a\mathbb{Z}$ for some $a > 0$ in $\mathbb{R}$, we say that ν is a *principal valuation*. For any $a > 1$ in $\mathbb{R}$, put

$$a^{-\infty} = 0 \quad \text{and} \quad -\log_a 0 = \infty$$

1.10. PROPOSITION. Let $a > 1$ be in $\mathbb{R}$ and, for any exponential valuation $\nu : F \longrightarrow \mathbb{R}_\infty$, define $v_\nu : F \to \mathbb{R}$ by

$$v_\nu(x) = a^{-\nu(x)} \qquad \text{for all } x \in F$$

The the map $\nu \longmapsto v_\nu$ is an equivalence preserving bijective correspondence between the exponential and non-archimedean valuations of F.

Proof. It is clear that v_ν is a non-archimedean valuation of F. If $v : F \longrightarrow \mathbb{R}$ is any non-archimedean valuation, define $\nu : F \longrightarrow \mathbb{R}_\infty$ by

$$\nu(x) = -\log_a v(x)$$

Then ν is an exponential valuation such that $v = v_\nu$. Because the given map is obviously injective, we conclude that it is in fact a bijection. The remaining assertion being obvious, the result follows. ■

For future use, we now classify all valuations of $\mathbb{Q}$.

1.11. PROPOSITION. If v is a nontrivial valuation of $\mathbb{Q}$, then either v is equivalent to the ordinary absolute value or it is equivalent to the p-adic valuation v_p for a unique rational prime p.

Proof. First assume that v is archimedean. Given integers $m,n > 1$, we may write

$$m = a_0 + a_1 n + \dots + a_r n^r \quad (a_i \in \mathbb{Z},\ a_r \neq 0,\ 0 \leqslant a_i < n)$$

Then $n^r \leqslant m$ and therefore $r \leqslant (\log m)/(\log n)$. Because $v(a_i) \leqslant a_i < n$, it follows that

$$\begin{aligned} v(m) &\leqslant v(a_0) + v(a_1)v(n) + \dots + v(a_r)v(n)^r \\ &\leqslant n(1+v(n) + \dots + v(n)^r) \\ &\leqslant n(r+1)\max\{1,v(n)^r\} \\ &\leqslant n(1+ (\log m)/(\log n))\ \max\{1,v(n)^{(\log m)/(\log n)}\} \end{aligned}$$

Now put $m = M^k$ with natural k, extract the k-th roots on both sides and let k tend to infinity with M fixed. In this way the inequality

$$v(M) \leqslant \max\{1,v(n)^{(\log M)/(\log n)}\} \tag{11}$$

follows, valid for every pair M,n of natural numbers larger than 1. Since v is archimedean, we may, by Proposition 1.7, choose M with $v(M) > 1$ and obtain the inequality

$$v(n)^{(\log M)/(\log n)} > 1$$

which shows that $v(n) > 1$ for all $n \geqslant 1$. Applying (ii), we conclude therefore that for all natural $M,n > 1$

$$v(M) \leqslant v(n)^{(\log M)/(\log n)}$$

and hence that

$$(\log v(M))/(\log M) \leqslant (\log v(n))/(\log n)$$

However we may interchange M and n to get the opposite inequality and finally we arrive at $(\log v(n))/(\log n) = c$, a constant independent of n, or equivalently, $v(n) = n^c$ for all natural $n > 1$. This obviously implies that $v(x) = |x|^c$ for all $x \in \mathbb{Q}$, proving that v is equivalent to the ordinary absolute value.

Let us now consider the case of non-archimedean v. Owing to Proposition 1.7, we have $v(n) \leqslant 1$ for all $n \in \mathbb{N}$. Put

$$X = \{n \in \mathbb{Z} | v(n) < 1\}$$

Then X is nonempty since otherwise $v(n) = 1$ for all $n \in \mathbb{Z}$ and v is trivial contrary to the assumed nontriviality of v. Furthermore, by (4), X is an ideal of $\mathbb{Z}$ which is obviously nonzero and, since $1 \notin X$, distinct from $\mathbb{Z}$. Therefore $X = p\mathbb{Z}$ with p being the smallest positive integer in X. But then p is a prime, because a factorization $p = rs$ with $1 < r,s < p$ would imply $1 > v(p) = v(r)v(s)$ and hence (say) $v(r) < 1$, which is impossible. If

$$v(p) = c$$

and $n = p^{\nu}b$ with $0 \neq n \in \mathbb{Z}$, $b \in \mathbb{Z}$, $(p,b) = 1$, then $v(n) = c^{\nu}$. Applying Lemma 1.1(i), we conclude that v is equivalent to the p-adic valuation. Since p is uniquely determined by the equivalence class of v, the result follows. ■

Let v be a valuation of F. We say that v is *discrete* if the value group of v is an infinite cyclic group. The following observation is an easy consequence of Propositions 1.7 and 1.11.

1.12. PROPOSITION. Every discrete valuation is non-archimedean.

Proof. Suppose that v is an archimedean valuation of F. Then, by Corollary 1.8, $\text{char}F = 0$ and hence $\mathbb{Q} \subseteq F$. The restriction of v to $\mathbb{Q}$ is archimedean by Proposition 1.7. Therefore, by Proposition 1.11, v is equivalent to $|x|$ and hence nondiscrete. ∎

The proof of the next result requires the following useful observation.

1.13. LEMMA. Let G be a nonzero additive subgroup of $\mathbb{R}$. Then G is either dense in $\mathbb{R}$ or $G = a\mathbb{Z}$ for some $a > 0$ (in particular, G is discrete and isomorphic to $\mathbb{Z}$).

Proof. First assume that G has a point of accumulation. Then, for any real $\varepsilon > 0$, there exists $x \neq 0$ in G such that $0 < x < \varepsilon$. The subset $\{nx | n \in \mathbb{Z}\}$ of G is therefore a set of equidistant points on the real line, each at a positive distance less than ε from the next. Hence G is dense in $\mathbb{R}$.

Now assume that G has no points of accumulation. Then there exists a smallest positive value a in G. Dividing by a, if necessary, we may assume that $a = 1$ in which case $\mathbb{Z} \subseteq G$. On the other hand, if x is an arbitrary element of G, then we find a unique $n \in \mathbb{Z}$ such that $x = n + \rho$, where

$$0 \leqslant \rho < 1$$

Thus $\rho = x - n \in G$ and so $\rho = 0$, by our choice of a. Consequently

$$x = n \in \mathbb{Z}$$

and the result follows. ∎

The equivalence of (i) and (iii) in the result below justifies the adjective "discrete".

1.14. PROPOSITION. Let v be a nontrivial valuation of F. Then the following conditions are equivalent:

(i) v is discrete

(ii) the corresponding exponential valuation is principal

(iii) $\{\log v(x) | x \in F^*\}$ is a discrete subset of $\mathbb{R}$

(iv) $\{\log v(x) | x \in F^*\} = a\mathbb{Z}$ for some $a > 0$ in $\mathbb{R}$

(v) v is equivalent to a valuation w such that $\{\log w(x) | x \in F^*\} = \mathbb{Z}$

Proof. (i) ⇒ (ii): By hypothesis, $\{v(x) | x \in F^*\} = \{a^n | n \in \mathbb{Z}\}$ for some $1 \neq a > 0$ in $\mathbb{R}$. Therefore

$$\{-\log v(x) \mid x \in F^*\} = \{\log v(x) \mid x \in F^*\} = (\log a)\mathbb{Z}$$

as required.

(ii) $\Rightarrow$ (iii): Obvious

(iii) $\Rightarrow$ (iv): Follows from Lemma 1.13

(iv) $\Rightarrow$ (v): Define $w(x) = v(x)^{1/a}$ for all $x \in F$. Then w is equivalent to v and

$$\{\log w(x) \mid x \in F^*\} = (1/a)\{\log v(x) \mid x \in F^*\} = \mathbb{Z}$$

as required.

(v) $\Rightarrow$ (i): There exists $r > 0$ in $\mathbb{R}$ such that $v(x) = w(x)^r$ for all $x \in F$. Thus

$$\{\log v(x) \mid x \in F^*\} = r\mathbb{Z}$$

which shows that the value group of v is the infinite cyclic group generated by e^r. ∎

Let R be a commutative ring. We say that R is a *discrete valuation ring* if it is a principal ideal domain that has a unique nonzero prime ideal.

Let P be the nonzero prime ideal of a discrete valuation ring R. Then $P = pR$ for a prime p of R determined uniquely up to multiplication by a unit. The nonzero ideals of R are of the form p^nR, $n \geqslant 0$. Therefore, if $x \neq 0$ is any element of R, one can write

$$x = p^n u \quad \text{with} \quad n \geqslant 0 \quad \text{and} \quad u \in U(R)$$

If F is the quotient field of R and $x = a/b$ is any element of F^*, one can again write x in the form $p^n u$, with $n \in \mathbb{Z}$ this time. Setting $\nu(x) = n$ and $\nu(0) = \infty$, one immediately verifies that the map $\nu : F \longrightarrow \mathbb{R}_\infty$ is a principal valuation. Any corresponding discrete valuation of F given by Proposition 1.10 is called a *valuation of F associated with R*.

Let F be a field. A subring R of F is said to be a *valuation ring* if for any $x \in F - R$ we have $x^{-1} \in R$. Of course, the latter implies that F is the quotient field of R.

1.15. LEMMA. Let R be a valuation ring in F. If M denotes the set of non-units in R, then M is the unique maximal ideal of R.

Proof. Given $x,y \in R$, if $xy \in U(R)$, then $x,y \in U(R)$. In particular, this says that M is closed under multiplication. Now assume that $x,y \in M$. If $x = 0$ or $y = 0$, then certainly $x - y \in M$, so we may assume that $x \neq 0, y \neq 0$. Then, by definition, either $x^{-1}y \in R$ or $y^{-1}x \in R$, and since

$$x - y = x(1-x^{-1}y) = y(y^{-1}x-1)$$

we conclude in either case that $x - y$ is a nonunit of R. Finally, any proper ideal of R consists of nonunits and is therefore contained in M. ■

1.16. PROPOSITION. Let F be an arbitrary field and let v be a non-archimedean valuation of F. Define R_v and P_v by

$$R_v = \{x \in F | v(x) \leq 1\}, \quad P_v = \{x \in F | v(x) < 1\}$$

(i) R_v is a valuation ring in F and P_v is a unique maximal ideal of R_v. In particular, $J(R_v) = P_v$ and $U(R_v) = \{x \in F | v(x) = 1\}$

(ii) If v is a discrete valuation, then R_v is a discrete valuation ring, P_v is a unique nonzero prime ideal of R_v and v coincides with a valuation on F associated with R_v.

Proof. (i) That R_v is a subring of F and P_v is an ideal of R_v is a consequence of (1), (2), (4) and Lemma 1.1(ii). Let $0 \neq x \in F$. Since

$$v(x^{-1}) = v(x)^{-1},$$

either $v(x) \leq 1$ or $v(x^{-1}) \leq 1$. Hence either $x \in R_v$ or $x^{-1} \in R_v$, proving that R_v is a valuation ring in F. Since $P_v \neq R_v$, it suffices to verify that any nonunit x of R_v belongs to P_v. But if $v(x) = 1$, then $v(x^{-1}) = 1$ and so $x \in U(R_v)$, a contradiction.

(ii) Choose $p \in F$ with $v(p) < 1$ and such that $v(F^*) = \langle v(p) \rangle$. We claim that the $p^n R_v$, $n \geq 0$, are all ideals of R_v; if sustained, it will follow that R_v is a discrete valuation ring with P_v being a unique nonzero prime ideal of R_v. To this end, note that given $x \in F^*$, we have $v(x) = v(p)^n = v(p^n)$ for some $n \in \mathbb{Z}$ and so $x = p^n u$ for some $u \in U(R_v)$, by virtue of (i). Because $v(p) < 1$, $x \in R_v$ if and only if $n \geq 0$. Given an ideal I of R_v, let m be the smallest nonnegative integer such that $p^m \in I$. Then $x \in I$ implies $n \geq m$

and so $x \in p^m R_v$, proving that $I = p^m R_v$.

To prove the final assertion, put $a = v(p)^{-1}$ so that $a > 1$ is in $\mathbb{R}$. If ν is the principal valuation associated with R_v, then $\nu(x) = n$ and

$$v_\nu(x) = a^{-\nu(x)} = v(p)^n = v(x),$$

as required.

We shall refer to R_v, P_v and R_v/P_v as the *valuation ring, maximal ideal* and *residue class field* of v, respectively. Owing to Proposition 1.5, both R_v and P_v are uniquely determined by the equivalence class of v, i.e. by the non-archimedean prime P of F containing v. It is therefore customary to refer to R_v, P_v and R_v/P_v as the *valuation ring, maximal ideal* and *residue class field* of P, respectively. We close by providing the following characterization of discrete valuation rings.

1.17. PROPOSITION. Let R be an integral domain with quotient field F. Then R is a discrete valuation ring if and only if R is the valuation ring of a discrete valuation of F.

Proof. If R is a discrete valuation ring, then R is the valuation ring of the valuation of F associated with R. The converse is true, by virtue of Proposition 1.16. ∎

2. VALUATION RINGS AND PLACES

Let F be a field and let R be a valuation ring on F. We know, from Lemma 1.15, that $J(R)$ is the unique maximal ideal of R. In particular, we have a natural homomorphism of R onto the field $K = R/J(R)$. It is often convenient to extend the map $R \longrightarrow K$ to all of F by defining

$$\psi_R(x) = \begin{cases} x + J(R) & \text{if } x \in R \\ \infty & \text{if } x \in F - R \end{cases} \tag{1}$$

where ∞ is just some additional point not contained in F. The constructed map $\psi : F \longrightarrow K \cup \{\infty\}$ is called a *place*. More formally, a place is a map

$$\psi : F \longrightarrow K \cup \{\infty\}$$

such that $R = \psi^{-1}(K)$ is a subring of F with $\psi : R \longrightarrow K$ a homomorphism into the field K and such that $\psi(x) = \infty$ if and only if $\psi(x^{-1}) = 0$. If ψ is a place and $\psi(x) \neq \infty$ for some $x \in F$, then ψ is said to be *finite* on x. Thus ψ is finite on x if and only if $x \in \psi^{-1}(K)$. The following observation shows that there is, in essence, a natural bijection between valuation rings in F and place maps.

2.1. LEMMA. (i) If R is a valuation ring in F, then the map ψ_R defined in (1) is a place

(ii) If $\psi : F \longrightarrow K \cup \{\infty\}$ is a place and R is the set of all elements of F on which ψ is finite, then R is a valuation ring in F. Moreover, $\psi = f\psi_R$ where ψ_R is defined by (1) and $f : R/J(R) \cup \{\infty\} \longrightarrow K \cup \{\infty\}$ is induced by an isomorphism of $R/J(R)$ into K.

Proof. (i) Only the last property in the definition of a place needs verification. By definition of ψ_R, $\psi_R(x) = \infty$ if and only if $x \notin R$, and $\psi_R(x) = 0$ if and only if $x \in J(R)$. But R is a valuation ring, hence $0 \neq x \in J(R)$ if and only if $x^{-1} \notin R$. Hence $\psi_R(x) = \infty$ if and only if $\psi_R(x^{-1}) = 0$, as required.
(ii) Since $R = \psi^{-1}(K)$, R is necessarily a subring of F, by definition of a place. Moreover, if $x \notin R$, then $\psi(x) = \infty$ and $x^{-1} \in R$. Thus R is a valuation ring in F. Finally, $\psi(x) = 0$ if and only if $x^{-1} \notin R$ so that $\psi^{-1}(0) = J(R)$ is the set of nonunits of R and the result follows. ∎

In what follows, we shall refer to the valuation ring R in Lemma 2.1(ii), as the *valuation ring corresponding to the place* ψ

2.2. THEOREM. (Extension theorem for places). Let S be a subring of a field F, let K be an algebraically closed field and let $f : S \longrightarrow K$ be a homomorphism. Then there exists a place $\psi : F \longrightarrow K \cup \{\infty\}$ such that the corresponding valuation ring R satisfies $R \supseteq S$ and $\psi|S = f|S$.

Proof. Consider the set T of all ordered pairs (R,ϕ) such that

$$F \supseteq R \supseteq S$$

and $\phi : R \longrightarrow K$ is a homomorphism extending f. If $(R_1,\phi_1),(R_2,\phi_2) \in T$, we write $(R_1,\phi_1) \leqslant (R_2,\phi_2)$ if and only if $R_1 \subseteq R_2$ and $\phi_1 = \phi_2|R_1$. Since

$(S,f) \in T$, we may choose, by Zorn's lemma, a maximal member, say (R,ϕ), in T under the given ordering. Our aim is to show that R is a valuation ring and that $\text{Ker}\phi = J(R)$ is a unique maximal ideal of R. Once this is accomplished, the required map ψ will be given by $\psi(x) = \phi(x)$ if $x \in R$ and $\psi(x) = \infty$ if $x \in F - R$.

For the sake of clarity, we divide the rest of the proof into three steps.

Step 1. *Here we show that* $\text{Ker}\phi$ *is a unique maximal ideal of* R.

Put $R^* = \{r/s \in F | r \in R,\ s \in R - \text{Ker}\phi\}$ and define $\phi^* : R^* \longrightarrow K$ by

$$\phi^*(r/s) = \phi(r)/\phi(s)$$

Then one immediately verifies that R^* is a ring containing R and that ϕ^* is a well-defined homomorphism extending ϕ. Thus, the maximality of (R,ϕ) ensures that $R = R^*$; in particular, if $r \in R - \text{Ker}\phi$, then $r^{-1} \in R$. Thus $\text{Ker}\phi$ consists of all nonunits of R and therefore $\text{Ker}\phi = J(R)$ is a unique maximal ideal of R.

Step 2. *It will be shown here that if* $\lambda \in F - R$, *then there exists a polynomial* $g(X) \in J(R)[X]$ *with* $g(\lambda) = 1$

Let I be the kernel of the natural homomorphism $R[X] \longrightarrow R[\lambda]$ given by

$$X \longmapsto \lambda$$

Then $I = \{f(X) \in R[X] | f(\lambda) = 0\}$ and ϕ extends naturally to the epimorphism $\bar{\phi} : R[X] \longrightarrow \phi(R)[X]$. We claim that the ideal $\bar{\phi}(I)$ of $\phi(R)[X]$ coincides with $\phi(R)[X]$; if sustained, it will follow that $1 \in \bar{\phi}(I)$, and hence there exists $g(X) \in J(R)[X] = \text{Ker}\bar{\phi}$ and $h(X) \in I$ with $1 = g(X) + h(X)$. But then setting $X = \lambda$ yields $g(\lambda) = 1$, as required.

Assume by way of contradiction that $\bar{\phi}(I) \neq \phi(R)[X]$. Then, since $\phi(R)[X]$ is a principal ideal domain, we have $\bar{\phi}(I) = (h(X))$ for some polynomial $h(X)$ of positive degree. Since K is algebraically closed, there exists $\mu \in K$ with $h(\mu) = 0$ and hence the combined homomorphism

$$R[X] \longrightarrow \phi(R)[X] \longrightarrow K$$

given by $\bar{\phi}$ followed by evaluation at $\mu \in K$ extends ϕ and has I in its kernel. But this maps factors through $R[X]/I \cong R[\lambda]$, and since $\lambda \notin R$, this contradicts the maximality of (R,ϕ).

Step 3. Here we complete the proof by showing that R is a valuation ring. Assume by way of contradiction that there exists $\lambda \in F$ with $\lambda \notin R$ and $\lambda^{-1} \notin R$. Then, by Step 2, there exist polynomials $g(X)$, $h(X) \in J(R)[X]$ with

$$g(\lambda) = h(\lambda^{-1}) = 1$$

We may assume that $g(X)$, $h(X)$ have been chosen with smallest possible degrees s and t, respectively, say $t \leqslant s$. If $h(X) = \sum_{i=0}^{t} a_i X^i$, then $1 = \sum_{i=0}^{t} a_i \lambda^{-i}$ so that

$$(1-a_0)\lambda^t = \sum_{i=0}^{t-1} a_{t-i}\lambda^i$$

Therefore, since $1 - a_0 \in R - J(R)$ is a unit in R, we can write λ^t as

$$\lambda^t = \sum_{i=0}^{t-1} (1-a_0)^{-1} a_{t-i}\lambda^i$$

a polynomial over $J(R)$ in λ of degree less than t. By using this expression for λ^t in the equation $1 = g(\lambda)$, we may eliminate all terms of degree $\geqslant t$. However, by the minimality of s, we cannot lower the degree of $g(X)$. Thus we must have $s < t$, a contradiction. ■

The rest of this section is devoted to exhibiting a number of applications of the preceding result.

2.3. THEOREM. Let S be a subring of a field F and let $\lambda_1,\lambda_2,\ldots,\lambda_n$ be finitely many elements of F. Then there is a nonzero element s_0 in S such that if $f : S \longrightarrow K$ is any homomorphism into an algebraically closed field K with $f(s_0) \neq 0$, then f extends to a place

$$\psi : F \longrightarrow K \cup \{\infty\}$$

such that the corresponding valuation ring R contains $S[\lambda_1,\lambda_2,\ldots,\lambda_n]$.

Proof. We have $F_0 \subseteq F_0(\lambda_1,\ldots,\lambda_n) \subseteq F$, where F_0 is the quotient field of S. We may harmlessly assume that $\lambda_1,\ldots,\lambda_k$, $k \leqslant n$ is a transcendency basis of $F_0(\lambda_1,\ldots,\lambda_n)$ over F_0. Hence $S_1 = S[\lambda_1,\ldots,\lambda_k] \cong S[X_1,\ldots,X_k]$ and

$$F_1 = F_0(\lambda_1,\ldots,\lambda_n)$$

is the quotient field of S_1. Since $\lambda_{k+1},\ldots,\lambda_n$ are algebraic over F_1, there exists a polynomial

$$f(X) = \sum_{i=0}^{t} a_i X^i \in S_1[X] \qquad (a_t \neq 0)$$

with $f(\lambda_i) = 0$, $1 \leqslant i \leqslant n$. Since $a_t \in S[\lambda_1,\ldots,\lambda_k]$ is a nonzero polynomial, we can choose $s_0 \in S$ to be one of its nonzero coefficients.

To prove that s_0 satisfies the required property, let $f : S \longrightarrow K$ be any homomorphism into an algebraically closed field K with $f(s_0) \neq 0$. We may extend f to a homomorphism

$$\bar{f} : S[\lambda_1,\ldots,\lambda_k] \longrightarrow K[X_1,\ldots,X_k]$$

which sends λ_i to X_i. Our choice of s_0 and the assumption that $f(s_0) \neq 0$ ensure that $\bar{f}(a_t) \neq 0$. Therefore, since K is an infinite field, there exist $\mu_1,\mu_2,\ldots,\mu_k$ in K such that $\bar{f}(a_t)(\mu_1,\ldots,\mu_k) \neq 0$. Consider the map

$$\begin{cases} S[\lambda_1,\ldots,\lambda_k] \xrightarrow{\alpha} K \\ g(\lambda_1,\ldots,\lambda_k) \longrightarrow \bar{f}g(\mu_1,\ldots,\mu_k) \end{cases}$$

Then α is a homomorphism extending f and $\alpha(a_t) \neq 0$. Applying Theorem 2.2 to the map $\alpha : S[\lambda_1,\ldots,\lambda_k] \longrightarrow K$, we may find a place $\psi : F \longrightarrow K \cup \{\infty\}$ which extends α and such that the corresponding valuation ring R contains

$$S[\lambda_1,\ldots,\lambda_k]$$

By the foregoing, we are left to verify that $R \supseteq S[\lambda_1,\ldots,\lambda_n]$. Assume by way of contradiction that $\lambda_j \notin R$ for some $j > k$. Then $\lambda_j^{-1} \in R$ and

$$\psi(\lambda_j^{-1}) = 0,$$

by the definition of R. Furthermore, $\lambda_j^{-t} f(\lambda_j) = 0$ and so

$$a_t + \sum_{i=1}^{t} a_{k-i}\lambda_j^{-i} = 0$$

Since all the terms in the above sum are in R, we find

$$0 = \psi(a_t) + \sum_{i=1}^{t} \psi(a_{k-i})\psi(\lambda_j^{-1})^i = \alpha(a_t)$$

since ψ extends α and $\psi(\lambda_j^{-1}) = 0$. This contradiction completes the proof of the theorem. ∎

2.4. COROLLARY. Let E/F be a field extension and let $\bar{F}$ be an algebraic closure of F. Then, for any nonzero elements $\lambda_1, \lambda_2, \dots, \lambda_n$ in E, the natural embedding $F \longrightarrow \bar{F}$ extends to a place $\psi : E \longrightarrow \bar{F} \cup \{\infty\}$ such that $\psi(\lambda_i) \neq 0$ and $\psi(\lambda_i) \neq \infty$ for all $i \in \{1,2,\dots,n\}$.

Proof. We apply Theorem 2.3 with $S = F$ and $F = E$, and with

$$\{\lambda_1, \lambda_2, \dots, \lambda_n, \lambda_1^{-1}, \lambda_2^{-1}, \dots, \lambda_n^{-1}\}$$

as the finite set of elements of E. Then we may choose a nonzero element s_0 in S such that the natural embedding $F \longrightarrow \bar{F}$ extends to a place

$$\psi : E \longrightarrow \bar{F} \cup \{\infty\}$$

such that $\lambda_i, \lambda_i^{-1} \in R$ for all i, where R is the corresponding valuation ring in E. Thus $\psi(\lambda_i) \neq 0$ and $\psi(\lambda_i) \neq \infty$, as required. ■

2.5. COROLLARY. Let F be a field of characteristic 0 and let $\lambda_1, \dots, \lambda_n$ be finitely many nonzero elements of F. Then there is an infinite set P of rational primes such that for any $p \in P$, there exists a place

$$\psi_p : F \longrightarrow \bar{\mathbb{F}}_p \cup \{\infty\}$$

with $\psi_p(\lambda_i) \neq 0$ and $\psi_p(\lambda_i) \neq \infty$ for all $i \in \{1,2,\dots,n\}$. Here $\bar{\mathbb{F}}_p$ denotes the algebraic closure of $\mathbb{F}_p$.

Proof. Since $\mathrm{char} F = 0$, we may apply Theorem 2.3 with $S = \mathbb{Z}$ and with

$$\{\lambda_1, \lambda_2, \dots, \lambda_n, \lambda_1^{-1}, \lambda_2^{-1}, \dots, \lambda_n^{-1}\}$$

as the finite set of elements in F. Let s_0 be chosen as in Theorem 2.3 and let P denote the set of primes p not dividing s_0. Then, for any given $p \in P$, the map $f_p : \mathbb{Z} \longrightarrow \mathbb{Z}/p\mathbb{Z} \subseteq \bar{\mathbb{F}}_p$ satisfies $f(s_0) \neq 0$. Hence, by Theorem 2.3, f_p extends to a place $\psi_p : F \longrightarrow \bar{\mathbb{F}}_p \cup \{\infty\}$ such that $\lambda_i, \lambda_i^{-1} \in R_p$ for all i, where R_p is the valuation ring in F corresponding to ψ_p. Thus

$$\psi_p(\lambda_i) \neq 0$$

and $\psi_p(\lambda_i) \neq \infty$ for all $i \in \{1,2,\dots,n\}$. ■

Our final application of Theorem 2.3 requires the following result, whose proof will be omitted.

2.6. PROPOSITION. Let $f(X) \in \mathbb{Z}[X]$ be a monic irreducible polynomial of degree > 1. Then there exist infinitely many primes p such that the polynomial

$$f(X) \bmod p$$

does not have all its roots in $\mathbb{F}_p$.

Proof. See Janusz(1973, p.139)). ∎

2.7. THEOREM. Let F be a field of characteristic 0 and let $\lambda_0, \lambda_1, \ldots, \lambda_n$ be finitely many elements of F with $\lambda_0 \notin \mathbb{Q}$. Then there is an infinite set P of rational primes and a family of valuation rings $\{R_p | p \in P\}$ in F such that for all $p \in P$, $R_p/J(R_p)$ is a field of characteristic p and

$$\lambda_0, \lambda_1, \ldots, \lambda_n \in R_p \quad \text{and} \quad \lambda_0 + J(R_p) \notin \mathbb{F}_p$$

(here $\mathbb{F}_p$ is the prime subfield of $R_p/J(R_p)$).

Proof. All we have to do is to exhibit an infinite set P of rational primes and a place $\psi_p : F \longrightarrow \bar{\mathbb{F}}_p \cup \{\infty\}$, for each $p \in P$, such that $\psi_p(\lambda_i) \neq \infty$ for all i and $\psi_p(\lambda_0) \notin \mathbb{F}_p$. We apply Theorem 2.3 with $S = \mathbb{Z}[\lambda_0]$ and with $\{\lambda_1, \lambda_2, \ldots, \lambda_n\}$ as the finite set of elements of F. Let s_0 be chosen as in Theorem 2.3. We divide the proof into two steps.

Step 1. Here we treat the case where λ_0 *is transcendental over* $\mathbb{Q}$

Then $S = \mathbb{Z}[\lambda_0]$ is identifiable with a polynomial ring over $\mathbb{Z}$ in the variable λ_0 and so $s_0 = s_0(\lambda_0)$ is a nonzero polynomial. For each prime p, there is a natural homomorphism

$$\begin{cases} \mathbb{Z}[\lambda_0] \xrightarrow{f_p} \mathbb{F}_p[X] \\ \lambda_0 \longmapsto X \end{cases}$$

It is clear that $f_p(s_0) \neq 0$ for all but finitely many primes p. Since $\bar{\mathbb{F}}_p$ is an infinite field, if $f_p(s_0) \neq 0$, then there is an element $\mu \in \bar{\mathbb{F}}_p - \mathbb{F}_p$ with $f_p(s_0)(\mu) \neq 0$. But then the map f_p followed by evaluation at $X = \mu$ yields a homomorphism of S into $\bar{\mathbb{F}}_p$ with s_0 not mapped to zero. The required assertion is therefore a consequence of Theorem 2.3.

Step 2. Here we assume that λ_0 *is algebraic over* $\mathbb{Q}$.

We may choose a nonzero $n \in \mathbb{Z}$ such that $\lambda = n\lambda_0$ is an algebraic integer, and we denote by $f(X)$ its minimal monic polynomial over $\mathbb{Z}$. Because $\lambda_0 \notin \mathbb{Q}$, $f(X)$ is irreducible of degree > 1, and moreover, $\mathbb{Z}[\lambda] \cong \mathbb{Z}[X]/(f(X))$. Since $s_0 \in \mathbb{Z}[\lambda_0]$ and λ_0 is algebraic over $\mathbb{Q}$, s_0 is also algebraic over $\mathbb{Q}$, say

$$a_t s_0^t + a_{t-1} s_0^{t-1} + \dots + a_0 = 0 \qquad (a_i \in \mathbb{Z},\ a_0 \neq 0)$$

Owing to Proposition 2.6, we may choose infinitely many primes p such that

$$f(X) \bmod p$$

does not have all its roots in $\mathbb{F}_p$ and such that $p \nmid a_0 n$. For any such p, we may therefore find a nonlinear irreducible factor $g_p(X) \in \mathbb{F}_p[X]$ of $f(X) \bmod p$ Consider the composite map

$$\mathbb{Z}[X] \longrightarrow \mathbb{F}_p[X] \longrightarrow \mathbb{F}_p[X]/(g(X))$$

Since $f(X)$ is mapped to zero, this map factors through $\mathbb{Z}[X]/(f(X)) \cong \mathbb{Z}[\lambda]$. This gives a homomorphism ϕ_p of $\mathbb{Z}[\lambda]$ onto a finite field of characteristic p with $\phi_p(\lambda) \notin \mathbb{F}_p$, since $g_p(X)$ has no roots in $\mathbb{F}_p$. Now consider the ring

$$V_p = \{m^{-1}x \mid x \in \mathbb{Z}[\lambda],\ m \in \mathbb{Z},\ p \nmid m\} \subseteq F$$

By setting $\phi_p(m^{-1}x) = \phi_p(m)^{-1}\phi_p(x)$, we may extend ϕ_p to V_p. Since $p \nmid n$ we have $\lambda_0 \in V_p$ and therefore $S = \mathbb{Z}[\lambda_0] \subseteq V_p$. Thus there exists a homomorphism $\phi_p : S \longrightarrow \bar{\mathbb{F}}_p$ with $\phi_p(\lambda_0) = \phi_p(n)^{-1}\phi_p(\lambda) \notin \mathbb{F}_p$. Furthermore, by applying ϕ_p to the equation satisfied by s_0, we derive

$$\phi_p(s_0)\{\sum_{i=1}^{t} \phi_p(a_i)\phi_p(s_0^{i-1}) + \phi_p(a_0) = 0$$

Since $p \nmid a_0$, $\phi_p(a_0) \neq 0$ and so $\phi_p(s_0) \neq 0$. Now apply Theorem 2.3. ■

The following application of Theorem 2.7 to group algebras is often useful,

2.8. COROLLARY. Let F be a field of characteristic 0, let G be an arbitrary group and let $x = \Sigma x_g g \in FG$. If $x_g \notin \mathbb{Q}$, then there is an infinite set P of rational primes and a family of valuation rings $\{R_p \mid p \in P\}$ in F such that for all $p \in P$

(a) $R_p/J(R_p)$ is a field of characteristic p

(b) $x \in R_pG$

(c) The image $\bar{x} = \Sigma \bar{x}_g g$ of x in $(R_p/J(R_p))G$ has the property that $\bar{x}_g$ is not contained in the prime subfield of $R_p/J(R_p)$.

Proof. Apply Theorem 2.7 to the case where $\{\lambda_0,\lambda_1,\ldots,\lambda_n\}$ is the set of all nonzero coefficients of x and $\lambda_0 = x_g$. ■

3. DEDEKIND DOMAINS

An integral domain R is called a *Dedekind domain* if it is noetherian, integrally closed and every nonzero prime ideal of R is maximal. The ring $\mathbb{Z}$ and, more generally, any principal ideal domain is a Dedekind domain.

Let R be an integral domain and let F be the quotient field of R. By a *fractional ideal* of R, we understand any nonzero R-submodule I of F for which there exists a nonzero d in R such that $d\cdot I \subseteq R$. Expressed otherwise this means that the elements of I have a "common denominator" $d \in R$. The ordinary nonzero ideals of R are fractional ideals (with $d = 1$). Any nonzero finitely generated R-submodule I of F is a fractional ideal. This follows from the fact that, if $I = Rx_1 + \ldots + Rx_n$ for some $x_i = r_i/d_i$ $(r_i, d_i \in R)$, then $d = d_1 d_2 \ldots d_n$ is a common denominator for the elements of I. Conversely, if R is noetherian, every fractional ideal I is a finitely generated R-submodule of F. This is so since $I \subseteq d^{-1}R$ for some $0 \neq d \in R$ and since the R-modules $d^{-1}R$ and R are isomorphic.

We define the product $I_1 \cdot I_2$ of two fractional ideals I_1 and I_2 as the set of all finite sums $\Sigma x_i y_i$, where $x_i \in I_1$ and $y_i \in I_2$. If d_1 and d_2 are the common denominators for I_1 and I_2, then $d_1 d_2$ is a common denominator for the R-submodule $I_1 \cdot I_2$ of F. Hence $I_1 \cdot I_2$ is a fractional ideal, and so the fractional ideals of R constitute a multiplicative monoid whose identity element is R. This monoid will be denoted by $I(R)$. The invertible elements of $I(R)$, also called *invertible fractional ideals* of R, constitute a group. Note that every *principal fractional ideal*, i.e. an ideal of the form Rx with $0 \neq x \in F$ is invertible. Moreover, the principal fractional ideals of R form a group which we denote by $P(R)$. It will be shown below that if R is a Dedekind domain then $I(R)$ is a free abelian group with the collection of nonzero prime ideals of R as free generators. The factor group

$$C(R) = I(R)/P(R)$$

is called the *class group* of a Dedekind domain R. Thus $C(R) = 1$ if and only if each nonzero prime ideal of R is principal. The following simple observation ties together Dedekind domains, UFD's and PID's.

3.1. PROPOSITION. The following conditions are equivalent:

(i) R is a Dedekind domain and $C(R) = 1$

(ii) R is a Dedekind domain and a UFD

(iii) R is a PID

Proof. (i) ⇒ (ii): Let $r \in R$ be a nonzero nonunit. Then Rr is a nontrivial ideal of R and so we may write $Rr = (Rp_1) \dots (Rp_n)$ for some prime elements $p_1, \dots, p_n$ of R. Hence $r = p_1 p_2 \dots p_n u$ for some unit u of R, proving that r is a finite product of primes. Thus, by Proposition 1.2.13, R is a UFD.

(ii) It is enough to prove that every nonzero prime ideal P of R is principal. Let $a \in P$, $a \neq 0$. Since a is a nonzero nonunit, it is a product of prime elements, $a = p_1 p_2 \dots p_r$. Because $a \in P$, it follows that $p_i \in P$ for some i, hence $Rp_i \subseteq P$. But Rp_i is a prime ideal of R, hence it must be maximal, since R is a Dedekind domain. Thus $P = Rp_i$ is a principal ideal.

(iii) ⇒ (i): If R is a PID, then R is a Dedekind domain in which any nonzero prime ideal is principal. Hence $C(R) = 1$ as required. ■

The following lemma will enable us to provide a large class of Dedekind domains.

3.2. LEMMA. Let $R \subseteq S$ be commutative rings

(i) If P is a prime ideal of S, then $R/(P \cap R)$ is identifiable with a subring of S/P and, in particular, $P \cap R$ is a prime ideal of R.

(ii) If S is integral over R and $P_1 \subseteq P_2$ are prime ideals of S such that $P_1 \cap R = P_2 \cap R$, then $P_1 = P_2$

(iii) If S is integral over R and P a prime ideal of S such that $P \cap R$ is a maximal ideal of R, then P is a maximal ideal of S.

Proof. (i) Note that $P \cap R$ is the kernel of the composite map

$$R \longrightarrow S \longrightarrow S/P$$

Hence $R/(P \cap R)$ admits an injective homomorphism into S/P.

(ii) Passing to the quotient by P_1, we may assume that $P_1 = 0$. If $P_1 \neq P_2$, then there is a nonzero $x \in P_2$. Let

$$x^n + r_{n-1}x^{n-1} + \ldots + r_0 = 0 \qquad (r_i \in R)$$

be its minimal equation over R. Then $r_0 \neq 0$ and r_0 belongs to the ideal of S generated by x. Thus $P_1 \cap R = P_2 \cap R$, a contradiction

(iii) Direct consequence of (ii). ■

3.3. THEOREM. Let R be a Dedekind domain with quotient field F, let E/F be a finite separable field extension and let S be the integral closure of R in E. Then

(i) S is a Dedekind domain and a finitely generated R-module

(ii) For any nonzero prime ideal P of S, the field S/P is a finite extension of $R/(P \cap R)$.

Proof. (i) By Corollary 2.6.16, S is noetherian and a finitely generated R-module. If $x \in E$ is integral over S, then x is integral over R by Proposition 1.3.7. Hence $x \in S$, proving that S is integrally closed. The fact that each nonzero prime ideal of S is maximal being a consequence of Lemma 3.2(iii), (i) follows.

(ii) Let P be any nonzero prime ideal of S. By (i), S/P is a field and, by Lemma 3.2(i), $R/(P \cap R)$ is a subfield. Since S is a finitely generated R-module, S/P is a finite extension of $R/(P \cap R)$, as required. ■

Remark. It can be shown that S is a Dedekind domain without the assumption that E/F is separable (see Bourbaki (1972))

To exhibit some further properties of Dedekind domains, we first provide some preliminary observations.

3.4. LEMMA. Let R be a Dedekind domain with quotient field F.

(i) Every nonzero ideal of R contains a product of nonzero prime ideals

(ii) For any nonzero ideal I of R, $I^{-1} = \{\lambda \in F | I\lambda \subseteq R\}$ is a fractional ideal of R with $I^{-1} \supseteq R$ and $I^{-1}I = R$. Furthermore, if $I \neq R$ then $I^{-1} \supset R$ and if $I_1 \supseteq I$, then $I_1^{-1} \subseteq I^{-1}$.

(iii) If $P_1 \ldots P_n \subseteq Q_1 \ldots Q_m$, where the $\{P_i\}$ and $\{Q_j\}$ are nonzero prime ideals of R, then $m \leqslant n$ and, after rearrangement of the $\{Q_j\}$ if necessary,

$$Q_1 = P_1, \ldots, Q_m = P_m$$

Proof. (i) Let S be the set of nonzero ideals of R for which (i) does not hold. If S were empty, then, since R is noetherian, S would contain a maximal element. Obviously $I \neq R$, I is not prime and every ideal of R containing I properly must contain a product of nonzero prime ideals. Since I is not prime, there exist $\alpha, \beta \in R$ such that $\alpha, \beta \notin I$ but $\alpha\beta \in I$. The ideals $I + \alpha R$, $I + \beta R$ both properly contain I and hence contain products of nonzero prime ideals. But

$$(I+\alpha R)(I+\beta R) \subseteq I \cdot I + \alpha I + \beta I + \alpha\beta R \subseteq I,$$

so I also contains a product of nonzero prime ideals, a contradiction.

(ii) If $0 \neq x \in I$, then $\lambda x \in R$ implies $\lambda \in Rx^{-1}$. Therefore I^{-1} is an R-submodule of the finitely generated R-module Rx^{-1}, and I^{-1} is also a finitely generated R-module. Hence I^{-1} is a fractional ideal of R, and clearly

$$I^{-1} \supseteq R$$

It is also clear that $I_1 \supseteq I$ implies $I_1^{-1} \subseteq I^{-1}$.

Assume that $I \neq R$ and choose $0 \neq x \in I$. Then, by (i), we can find prime ideals $P_1, \ldots, P_n$ such that

$$P_1 \ldots P_n \subseteq xR \subseteq I$$

Choose such a set of prime ideals with minimal possible n. Because $I \neq R$, the fact that the ideals of R satisfy the ascending chain condition implies that I is contained in some maximal ideal P. Hence

$$P_1 \ldots P_n \subseteq P$$

If no $P_i = P$, each P_i contains an element $x_i \notin P$, and $x_1 \ldots x_n \in P$, which is impossible. Thus P coincides with some P_i, say $P = P_1$. We now have

$$PP_2 \ldots P_n \subseteq xR \subseteq I \subseteq P$$

By the minimality of n, we know that $P_2 \ldots P_n \not\subseteq xR$. Choose $y \in P_2 \ldots P_n$ such that $y \notin xR$, and set $z = x^{-1}y$. Then $z \notin R$, and we have

$$zR = x^{-1}yI \subseteq x^{-1}yP \subseteq x^{-1}PP_2 \ldots P_n \subseteq R,$$

so $z \in I^{-1}$ and $I^{-1} \supset R$.

Finally, set $J = I^{-1}I$. Then J is a nonzero ideal of R. Therefore

$$II^{-1}J^{-1} = JJ^{-1} \subseteq R$$

so $I^{-1}J^{-1} \subseteq I^{-1}$. Thus for any $x \in J^{-1}$ we have $I^{-1}x \subseteq I^{-1}$, and so

$$I^{-1}x^m \subseteq I^{-1}$$

for all m. But then $I^{-1}[x]$ is an R-submodule of I^{-1} and is therefore finitely generated. On the other hand, $R[x]$ being a submodule of $I^{-1}[x]$ is also finitely generated. Since R is integrally closed, it follows that $x \in R$. This shows that $J^{-1} \subseteq R$ and therefore $J^{-1} = R$. Hence, by the foregoing, $J = R$, as required.

(iii) We have $P_1 \ldots P_n \subseteq Q_1 \ldots Q_n \subseteq Q_1$, whence, as in (ii), Q_1 must equal some P_i, say $Q_1 = P_1$. Then

$$P_1^{-1}P_1 \ldots P_n \subseteq Q_1^{-1}Q_1 \ldots Q_m$$

so $P_2 \ldots P_n \subseteq Q_2 \ldots Q_m$, and so on. ∎

3.5. THEOREM. Let R be a Dedekind domain. Then any nontrivial ideal I of R can be written uniquely, up to the order of occurence, as a product of prime ideals.

Proof. By Lemma 3.4(i), I contains a product of nonzero prime ideals

$$P_1 \ldots P_n \subseteq I$$

Choose a maximal ideal P such that $I \subseteq P$. As in the proof of Lemma 3.4(ii), P coincides with some P_i, say $P = P_1$. Then

$$P^{-1}PP_2 \ldots P_n = P_2 \ldots P_n \subseteq P^{-1}I \subseteq R$$

Assume that we have shown that a nontrivial ideal of R which contains a product of fewer than n prime ideals is expressible as a product of prime ideals (a fact which is trivially true for $n = 2$ since prime ideals are maximal). We then conclude that

$$P^{-1}I = Q_1 \ldots Q_r \qquad Q_j \text{ prime}$$

and so

$$I = PP^{-1}I = PQ_1 \ldots Q_r$$

Thus, by induction, I is a product of prime ideals. The uniqueness of the factorization is a consequence of Lemma 3.4(iii). ■

We next provide the following characterization of Dedekind domains.

3.6. THEOREM. An integral domain R is a Dedekind domain if and only if every fractional ideal of R is invertible.

Proof. Assume that every fractional ideal of R is invertible. If I is any nonzero ideal of R, then since $II^{-1} = R$ there exist elements

$$a_i \in I,\ b_i \in I^{-1},\ 1 \leqslant i \leqslant m,$$

such that $a_1b_1 + \ldots + a_mb_m = 1$. Now, if $x \in I$, then

$$x = (xb_1)a_1 + \ldots + (xb_m)a_m$$

and $xb_j \in R$, whence $I = Ra_1 + \ldots + Ra_m$ is finitely generated. Thus R is noetherian.

Let P be a nonzero prime ideal of R and let M be a maximal ideal of R containing P. We have $PM^{-1} \subseteq MM^{-1} = R$ and so PM^{-1} is an ideal of R. As $(PM^{-1})M = P$, we must have either $PM^{-1} \subseteq P$ or $M \subseteq P$, since P is a prime ideal. The first possibility gives $M^{-1} = P^{-1}PM^{-1} \subseteq P^{-1}P = R$, which implies $M^{-1} = R$ and $M = R$, a contradiction. Thus $M = P$, proving that P is maximal.

To prove that R is integrally closed, choose any element x in the quotient field F of R such that x is integral over R. Then, by Proposition 1.3.6, $R[x]$ is a finitely generated R-module and hence a fractional ideal of R. Moreover $R[x]^2 = R[x]$ and we obtain

$$R[x] = R\,R[x] = R[x]R[x]R[x]^{-1} = R$$

which implies $x \in R$. Thus R is integrally closed and, by the foregoing, is a Dedekind domain.

Conversely, assume that R is a Dedekind domain and let I be a fractional ideal of R. Choose $d \in R$ such that $dI \subseteq R$. Then by Theorem 3.5 we may express dI as a product of prime ideals.

$$dI = P_1 \dots P_n$$

(if $dI = R$, set $n = 0$, and interpret a vacuous product of prime ideals as being equal to R). If $dR = Q_1 \dots Q_m$ with the $\{Q_i\}$ prime ideals, we have

$$I = d^{-1}P_1 \dots P_n = P_1 \dots P_n Q_1^{-1} \dots Q_m^{-1}$$

Hence every fractional ideal can be expressed as a product of prime ideals and inverses of prime ideals. Thus I is invertible, as required. ■

3.7. COROLLARY. Let R be a Dedekind domain. Then $I(R)$ is a free abelian group with the collection of nonzero prime ideals of R as free generators.

Proof. As we have seen in the proof of Theorem 3.6, every fractional ideal of R can be expressed as a product of prime ideals of R and their inverses. Hence the nonzero prime ideals of R generate $I(R)$, and Theorem 3.5 shows that they generate it freely. ■

Let R be a Dedekind domain and let A,B be nonzero ideals of R. We say that A is *divisible* by B or that B *divides* A if $A = BC$ for some ideal C of R. The *greatest common divisor* of A and B is defined to be an ideal which divides both A and B and is divisible by every other ideal with this property. Similarly, the *least common multiple* of A,B is defined as an ideal divisible by A and B and dividing any other such ideal. We say that A and B are *relatively prime* if they have no common prime factor. As we shall see below, g.c.d. $\{A,B\}$ and ℓ.c.m. $\{A,B\}$ both exist and are unique.

3.8. LEMMA. Let A,B be nonzero ideals of a Dedekind domain R. Then B divides A if and only if B contains A.

Proof. If B divides A, then $A = BC$ for some ideal C of R and hence $B \supseteq A$. Conversely, assume that $B \supseteq A$. If $B = R$, then $A = BA$ and B divides

A. If $B \neq R$, then by Theorem 3.5 $B = Q_1 \ldots Q_m$ and $A = P_1 \ldots P_n$, where the $\{P_i\}$ and $\{Q_j\}$ are nonzero prime ideals of R. Hence, by Lemma 3.4(iii), B divides A. ■

We are now ready to prove

3.9. PROPOSITION. Let A,B be nonzero ideals of a Dedekind domain R, and let

$$A = P_1^{a_1} \ldots P_n^{a_n}, \quad B = P_1^{b_1} \ldots P_n^{b_n} \qquad (a_i, b_i \geq 0)$$

where the $\{P_i\}$ are distinct prime ideals, and P_i^0 is taken to be R.

(i) g.c.d. $\{A,B\} = A + B = \prod_{i=1}^{n} P_i^{\min(a_i,b_i)}$

(ii) l.c.m. $\{A,B\} = A \cap B = \prod_{i=1}^{n} P_i^{\max(a_i,b_i)}$

(iii) $AB = (A+B)(A \cap B)$

(iv) A,B are relatively prime if and only if $A + B = R$

(v) The nonzero ideals $A_1,\ldots,A_n$ of R are pairwise relatively prime if and only if

$$A_i + \prod_{j \neq i} A_j = R \qquad (1 \leq i \leq n)$$

(vi) If the nonzero ideals $A_1,\ldots,A_n$ of R are pairwise relatively prime, then $\prod_{i=1}^{n} A_i = \bigcap_{i=1}^{n} A_i$

Proof. We first note that (iii) follows from (i) and (ii), (iv) from (i), (v) from (iv) and (vi) from (v) and (iii).

That g.c.d. $\{A,B\} = A + B$ is a consequence of Lemma 3.8. This proves (i) upon noting that A divides B if and only if $a_i \leq b_i$ for all i. A similar argument proves (ii). ■

3.10. THEOREM. (Chinese Remainder Theorem). Let R be a Dedekind domain and let the nonzero ideals $A_1, A_2, \ldots, A_n$ of R be pairwise relatively prime. Then, for any $\alpha_1, \alpha_2, \ldots, \alpha_n$ in R, there exists an element $\alpha \in R$ (uniquely determined mod $A_1 \ldots A_n$) such that

$$\alpha \equiv \alpha_i \pmod{A_i} \qquad (1 \leq i \leq n)$$

Proof. Owing to Proposition 3.9(v), there exist elements $\beta_i \in A_i$ and $\beta_i' \in \prod_{j \neq i} A_j$ such that $\beta_i + \beta_i' = 1$, $1 \leq i \leq n$. Setting $\alpha = \beta_1'\alpha_1 + \ldots + \beta_n'\alpha_n$, we have $\alpha \equiv \beta_i'\alpha_i (\mathrm{mod}\ A_i)$. Because $\beta_i + \beta_i' = 1$, we have $\beta_i' \equiv 1 (\mathrm{mod}\ A_i)$ and thus $\alpha \equiv \alpha_i (\mathrm{mod}\ A_i)$. Now assume that $\alpha' \equiv \alpha_i (\mathrm{mod}\ A_i)$, $1 \leqslant i \leqslant n$. Then, by Proposition 3.9(vi),

$$\alpha - \alpha' \in \bigcap_{i=1}^{n} A_i = \prod_{i=1}^{n} A_i,$$

as required. ∎

3.11. COROLLARY. Let R be a Dedekind domain and let the nonzero ideals

$$A_1, \ldots, A_n$$

of R be pairwise relatively prime. Then

$$R/(A_1 \ldots A_n) \cong (R/A_1) \times (R/A_2) \times \ldots \times (R/A_n)$$

Proof. Consider the ring homomorphism

$$R \longrightarrow (R/A_1) \times \ldots \times (R/A_n),\ r \longmapsto (r+A_1, \ldots, r+A_n)$$

The kernel of this map is $\bigcap_{i=1}^{n} A_i = \prod_{i=1}^{n} A_i$, by Proposition 3.9(vi). This map is surjective, by Theorem 3.10, hence the result. ∎

3.12. COROLLARY. Let I be a nontrivial ideal of a Dedekind domain R, say $I = P_1^{n_1} \ldots P_k^{n_k}$, $n_i > 0$, where $P_1, \ldots, P_k$ are distinct prime ideals of R. Then

$$R/I \cong \prod_{i=1}^{k} (R/P_i^{n_i})$$

Proof. The ideals $A_i = P_i^{n_i}$, $1 \leqslant i \leqslant k$, are pairwise relatively prime. Now apply Corollary 3.11. ∎

For future use, we next record

3.13. THEOREM. Let R be a Dedekind domain and let A, B be nonzero ideals of R. Then there exists a nonzero ideal C of R such that C is relatively prime to B and AC is a principal ideal.

Proof. We may write

$$A = P_1^{a_1} \ldots P_n^{a_n},\ B = P_1^{b_1} \ldots P_n^{b_n} \qquad (a_i \geqslant 0,\ b_i \geqslant 0)$$

where $P_1, \ldots, P_n$ are distinct prime ideals. Choose $\alpha_i \in P_i^{a_i} - P_i^{a_i+1}$, $1 \leqslant i \leqslant n$.

Then, by Theorem 3.10, there exists $\alpha \in R$ such that

$$\alpha \equiv \alpha_i \pmod{P_i^{a_i+1}} \qquad (1 \leqslant i \leqslant n)$$

Then $\alpha \in P_i^{a_i} - P_i^{a_i+1}$, so $\alpha \in \bigcap_{i=1}^{n} P_i^{a_i} = A$ (Proposition 3.9(vi)) and $\alpha R + AB \subseteq A$. Hence, by Lemma 3.8, $\alpha R + AB = P_1^{c_1} \ldots P_n^{c_n}$ with $c_i \geqslant a_i$. But $\alpha \notin P_i^{a_i+1}$, so $c_i = a_i$ and therefore $\alpha R + AB = A$. Since $\alpha R \subseteq A$, we have $\alpha R = AC$ for some ideal C of R (Lemma 3.8). Therefore $AC + AB = A$ and multiplying by A^{-1} yields $C + B = R$, as required. ■

3.14. COROLLARY. Every ideal in a Dedekind domain R is generated as an R-module by at most two elements.

Proof. Let A be a nonzero ideal of R. It suffices to show that

$$A = r_1R + r_2R$$

for some $r_1, r_2 \in R$. Let r_1 be any nonzero element of R. By Theorem 3.13, we may choose a nonzero ideal B of R relatively prime to r_1R such that AB is principal, say $AB = r_2R$. Then we have

$$r_1R + r_2R = r_1R + AB \subseteq A$$

On the other hand, since $B + r_1R = R$, $1 = \beta + r_1\alpha$ for some $\beta \in B$, $\alpha \in R$. Hence, for each $a \in A$, we have

$$a = ar_1\alpha + a\beta \in r_1R + AB$$

and so $A \subseteq r_1R + r_2R$, as required. ■

3.15. COROLLARY. Let A, B be nonzero ideals of a Dedekind domain R. Then the additive groups of R/A and B/AB are isomorphic. In particular, for any $n \geqslant 1$, R/A and A^n/A^{n+1} have isomorphic additive groups.

Proof. By Theorem 3.13, we may find a nonzero ideal C of R prime to A such that $BC = rR$ for some $r \in R$. Then the map $R \xrightarrow{\phi} B/AB$, $x \longmapsto rx + AB$ is an additive homomorphism, which is surjective, since

$$rR + AB = CB + AB = (C+A)B = RB = B$$

Assume that $x \in \operatorname{Ker}\phi$. Then $rx \in AB$ and $rxC \subseteq ABC = rA$, whence $xC \subseteq A$. Since $A + C = R$, we have $a + c = 1$ for some $a \in A$, $c \in C$. Then

$x = xa + xc \in A$. Conversely, if $x \in A$, then $\phi(x) = rx + AB = AB$, proving that $\mathrm{Ker}\phi = A$. ∎

3.16. COROLLARY. Let R be a Dedekind domain. Then the following conditions are equivalent:

(i) For every nonzero prime ideal P of R, R/P is finite

(ii) For every nonzero ideal I of R, R/I is finite

Proof. It is clear that (ii) implies (i). Assume that (i) holds. Then, by looking at the chain $P^n \subset P^{n-1} \subset \ldots \subset P \subset R$ and applying Corollary 3.15, we see that R/P^n is finite. The desired conclusion is therefore a consequence of Corollary 3.12. ∎

Assume that R is a Dedekind domain such that for every nonzero prime ideal P of R, R/P is finite. Then we put $N(P) = |R/P|$ and refer to $N(P)$ as the *norm* of P. Note that since R/P is a finite field, $N(P)$ is a positive power of a rational prime p. If I is any fractional ideal of R, say

$$I = P_1^{n_1} P_2^{n_2} \ldots P_k^{n_k} \qquad (n_i \in \mathbb{Z})$$

where $P_1,\ldots,P_k$ are distinct prime ideals of R, then the *norm* of I is defined by

$$N(I) = N(P_1)^{n_1} \ldots N(P_k)^{n_k}$$

Note that if I is a nonzero ideal of R (equivalently, if all $n_i \geqslant 0$), then by Corollaries 3.12 and 3.15,

$$N(I) = |R/I|$$

3.17. COROLLARY. Let R be a Dedekind domain such that for every nonzero prime ideal P of R, R/P is finite. Then the map

$$\begin{cases} I(R) \longrightarrow \mathbb{Q}^* \\ I \longmapsto N(I) \end{cases}$$

is a homomorphism.

Proof. Apply Corollary 3.7. ∎

Assume that F is the quotient field of a Dedekind domain R which is not a field. Let P be a nonzero prime ideal of R. We know, from Corollary 3.7 that the fractional ideals of R form a free abelian group with the collection of nonzero prime ideals of R as free generators. Hence, for any nonzero $x \in F$, we may factor the principal fractional ideal Rx into a product of powers of prime ideals. Let $\nu_P(x)$ denote the exponent to which P occurs in this factorization. If P does not occur, we set $\nu_P(x) = 0$. Also, we put $\nu_P(0) = \infty$. Then the map

$$\begin{cases} F \longrightarrow \mathbb{R}_\infty \\ x \longmapsto \nu_P(x) \end{cases}$$

is easily seen to be an exponential valuation. Hence, by Proposition 1.10, for any $a > 1$ in $\mathbb{R}$, the map $v_P : F \longrightarrow \mathbb{R}$ defined by

$$v_P(x) = a^{-\nu_P(x)} \qquad (x \in F) \tag{1}$$

is a non-archimedean valuation of F. Moreover, since the value group of v_P is the infinite cyclic group generated by a, v_P is in fact discrete. We refer to ν_P as the *exponential valuation associated with* P and to v_P as the *P-adic valuation*. If the factor ring R/P is finite, we put $N(P) = |R/P|$ and as usual refer to $N(P)$ as the *norm* of P. Of course, since R/P is a finite field, $N(P)$ is a power of a unique prime contained in P. The P-adic valuation v_P given by (1) is called *normalized* if $a = N(P)$.

3.18. THEOREM. Let R be a Dedekind domain which is not a field and let F be the quotient field of R. For any nonzero prime ideal P of R, let v_P be the P-adic valuation of F and let R_P be the localization of R at P.

(i) R_P is the valuation ring of v_P (in particular, R_P is a discrete valuation ring)

(ii) If $Q \neq P$ is another nonzero prime ideal of R, then v_P and v_Q are nonequivalent

(iii) $R/P^n \cong R_P/P^nR_P$ for all $n \geqslant 1$

(iv) The intersection of all R_P taken over all nonzero prime ideals P of R equals R

(v) $U(R) = \{x \in F | v_P(x) = 1$ for all nonzero prime ideals P of $R\}$

Proof. (i) Straightforward.

(ii) If v_P and v_Q are equivalent, then they have the same valuation ring. Hence by (i), $R_P = R_Q$ and so $P = Q$

(iii) Let $\nu = \nu_P$ so that by (i), $R_P = \{x \in F | \nu(x) \geqslant 0\}$ and

$$PR_P = \{x \in F | \nu(x) > 0\}$$

If $r \in R \cap P^n R_P$, then $\nu(r) \geqslant n$ and so $r \in P^n$. Since $P^n \subseteq R \cap P^n R_P$, we deduce that

$$P^n = R \cap P^n R_P \qquad \text{for all } n \geqslant 1$$

Therefore the map $R \longrightarrow R_P/P^n R_P$, $r \longmapsto r + P^n R_P$ induces an injective homomorphism

$$R/P^n \longrightarrow R_P/P^n R_P$$

Let $a = a_1/a_2$ with $a_1 \in R$, $a_2 \in R - P$ and let $m = \nu(a_2)$. Then $\nu(a_1) \geqslant m$ and so a_1 lies in the ideal $P^{n+m} + a_2 R = P^m$. Thus the congruence

$$a_2 x \equiv a_1 (\text{mod } P^{n+m})$$

has a solution in R, and for this solution we have

$$\nu(a-x) = \nu(a_1 - a_2 x) - \nu(a_2) \geqslant n,$$

i.e. $a - x \in P^n R_P$. This proves that the given map is in fact an isomorphism.

(iv) It suffices to show that $\bigcap_P R_P \subseteq R$. So assume that $x \in \bigcap_P R_P$ so that $\nu_P(x) \geqslant 0$ for all nonzero prime ideals P of R. Then the principal fractional ideal xR is an ideal of R and thus $x \in R$.

(v) Assume that $v_P(x) = 1$ for all P. Then $v_P(x^{-1}) = 1$ for all P and so, by (iv), both x and x^{-1} are in R, proving that $x \in U(R)$. Conversely, if $x \in U(R)$, then $x \in U(R_P)$ for all P. Hence, by Proposition 1.16(i), $v_P(x) = 1$ for all P as required. ■

4. COMPLETION OF A FIELD

Let v be a valuation of a field F. A *Cauchy sequence* is a sequence $\{x_n\}$ of elements of F such that for each $\varepsilon > 0$ there exists an integer N such that

$$v(x_m - x_n) < \varepsilon \qquad \text{for all } m,n > N$$

The following properties are easy consequences of the definitions:

(i) Every convergent sequence is a Cauchy sequence

(ii) If a subsequence of a Cauchy sequence converges, then so does the sequence, and to the same limit.

(iii) If $\{x_n\}$ and $\{y_n\}$ are Cauchy sequences, so are

$$\{x_n+y_n\},\ \{x_n-y_n\},\ \{x_ny_n\} \text{ and } \{\lambda x_n\}$$

where $\lambda \in F$.

(iv) If $\{x_n\}$ is a Cauchy sequence which does not converge to 0, then $x_n \neq 0$ for some n_0 on. If we form $\{x_n^{-1}\}$ starting with $x_{n_0}^{-1}$ as first term, then $\{x_n^{-1}\}$ is also a Cauchy sequence.

Of course, the converse of (i) need not be true, as is evident from the case $F = \mathbb{Q}$ and v the ordinary absolute value. A field F is said to be *complete* with respect to v if every Cauchy sequence relative to v converges to an element of F. For example, if $F = \mathbb{R}$ or $\mathbb{C}$ and v is the ordinary absolute value, then F is complete with respect to v.

Let v be a valuation of a field F. By a *completion* of F at v (or at the prime of F containing v) we mean a pair $(\tilde{F},\tilde{v})$ consisting of a field $\tilde{F}$ and a valuation $\tilde{v}$ of $\tilde{F}$ with the following properties:

(a) $\tilde{F}$ is complete with respect to $\tilde{v}$

(b) F is a subfield of $\tilde{F}$ with $v(x) = \tilde{v}(x)$ for all $x \in F$

(c) F is dense in $\tilde{F}$ (i.e. every element of $\tilde{F}$ is a limit of a convergent sequence of elements of F)

Let v_i be a valuation of a field F_i, $i = 1,2$. A map

$$f : F_1 \longrightarrow F_2$$

is said to be an *isometric isomorphism* if f is an isomorphism of fields such that

$$v_2(f(x)) = v_1(x) \qquad \text{for all } x \in F_1$$

Of course, if F_1 and F_2 are isometrically isomorphic, then they are isomorphic as valuation fields.

Assume that $\{x_n\}$ is a Cauchy sequence of elements of F. Then $\{v(x_n)\}$ is a Cauchy sequence of $\mathbb{R}$ since by Lemma 1.1(iii)

$$|v(x_m) - v(x_n)| \leqslant v(x_m - x_n)$$

Therefore, by the completeness of $\mathbb{R}$,

$$\lim_{n \to \infty} v(x_n) \quad \text{always exists}$$

This fact will play a fundamental role in the ensuing discussion.

It will next be shown how to construct a completion of F at v. Given Cauchy sequences $\{x_n\}$ and $\{y_n\}$, we write $\{x_n\} \sim \{y_n\}$ to mean that $\lim_{n \to \infty} v(x_n - y_n) = 0$. Then $\sim$ is obviously an equivalence relation on the set of Cauchy sequences of F. We write $\{x_n\}^*$ for the equivalence class of $\{x_n\}$ and by $\widetilde{F}$ the set of all such $\{x_n\}^*$. For any $x \in F$, the sequence $\{x_n\}$ defined by $x_1 = x_2 = \ldots = x$ is obviously a Cauchy sequence; let us denote this $\{x_n\}^*$ by x^*. Then $x^* = y^*$ if and only if $x = y$. One immediately verifies that the following addition and multiplication on $\widetilde{F}$ are well-defined:

$$\{x_n\}^* + \{y_n\}^* = \{x_n + y_n\}^*$$

$$\{x_n\}^* \{y_n\}^* = \{x_n y_n\}^*$$

We are now in a position to establish the following fundamental result:

4.1. THEOREM. Let v be a valuation of a field F, let $\widetilde{F}$ be defined as above and let $\widetilde{v} : \widetilde{F} \longrightarrow \mathbb{R}$ be defined by

$$\widetilde{v}(\{x_n\}^*) = \lim_{n \to \infty} v(x_n)$$

(i) $(\widetilde{F}, \widetilde{v})$ is a completion of F at v (we regard F as embedded in $\widetilde{F}$ by means of $x \longmapsto x^*$). In particular, by Proposition 1.7, v is archimedean if and only if so is $\widetilde{v}$.

(ii) $\widetilde{F}$ is uniquely determined by F and v, up to isometric isomorphism which is the identity map on F.

(iii) Assume that v is non-archimedean and let R_v, $R_{\tilde{v}}$ and P_v, $P_{\tilde{v}}$ be the valuation rings and maximal ideals of v and $\tilde{v}$, respectively. Then

(a) The value groups of v and $\tilde{v}$ are the same (in particular, v is discrete if and only if so is $\tilde{v}$)

(b) R_v is dense in $R_{\tilde{v}}$, P_v is dense in $P_{\tilde{v}}$ and the map $R_v/P_v \longrightarrow R_{\tilde{v}}/P_{\tilde{v}}$ induced by the embedding $F \longrightarrow \tilde{F}$ is an isomorphism.

Proof. (i) It is straightforward to verify that $\tilde{F}$ is a field, $F \longrightarrow \tilde{F}$, $x \longmapsto x^*$ is an injective homomorphism and that $\tilde{v}$ is well defined. Clearly $\tilde{v}$ extends v. Let $\{\alpha_m\}$ be a Cauchy sequence of elements of $\tilde{F}$. For each m, we may find a Cauchy sequence $\{x_n^{(m)}, n = 1,2,\ldots\}$ of elements of F such that

$$\alpha_m = \{x_n^{(m)}\}^* \qquad (m = 1,2,\ldots)$$

One proves easily that $\{x_n^{(n)}, n = 1,2,\ldots\}$ is a Cauchy sequence from F and hence it determines an element $\beta = \{x_n^{(n)}\}^* \in \tilde{F}$. Then we have

$$\lim_{m\to\infty} v(\alpha_m-\beta) = \lim_{m\to\infty}\ \lim_{n\to\infty} v(x_n^{(m)}-x_n^{(n)}) = 0$$

so that $\{\alpha_m\}$ converges to β in $\tilde{F}$. Hence $\tilde{F}$ is complete with respect to $\tilde{v}$.

Assume that $\{x_k\}^*$ is an element of $\tilde{F}$. We claim that

$$\lim_{k\to\infty} x_k^* = \{x_k\}^* \tag{1}$$

which will prove that F is dense in $\tilde{F}$. Indeed, given $\varepsilon > 0$, we can find N such that $v(x_m-x_n) < \varepsilon$ if $m,n \geqslant N$. Then $\lim_{n\to\infty} v(x_m-x_n)$ exists and this is $\leqslant \varepsilon$. On the other hand,

$$\tilde{v}(\{x_k\}^* - x_m^*) = \lim_{n\to\infty} v(x_m-x_n)$$

so $\tilde{v}(\{x_k\}^* - x_m^*) \leqslant \varepsilon$ if $m \geqslant N$. This proves (1) and hence F is dense in $\tilde{F}$.

(ii) Let (F_1,v_1) and (F_2,v_2) be two completions of F at v. Let $x \in F_1$ and let $\{x_k\}$ be a sequence of elements of F such that $\lim x_k = x$. Regarding $\{x_k\}$ as a Cauchy sequence of F_2, we also have $\lim x_k = f(x)$ for a unique $f(x) \in F_2$. Then the map $F_1 \longrightarrow F_2$, $x \longmapsto f(x)$ is obviously well-defined, restricts to the identity map on F and a routine verification shows that it is in fact an isometric isomorphism.

(iii) Because v is non-archimedean, so is $\tilde{v}$ by virtue of (i). Let $\alpha = \{\alpha_n\}^*$ be a nonzero element of $\tilde{F}$. We claim that there is a positive integer n_0 such that

$$v(\alpha) = \tilde{v}(\alpha_n^*) \qquad \text{for all } n > n_0 \tag{2}$$

As we have seen in (i), $\alpha = \lim_{n\to\infty} \alpha_n^*$. Therefore, for a sufficiently large n, we have

$$\tilde{v}(\alpha_n^*) = \tilde{v}(\alpha+(\alpha_n^*-\alpha)) \leqslant \max\{\tilde{v}(\alpha), \tilde{v}(\alpha_n^*-\alpha) = \tilde{v}(\alpha) \tag{3}$$

since $\tilde{v}(\alpha) \neq 0$ and $\tilde{v}(\alpha_n^*-\alpha)$ can be made as small as we please, in particular less than $\tilde{v}(\alpha)$, by selecting n large enough. Because $\lim_{n\to\infty} v(\alpha_n) = \tilde{v}(\alpha) \neq 0$, it follows that $\tilde{v}(\alpha_n^*) \geqslant \tilde{v}(\alpha)/2 > 0$ for n sufficiently large. Therefore, for sufficiently large n, $\tilde{v}(\alpha_n^*-\alpha) < \tilde{v}(\alpha_n^*)$. But then

$$\tilde{v}(\alpha) = \tilde{v}(\alpha_n^*+(\alpha-\alpha_n^*)) \leqslant \max\{\tilde{v}(\alpha_n^*), \tilde{v}(\alpha-\alpha_n^*)\} = \tilde{v}(\alpha_n^*) \tag{4}$$

by choosing n large enough. This substantiates our claim, by applying (3) and (4).

Invoking (2), we infer that v and $\tilde{v}$ have the same value group. Moreover, by (2), we may also assume that for each $\alpha = \{\alpha_n\}^* \in \tilde{F}$

$$\tilde{v}(\alpha) = \tilde{v}(\alpha_n^*) \qquad \text{for all } n \geqslant 1$$

In particular, $\tilde{v}(\alpha) \leqslant 1$ (respectively, $\tilde{v}(\alpha) < 1$) if and only if $\tilde{v}(\alpha_n^*) \leqslant 1$ (respectively, $\tilde{v}(\alpha_n^*) < 1$) for all $n \geqslant 1$. However, $\alpha = \lim_{n\to\infty} \alpha_n^*$ and so R_v is dense in $R_{\tilde{v}}$ and P_v is dense in $P_{\tilde{v}}$. Because $R_v \cap P_{\tilde{v}} = P_v$, the map

$$R_v/P_v \longrightarrow R_{\tilde{v}}/P_{\tilde{v}}$$

is an injective homomorphism. Since R_v is dense in $R_{\tilde{v}}$, given any $\alpha \in R_{\tilde{v}}$, we can find $a \in R_v$ such that $\tilde{v}(\alpha-a) < 1$. Then $a + P_{\tilde{v}} = \alpha + P_{\tilde{v}}$ and so the map is surjective. This completes the proof of the theorem. ■

Let R be a Dedekind domain with quotient field F and let $R \neq F$. Then every nonzero prime ideal P of R determines the P-adic valuation v_P of F introduced in the previous section. Owing to Theorem 2.1, there exists a completion, say $(F_P, \tilde{v}_P)$ of F at v_P. Moreover, the field F_P is determined uniquely up to isometric isomorphism. We shall refer to F_P as the *P-adic field*. The valuation ring of $\tilde{v}_P$ is called the ring of *P-adic integers* , or the *ring of integers* of F_P. In the case $R = \mathbb{Z}$ and $F = \mathbb{Q}$ we will not distinguish between the prime p and the prime ideal $P = p\mathbb{Z}$. Therefore we will write $\mathbb{Q}_p$ instead of $\mathbb{Q}_P$. The field $\mathbb{Q}_p$ is called the (*rational*) *p-adic field*, and its ring of integers, denoted by $\mathbb{Z}_p$, is called the *ring of p-adic integers* (although the same symbol $\mathbb{Z}_p$ will be used to denote a cyclic group of order p, the context should preclude any possible confusion).

Our next aim is to describe topologies on finite-dimensional vector spaces over a complete field which are consistent with the topology of that field. We digress briefly to introduce the necessary definitions.

Let V be a vector space over a field F and let v be a valuation of F. A *norm* on V associated with v is a real-valued function $\| \ \|_v$ with the following properties:

(i) $\|x\|_v \geq 0$ and $\|x\|_v = 0$ if and only if $x = 0$

(ii) $\|\lambda x\| = v(\lambda)\|x\|_v$ for all $\lambda \in F$, $x \in V$

(iii) $\|x+y\|_v \leq \|x\|_v + \|y\|_v$ for all $x,y \in V$

We can introduce a distance function on V by defining $\|x-y\|_v$ to be the distance between two typical points x and y of V; this makes V into a metric space and the various topological concepts are thereby introduced into V. The topology defined by $\| \ \|_v$ is called a *norm topology* associated with v.

A typical example of a norm is as follows. Let E/F be a field extension, let v be a valuation of F and assume that v can be extended to a valuation w of E. If we define $\|x\| = w(x)$ for all $x \in E$, then $\| \ \|$ is obviously a norm associated with v. Furthermore, the corresponding norm topology on E coincides with the topology induced by w. If we further assume that E/F is a finite field extension and F is complete, then it satisfies the hypotheses of the following theorem.

4.2. THEOREM. Let F be a complete field with respect to a valuation v, let

E be a finite-dimensional space over F and let $\| \ \|_v$ be a norm on E associated with v. Then the map

$$\begin{cases} F^n \longrightarrow E \\ (\lambda_1,\dots,\lambda_n) \longmapsto \sum_{i=1}^{n} \lambda_i a_i \end{cases}$$

where $\{a_1,\dots,a_n\}$ is an F-basis of E, is a topological isomorphism .

Proof. Because f is obviously a continuous isomorphism, it suffices to verify the following property:

If $\|x\|$ denotes the norm on E, then there exists a positive constant C such that

$$v(\lambda_i) \leqslant C \, \| \sum_{i=1}^{n} \lambda_i a_i \| \quad \text{for all } i \in \{1,2,\dots,n\},\ \lambda_i \in F \tag{5}$$

Denote by E_r the subset of E consisting of all $x \in F$ such that in

$$x = \sum_{i=1}^{n} \lambda_i a_i$$

at most r of the λ_i's are nonzero. We prove (5) by induction on r successively for the sets $E_1, E_2, \dots, E_n = E$. Because the case $r = 1$ is included in our hypotheses, we may assume that (5) is proved for $E_1,\dots,E_{r-1}$. Hence for $x \in E_{r-1}$ we have

$$v(\lambda_i) \leqslant C_1 \| \sum_{i=1}^{n} \lambda_i a_i \| \qquad (1 \leqslant i \leqslant n)$$

for some positive constant C_1. If our statement were false in E_r, we could find a sequence $x^{(n)} \neq 0$ with

$$x^{(n)} = \lambda_{i_1}^{(n)} a_{i_1} + \dots + \lambda_{i_r}^{(n)} a_{i_r}$$

such that the quotient

$$\frac{\|x^{(n)}\|}{v(\lambda_{i_1}^{(n)})}$$

tends to zero. For $y^{(n)} = x^{(n)}/\lambda_{i_1}^{(n)} = a_{i_1} + y_{i_2}^{(n)} a_{i_2} + \dots + y_{i_r}^{(n)} a_{i_r}$ we thus have $\|y^{(n)}\| \longrightarrow 0$, and hence $y^{(n)}$ tends to zero. That this leads to a contradiction can be seen as follows.

The differences $y^{(m)} - y^{(n)}$ lie in E_{r-1}, so by the inductive hypothesis

we have

$$v(y_{i_j}^{(m)}-y_{i_j}^{(n)}) \leq C_1\|y^{(n)}-y^{(m)}\| \longrightarrow 0 \qquad (2 \leq j \leq r)$$

Hence the $\{y_{i_j}^{(n)}\}$ are Cauchy sequences, $2 \leq j \leq r$, so the completeness of F guarantees that they have limits, say $z_j = \lim y_{i_j}^{(n)}$. But this shows that $y^{(n)}$ tends to $a_{i_1} + z_{i_2}a_{i_2} + \ldots + z_{i_r}a_{i_r}$ which is obviously nonzero, a contradiction. ■

4.3. COROLLARY. Let F be a complete field with respect to a valuation v and let E/F be an algebraic field extension such that v extends to a valuation of E. Then

(i) If E/F is a finite field extension of degree n, then E is topologically isomorphic to F^n and, in particular, E is complete.

(ii) Any two extensions of v to a valuation of E coincide.

Proof. (i) Direct consequence of Theorem 4.2.

(ii) Let v_1 and v_2 be two valuations of E extending v. We may assume without loss of generality that E/F is a finite extension. By (i), v_1 and v_2 determine the same topology on E and hence are equivalent. If v is trivial, then F is discrete, hence so is E, by virtue of (i). Thus both v_1 and v_2 are trivial. Otherwise, we have $v_2(x) = v_1(x)^r$ for some $r \in \mathbb{R}$, and there exists $\lambda \in F$ such that $v(\lambda) \neq 0, 1$. Taking $x = \lambda$, we see that $r = 1$, hence $v_1 = v_2$. ■

We close this section by determining all complete archimedean fields. The field $\mathbb{R}$ being the completion of $\mathbb{Q}$ at the ordinary absolute value is one of such fields. Another example of a complete archimedean field is $\mathbb{C}$. Indeed, the valuation on $\mathbb{C}$ given by $|a + bi| = \sqrt{a^2+b^2}$ extends the absolute value on $\mathbb{R}$ and hence, by Corollary 4.3(i), $\mathbb{C}$ is complete. It turns out that $\mathbb{C}$ and $\mathbb{R}$ are the only examples of complete archimedean fields. Before proving this, let us make the following general observation.

Let E/F be a field extension and let w be a valuation of E such that E is complete with respect to w. Let $\tilde{F}$ denote the subfield of E consisting of limits of all Cauchy sequences of elements of F. Then $\tilde{F}$ is a closed subspace of the complete metric space E and hence $\widetilde{F}$ is complete with respect to $\widetilde{v} = w|\widetilde{F}$. Thus if $v = w|F$, then $(\widetilde{F},\widetilde{v})$ is the completion of F at v with $F \subseteq \widetilde{F} \subseteq E$. We shall refer to $\widetilde{F}$ as the *completion* of F in E.

The following lemma contains some burden of the proof that $\mathbb{C}$ and $\mathbb{R}$ are the only complete archimedean fields.

4.4. LEMMA. Let F be a complete valuation field with valuation v, and let E be a quadratic extension of F. Then a real-valued function $\widetilde{v} : E \longrightarrow \mathbb{R}$ defined by

$$\widetilde{v}(\lambda) = \sqrt{v(N(\lambda))}$$

where N is the norm $N_{E/F}$, is a valuation of E

Proof. It is clear that $\widetilde{v}(\lambda) \geqslant 0$ for all $\lambda \in E$, that $\widetilde{v}(\lambda) = 0$ if and only if $\lambda = 0$, and that $\widetilde{v}(\lambda_1\lambda_2) = \widetilde{v}(\lambda_1)\widetilde{v}(\lambda_2)$ for all $\lambda_1,\lambda_2 \in E$. To prove that $\widetilde{v}(x+y) \leqslant \widetilde{v}(x) + \widetilde{v}(y)$, it suffices to verify

$$\widetilde{v}(\lambda-1) \leqslant 1 + \widetilde{v}(\lambda) \qquad \text{for all } \lambda \in E \tag{6}$$

Indeed, if (6) is true, then for elements $x,y \in E$, $y \neq 0$,

$$\widetilde{v}(-x/y - 1) \leqslant 1 + \widetilde{v}(-x/y)$$

whence $\widetilde{v}(y)^{-1}\widetilde{v}(x+y) \leqslant 1 + \widetilde{v}(x)\widetilde{v}(y)^{-1}$ and so $\widetilde{v}(x+y) \leqslant \widetilde{v}(x) + \widetilde{v}(y)$.

Assume by way of contradiction that there exists $\lambda \in E-F$ such that

$$\widetilde{v}(\lambda-1) > 1 + \widetilde{v}(\lambda)$$

Denote by $\bar{\lambda}$ the conjugate of λ (with respect to F), and set

$$(X-\lambda)(X-\bar{\lambda}) = X^2 + bX + c \qquad (b,c \in F)$$

We now have, by our assumption,

$$\sqrt{v(1+b+c)} > 1 + \sqrt{v(c)}$$

whence

$$v(1+b+c) > 1 + 2\sqrt{v(c)} + v(c)$$

$$1 + v(b) + v(c) > 1 + 2\sqrt{v(c)} + v(c)$$

and so $v(b)^2 > 4v(c)$. Because $c \neq 0$, we have $v(b) > 0$. We then set $a_0 = b$, and construct a sequence $a_1, a_2, \ldots$ by

$$a_{n+1} = -b - c/a_n$$

We claim that a_n is never 0, so that the above definition makes sense, and that $\{a_n\}$ is a Cauchy sequence; if sustained, it will follow that there exists an element $a \in F$ to which $\{a_n\}$ converges, and hence we have $a = -b - c/a$ or $c^2 + ba + c = 0$, which implies that $\lambda \in F$, a contradiction.

To prove that $a_n \neq 0$, it suffices to show that $v(a_n) \geqslant (1/2)v(b)$. We have of course, $v(a_0) \geqslant (1/2)v(b)$. Assume that we have $v(a_n) \geqslant (1/2)v(b)$. Then

$$\begin{aligned} v(a_{n+1}) &\geqslant v(b) - v(c)/v(a_n) \geqslant v(b) - v(c)/(1/2)v(b) \\ &> v(b) - (1/2)v(b) = (1/2)v(b), \end{aligned}$$

proving that $a_n \neq 0$.

To prove that $\{a_n\}$ is a Cauchy sequence, we note that

$$\begin{aligned} v(a_{n+1}-a_n) &= v(c/a_n - c/a_{n-1}) = (v(c)/v(a_n)v(a_{n-1}))v(a_n - a_{n-1}) \\ &\leqslant (4v(c)/v(b)^2)v(a_n - a_{n-1}) \\ &= \rho v(a_n - a_{n-1}) \qquad (\rho < 1) \end{aligned}$$

and hence the series $\sum_{n=1}^{\infty} v(a_n - a_{n-1})$ converges. Therefore $\{a_n\}$ is a Cauchy sequence, as required. ■

We are now ready to prove

4.5. THEOREM. (Ostrowski). Let v be an archimedean valuation of a field F. Then there is an injective homomorphism $\psi : F \longrightarrow \mathbb{C}$ and a positive real number r such that

$$v(x) = |\psi(x)|^r \qquad \text{for all } x \in F$$

Furthermore, if F is complete, then $\psi(F) = \mathbb{C}$ or $\mathbb{R}$.

Proof. By passing to the completion of F, if necessary, we may assume that F is complete. By Corollary 1.8, char$F = 0$ and so $\mathbb{Q} \subseteq F$. By Proposition 1.11, any archimedean valuation of $\mathbb{Q}$ is equivalent to the ordinary absolute value. By looking at the completion of $\mathbb{Q}$ in F, we may assume by Theorem 4.1, that F contains $\mathbb{R}$ and $v(x) = |x|$ for every $x \in \mathbb{R}$. We now claim that the result will follow provided we show that

(a) If $\sqrt{-1} \in F$ and if $v(x) = |x|$ for all $x \in \mathbb{C}$, then $F = \mathbb{C}$

(b) If $\sqrt{-1} \notin F$, then v is uniquely extended to a valuation of $F(\sqrt{-1})$ and $F(\sqrt{-1})$ is complete.

Indeed if $\sqrt{-1} \in F$, then $|\ |$ in $\mathbb{R}$ can be extended to $|\ |$ in $\mathbb{C}$; hence by (b) the restriction of v to $\mathbb{C}$ must be $|\ |$ and $F = \mathbb{C}$, by (a). If $\sqrt{-1} \notin F$, then $F(\sqrt{-1}) = \mathbb{C}$, by (a) and $F = \mathbb{R}$ since $F \supseteq \mathbb{R}$.

Property (b) being a consequence of Lemma 4.4 and Corollary 4.3, we are left to verify (a). In what follows we assume that the reader is familiar with the following standard fact: If f is a real-valued continuous function defined on a bounded closed set K on the real plane, then $f(K)$ has the least value.

Assume by way of contradiction that $F \neq \mathbb{C}$ and fix an element $a \in F - \mathbb{C}$. Let $f(z) = v(z-a)$ for $z \in \mathbb{C}$. Then f is a continuous function and $f(z) > 0$ for all $z \in \mathbb{C}$. Now $f(z) = v(z-a) \geqslant v(z) - v(a) = |z| - v(a)$, and so if $|z|$ is sufficiently large, then $f(z)$ is large. Let N be a sufficiently large number, and consider f on $K = \{z \in \mathbb{C} \,|\, |z| \leqslant N\}$. Then the least value $f(k)$ $(k \in K)$ in $f(K)$ is the least value in $f(\mathbb{C})$. We now set $r = f(k)$ and claim that the following assertion will give the desired contradiction:

If $x,y \in \mathbb{C}$, $f(x) = r$ and $|y| < r$, then $f(x-y) = r$ (7)

Indeed, take $z \in \mathbb{C}$ such that $|z|$ is sufficiently large, and take a natural number m such that $|z/m| < r$. By (7), $f(k-j(z/m)) = r$ for $j = 0,1,\ldots$ (by induction on j). By the case $j = m$, we have $f(k-z) = r$. On the other hand,

$$f(k-z) = v(k-z-a) \geqslant v(z) - v(k-a) \geqslant |z| - |k| - v(a)$$

and thus

$$r \geqslant |z| - |k| - v(a),$$

contradicting to the fact that $|z|$ is large.

To prove (7), set $b = x-a$ so that $v(b) = f(x) = r$. For each natural number n, we consider $b^n - y^n$. If ε is a primitive n-th root of unity, then

$$b^n - y^n = (b-y)(b-\varepsilon y) \ldots (b-\varepsilon^{n-1}y)$$

and so

$$v(b^n-y^n) = v(b-y)\ v(b-\varepsilon y) \ldots v(b-\varepsilon^{n-1}y)$$

Because of the minimality of r, we have $v(b^n-y^n) \geqslant v(b-y)r^{n-1}$ and thus

$$v(b - (y^n/b^{n-1})) \geqslant v(b-y)$$

It follows that

$$v(b) + v(y)^n/v(b)^{n-1} \geqslant v(b-y) = f(x-y)$$

Since $v(y) < r = v(b)$, $v(y)^n/v(b)^{n-1}$ tends to 0 if n tends to infinity. Thus $r = v(b) \geqslant f(x-y)$. By the minimality of r, we have $f(x-y) = r$, as required. ■

We close by providing an application of Corollary 4.3.

4.6. THEOREM. Let F be a complete field with respect to a discrete valuation v, let R be the valuation ring of v, and let E/F be a finite separable field extension of degree n. If S is the integral closure of R in E, then

(i) S is a discrete valuation ring and is a free R-module of rank n

(ii) E is complete with respect to the valuation associated with S

(iii) S is the valuation ring of the extension of v to E.

Proof. By Corollary 2.6.17, S is a free R-module of rank n. By Theorem 3.3, S is a Dedekind domain. Let P be any nonzero prime ideal of S and let v_P be the P-adic valuation of E. By Theorem 3.18, S_P is the valuation ring of v_P. We claim that it suffices to show that the restriction w of v_P to F is equivalent to v. Indeed, if this is the case, then by Corollary 4.3, the equivalence class of v_P is independent of the choice of P. Hence P is a unique nonzero prime ideal of S and therefore $S = S_P$ is a discrete valuation ring, proving (i) Furthermore, since in this case v_P is the valuation of E associated with S, (ii) follows from Corollary 4.3(i). Finally, (iii) is a consequence of the fact that S is the valuation ring of v_P.

To substantiate our claim, note that w is nontrivial, since $v_P(s) < 1$ for all nonzero $s \in P$ and since, by Lemma 3.2(ii), $P \cap R \neq 0$. Assume that $w(\lambda) < 1$ for some $\lambda \in F$. If $\lambda \in R$, then λ is a nonunit, hence $v(\lambda) < 1$. If $\lambda \notin R$, then $\lambda^{-1} \in R \subseteq S$, so $w(\lambda^{-1}) \leqslant 1$ which is impossible, since $w(\lambda) < 1$. Hence, by Proposition 1.5, w and v are equivalent, as required. ∎

5. EXTENSIONS OF VALUATIONS

In this section we examine the ways of extending a valuation of a given field to a larger field. Our first aim is to complement Corollary 4.3 by proving that the valuation of E extending v always exists. This will be achieved with the aid of an important result known as Hensel's lemma.

Let I be an ideal of a cummutative ring R and let $- : R \longrightarrow R/I$ be the natural homomorphism. Then the map

$$\begin{cases} R[X] \longrightarrow (R/I)[X] \\ \Sigma r_i X^i \longmapsto \Sigma \bar{r}_i X^i \end{cases}$$

is obviously a surjective homomorphism. In what follows, we write $\bar{f}(X)$ (or simply $\bar{f}$) for the image of $f(X) \in R[X]$ in $(R/I)[X]$, where R is the valuation ring of a valuation v and I is the maximal ideal of R.

5.1. LEMMA. (Hensel's lemma). Let F be a complete field with respect to a non-archimedean valuation v, let R be the valuation ring of v and let $f \in R[X]$ be such that $\bar{f} \neq 0$. Assume that $g_0, h_0 \in R[X]$ are such that $\bar{g}_0$ and $\bar{h}_0$ are relatively prime and $\bar{f} = \bar{g}_0\bar{h}_0$. Then there exist $g,h \in R[X]$ with

$$f = gh,\ \bar{g} = \bar{g}_0,\ \bar{h} = \bar{h}_0 \text{ and } \deg g = \deg g_0$$

Proof. We may harmlessly assume that the leading coefficient of g_0 and h_0 are not in I, in which case they are invertible. Hence there is no loss of generality in assuming that g_0 is a monic polynomial. Let $n = \deg f$, $r = \deg g$, $s = \deg h_0$ and b the leading coefficient of h_0. Then the leading coefficient of g_0h_0 is b and $\deg(g_0h_0) = r + s \leqslant n$.

Let ν be the exponential valuation of F associated with v. Since

$$f - g_0h_0 \in I[X]$$

and $I = \{x \in F | \nu(x) > 0\}$, any coefficient c of $f - g_0h_0$ with minimal $\nu(c)$ satisfies $\nu(c) > 0$. We set $\delta_1 = \nu(c)$ and note that if $\delta_1 = \infty$ then $f = g_0h_0$ and there is nothing to prove. From now on, we may therefore assume that $\delta_1 > 0$ is in $\mathbb{R}$.

Because $(\bar{g}_0, \bar{h}_0) = 1$, there exist $\ell, m \in R[X]$ such that

$$\bar{\ell}\, \bar{g}_0 + \bar{m}\, \bar{h}_0 = \bar{1}$$

The least of the values of ν on the coefficients of $\ell g_0 + mh_0 - 1$ is some $\delta_2 > 0$. Set $\varepsilon = \min(\delta_1, \delta_2)$ and let $\pi \in R$ be such that $\nu(\pi) = \varepsilon$. Then

$$f \equiv g_0h_0 \pmod{\pi} \tag{1}$$

$$\ell g_0 + mh_0 \equiv 1 \pmod{\pi} \tag{2}$$

We now construct g as the limit of a certain sequence $\{g_k\}$ of polynomials of degree r with $g_1 = g_0$ and h as the limit of a certain sequence $\{h_k\}$ of polynomials of degree $\leqslant n-r$ with $h_1 = h_0$. Assume that g_k and h_k satisfy

$$f \equiv g_kh_k \pmod{\pi^{k+1}} \tag{3}$$

$$g_k \equiv g_0 \pmod{\pi} \tag{4}$$

$$h_k \equiv h_0 \pmod{\pi} \tag{5}$$

and that g_k is a monic polynomial. In order to determine g_{k+1} and h_{k+1}, we write

$$g_{k+1} = g_k + \pi^{k+1}x \tag{6}$$

$$h_{k+1} = h_k + \pi^{k+1}y \tag{7}$$

Then we have

$$g_{k+1}h_{k+1} - f = g_kh_k - f + \pi^{k+1}(g_ky + h_kx) - \pi^{2k+2}xy$$

By (3), we may write $f - g_kh_k = \pi^{k+1}t$ and therefore

$$g_{k+1}h_{k+1} - f \equiv \pi^{k+1}(g_ky + h_kx - t) \pmod{\pi^{k+2}}$$

which implies that the left-hand side is divisible by π^{k+2} provided

$$g_ky + h_kx \equiv t \pmod{\pi} \tag{8}$$

To achieve this, we multiply (2) by t:

$$t\ell g_0 + tmh_0 \equiv t(\text{mod } \pi) \tag{9}$$

and write

$$tm = qg_0 + x \tag{10}$$

where $\deg x < r$. Then (9) can be rewritten as

$$(t\ell + qh_0)g_0 + xh_0 \equiv t(\text{mod } \pi)$$

and replacing by 0 those coefficients of $t\ell + qh_0$ which are divisible by π, we obtain

$$yg_0 + xh_0 \equiv t(\text{mod } \pi) \tag{11}$$

This proves (8), by applying (11), (4) and (5). Moreover, $\deg x < r$ so that by (6), g_{k+1} and g_k have the same degree and the same leading coefficients. It therefore suffices to verify that $\deg y \leqslant n-r$. Assume the contrary. Then the leading term of yg_0 has degree $> n$ and there are no terms of this degree in xh_0 and t. Applying (11), we see that the leading coefficient of yg_0 is divisible by π. Thus the leading coefficient of y is divisible by π, contrary to our choice of y.

As we have seen above, it follows from (8) that

$$f \equiv g_{k+1}h_{k+1}(\text{mod } \pi^{k+2})$$

By (6), the coefficients of $g_{k+1} - g_k$ are divisible by π^{k+1} and therefore they tend to 0 as $k \longrightarrow \infty$. Bearing in mind that F is complete, we see that $\lim_{k\to\infty} g_k$ exists, say equal to a monic polynomial $g \in R[X]$. Similarly, $\lim_{k\to\infty} h_k$ exists and we denote it by h. Finally, taking the limits of both sides of (3), we derive $f = gh$ and, by applying (4) and (5), we also have $\bar{g} = \bar{g}_0$ and $\bar{h} = \bar{h}_0$. So the lemma is true. ■

We now turn to ways of extending a valuation of a field to a larger field. The following easy consequence of Hensel's lemma will clear our path.

5.2. COROLLARY. Let F be a complete field with respect to a non-archimedean valuation v. If

$$f(X) = a_0 + a_1 X + \dots + a_n X^n \in F[X]$$

is irreducible, then

$$\min\{\nu(a_0), \nu(a_1), \dots, \nu(a_n)\} = \min\{\nu(a_0), \nu(a_n)\}$$

where ν is the exponential valuation associated with v.

Proof. We keep the notation of the preceding lemma. Let a_k be such that $\nu(a_k) = \{\min\nu(a_i) | 0 \leqslant i \leqslant n\}$. Dividing each coefficient of $f(X)$ by a_k, if necessary, we may assume that $f \in R[X]$, $\bar{f} \neq 0$ and that $\nu(a_k) = 0$. Assume by way of contradiction that $\nu(a_0) > 0$ and $\nu(a_n) > 0$. Then there exists $r \in \mathbb{N}$ with $r < n$ for which $\nu(a_r) = 0$ and $\nu(a_i) > 0$ for $i = r+1, \dots, n$. Setting

$$g_0 = a_0 + a_1 X + \dots + a_r X^r \quad \text{and} \quad h_0 = 1$$

it follows that $\bar{f} = \bar{g}_0 \bar{h}_0$. Therefore, by Lemma 5.1, f is reducible, a contradiction. ■

We are now ready to prove

5.3. THEOREM. Let F be a complete field with respect to a valuation v and let E be an algebraic extension of F. Then the map $w : E \longrightarrow \mathbb{R}$ given by

$$w(\lambda) = [v(N_{F(\lambda)/F}(\lambda))]^{1/n}$$

where $n = (F(\lambda):F)$, is a unique extension of v to a valuation of E. Furthermore, if E/F is a finite extension of degree m, then E is complete and

$$w(\lambda) = [v(N_{E/F}(\lambda))]^{1/m} \qquad \text{for all } \lambda \in E$$

Proof. Owing to Lemma 2.6.2(ii), both expressions for w coincide if the extension E/F is finite. Moreover, it is obvious that w extends v. Hence, by Corollary 4.3, we are left to verify that w is a valuation of E.

We first assume that v is non-archimedean and denote by ν the exponential valuation associated with v so that $\nu(x) = -\log v(x)$ for all $x \in F$. Given $\lambda \in E$, let

$$X^n + a_{n-1}X^{n-1} + \dots + a_0 \in F[X]$$

be the minimal polynomial for λ. Define $\nu' : E \longrightarrow \mathbb{R}_\infty$ by

$$\nu'(\lambda) = (1/n)\nu(a_0)$$

Then ν' extends ν and we assert that ν' is an exponential valuation of E. Indeed, let $\mu \in E$ and let $K = F(\lambda,\mu)$ have degree t over F. Then, by Lemma 2.6.2(ii),

$$N_{K/F}(\lambda) = (-1)^t a_0^r \quad , r = t/n$$

and thus

$$\nu(N_{K/F}(\lambda)) = \nu(a_0^r) = r\nu(a_0)$$

$$\nu'(\lambda) = (1/n)\nu(a) = (1/t)\nu(N_{K/F}(\lambda))$$

Since $N_{K/F}(\lambda\mu) = N_{K/F}(\lambda)N_{K/F}(\mu)$, we infer that

$$\nu'(\lambda\mu) = \nu'(\lambda) + \nu'(\mu)$$

To prove that $\nu'(\lambda+\mu) \geqslant \min\{\nu'(\lambda),\nu'(\mu)$, we may assume that $\mu = 1$. This is so since

$$\nu'(\lambda+\mu) = \nu'(\mu) + \nu'(1+\lambda/\mu)$$

and

$$\min\{\nu'(\lambda),\nu'(\mu) = \nu'(\mu) + \min\{\nu'(\lambda/\mu),0\}$$

The minimal polynomial for $\lambda + 1$ is

$$X^n + \ldots + (a_0 - a_1 + a_2 - \ldots + (-1)^{n-1}a_{n-1} + (-1)^n)$$

Invoking Corollary 5.2, we have

$$\begin{aligned}\nu'(\lambda+1) &= (1/n)\nu(a_0-a_1+\ldots) \\ &\geqslant (1/n)\min\{\nu(a_0),\nu(a_1),\ldots,\nu(a_{n-1}),\nu(1)\} \\ &= (1/n)\min\{\nu(a_0),\ \nu(1)\} \\ &= \min\{\nu'(\lambda),\ \nu'(1)\},\end{aligned}$$

proving that ν' is an exponential valuation extending ν. This establishes the nonarchimedean case since the valuation of E corresponding to ν' coincides with w.

If v is archimedean, it suffices, by virtue of Theorem 4.5, to consider the case where $F = \mathbb{R}$, $E = \mathbb{C}$ and $v(\lambda) = |\lambda|^r$ for some $r > 0$ in $\mathbb{R}$. Then the map

$w : \mathbb{C} \longrightarrow \mathbb{R}$ given by $w(x) = |x|^r$ is a valuation of $\mathbb{C}$ extending v. However, for $x = a + bi$, where $a,b \in \mathbb{R}$, we have

$$|x| = \sqrt{a^2+b^2} = \sqrt{N(x)} = \sqrt{|N(x)|}$$

so that

$$w(x) = |x|^r = \sqrt{v(N(x))}$$

as we wished to show. ∎

The above result represents a turning point in our discussion. Before putting it to use we shall bring in certain additional facts.

5.4. LEMMA. Let R be a commutative ring and $I_1,\ldots,I_n$ ideals of R such that $I_i + I_j = R$ for $i \neq j$. Then the natural map

$$\begin{cases} R \longrightarrow \prod_{i=1}^{n} (R/I_i) \\ r \longmapsto (r+I_1,\ldots,r+I_n) \end{cases}$$

induces an isomorphism

$$R/(\bigcap_{i=1}^{n} I_i) \cong \prod_{i=1}^{n} (R/I_i)$$

Proof. It clearly suffices to verify that, given $r_1,\ldots,r_n \in R$ there exists $r \in R$ such that $r \equiv r_i (\text{mod } I_i)$ for all i. We argue by induction on n. If $n = 2$, we have an expression $1 = a_1 + a_2$ with $a_i \in I_i$, and we let $r = r_2 a_1 + r_1 a_2$. Assume the result holds for $n-1$ ideals. For each $i \geq 2$, we can find elements $a_i \in I_1$ and $b_i \in I_i$ such that $a_i + b_i = 1$, $i \geq 2$. The product $\prod_{i=2}^{n} (a_i+b_i)$ is equal to 1 and is contained in $I_1 + \prod_{i=2}^{n} I_i$, and so

$$I_1 + \prod_{i=2}^{n} I_i = R$$

Applying the case $m = 2$, we can find an element $y_1 \in R$ such that

$$y_1 \equiv 1 (\text{mod } I_1)$$
$$y_1 \equiv 0 (\text{mod } \prod_{i=2}^{n} I_i)$$

We find similarly elements $y_2,\ldots,y_n$ such that

$$y_j \equiv 1 (\text{mod } I_j) \quad \text{and} \quad y_j \equiv 0 (\text{mod } I_i) \quad \text{for } i \neq j$$

Then $r = r_1y_1 + \ldots + r_ny_n$ satisfies our requirements. ■

Let L/F and E/F be field extensions. By a *composite* of L and E over F we understand a triple (K,ϕ,ψ), where K/F is a field extension and

$$\phi : L \to K,\ \psi : E \to K$$

are F-homomorphisms such that K is generated (as a field) by the images of L and E. In the case ϕ and ψ are not pertinent to the discussion, we refer to K itself as a composite of E and L. Two such composites K and K' are said to be *equivalent* if there is an F-isomorphism $f : K \to K'$ which renders commutative the following diagram:

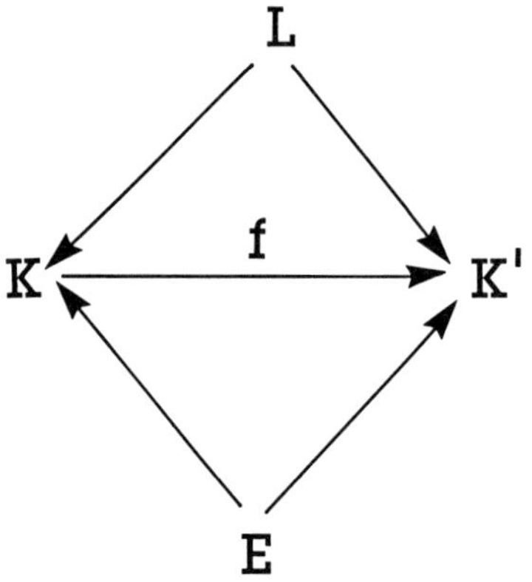

Now assume that E/F is finite. Then $L \otimes_F E$ is a finite-dimensional commutative L-algebra. Therefore $L \otimes_F E$ has only finitely many maximal ideals

$$M_1,\ldots,M_n\ ,$$

say, and we put

$$K_i = (L \otimes_F E)/M_i \qquad (1 \leqslant i \leqslant n)$$

Then each K_i is a composite of L and E via the natural embeddings. Moreover, the following result holds:

5.5. LEMMA. Let L/F and E/F be field extensions, let $(E:F) < \infty$ and let the K_i, $1 \leqslant i \leqslant n$, be defined as above.

(i) $K_1,\ldots,K_n$ are all nonequivalent composites of L and E and, for any full set $A_1,\ldots,A_n$ of such nonequivalent composites

$$(L \underset{F}{\otimes} E)/J(L \underset{F}{\otimes} E) \cong A_1 \times \dots \times A_n \qquad \text{(as } L\text{-algebras)} \qquad (12)$$

In particular

$$(E:F) \geqslant \sum_{i=1}^{n} (A_i:L)$$

with equality if and only if $J(L \underset{F}{\otimes} E) = 0$

(ii) If $E = F(\theta)$ and f is the minimal polynomial of θ over F, then f has precisely n distinct irreducible factors in $L[X]$, say $f_1, f_2, \dots, f_n$. Furthermore, if α_i is a root of f_i, then with respect to the natural embedding $L \longrightarrow L(\alpha_i)$ and the F-homomorphism $F(\theta) \longrightarrow L(\alpha_i)$, $\theta \longmapsto \alpha_i$, the

$$L(\alpha_i),\ 1 \leqslant i \leqslant n,$$

are all nonequivalent composites of L and E and, provided E/F is separable, we have

$$(E:F) = \sum_{i=1}^{n} (A_i:L) \qquad (13)$$

Proof. (i) Because $J(L \underset{F}{\otimes} E) = \bigcap_{i=1}^{n} M_i$, it follows from Lemma 5.4 that (12) holds for the $K_1, \dots, K_n$. But if K_i is equivalent to A_i, then $K_i \cong A_i$ as L-algebras. Hence (12) will hold in the general case, provided we show that $K_1, \dots, K_n$ are all nonequivalent composites of L and E.

Let (K, α, β) be any composite of L and E. Then there is a homomorphism $\gamma : L \underset{F}{\otimes} E \longrightarrow K$ given by $\gamma(x \otimes y) = \alpha(x)\beta(y)$, $x \in L$, $y \in E$, and $\mathrm{Ker}\gamma$ is a maximal ideal, say M_i, of $L \underset{F}{\otimes} E$. Hence γ induces an F-homomorphism

$$f : K_i \longrightarrow K,$$

which in view of our choice of γ, renders the corresponding diagram commutative. Hence K is equivalent to K_i.

Finally, assume that K_i and K_j are equivalent. Then the commutativity of the corresponding diagram shows that the map $x \otimes y + M_i \longmapsto x \otimes y + M_j$ is an isomorphism of K_i and K_j. Since the latter is possible only if $i = j$, the required assertion follows.

(ii) Since $F(\theta) \cong F[X]/(f)$, we have an exact sequence

$$0 \longrightarrow (f) \longrightarrow F[X] \longrightarrow F(\theta) \longrightarrow 0$$

Tensoring with L, we obtain the exact sequence

$$0 \longrightarrow (f) \longrightarrow L[X] \longrightarrow L \otimes_F F(\theta) \longrightarrow 0$$

and thus

$$L \otimes_F E \cong L[X]/(f) \tag{14}$$

Let $f = f_1^{e_1} \ldots f_r^{e_r}$ be the decomposition of f over L, where the f_i are irreducible and distinct. Then, by (14) and Proposition 1.2.17,

$$(L \otimes_F E)/J(L \otimes_F E) \cong \prod_{i=1}^{r} L[X]/(f_i) \cong \prod_{i=1}^{r} L(\alpha_i)$$

and hence $r = n$. Moreover, if $F(\theta)/F$ is separable, then $e_1 = e_2 = \ldots = e_n = 1$ and so, by Proposition 1.2.17, $J(L \otimes_F E) = 0$. Applying (i), we therefore conclude that (13) holds.

Finally, assume that $L(\alpha_i)$ is equivalent to $L(\alpha_j)$. Then the commutativity of the corresponding diagram shows that there is an L-isomorphism

$$f : L(\alpha_i) \longrightarrow L(\alpha_j)$$

with $f(\alpha_i) = \alpha_j$. Since the latter obviously implies that $i = j$, the result follows. ■

Let E/F be a finite field extension and let v be a valuation of F. We now take up the problem of determining all the extensions of v to E. The case in which F is complete has been treated in Theorem 5.3. In what follows we shall use this theorem to treat the general case.

Let $(\widetilde{F},\widetilde{v})$ be the completion of F at v and let (K,ϕ,ψ) be a composite of $\widetilde{F}$ and E over F. Since $(E:F) = m < \infty$, we have $(K:\phi(\widetilde{F})) \leqslant m < \infty$. The valuation $\widetilde{v}$ of $\widetilde{F}$ can be transferred to $\phi(\widetilde{F})$ by defining $\widetilde{v}_\phi(\phi(x)) = \widetilde{v}(x)$ for all $x \in \widetilde{F}$. Clearly $\widetilde{v}_\phi$ coincides with v on F. Since $\widetilde{F}$ is complete relative to $\widetilde{v}$, it is clear that $\phi(\widetilde{F})$ is complete relative to $\widetilde{v}_\phi$. Since K is a finite extension of $\phi(\widetilde{F})$, $\widetilde{v}_\phi$ has a unique extension to a valuation $\widetilde{w}_\phi$ of K. Hence the map $w_\phi : E \longrightarrow \mathbb{R}$ defined by

$$w_\phi(x) = \widetilde{w}_\phi(\psi(x))$$

is a valuation which extends v. We shall refer to w_ϕ as the *valuation of* E

associated with the composite (K,ϕ,ψ). Note that, by Lemma 5.5, there are only finitely many nonequivalent composites of $\widetilde{F}$ and E. We are at last in a position to attain our main objective, which is to prove the following result.

5.6. THEOREM. Let E/F be a finite field extension, let v be a valuation of F and let $(\widetilde{F},\widetilde{v})$ be the completion of F at v. Let (K_i,ϕ_i,ψ_i), $1 \leqslant i \leqslant n$, be all nonequivalent composites of $\widetilde{F}$ and E over F and, for each i, let v_i be the valuation of E associated with (K_i,ϕ,ψ_i). Then $v_1,v_2,\ldots,v_n$ are all distinct valuations of E extending v (in fact, all the v_i are nonequivalent). Moreover, if E_i is the completion of E with respect to v_i, then

$$(E:F) \geqslant \sum_{i=1}^{n} (E_i:\widetilde{F})$$

with equality if E/F is separable.

Proof. First, suppose the two composites (K,α,β) and (K',γ,σ) of $\widetilde{F}$ and E are equivalent. Then we have an F-isomorphism $f : K \longrightarrow K'$ such that

$$f\alpha(x) = \gamma(x),\ x \in \widetilde{F},$$

and $f\beta(y) = \sigma(y)$, $y \in E$. For the valuations on $\alpha(\widetilde{F})$ and $\gamma(\widetilde{F})$ we have

$$\widetilde{v}_\gamma(\gamma(x)) = \widetilde{v}(x) = \widetilde{v}_\alpha(\alpha(x))$$

Hence

$$\widetilde{v}_\gamma(f\alpha(x)) = \widetilde{v}_\alpha(\alpha(x)) \qquad \text{for all } x \in \widetilde{F}$$

Let $\widetilde{w}_\alpha$ and $\widetilde{w}_\gamma$ be the valuations of K and K', respectively, which extend $\widetilde{v}_\alpha$ and $\widetilde{v}_\gamma$. Now $\phi(f(y)) = \widetilde{w}_\alpha(y)$, $y \in K$, defines a valuation of K' such that

$$\phi(f\alpha(x)) = \widetilde{w}_\alpha(\alpha(x)) = \widetilde{v}_\alpha(\alpha(x)) = \widetilde{v}_\gamma(f\alpha(x)) \qquad \text{for all } x \in \widetilde{F}$$

Thus ϕ is an extension of the valuation $\widetilde{v}_\gamma$ of $\gamma(\widetilde{F})$. Since $\gamma(\widetilde{F})$ is complete, this extension is unique and therefore it coincides with $\widetilde{w}_\gamma$. Hence we have $\widetilde{w}_\alpha(y) = \widetilde{w}_\gamma(f(y))$ for all $y \in K$, which in turn implies that

$$w_\alpha(x) = \widetilde{w}_\alpha(\beta(x)) = \widetilde{w}_\gamma(f\beta(x)) = \widetilde{w}_\gamma(\sigma(x)) = w_\gamma(x) \qquad \text{for all } x \in E$$

Thus equivalent composites give the same valuation.

Conversely, assume that $w_\alpha = w_\gamma$. Observe that K,K' are the closures of

$\beta(E), \sigma(E)$, respectively, in the topology defined by the valuations of K, K'. Hence K, K' are completions of $\beta(E)$, $\delta(E)$, respectively, with respect to the restrictions of $\tilde{w}_\alpha$, $\tilde{w}_\gamma$ to $\beta(E)$, $\delta(E)$. By the uniqueness argument of Theorem 4.1, the isometric isomorphism $\beta(E) \longrightarrow \delta(E)$, $\beta(x) \longmapsto \delta(x)$, has a unique extension to an isometric isomorphism $f : K \to K'$. We have $\tilde{w}_\alpha(y) = \tilde{w}_\gamma(f(y))$ for all $y \in K$ and $f\beta(x) = \delta(x)$ for all $x \in E$. Since $\alpha(\tilde{F})$, $\gamma(\tilde{F})$ are the closures of F in K, K', respectively, and since f is the identity on F, it is clear that f maps $\alpha(\tilde{F})$ onto $\gamma(\tilde{F})$. Hence the restriction of f to $\alpha(\tilde{F})$ is an isometric isomorphism which is the identity on F. On the other hand, the map $\alpha(x) \longmapsto \gamma(x)$ has the same properties since $\tilde{v}_\alpha(\alpha(x)) = \tilde{v}(x) = \tilde{v}_\gamma(\gamma(x))$. Hence, by Theorem 4.1, $\alpha(x) \longmapsto \gamma(x)$ coincides with f. Thus $f\alpha(x) = \gamma(x)$ for all $x \in \tilde{F}$, proving that K and K' are equivalent.

It will next be shown that every valuation w of E which is an extension of v can be obtained from a composite in the manner indicated. To see this, let $\tilde{E}$ be the completion of E relative to w and let $\psi : E \longrightarrow \tilde{E}$ be the canonical embedding. Now we have an isomorphism ϕ of $\tilde{F}$ into the closure of F in $\tilde{E}$. The subfield of $\tilde{E}$ generated by $\phi(\tilde{F})$ and $\psi(E)$ is a finite extension of $\phi(\tilde{F})$, so it is complete relative to the valuation obtained from $\tilde{E}$. It follows that this coincides with $\tilde{E}$. Thus we have a composite $(\tilde{E}, \phi, \psi)$ of $\tilde{F}$ and E and one immediately verifies that the valuation of E obtained from this composite coincides with w. This proves that $v_1, \ldots, v_n$ are distinct valuations of E extending v. If v is trivial, then $\tilde{F} = F$ and hence $n = 1$. If v is non-trivial and v_i, v_j are equivalent, say $v_i(x) = v_j(x)^r$ for all $x \in E$ and some $r > 0$ in $\mathbb{R}$, then $v(y) = v(y)^r$ for all $y \in F$ and hence $r = 1$. This proves the first assertion and the second assertion follows by virtue of Lemma 5.5. ∎

The special case in which $E = F(\Theta)$ is a simple algebraic extension of F deserves a separate mention. This is due to the fact that, by Lemma 5.5, one can choose a distinguished family of all nonequivalent composites of $\tilde{F}$ and E and to determine their number exclusively in terms of the minimal polynomial of Θ over F.

5.7. COROLLARY. Let $E = F(\theta)$ be a simple algebraic extension of F, let v be a valuation of F and let $(\tilde{F},\tilde{v})$ be the completion of F at v. Let f be the minimal polynomial of θ over F, let $f_1, f_2, \ldots, f_n$ be all distinct irreducible factors of f in $\tilde{F}[X]$ and let α_i be a root of f_i, $1 \leqslant i \leqslant n$. For each $i \in \{1,2,\ldots,n\}$, let w_i be the unique extension of $\tilde{v}$ to $\tilde{F}(\alpha_i)$ and let $\psi_i : F(\theta) \longrightarrow \tilde{F}(\alpha_i)$ be the F-homomorphism which sends θ to α_i. Then the maps $v_i : E \longrightarrow \mathbb{R}$, $1 \leqslant i \leqslant n$, given by

$$v_i(x) = w_i(\psi_i(x))$$

are all distinct valuations of E extending v (in fact, all the v_i are non-equivalent)

Proof. Direct consequence of Lemma 5.5 and Theorem 5.6. ■

5.8. COROLLARY. Let E/F be a finite field extension and let v be a valuation of E such that the restriction w of v to F is nontrivial. Then v is discrete if and only if so is w.

Proof. If v is discrete, then certainly so is w. The converse is a consequence of Theorems 4.1, 5.3 and 5.6. ■

6. VALUATIONS OF ALGEBRAIC NUMBER FIELDS

The general theory of the preceding sections is well illustrated by the example of an algebraic number field.

Let $F = \mathbb{Q}(\theta)$ be an algebraic number field, that is, a finite extension of the rational number field $\mathbb{Q}$ generated by adjunction of a primitive element θ. Let $f(X) \in \mathbb{Q}[X]$ be the minimal polynomial of θ and let $\alpha_1, \alpha_2, \ldots, \alpha_{r_1}$ (respectively, $\{\beta_1,\bar{\beta}_1\}, \ldots, \{\beta_{r_2},\bar{\beta}_{r_2}\}$) be all real roots (respectively, all pairs of conjugate complex roots) of $f(X)$. Then $f(X)$ splits into $r_1 + r_2$ real irreducible factors and

$$(F:\mathbb{Q}) = r_1 + 2r_2$$

The pair $[r_1, r_2]$ is called the *signature* of F. Those fields F for which $r_2 = 0$ are called *totally real*, and those with $r_1 = 0$ are called *totally complex.*

Let $\lambda_1, \lambda_2, \ldots, \lambda_{r_1}$ be the real embeddings $F \longrightarrow \mathbb{R}$ defined by

$$\lambda_i(\theta) = \alpha_i \qquad (1 \leqslant i \leqslant r_1)$$

and let $\mu_1, \mu_2, \ldots, \mu_{r_2}$ be the complex embeddings $F \longrightarrow \mathbb{C}$ defined by

$$\mu_j(\theta) = \beta_j \qquad (1 \leqslant j \leqslant r_2)$$

Let R be the ring of integers of F and, for each nonzero prime ideal P of R, let v_P be the P-adic valuation of F defined by

$$v_P(x) = a^{-\nu_P(x)} \qquad (x \in F)$$

where $\nu_P(x)$ is the exponent to which P occurs in the factorization of Rx into a product of integral powers of prime ideals of R, and $a > 1$ is any fixed element in $\mathbb{R}$. Recall that v_P is called normalized if $a = N(P)$.

6.1. THEOREM (Ostrowski(1935)). Further to the notation above, let Ω be the set of all nonzero prime ideals of R and let $v_i : F \longrightarrow \mathbb{R}$ and $w_j : F \longrightarrow \mathbb{R}$ be defined by

$$v_i(x) = |\lambda_i(x)| \qquad (1 \leqslant i \leqslant r_1)$$
$$w_j(x) = |\mu_j(x)| \qquad (1 \leqslant j \leqslant r_2)$$

Then

$$\{v_i | 1 \leqslant i \leqslant r_1\} \cup \{w_j | 1 \leqslant j \leqslant r_2\} \cup \{v_P | P \in \Omega\}$$

is a full set of nonequivalent nontrivial valuations of F. In particular, F has precisely $r_1 + r_2$ archimedean primes and all other primes are discrete.

Proof. Let v be a nontrivial valuation of F. For the sake of clarity, we divide the proof into two steps.

Step 1. Here we treat the case where v is archimedean.

The restriction of v to $\mathbb{Q}$ is archimedean, by virtue of Proposition 1.7. Hence by Proposition 1.11, we may assume that v restricts to the ordinary absolute value of $\mathbb{Q}$. If $(\tilde{\mathbb{Q}}, \tilde{v})$ is the completion of $\mathbb{Q}$ at $|\ |$, then $\tilde{\mathbb{Q}} = \mathbb{R}$ and $\tilde{v}$ is the ordinary absolute value of $\mathbb{R}$. Thus all nonequivalent extensions of $|\ |$ to F provide a full set of nonequivalent archimedean valuations of F. Now

$\mathbb{R}(\alpha_i) = \mathbb{R}$ and $\mathbb{R}(\beta_j) = \mathbb{C}$, $1 \leqslant i \leqslant r_1$, $1 \leqslant j \leqslant r_2$. Hence the unique extension of $\tilde{v}$ to $\mathbb{R}(\alpha_i)$ and $\mathbb{R}(\beta_j)$ is again the ordinary absolute value. The desired conclusion is therefore a consequence of Corollary 5.7.

Step 2. Now we treat the case where v is non-archimedean.

To begin with, note that the v_P, $P \in \Omega$, are nonequivalent, by virtue of Theorem 3.18(ii). Let ν be the exponential valuation corresponding to v. We assert that $\nu(r) \geqslant 0$ for all $r \in R$. Assume the contrary, say $\nu(r) < 0$ for some $r \in R$. Because r is integral over $\mathbb{Z}$, we have

$$r^n = z_1 r^{n-1} + \dots + z_n$$

where the z_i are suitable elements of $\mathbb{Z}$. Then $\nu(z_i) \geqslant 0$ and

$$n\nu(r) = \nu(r^n) \geqslant \min\{\nu(z_1 r^{n-1}), \dots, \nu(z_n)\}$$

so that $n\nu(r) \geqslant \nu(z_i r^{n-i}) \geqslant (n-i)\,\nu(r)$ for some $i \in \{1,2,\dots,n\}$, which is clearly impossible.

The set of all x in R with $\nu(x) > 0$ is a prime ideal, say P, in R. Let ν_P be the exponential valuation corresponding to v_P. Note that if

$$\nu_P(x) = 0$$

for $x \in R$, then $x \notin P$ and thus $\nu(x) = 0$. We now show that ν and ν_P are equivalent which will finish the proof. Since F is the quotient field of R, it suffices to verify that there exists $c > 0$ in $\mathbb{R}$ such that $\nu(x) = c\nu_P(x)$ for all $x \in R$.

To this end, fix $x_0 \in P - P^2$ and put $c = \nu(x_0)$. Then $\nu_P(x_0) = 1$ and c is a positive real number. Furthermore, given $x \in R$, there exists an ideal Q in R which is not divisible by P and such that

$$Rx = P^r Q$$

where $r = \nu_P(x)$. In Q there is an element y with $\nu_P(y) = 0$ and hence with $\nu(y) = 0$. Write $zx = x_0^r y$ for some $z \in R$. Then $\nu_P(z) = 0$ since

$$\nu_P(x) = \nu_P(x_0^r) = r$$

and $\nu_P(y) = 0$. Hence $\nu(z) = 0$ and thus $\nu(x) = r\nu(x_0) = c\nu_P(x)$, as required. ■

The valuations of F exhibited in Theorem 6.1, where each v_P is normalized, are called *canonical*. The valuations $v_i (1 \leq i \leq r_1)$ and $w_j (1 \leq j \leq r_2)$ are called *real* and *complex* , respectively. As is customary, we adopt the convention of calling each generalized valuation w_j^2 a valuation, although it need not satisfy the triangular inequality. This convention allows us to define the set of all normalized valuations of F by

$$\{v_i |\ 1 \leq i \leq r_1\} \cup \{w_j^2 |\ 1 \leq j \leq r_2\} \cup \{v_P | P \in \Omega\}$$

where Ω is the set of all nonzero prime ideals of R and each v_P is normalized.

Let K be an arbitrary field, let $x \in K$ and let p be a prime of K. If $v(x) = 1$ for some v in p, then $v(x) = 1$ for all v in p. We may therefore write unambiguously $p(x) = 1$ to mean $v(x) = 1$ for some (hence all) v in p.

6.2. COROLLARY. Let R be the ring of integers of an algebraic number field F and let $\{p_i | i \in I\}$ be the set of all non-archimedean primes of F. Then

$$U(R) = \{x \in F | p_i(x) = 1 \text{ for all } i \in I\}$$

Proof. By Theorem 6.1, we may choose v_i in p_i such that

$$\{v_i | i \in I\} = \{v_P | P \in \Omega\},$$

where Ω is the set of all nonzero prime ideals of R. The desired conclusion is therefore a consequence of Theorem 3.18(v). ∎

7. RAMIFICATION INDEX AND RESIDUE DEGREE

In what follows E/F denotes a finite field extension of degree n, v an exponential valuation on F, with an extension w to E. We shall write R_v, P_v, Γ_v for the valuation ring, maximal ideal and the value group of v, and R_w, P_w, Γ_w for the corresponding entities for w. In case both v and w are principal, we write p and π for the prime elements of R_v and R_w (i.e. the elements of least positive value). We say that E/F is an *extension of complete discrete valued fields* if v and w are principal and complete.

Since w restricts to v on F, we have $\Gamma_v \subseteq \Gamma_w$. The index

$$e = e_{E/F} = (\Gamma_w : \Gamma_v)$$

is called the *ramification index* of w (in the extension E/F). Since

$$R_v \cap P_w = P_v,$$

the natural homomorphism $R_v/P_v \longrightarrow R_w/P_w$ is an embedding, so denoting the residue class fields K_v, K_w, respectively, we see that K_w is an extension of K_v. The degree

$$f = f_{E/F} = (K_w : K_v)$$

is called the *residue degree* of w (in the extension E/F). In the special case where $F = \mathbb{Q}_p$, the rational p-adic field, we refer to e and f as the *absolute ramification index* and *absolute residue degree*, respectively. Both the residue degree and ramification index are finite and, in fact, we have

7.1. THEOREM. (i) $ef \leqslant n$

(ii) If v is principal and complete, then

(a) w is principal and complete, $e = w(p)/w(\pi)$, $n = ef$ and $R_w p = P_w^e$

(b) R_w is a free R_v-module of rank n with basis $\{u_i\pi^j \mid 1 \leqslant i \leqslant f, 0 \leqslant j \leqslant e-1\}$, where $u_1,\dots,u_f$ are any elements of R_w such that their images in K_w form a basis of K_w over K_v. Furthermore, the $u_i\pi^j$ form a basis of E over F.

Proof. For the sake of clarity, we divide the proof into three steps.

Step 1. Let $u_1,\dots,u_r \in R_w$ be such that their images $\bar{u}_1,\dots,\bar{u}_r$ in K_w are linearly independent over K_v. We claim that for any $\lambda_1,\dots,\lambda_r \subset F$,

$$w(\lambda_1 u_1 + \dots + \lambda_r u_r) = \min\{v(\lambda_1),\dots,v(\lambda_r)\} \tag{1}$$

To prove (1), we may assume, after appropriate renumbering that

$$v(\lambda_1) = \dots = v(\lambda_k) < v(\lambda_i) \qquad (i > k)$$

Because $\bar{u}_i \neq 0$, we have $w(u_i) = 0$ and so the left-hand side of (1) has value $\geqslant v(\lambda_1)$. If the value is $> v(\lambda_1)$, then after dividing by λ_1 and reducing mod P_w we would get

$$\bar{u}_1 + (\overline{\lambda_2/\lambda_1})\bar{u}_2 + \dots + (\overline{\lambda_r/\lambda_1})\bar{u}_r = 0$$

contrary to the linear independence of the $\bar{u}_i$ over K_v. This establishes (1).

Step 2. Let $x_1,\ldots,x_s \in E$ be such that the $w(x_j)$ are incongruent mod Γ_v. We assert that the rs elements $u_i x_j$, $1 \leq i \leq r$, $1 \leq j \leq s$, are F-linearly independent. This will prove that $rs \leq n$; in particular, taking $s = e$, $r = f$ we obtain (i).

Assume that there is a relation

$$\Sigma\lambda_{ij}u_i x_j = 0 \qquad (\lambda_{ij} \in F) \tag{2}$$

Setting $a_j = \Sigma u_i \lambda_{ij}$, we then have $\Sigma a_j x_j = 0$. It follows that

$$w(a_h x_h) = w(a_k x_k)$$

for some h,k, $1 \leq h < k \leq s$. This means that

$$w(x_h/x_k) = w(x_h) - w(x_k) = w(a_k) - w(a_h) \tag{3}$$

But the right-hand side of (3) is in Γ_v, by (1), and the left-hand side is not, by our choice of the x_i. Hence all the a_j vanish and so, again by (1), the λ_{ij} must also vanish, which shows that the $u_i x_j$ are F-linearly independent.

Step 3. Completion of the proof.

Assume that v is prinicpal and complete. Then w is principal and complete, by Corollary 5.8 and Theorem 5.3. Since both v and w are principal,

$$\Gamma_v = v(p)\mathbb{Z} \text{ and } \Gamma_w = w(\pi)\mathbb{Z}$$

Let $k \geq 1$ be such that $R_w p = P_w^k$. Because $P_w^k = (\pi^k)$, we have $w(p) = kw(\pi)$. Hence

$$e = (\Gamma_w : \Gamma_v) = w(p)/w(\pi) = k$$

We now claim that every element of R_w is an R_v-linear combination of the elements $u_i\pi^j$, $1 \leq i \leq f$, $0 \leq j \leq e-1$. If sustained, it will follow (by multiplying by a negative power of π) that every element of E is an F-linear combination of the $u_i\pi^j$. Hence $n \leq ef$ and, by (i) $n = ef$. Furthermore, it will follow that the $u_i\pi^j$ form an F-basis of E and hence an R_v-basis of R_w, which will complete the proof.

Choose $x \in R_w$. Then $w(x) = w(\pi^k)$ for some $k \geq 0$. We can write

$$k = m_1 e + j_1$$

with $m_1 \geqslant 0$, $0 \leqslant j_1 \leqslant e-1$. Then $w(x) = w(p^{m_1}\pi^{j_1})$ (since $e = w(p)/w(\pi)$) so $y = (p^{m_1}\pi^{j_1})^{-1}x$ satisfies $w(y) = 0$. Then the definition of the u_i shows that there exist elements $\alpha_{1i} \in R_v$ such that $y - \sum_{i=1}^{f} \alpha_{1i}u_i \in P_w$. Then

$$w(\sum_{i=1}^{f} \alpha_{1i}u_i) = w(y) = 0$$

and, if $x = p^{m_1}\pi^{j_1}(y - \sum_{i=1}^{f} \alpha_{1i}u_i)$, then $w(x_1) > w(x)$. We have

$$x = p^{m_1}\pi^{j_1}y = p^{m_1}\pi^{j_1}(\sum_{i=1}^{f} \alpha_{1i}u_i) + x_1 \tag{4}$$

We may repeat this argument with x_1 and obtain a sequence $x_1, x_2, \ldots$ such that

$$x_{k-1} = p^{m_k}\pi^{j_k}(\sum_{i=1}^{f} \alpha_{ki}u_i) + x_k \qquad (\alpha_{ki} \in R_v, m_k \geqslant 0,\ 0 \leqslant j_k \leqslant e-1) \tag{5}$$

where

$$w(\sum_{i=1}^{f} \alpha_{ki}u_i) = 0 \quad \text{and} \quad w(x_k) > w(x_{k-1})$$

Then (5) implies that $w(x_{k-1}) = w(p^{m_k}\pi^{j_k})$. It follows that $x_k \longrightarrow 0$, $p^{m_k} \longrightarrow 0$, and $(\sum_{i=1}^{f} \alpha_{ki}u_i)p^{m_k} \longrightarrow 0$. The latter implies that every infinite series whose terms form a subsequence of the sequence $(\sum_{i=1}^{f} \alpha_{ki}u_i)p^{m_k}$, $k = 1,2,\ldots$, converges. By (4) and (5), we have

$$x = p^{m_1}\pi^{j_1}(\Sigma\alpha_{1i}u_i) + \ldots + p^{m_k}\pi^{j_k}(\Sigma\alpha_{ki}u_i) + x_k \tag{6}$$

Because $x_k \longrightarrow 0$ and the coefficients of the various powers of π^j, $0 \leqslant j \leqslant e-1$, in (6) converge, we obtain from (6) that

$$x = \Sigma\beta_{ij}u_i\pi^j \qquad (0 \leqslant j \leqslant e-1)$$

for some $\beta_{ij} \in R_v$. This completes the proof of the theorem. ∎

7.2. REMARK. Let E/F be an extension of complete discrete valued fields. Then the proof of Step 3 shows that if $f_{E/F}$ is finite, then so is E/F.

Let E/F be an extension of complete discrete valued fields. We say that E/F is *unramified* if $e = 1$ (equivalently, if $f = n$). In case $e = n$ (equivalently, $f = 1$), E/F is said to be *totally ramified*

Let E/F be a field extension and let v_1, v_2 be two valuations of E. Then v_1 and v_2 are said to be *F-conjugate* if $v_1 = v_2^{\sigma}$ for some $\sigma \in \mathrm{Gal}(E/F)$, where

$$v_2^{\sigma}(x) = v_2(\sigma(x)) \qquad \text{for all} \quad x \in E$$

It is clear that conjugate valuations restrict to the same valuation of F. The following result shows that the converse is also true, provided E/F is a finite Galois extension.

7.3. THEOREM. Let E/F be a finite field extension of degree n, let v be an exponential valuation of F and let $v_1,\dots,v_r$ be all distinct valuations of E which extend v. Let e_i, f_i be the ramification index and residue degree of v_i in the extension E/F, $1 \leqslant i \leqslant r$.

(i) $\sum_{i=1}^{r} e_if_i \leqslant n$

(ii) $\sum_{i=1}^{r} e_if_i = n$, if E/F is separable and v is principal

(iii) If E/F is Galois, then all the v_i are F-conjugate and hence $e_i = e$, $f_i = f$ independently of i. In particular, $efr \leqslant n$ and, if E/F is Galois and v is principal, then $efr = n$.

Proof. (i) and (ii). Let E_i be the completion of E at v_i and let $\tilde{F}$ be the completion of F at v. Owing to Theorem 5.6, we have

$$n \geqslant \sum_{i=1}^{r} (E_i:\tilde{F})$$

with equality if E/F is separable. By Theorem 4.1, E_i and E have the same value group relative to v_i, and $\tilde{F}$ and F have the same value group relative to v. Hence the ramification index e_i of E over F relative to v_i is the same as that of E_i over $\tilde{F}$. Similarly, Theorem 5.6 and the definitions show that the residue degree f_i of E/F relative to v_i is the same as that of E_i over $\tilde{F}$. By Theorem 7.1, we have

$$e_if_i \leqslant (E_i:\tilde{F}) \quad \text{and} \quad e_if_i = (E_i:\tilde{F}) \quad \text{if} \quad v \quad \text{is principal}$$

Hence

$$\sum_{i=1}^{r} e_if_i \leqslant \sum_{i=1}^{r} (E_i:\tilde{F}) \leqslant n$$

in every case with equalities if E/F is separable and v is principal.

(iii) Fix $i,j \in \{1,2,\dots,r\}$ and consider the valuations of E defined by

$$x \longmapsto v_i(\sigma(x)),\ x \longmapsto v_j(\sigma(x)) \qquad \text{for all} \quad \sigma \in G = \mathrm{Gal}(E/F)$$

If these two sets of valuations are not disjoint, we have $v_i^\sigma = v_j^\tau$ for some $\sigma,\tau \in G$ and so v_i and v_j are F-conjugate. If they are disjoint, then by Theorem 1.9, there exists $\lambda \in E$ such that $v_i(\sigma(\lambda)) > 0$, $v_j(\sigma(\lambda) - 1) > 0$ for all $\sigma \in G$. Hence

$$v_j(\sigma(\lambda)) = 0 \quad \text{and} \quad v(N(\lambda)) = \sum_\sigma v_j(\sigma(\lambda)) = 0$$

but also $v(N(\lambda)) = \sum_\sigma v_i(\sigma(\lambda)) > 0$, which is a contradiction. ■

We now apply the foregoing to extensions of Dedekind domains. Let R be a Dedekind domain with quotient field F, let E/F be a finite separable field extension and let S be the integral closure of R in E. We know, from Theorem 3.3, that S is a Dedekind domain and a finitely generated R-module. If Q is a nonzero prime ideal of S and if $P = Q \cap R$, we will say that Q *divides* P or that Q *lies over* P, and we will write $Q|P$. This relation is also equivalent to saying that $Q \supseteq PS$ Denote by e_Q the exponent of Q in the decomposition of PS into prime ideals. Thus

$$PS = \prod_{Q|P} Q^{e_Q} \tag{7}$$

The integer e_Q is called the *ramification index* of Q in the extension E/F. If Q is a nonzero prime ideal of S and $P = Q \cap R$, then the field S/Q is a finite extension of R/P. The degree of this extension is called the *residue degree* of Q in the extension E/F, and is denoted by f_Q. Thus

$$f_Q = (S/Q:R/P)$$

In case one wants to specify P, it is customary to write $e_{Q/P}$ and $f_{Q/P}$ instead of e_Q and f_Q, respectively. When there is only one prime ideal Q dividing P and $f_Q = 1$, one says that E/F is *totally ramified* at P. If $e_Q = 1$ and S/Q is separable over R/P, then we say that E/F is *unramified* at Q.

In what follows, for any nonzero prime ideal L of R or S, we write ν_L for the exponential valuation associated with L.

7.4. LEMMA. Let P be a nonzero prime ideal of R.

(i) $\{(1/e_Q)\nu_Q | Q|P\}$ is the set of all distinct valuations of E extending ν_P

(ii) For each prime ideal Q of S lying over P, e_Q and f_Q are the ramification index and residue degree of $(1/e_Q)\nu_Q$, respectively.

Proof. (i) By (7), $\prod_{Q|P} Q^{e_Q}$ is the greatest common divisor of all xS, $x \in P$. Hence we may choose $\lambda \in P$ such that $\lambda \in Q^{e_Q} - Q^{e_Q+1}$. Since $P^2 \subseteq \prod_{Q|P} Q^{2e_Q}$, we must also have $\lambda \notin P^2$. This shows that $\nu_P(\lambda) = 1$ and $\nu_Q(\lambda) = e_Q$. Now, if $r \in R$, then $r = \lambda^n u$ for some $n \geq 1$ and some unit u of $R_P \subseteq S_Q$. Hence

$$\nu_Q(r) = ne_Q = e_Q \nu_P(r)$$

and therefore $\nu_Q(x) = e_Q \nu_P(x)$ for all $x \in F$.

Finally, let w be any valuation of E extending ν_P. Since ν_P is discrete, so is w by Corollary 5.8. Let R_w and P_w be the valuation ring and maximal ideal of w, respectively. Then R_w is integrally closed in its quotient field E and contains R, hence $R_w \supseteq S$. Let $Q = P_w \cap S$. Obviously $Q \cap R = P$, so Q divides P. The ring R_w thus contains S_Q. But one checks immediately that every discrete valuation ring is a maximal subring of its quotient field. Hence $R_w = S_Q$ so that w is equivalent to ν_Q, say

$$w(x) = n\nu_Q(x)$$

for all $x \in E$. Since

$$w(x) = n\nu_Q(x) = \nu_P(x) = (1/e_Q)\nu_Q(x) \qquad \text{for all } x \in F$$

we see that $w = (1/e_Q)\nu_Q$, as required.

(ii) The residue degree of $(1/e_Q)\nu_Q$ is $(S_Q/QS : R_P/PR_P)$ which is also

$$(S/Q:R/P) = f_Q,$$

by Theorem 3.18(iii). The value group of $(1/e_Q)\nu_Q$ is $(1/e_Q)\mathbb{Z}$, while the value group of ν_P is $\mathbb{Z}$. Hence the corresponding ramification index is $((1/e_Q)\mathbb{Z}:\mathbb{Z}) = e_Q$, as required. ■

7.5. COROLLARY. Let R be a Dedekind domain with quotient field F, let E/F be a finite separable field extension of degree n and let S be the integral closure of R in E.

(i) For any nonzero prime ideal P of R,

$$n = \sum_{Q|P} e_Q f_Q$$

Furthermore, if E/F is Galois, then $e_Q = e$, $f_Q = f$ independently of $Q|P$ and, in particular, $n = efr$ where r the number of Q dividing P.

(ii) The number of prime ideals Q of S which divide a nonzero prime ideal P of R is at least 1 and at most n.

Proof. (i) Direct consequence of Lemma 7.4 and Theorem 7.3(ii), (iii).
(ii) Direct consequence of (i). ∎

8. STRUCTURE OF COMPLETE DISCRETE VALUED FIELDS

A. Notation and terminology.

Since we shall deal exclusively with non-archimedean valuations, it will be more convenient to write these valuations additively. Thus, in accordance with our previous conventions, we use the term "principal valuation" instead of "discrete valuation". Throughout, F denotes a *complete discrete valued field*. This means that we are given a fixed principal valuation v of F with respect to which F is complete. By a *prime* of F we understand an element of least positive value. We write R,P and $\bar{F}$ for the valuation ring, the unique maximal ideal and residue class field of v, respectively. Thus, by definition,

$$R = \{x \in F | v(x) \geqslant 0\}, \; P = \{x \in F | v(x) > 0\} \quad \text{and} \quad \bar{F} = R/P$$

Since v is fixed, we refer to R and $\bar{F}$ as the *valuation ring and residue class field of* F, respectively.

Assume that char$F \neq$ char$F$, i.e. char$F = 0$ and char$\bar{F} = p > 0$. Then we say that F is *absolutely unramified* if p is a prime element of F.

By a *residue system* of $\bar{F}$ in F we mean any set S of elements of R such that $s + P$, $s \in S$, are all distinct elements of $\bar{F}$. A residue system S is said to be multiplicative if S is closed under multiplication. Of course, in general, a multiplicative system need not exit . But if it does exist, then $0 \in S$ and the set S^* of nonzero elements of S forms a multiplicative group isomorphic to $\bar{F}^*$. In our treatment of F we shall distinguish two cases

(i) char$F$ = char$\bar{F}$ (the equal characteristic case)

(ii) char$F \neq$ char$\bar{F}$, i.e. char$F = 0$, char$\bar{F} = p > 0$ (the unequal characteristic case).

Let E/F be a finite field extension. Then there is a unique principal valuation w of E which extends v. Since E is complete with respect to w, we refer to E/F as a *finite extension of complete discrete valued fields*. Under these circumstances, we write R_F and P_F instead of R and P, and denote by R_E, P_E the corresponding quantities for E. The notation for residue class fields will remain unchanged, i.e.

$$\bar{F} = R_F/P_F \quad \text{and} \quad \bar{E} = R_E/P_E$$

B. The equal characteristic case.

Our aim is to investigate the structure of F in the equal characteristic case. In what follows, we write π for a prime element of F.

8.1. LEMMA. Let S be a residue system of $\bar{F}$ in F. Then

(i) Every element $r \in R$ can be written uniquely as a convergent series

$$r = \sum_{i=0}^{\infty} r_i \pi^i \qquad (r_i \in S)$$

and, conversely, every such series converges to an element of R.

(ii) Every element $x \in F$ can be written uniquely as a convergent series

$$x = \sum_{i>>-\infty} x_i \pi^i \qquad (x_i \in S)$$

and, conversely, every such series converges to an element of F. Here $i>>-\infty$ indicates that the summation is formally over all integers i, but for $i < 0$ only finitely many terms are nonzero.

Proof. Property (ii) follows from (i) by multiplying by a suitable negative power of π. If $r \in R$, then by definition of S, there is an $r_0 \in S$ such that $r \equiv r_0 (\text{mod } P)$; if one writes $r = r_0 + \pi x_1$ and applies the same procedure to x_1, one obtains an $r_1 \in S$ such that $r = r_0 + r_1\pi + x_2\pi^2$ and so on. Thus r defines a unique infinite series $\sum_{i=0}^{\infty} r_i \pi^i$ with $r_i \in S$ for all i. The series converges to r and one sees easily that it is unique. Conversely, every series of the form $\sum_{i=0}^{\infty} r_i \pi^i$ converges to an element of R, since its general term converges to zero and F is complete. ∎

8.2. LEMMA. Assume that $\text{char}\bar{F} = p > 0$. Then, for all $n,m \in \mathbb{N}$ and $a,b \in R$,

$$a \equiv b(\text{mod } P^n) \text{ implies } a^{p^m} \equiv b^{p^m}(\text{mod } P^{n+m})$$

Proof. We may clearly assume that $m = 1$. If $a = b + \lambda\pi^n$ with $\lambda \in R$, then

$$\begin{aligned} a^p &= b^p + \binom{p}{1}b^{p-1}\lambda\pi^n + \ldots + \binom{p}{p-1}b\lambda^{p-1}\pi^{(p-1)n} + \lambda^p\pi^{p^n} \\ &\equiv b^p + pb^{p-1}\lambda\pi^n(\text{mod } P^{n+1}) \end{aligned}$$

Because $\text{char}\bar{F} = p$, we have $p \equiv 0(\text{mod } P)$ and so $p\pi^n \equiv 0(\text{mod } P^{n+1})$. Thus

$$a^p \equiv b^p(\text{mod } P^{n+1}),$$

as required. ∎

Recall that a field F is perfect if either $\text{char}F = 0$ or $\text{char}F = p > 0$ and $F^p = F$. In the latter case, the map $F \longrightarrow F$, $\lambda \longmapsto \lambda^p$ is an automorphism. Hence, for any $n \geqslant 1$ and any $\lambda \in F$, there is a unique element $\lambda^{p^{-n}} \in F$ such that

$$(\lambda^{p^{-n}})^{p^n} = \lambda$$

8.3. THEOREM. Assume that $\bar{F}$ is perfect.

(i) If $\text{char}\bar{F} = \text{char}F$, then there is a unique residue system of $\bar{F}$ in F which is a field

(ii) If $\text{char}\bar{F} = p > 0$ and $\text{char}F = 0$, then there is a unique multiplicative residue system of $\bar{F}$ in F (this system consists precisely of those elements r of R for which r is a p^n-th power for all $n \geqslant 0$).

Proof. We divide the proof into two steps.

Step 1. Here we consider the case where $\text{char}\bar{F} = 0$ (and hence $\text{char}F = 0$). As $\mathbb{Z} \longrightarrow R \longrightarrow \bar{F}$ is injective, the homomorphism $\mathbb{Z} \longrightarrow R$ extends to $\mathbb{Q}$, and we see that R contains $\mathbb{Q}$. By Zorn's lemma, there exists a maximal subfield L of R. If S denotes the image of L in $\bar{F}$, we shall show that $S = \bar{F}$. This will prove the required assertion.

We first claim that $\bar{F}$ is algebraic over S. Indeed, otherwise there would exist an $r \in R$ whose image $\bar{r}$ in $\bar{F}$ is transcendental over S. Hence the subring $L[r]$ of R maps into $S[r]$, so is isomorphic to $L[X]$, and

$$L[r] \cap P = 0$$

But then R contains the field $L(r)$ of rational functions in r, contradicting the maximality of L.

By the above, any $\lambda \in \bar{F}$ has a minimal polynomial $\bar{f}(X)$ over S. Since $\mathrm{char}\bar{F} = 0$, λ is a simple root of $\bar{f}$. Let $f \in L[X]$ be the polynomial corresponding to $\bar{f}$ under the isomorphism $S \longrightarrow L$. By Lemma 5.1, there is an $x \in R$ such that $\bar{x} = \lambda$ and $f(x) = 0$, and one can lift $S[\lambda]$ into R by sending λ to x. Invoking the maximality of L, we deduce that $\lambda \in S$. Thus $S = \bar{F}$ as we wished to show.

Step 2. Here we treat the case where $\mathrm{char}\bar{F} = p > 0$.

Fix $\lambda \in \bar{F}$ and, for all $n \geq 0$, let λ_n be a fixed inverse image of $\lambda^{p^{-n}}$ in R. Then the powers $\lambda_n^{p^n}$ all belong to the original residue class λ. Let us show that the sequence $\lambda_n^{p^n}$, $n \geq 0$, converges. Because λ_{n+1}^p and λ_n both belong to $\lambda^{p^{-n}}$, we have

$$\lambda_{n+1}^p \equiv \lambda_n (\text{mod } P)$$

and so, by Lemma 8.2,

$$\lambda_{n+1}^{p^{n+1}} \equiv \lambda_n^{p^n} (\text{mod } P^{n+1})$$

This proves the convergence. Since F is complete, the limit

$$f(\lambda) = \lim_{n \to \infty} \lambda_n^{p^n} = \lambda_0 + (\lambda_1^p - \lambda_0) + (\lambda_2^{p^2} - \lambda_1^p) + \ldots$$

exists in R and we have

$$f(\lambda) \equiv \lambda_0 (\text{mod } P)$$

Hence $f(\lambda)$ belongs to λ, proving that $S = f(\bar{F})$ is a residue system of $\bar{F}$ in R.

It will next be shown that $f(\lambda)$ is independent of the choice of the sequence λ_n. Indeed, let λ_n' denote another such sequence. Then

$$\lambda_n' \equiv \lambda_n (\text{mod } P)$$

and so, by Lemma 8.2,

$$(\lambda_n')^{p^n} \equiv \lambda_n^{p^n} \pmod{P^{n+1}}$$

proving that

$$\lim_{n\to\infty} (\lambda_n')^{p^n} = \lim_{n\to\infty} \lambda_n^{p^n}$$

Fix $x, y \in \bar{F}$ and let $f(x) = \lim x_n^{p^n}$, $f(y) = \lim y_n^{p^n}$. Then as $(xy)_n$ we can choose $x_n y_n$ and, because $(x_n y_n)^{p^n} = x_n^{p^n} y_n^{p^n}$, we derive

$$f(xy) = \lim(x_n^{p^n} y_n^{p^n}) = (\lim x_n^{p^n})(\lim y_n^{p^n}) = f(x)f(y)$$

This shows that S is a multiplicative residue system of $\bar{F}$ in F. In particular, we have $f(\lambda^p) = f(\lambda)^p$ for all $\lambda \in \bar{F}$. Because $\bar{F}$ is perfect, the latter implies that $S^p = S$. Let S_1 be any residue system of $\bar{F}$ in F with $S_1^p = S_1$. For all $n \geqslant 0$, let α_n denote a representative of $\lambda^{p^{-n}}$ in S_1. Because $S_1^p = S_1$, we then have $\alpha_n^{p^n} = \alpha_0$. Therefore, we also have $\lim_{n\to\infty} \alpha_n^{p^n} = \alpha_0$. On the other hand, α_n can be chosen as a sequence for the calculation of the representative of λ in S. Thus we have $\alpha_0 = f(\lambda) \in S$ and therefore $S = S_1$, proving that S is a unique multiplicative residue system of $\bar{F}$ in F.

If $r \in R$ is a p^n-th power for all $n \geqslant 0$, then r is the limit of an appropriate sequence $r_n^{p^n}$ and hence $r \in S$. Conversely, since $S^p = S$, each $r \in S$ is a p^n-th power for all $n \geqslant 0$, proving that S consists precisely of those elements r of R for which r is a p^n-th power for all $n \geqslant 0$. Finally if R itself is of characteristic p, then $(a+b)^{p^n} = a^{p^n} + b^{p^n}$ which implies that $f(x+y) = f(x) + f(y)$ for all $x, y \in \bar{F}$. Hence S is a field and the result follows. ∎

Let A be a commutative ring and let $A[[X]]$ be the set of all formal sums

$$\sum_{n=0}^{\infty} a_n X^n \qquad (a_n \in A)$$

Given two elements of $A[[X]]$, say

$$\sum_{n=0}^{\infty} a_n X^n \quad \text{and} \quad \sum_{n=0}^{\infty} b_n X^n$$

define their sum and product to be respectively

$$\sum_{n=0}^{\infty}(a_n+b_n)X^n \text{ and } \sum_{n=0}^{\infty} c_n$$

where

$$c_n = \sum_{i+j=n} a_i b_j$$

Then we see that $A[[X]]$ becomes a ring, to which we refer as the ring of *formal power series* in one variable over A. Let

$$f = \sum_{n=0}^{\infty} a_n X^n$$

be a nonzero element of $A[[X]]$. Then the smallest n for which $a_n \neq 0$ is called the *order* of f and will be denoted by $o(f)$. We agree to attach the order $+\infty$ to the zero element of $A[[X]]$. The following properties are easy consequences of the definitions:

$$o(f+g) \geqslant \min\{o(f),o(g)\}, \quad o(fg) \geqslant o(f) + o(g) \tag{1}$$

$$o(fg) = o(f) + o(g) \quad \text{if } A \text{ is an integral domain} \tag{2}$$

$$f \text{ is a unit of } A[[X]] \text{ if and only if } a_0 \text{ is a unit of } A \tag{3}$$

Now assume that A is a field. Then, by (3), each nonzero element f of $A[[X]]$ of order n is of the form $f = X^n g$ where g is a unit. Furthermore, by (2), $A[[X]]$ is an integral domain. It follows that $P = (X)$ is a prime ideal of $A[[X]]$ and all other ideals are powers of P. Thus $A[[X]]$ is a *discrete valuation ring with residue class field equal to* A. We denote by $A((X))$ the quotient field of $A[[X]]$ and refer to $A((X))$ as the *field of formal power series* in one variable over A. Each nonzero element of $A((X))$ can be uniquely written in the form

$$f = \sum_{n \geqslant n_0} a_n X^n \qquad (n_0 \in \mathbb{Z},\ a_{n_0} \neq 0)$$

and one defines the order $o(f)$ of f to be the integer n_0. Then one obtains a principal valuation of $A((X))$ whose valuation ring is $A[[X]]$ and the residue class field is A. An easy verification shows that the given valuation of $A((X))$ is complete. Hence $A((X))$ is a complete discrete valued field with residue class field A. Of course, since A is a subfield of $A((X))$, A and $A((X))$ have the same characteristic. It turns out that the latter property guarantees the converse, namely

8.4. THEOREM. Assume that $\mathrm{char}F = \mathrm{char}\bar{F}$. Then

$$R \cong \bar{F}[[X]]$$

and

$$F \cong \bar{F}((X))$$

Proof. Because F is the quotient field of R and $\bar{F}((X))$ is the quotient field of $\bar{F}[[X]]$, it suffices to prove that $R \cong \bar{F}[[X]]$. If the field $\bar{F}$ is perfect, then the required assertion is a consequence of Lemma 8.1(i) and Theorem 8.3(i). For the case where $\bar{F}$ is not necessarily perfect we refer to Cohen(1946) and Roquette(1959). ■

C. The unequal characteristic case.

Here we assume that $\mathrm{char}F = 0$, $\mathrm{char}\bar{F} = p > 0$, $\bar{F}$ is perfect and examine the structure of F in case F is absolutely unramified. It turns out that F can be identified with the quotient field of the Witt ring $W(\bar{F})$ of $\bar{F}$ (hence, there is precisely one such F up to isomorphism).

8.5. LEMMA. Let A be a discrete valuation ring with quotient field B and residue class field $\bar{B}$, let α be a prime element of B and let S be a residue system of $\bar{B}$ in B. Then B is complete if and only if the elements of A are given by the collection of all series of the form

$$a = \sum_{i=0}^{\infty} a_i \alpha^i \qquad (a_i \in S)$$

Proof. Let $\widetilde{B}$ be a completion of B and let $\widetilde{A}$ and K be the valuation ring and residue class field of $\widetilde{B}$, respectively. Then B is complete if and only if $\widetilde{A} = A$ (equivalently, $\widetilde{B} = B$). By Theorem 4.1(iii), S is a residue system of K in $\widetilde{B}$. Assume that F is complete. Then, by Lemma 8.1(i), the elements of A are given by the collection of all series of the required form. Conversely, assume that the given condition holds. The element α is obviously a prime element of $\widetilde{B}$. Hence, by Lemma 8.1(i), $A = \widetilde{A}$ as required. ■

Before proving our next result, it will be convenient to recall the following properties of Witt rings. Let K be a perfect field of characteristic $p > 0$ and let $W(K)$ be the Witt ring of K. Then

(a) $W(K)$ is an integral domain of characteristic 0 with unique maximal ideal I given by $I = \{(0,x_1,x_2,\dots) \in W(K) | x_i \in K\}$ and with $W(K)/I \cong K$

(b) Each nonzero element of $W(K)$ can be uniquely written in the form $p^k c$ with $k \geq 0$ and $c \notin I$. Here k is the smallest integer with $x_k \neq 0$.

(For (a) and (b) refer to Corollary 2.13.12).

(c) For any $x \in W(K)$, $p^i x = (\underbrace{0,\dots,0}_{i},x_0^{p^i},\dots,x_n^{p^i},\dots)$ (Theorem 2.13.6(ii))

(d) If $x,y \in W(K)$ are such that $x_i = 0$ or $y_i = 0$ for all $i \geq 1$, then

$$x + y = (x_0+y_0,x_1+y_1,\dots,x_{n-1}+y_{n-1},\dots)$$

(Theorem 2.13.4(iv)).

8.6. THEOREM. Let K be a perfect field of characteristic $p > 0$ and let L be the quotient field of $W(K)$. Then L is an absolutely unramified complete discrete valued field of characteristic 0 with valuation ring $W(K)$ and residue class field K.

Proof. It is a consequence of (a) and (b) that $W(K)$ is a discrete valuation ring of characteristic 0 with prime element p and residue class field K. Hence, by Lemma 8.5, it suffices to verify that the elements of $W(K)$ are given by the collection of all series of the form $\sum_{i=0}^{\infty} s_i p^i$ where each s_i belongs to a fixed residue system S of K in $W(K)$.

Since the map $W(K) \longrightarrow K$, $(x_i) \longmapsto x_0$ is a surjective homomorphism with kernel I (Theorem 2.13.3), we may choose as S the set of all elements of the form $(x_0,0,0,\dots)$. Note that, by the definition of multiplication in $W(K)$ the residue system S is multiplicative. Since K is perfect, a typical element of $W(K)$ is $x = (x_0,x_1^p,x_2^{p^2},\dots)$. Applying (c) and (d), we have

$$\begin{aligned} x = {} & (x_0,0,\dots\dots\dots\dots) \\ & + (0,x_1^p,0,\dots\dots\dots\dots) \\ & \dots\dots\dots\dots\dots\dots \end{aligned}$$

$$+ (0,\ldots\ldots\ldots\ldots\ldots\ldots,0,x_{n-1}^{p^{n-1}},0,\ldots\ldots\ldots)$$
$$+ (0,\ldots\ldots\ldots\ldots\ldots\ldots,0,0,x_n^{p^n},x_{n+1}^{p^{n+1}},\ldots\ldots)$$
$$= s_0 + ps_1 + \ldots + p^{n-1}s_{n-1} + p^n r$$

where $s_i = (x_i,0,\ldots)$ and $r \in W(K)$. Hence

$$x \equiv s_0 + ps_1 + \ldots + p^{n-1}s_{n-1} \pmod{p^n}$$

By taking the limit with respect to the valuation $w(p^k c) = k$ of $W(K)$, we deduce that the elements of $W(K)$ are given by the collection of all series of the form

$$r = \sum_{i=0}^{\infty} s_i p^i \qquad (s_i \in S)$$

as required. ■

We are now ready to prove the main result.

8.7. THEOREM. Assume that char$F = 0$ and that $\bar{F}$ is perfect of characteristic $p > 0$. Then

(i) F contains exactly one absolutely unramified complete subfield F_0 having the same residue class field $\bar{F}$.

(ii) F_0 consists precisely of elements of the form

$$x = \sum_{i>>-\infty} x_i p^i \qquad (x_i \in S)$$

where S is a unique multiplicative residue system of $\bar{F}$ in F.

(iii) F_0 is isomorphic (as valuation field) to the quotient field of $W(\bar{F})$.

Proof. Assume that F_0 is an absolutely unramified subfield of F having the same residue class field $\bar{F}$. Then the multiplicative residue system of $\bar{F}$ in F_0 is also a multiplicative residue system of $\bar{F}$ in F, hence it coincides with S. Moreover, since F_0 is absolutely unramified, p is a prime element of F_0. Therefore, by Lemma 8.1, F_0 consists precisely of elements of the form

$$x = \sum_{i>>-\infty} x_i p^i \qquad (x_i \in S)$$

Thus, if such F_0 does exist, then it is unique within F.

Let $R_1 = W(\bar{F})$, let R_0 consist of all elements of F of the form

$$\sum_{i=0}^{\infty} x_i p^i, \quad x_i \in S$$

and let F_0 be the quotient field of R_0. Let v_0 be defined by

$$v_0(x) = [1/v(p)]v(x)$$

for all $x \in F_0$. By the foregoing, it suffices to exhibit a bijection

$$f : R_1 \longrightarrow R_0$$

which preserves sums and products and satisfies $w(r) = v_0(f(r))$ for all $r \in R_1$, where w is the valuation associated with R_1.

A typical element r of R_1 is

$$r = (r_0, r_1^p, r_2^{p^2}, \dots) \qquad (r_i \in \bar{F}) \tag{1}$$

and we define

$$f(r) = \sum_{i=0}^{\infty} x_i p^i$$

with $x_i \in S$ representing r_i $(i = 0,1,2,\dots)$. Then obviously f is a bijection which satisfies $w(r) = v_0(f(r))$. It therefore remains to show that

$$f(r) \circ f(t) = f(r \circ t)$$

for all $r, t \in R_1$, where $\circ$ denotes addition or multiplication.

Since $S \cong \bar{F}^*$ and $\bar{F}$ is perfect, for each $s \in S$ and each $k \geqslant 0$ there exists a unique $s^{p^{-k}} \in S$ such that $(s^{p^{-k}})^{p^k} = s$. Let

$$x = (x_0, x_1^p, x_2^{p^2}, \dots) \in W(F) \qquad (x_i \in S)$$

and let $r_i \in \bar{F}$ be represented by $x_i (i = 0,1,2,\dots)$. We can now define

$$x^{p^{-n}} \in W(F)$$

by

$$x^{p^{-n}} = (x_0^{p^{-n}}, x_1^{p^{-n+1}}, x_2^{p^{-n+2}}, \dots)$$

Then we have

$$(x^{p^{-n}})^{(n)} = \sum_{i=0}^{n} (x_i^{p^{i-n}})^{p^{n-i}} p^i = \sum_{i=0}^{n} x_i p^i$$

(see formula (2) of Sec. 13 of Chapter 2), which is just the n-th partial sum of $f(r)$ where r is as in (1). This shows that

$$f(r) = \lim_{n \to \infty} (x^{p^{-n}})^{(n)} \tag{2}$$

Write $t = (t_0, t_1^p, t_1^{p^2}, \dots)$, $f(t) = \sum_{i=0}^{\infty} y_i p^i$, $r \circ t = (\gamma_0, \gamma_1^p, \gamma_2^{p^2}, \dots)$ and

$$f(r \circ t) = \sum_{i=0}^{\infty} z_i p^i$$

Setting

$$r^{p^{-n}} = (r_i^{p^{i-n}}), \quad t^{p^{-n}} = (t_i^{p^{i-n}}) \quad \text{and} \quad (r \circ t)^{p^{-n}} = (\gamma_i^{p^{i-n}}) \qquad (i=0,1,2,\dots)$$

it follows from Theorem 2.13.6(i) that

$$r^{p^{-n}} \circ t^{p^{-n}} = (r \circ t)^{p^{-n}}$$

Hence

$$(r^{p^{-n}} \circ t^{p^{-n}})_k = [(r \circ t)^{p^{-n}}]_k \qquad (0 \leqslant k \leqslant n)$$

and therefore

$$(x^{p^{-n}} \circ y^{p^{-n}})_k \equiv (z^{p^{-n}})_k \pmod P$$

where $y = (y_i^{p^i})$ and $z = (z_i^{p^i})$. It follows, from Lemma 2.13.1, that

$$(x^{p^{-n}})^{(n)} \circ (y^{p^{-n}})^{(n)} \equiv (z^{p^{-n}})^{(n)} \pmod{P^{n+1}}$$

If we now let n tend to ∞, then by (2) we obtain

$$f(r) \circ f(t) = f(r \circ t)$$

as required. ∎

8.8. COROLLARY. Assume that char$F = 0$ and that $\bar{F}$ is perfect of characteristic $p > 0$. If F is absolutely unramified, then F is isomorphic (as valuation field) to the quotient field of the Witt ring of $\bar{F}$.

Proof. Let F_0 be as in Theorem 8.7. Then $F = F_0$ and the result follows. ∎

D. The inertia field.

In what follows, E/F denotes a finite extension of complete discrete valued fields. We write π_F and π_E for the prime elements of F and E, respectively. In case char$F \neq$ char$\bar{F}$ and $\bar{F}$ is perfect, we write F_0 (respectively, E_0) for a unique absolutely unramified complete subfield of F (respectively, E) having the residue class field $\bar{F}$ (respectively, $\bar{E}$). In case char$F$ = char$\bar{F}$

and $\bar{F}$ is perfect, we identify $\bar{F}$ with a unique residue system of $\bar{F}$ in F which is a field. The rest of the notation is taken from Sec. A.

Put $n = (E:F)$ and recall, from Theorem 7.1, that

$$n = e_{E/F}f_{E/F} \quad \text{and} \quad e_{E/F} = w(\pi_F)/w(\pi_E)$$

Recall also that E/F is said to be unramified if $e_{E/F} = 1$ (equivalently, if π_F is a prime of E), while if $f_{E/F} = 1$ (equivalently, if $\bar{F} = \bar{E}$), then we say that E/F is totally ramified.

8.9. LEMMA. (i) If E/F is totally ramified of degree e, then $E = F(\pi_E)$ and π_E satisfies an equation of the form

$$x^e + a_1\pi_F x^{e-1} + \dots + a_{e-1}\pi_F x + a_e\pi_F = 0 \qquad (a_i \in R_F,\ 1 \leqslant i \leqslant e-1,\ a_e \in U(R_F)).$$

(Such an equation is called an Eisenstein equation of degree e)

(ii) If $E = F(\pi_0)$ where π_0 satisfies an Eisenstein equation of degree e, then E/F is totally ramified of degree e and π_0 is a prime element of E.

Proof. (i) Assume that E/F is totally ramified of degree e. Then, by Theorem 7.1(ii), R_E is a free R_F-module with basis $\{\pi_E^j \mid 0 \leqslant j \leqslant e-1\}$ (which is also an F-basis of E) and $\pi_E^e = \pi_F u$ for some unit u of R_E. Hence $E = F(\pi_E)$ and

$$\pi_E^e = \pi_F(a_1\pi_E^{e-1} + \dots + a_{e-1}\pi_E + a_e)$$

for some $a_i \in R_F$, $1 \leqslant i \leqslant e$. Furthermore, since u is a unit of R_E, a_e is a unit of R_F, which proves the required assertion.

(ii) Put $n = (E:F)$, $e^* = e_{E/F}$ and $f^* = f_{E/F}$. Then $n = e^*f^*$ and we must show that $n = e^* = e$ and that $w(\pi_0) = w(\pi_E)$. Because π_0 satisfies an equation of degree e, we have $n \leqslant e$ and thus $e^* \leqslant n \leqslant e$. On the other hand, note that $w(\pi_0) > 0$. This is so since otherwise, $1/\pi_0 \in R_E$ and the Eisenstein equation, rewritten in $1/\pi_0$, would yield a contradiction:

$$0 = 1 + a_1\pi_F(1/\pi_0) + \dots + a_{e-1}\pi_F(1/\pi_0)^{e-1} + a_e\pi_F(1/\pi_0)^e \equiv 1 (\text{mod } P_F)$$

Hence, for any $x \in P_F$, $w(\pi_0 x) > w(x) \geqslant w(\pi_F)$. Because $w(-a_e\pi_F) = w(\pi_F)$, we have $w(-a_e\pi_F + \pi_0 x) = w(\pi_F)$. But, by hypothesis,

$$\pi_0^e \equiv -a_e \pi_F (\mathrm{mod}\ \pi_0 P_F)$$

and hence $e w(\pi_0) = w(\pi_F)$. Since, by Theorem 7.1(ii), $e^* = w(\pi_F)/w(\pi_E)$ and, by definition of π_E, $w(\pi_0) \geqslant w(\pi_E)$ we deduce that $e^* \geqslant e$. Hence $n = e^* = e$ and $w(\pi_0) = w(\pi_E)$, as required. ■

8.10. COROLLARY. Assume that $\bar{F}$ is perfect and $\mathrm{char}F \neq \mathrm{char}\bar{F}$. Then

$$F = F_0(\pi_F)$$

Proof. One immediately verifies that the extension F/F_0 is finite. Since F_0 and F have the same residue class fields, F/F_0 is totally ramified. Now apply Lemma 8.9. ■

8.11. THEOREM. Assume that $\bar{F}$ (hence $\bar{E}$) is perfect. Then there exists a unique intermediate field T of E/F such that

(i) T/F is unramified of degree $f = f_{E/F}$.

(ii) E/T is totally ramified of degree $e = e_{E/F}$.

Furthermore, the field T satisfies

(iii) $T = E_0(\pi_F)$ if $\mathrm{char}F \neq \mathrm{char}\bar{F}$

(iv) $T = \bar{E}((\pi_F))$ if $\mathrm{char}F = \mathrm{char}\bar{F}$.

Proof. We first assume that $\mathrm{char}F = \mathrm{char}\bar{F}$. Then $E = \bar{E}((\pi_E))$ and

$$F = \bar{F}((\pi_F)),$$

so $T = \bar{E}((\pi_F))$ obviously satisfies (i) and (ii). Conversely, assume that T satisfies (i) and (ii). Since T/F is unramified, there exists an intermediate field L of $\bar{E}/\bar{F}$ such that $T = L((\pi_F))$. Since $(T:F) = f = (\bar{E}:\bar{F})$ and, on the other hand, $(T:F) = (L:\bar{F})$, we have $L = \bar{E}$. Thus $T = \bar{E}((\pi_F))$ as required.

Now assume that $\mathrm{char}F \neq \mathrm{char}\bar{F}$. By Corollary 8.10, $F = F_0(\pi_F)$ and $E = E_0(\pi_E)$. Setting $T = E_0(\pi_F)$, we then obviously have $F \subseteq T \subseteq E$. Since E/E_0 is totally ramified, so is E/T. Since F/F_0 is totally ramified, π_F satisfies an Eisenstein equation with respect to F_0 and hence to E_0 (since p is a prime element of both F_0 and E_0). But then π_F is a prime element of T, by Lemma 8.9(ii). Hence T/F is unramified, which clearly implies (i) and (ii). Finally, if T satisfies (i) and (ii), then T has the same multiplicative residue system, say S, as E and π_F is a prime element of T. Hence, by Lemma 8.1(ii), T consists precisely of elements of the form

$$x = \sum_{i>>-\infty} x_i \pi_F^i \qquad (x_i \in S)$$

and therefore is unique. ∎

We shall refer to the field T constructed in Theorem 8.11 as the *inertia field* of E/F.

8.12. COROLLARY. Every finite extension E/F of complete discrete valued fields with perfect residue class fields may be obtained by taking an unramified extension followed by a totally ramified extension.

Proof. Consider the chain $F \subseteq T \subseteq E$ and apply Theorem 8.11. ∎

From now on, we will be interested only in the unequal charactersitic case.

8.13. THEOREM. Assume that char$F$ = 0 and that $\bar{F}$ is perfect of characteristic $p > 0$. Then for any finite extension $K/\bar{F}$, there exists a finite unramified extension E/F such that $\bar{E} = K$.

Proof. By Corollary 8.10, $F = F_0(\pi_F)$ and, by Theorem 8.7(iii), F_0 can be identified with the quotient field of $W(\bar{F})$. Let $E = L(\pi_F)$ where L is the quotient field of $W(K)$. Since $K \supseteq \bar{F}$, we have $W(K) \supseteq W(\bar{F})$ and so $L \supseteq F_0$. Thus $E = L(\pi_F) \supseteq F_0(\pi_F) = F$. Note that $L = L_0 \subseteq E_0$ and so $E = E_0(\pi_F)$. Now, since $K/\bar{F}$ is finite, so is L/F_0 (by Remark 7.2). Hence E/F is also finite. Therefore, by Theorem 8.11, E/F is unramified. Finally, $\bar{L} = K$ (by Theorem 8.6) and $\bar{E} = \bar{L}$ (by Lemma 8.9(ii)), hence $\bar{E} = K$ as required. ∎

We close by proving the following results.

8.14. PROPOSITION. Assume that E/F is unramified, that char$F$ = 0 and that the residue class field $\bar{F}$ is finite, say of p^m elements, where p = char$\bar{F}$.

(i) $E = F(\varepsilon)$ where ε is a primitive $(p^{mf}-1)$-th root of unity and $f = f_{E/F}$.

(ii) Every p'-root of unity of E (i.e. a root of unity whose order is not divisible by p) is contained in $\langle\varepsilon\rangle$.

(iii) If $t \geqslant 1$ and δ is a primitive (p^t-1)-th root of unity, then $F(\delta)/F$ is unramified.

Proof. (i) Since $(\bar{E}:\bar{F}) = f$ and $|\bar{F}| = p^m$, we have $|\bar{E}| = p^{mf}$. Let S be a unique multiplicative residue system of $\bar{E}$ in E. Then S is a subgroup of E^* which is isomorphic to $\bar{E}^*$ and hence to the cyclic group of order $p^{mf}-1$. Thus S is a cyclic group generated by a primitive $(p^{mf}-1)$-th root of unity, say ε. Let $u_1 = \varepsilon, u_2, \ldots, u_f$ be elements of S such that their images in $\bar{E}$ form a basis of $\bar{E}$ over $\bar{F}$. Then, by Theorem 7.1(ii), $u_1, u_2, \ldots, u_f$ form a basis of E over F and thus $E = F(\varepsilon)$.

(ii) Assume that δ is a primitive p'-root of unity in E. Then δ is a p^n-th power for all $n \geqslant 0$, hence by Theorem 8.3, $\delta \in S = \langle\varepsilon\rangle$ as required.

(iii) Let $K/\bar{F}$ be a finite extension of degree t. By Theorem 8.13, there exists a finite unramified extension E/F such that $\bar{E} = K$. Then $f = f_{E/F} = t$ and so, by (i) $E = F(\varepsilon)$ where ε is a primitive $(p^{mt}-1)$-th root of unity. Since $(p^t-1)|(p^{mt}-1)$, we have $F(\delta) \subseteq F(\varepsilon)$. Since E/F is unramified, so is $F(\delta)/F$ as required. ∎

Before recording our last result, let us make the following observation. Let E/F be a finite extension of complete discrete valued fields, let $\operatorname{char} F = 0$ and let $\bar{F}$ be perfect of characteristic $p > 0$. If F is absolutely unramified (i.e. if p is a prime element of F), then by Theorem 8.11(iii), the inertia field T of E/F coincides with E_0. In particular, this applies to $F = \mathbb{Q}_p$, and for this reason the inertia field of $E/\mathbb{Q}_p$ is E_0. Of course, if $E/\mathbb{Q}_p$ is unramified, then $E = E_0$.

8.15. PROPOSITION. Let $E/\mathbb{Q}_p$ be a finite field extension, let R and P be the valuation ring and maximal ideal attached to the valuation of E and let

$$e = e_{E/\mathbb{Q}_p} \quad \text{and} \quad f = f_{E/\mathbb{Q}_p}$$

(i) $(E:\mathbb{Q}_p) = ef$

(ii) $Rp = P^e$

(iii) R is a free $\mathbb{Z}_p$-module of rank ef

(iv) $E_0 = \mathbb{Q}_p(\varepsilon)$, where ε is a primitive (p^f-1)-th root of unity

Proof. Apply Theorem 7.1(ii) and Proposition 8.14(i). ∎

A field F is said to be *local* if there is a complete discrete valuation v

of F such that the residue class field $\bar{F}$ of v is finite.

8.16. PROPOSITION. A field F is local if and only if it is one of the following types:

(i) A finite extension of the rational p-adic field $\mathbb{Q}_p$

(ii) A formal power series field in one variable over a finite field

Proof. We have already observed, in our discussion prior to Theorem 8.4 that if F satisfies (ii), then F is local. By Theorem 7.1, F is local if it satisfies (i). Conversely, assume that F is local and let $R,\bar{F}$ be the valuation ring and residue class field of F. If $\mathrm{char}F = \mathrm{char}\bar{F}$, then F satisfies (ii) by virtue of Theorem 8.4. So assume that $\mathrm{char}F = 0$ and $\mathrm{char}\bar{F} = p > 0$. Then the injection $\mathbb{Z} \longrightarrow R$ extends by continuity to an injection of the ring $\mathbb{Z}_p$ of p-adic integers into R and therefore $\mathbb{Q}_p \subseteq F$. By hypothesis, $\bar{F}$ is finite and so, by Remark 7.2, $F/\mathbb{Q}_p$ is also finite. ∎

E. Cyclotomic extensions of p-adic fields

Let $\mathbb{Q}_p$ be the rational p-adic field. A finite extension F of $\mathbb{Q}_p$ is said to be a *P-adic field* if P is the maximal ideal of the valuation ring of F. In case $F/\mathbb{Q}_p$ is unramified (equivalently, if P is generated by p), we say that F is a p-adic field. Thus a finite extension F of $\mathbb{Q}_p$ is a p-adic field if and only if p is a prime element of F. Any p-adic field F is completely determined by its degree f over $\mathbb{Q}_p$. Indeed, by Proposition 8.15(iv), $F = \mathbb{Q}_p(\varepsilon)$ where ε is a primitive (p^f-1)-th root of unity.

For any natural number n and any given field F, with $\mathrm{char}F \nmid n$, we denote by ε_n a primitive n-th root of unity over F. Our main objective is to investigate the structure of $F(\varepsilon_n)$ in case F is a p-adic field. The following general observation will clear our path.

8.17. PROPOSITION. Let $F/\mathbb{Q}_p$ be a finite field extension of absolute residue degree f, let $n \not\equiv 0 \pmod p$ and let m be the order of p^f modulo n, i.e. the smallest exponent such that

$$p^{fm} \equiv 1 \pmod n$$

Then the extension $F(\varepsilon_n)/F$ is unramified of degree m.

Proof. Since n divides $p^{fm}-1$, we have $F(\varepsilon_n) \subseteq F(\varepsilon_{p^{fm}-1})$. Hence, by Proposition 8.14(iii), $F(\varepsilon_{p^{fm}-1})/F$ is unramified and therefore so is $F(\varepsilon_n)/F$. By Proposition 8.14(i), (ii), we have

$$F(\varepsilon_n) = F(\varepsilon_{p^{fk}-1}), \quad <\varepsilon_n> \subseteq <\varepsilon_{p^{fk}-1}>$$

where $k = (F(\varepsilon_n):F)$. The containment $F(\varepsilon_{p^{fk}-1}) \subseteq F(\varepsilon_{p^{fm}-1})$ shows that $k|m$, while the containment $<\varepsilon_n> \subseteq <\varepsilon_{p^{fk}-1}>$ implies that n divides $p^{fk}-1$, i.e. $p^{fk} \equiv 1 \pmod{n}$. Hence $m = k$ and the result follows. ■

By the *p'-component* (or *p'-part*) of a torsion abelian group A we understand the (restricted) direct product of all q-components of A for which the prime q is distinct from p. Thus the p'-component of the torsion subgroup $t(F^*)$ of F^* is just the subgroup of $t(F^*)$ generated by all p'-roots of unity in F.

8.18. COROLLARY. Let $F/\mathbb{Q}_p$ be a finite field extension of absolute residue degree f. Then $<\varepsilon_{p^f-1}>$ is the p'-component of $t(F^*)$.

Proof. Let $n \not\equiv 0 \pmod{p}$ and let $\varepsilon_n \in F$. Since $F(\varepsilon_n) = F$, it follows from Proposition 8.17 that $p^f \equiv 1 \pmod{n}$. Thus n divides p^f-1 and hence $\varepsilon_n \in <\varepsilon_{p^f-1}>$, as required. ■

8.19. COROLLARY. Let F be a p-adic field of degree f over $\mathbb{Q}_p$. Then

(i) For any $n \not\equiv 0 \pmod{p}$, $F(\varepsilon_n)/F$ is unramified of degree m, where m is the order of p^f modulo n.

(ii) $<\varepsilon_{p^f-1}>$ is the p'-component of $t(F^*)$.

Proof. Since F is a p-adic field, f coincides with the absolute residue degree of F. Now apply Proposition 8.17 and Corollary 8.18. ■

So far we have considered only adjoining of p'-roots of unity to F. The next result examines the opposite case.

8.20. PROPOSITION. Let F be a p-adic field. Then, for any $n \geqslant 1$, $F(\varepsilon_{p^n})/F$ is totally ramified of degree $\phi(p^n) = p^{n-1}(p-1)$ and

$$\pi_n = 1 - \varepsilon_{p^n}$$

is a prime element of $F(\varepsilon_{p^n})$.

Proof. Put $\varepsilon_{p^{n-1}} = \varepsilon_{p^n}^p$, $\varepsilon_{p^{n-2}} = \varepsilon_{p^{n-1}}^p, \ldots, \varepsilon_p = \varepsilon_{p^2}^p$. Using induction on n, it suffices to show that

(a) $F(\varepsilon_p)/F$ is totally ramified of degree $p - 1$ with $\pi_1 = 1 - \varepsilon_p$ as a prime element

(b) $F(\varepsilon_{p^n})/F(\varepsilon_{p^{n-1}})$, $n > 1$, is totally ramified of degree p with $\pi_n = 1 - \varepsilon_{p^n}$ as a prime element, provided $\pi_{n-1} = 1 - \varepsilon_{p^{n-1}}$ is a prime element of $F(\varepsilon_{p^{n-1}})$.

To prove (a), note that $F(\varepsilon_p) = F(\pi_1)$ and

$$\frac{1-\varepsilon_p^p}{1-\varepsilon_p} = 1 + \varepsilon_p + \ldots + \varepsilon_p^{p-1} = 0$$

The latter implies that

$$\frac{1-(1-\pi_1)^p}{\pi_1} = p - \binom{p}{2}\pi_1 + \ldots + (-1)^{p-1}\pi_1^{p-1} = 0$$

and thus π_1 satisfies an Eisenstein equation over F of degree $p-1$. This proves (a), by applying Lemma 8.9(ii).

To prove (b), we first observe that $F(\varepsilon_{p^n}) = F(\varepsilon_{p^{n-1}})(\varepsilon_{p^n})$ and $\varepsilon_{p^n}^p = \varepsilon_{p^{n-1}}$. Hence $F(\varepsilon_{p^n}) = F(\varepsilon_{p^{n-1}})(\pi_n)$ and

$$(1-\pi_n)^p - (1-\pi_{n-1}) = \pi_{n-1} - \binom{p}{1}\pi_n + \binom{p}{2}\pi_n^2 + \ldots + (-1)^p\pi_n^p = 0$$

Therefore, π_n satisfies an Eisenstein equation over $F(\varepsilon_{p^{n-1}})$ of degree p which yields the assertion, by applying Lemma 8.9(ii). ■

By combining Corollary 8.19 and Proposition 8.20, we derive

8.21. COROLLARY. Let $n = kp^s$ with $k \not\equiv 0 \pmod p$ and let F be a p-adic field of degree f over $\mathbb{Q}_p$. Then $F(\varepsilon_k)$ is the inertia field of the extension $F(\varepsilon_n)/F$ with

$$e_{F(\varepsilon_n)/F} = \phi(p^s) \quad \text{and} \quad f_{F(\varepsilon_n)/F} = m$$

where m is the order of p^f modulo k.

Proof. By Corollary 8.19, $F(\varepsilon_k)/F$ is unramified of degree m, while by Proposition 8.20, the field $F(\varepsilon_n) = F(\varepsilon_k)F(\varepsilon_{p^s}) = F(\varepsilon_k)(\varepsilon_{p^s})$ is totally ramified of degree $\phi(p^s)$ over $F(\varepsilon_k)$. ■

4
Multiplicative groups of fields

This chapter is devoted exclusively to the study of multiplicative groups of fields. After establishing some preliminary results, we investigate the isomorphism class of F^*, where F is a distinguished field such as local or global. Concentrating on field extensions E/F, we then prove Brandis's theorem which asserts that if F is infinite and $E \neq F$, then E^*/F^* is not finitely generated. The rest of the chapter is based on works of May who made fundamental contributions to the subject. Numerous examples are given to illustrate that the results obtained are the best possible. Special attention is drawn to the study of fields whose multiplicative groups are free modulo torsion, since the corresponding results have a number of important applications in the study of unit groups of group algebras and the isomorphism problem.

1. SOME GENERAL OBSERVATIONS

Our aim here is to present some general observations required for subsequent investigations. We begin with

1.1. LEMMA. Let F be the quotient field of a unique factorization domain R, and let S be a complete set of nonassociated prime elements of S. Then F^* is a direct product of $U(R)$ and the free abelian group of rank $|S|$.

Proof. Since R is a UFD, each nonzero element of R is either a unit or of the form

$$up_1p_2\cdots p_n \qquad u \in U(R),\ p_i \in S,\ n \geqslant 1$$

This shows that $F^* = U(R)\cdot\langle S\rangle$. Suppose that $\lambda_1, \lambda_2, \ldots, \lambda_n$ are nonnegative integers such that

$$p_1^{\lambda_1} p_2^{\lambda_2} \cdots p_k^{\lambda_k} p_{k+1}^{-\lambda_{k+1}} \cdots p_n^{-\lambda_n} = u \in U(R)$$

Then we have

$$p_1^{\lambda_1} p_2^{\lambda_2} \cdots p_k^{\lambda_k} = u p_{k+1}^{\lambda_{k+1}} \cdots p_n^{\lambda_n}$$

which is possible only in the case $\lambda_i = 0$, $1 \leqslant i \leqslant n$, because R is a *UFD* and the primes $p_1, \ldots, p_n$ are nonassociated. Therefore $F^* = U(R) \times \langle S \rangle$ and, by taking $u = 1$, we see that $\langle S \rangle$ is free of rank $|S|$, as required. ■

Turning to Dedekind domains, we now prove

1.2. LEMMA. Let F be the quotient field of the Dedekind domain R, and let $P(R)$ be the group of all principal fractional ideals of R. Then $P(R)$ is free and

$$F^* \cong U(R) \times P(R)$$

In particular, if the class group of R is a torsion group, then F^* is a direct product of $U(R)$ and a free abelian group whose rank is equal to the cardinality of the set of nonzero prime ideals of R.

Proof. Let $I(R)$ be the group of all fractional ideals of R. We know, from Corollary 3.3.7, that $I(R)$ is a free abelian group with the collection of nonzero prime ideals of R as free generators. Therefore $P(R)$ is a free abelian group. Moreover, if $C(R)$ is torsion, then the ranks of $I(R)$ and $P(R)$ are the same. It therefore suffices to verify that $F^* \cong U(R) \times P(R)$.

To this end, observe that the map

$$\begin{cases} F^* \longrightarrow P(R) \\ a \longmapsto Ra \end{cases}$$

is a surjective homomorphism. Because R is the identity element of $P(R)$, the kernel of this map is $U(R)$. Hence

$$1 \longrightarrow U(R) \longrightarrow F^* \longrightarrow P(R) \longrightarrow 1$$

is an exact sequence which splits since $P(R)$ is free. This proves that $F^* \cong U(R) \times P(R)$, as we wished to show. ■

We next recall certain facts pertaining to cardinal arithmetic. Let

$$(a_i)_{i\in I}$$

be a family of cardinals, say $a_i = |A_i| (i \in I)$, and assume that

$$A_i \cap A_j = \emptyset \qquad \text{for} \quad i \neq j$$

Then we define

$$\sum_{i\in I} a_i = |\bigcup_{i\in I} A_i|$$

Similarly, we put

$$\prod_{i\in I} a_i = |\prod A_i|$$

A standard verification shows that

(1) If a and b are nonzero cardinals and at least one of a,b is infinite, then $a + b = ab = \max\{a,b\}$. In particular, if a is infinite, then $a^n = a$ for any natural number n

(2) If $(a_i)_{i\in I}$ is a family of cardinals, where $a_i = a$ for every $i \in I$ and at least one of a, $|I|$ is infinite, then

$$\sum_{i\in I} a_i = \max\{a,|I|\}$$

(3) If $(a_i)_{i\in I}$ and $(b_i)_{i\in I}$ are two families of cardinals indexed by the same set I and $a_i \leqslant b_i$ for all $i \in I$, then

$$\sum_{i\in I} a_i \leqslant \sum_{i\in I} b_i$$

(4) If A is the set of all finite subsets of an infinite set S, then

$$|S| = |A|$$

Invoking properties (1) - (4), we now prove the following useful observation .

1.3. LEMMA. Let $A \neq 1$ be a group, λ any cardinal and A^λ the (restricted) direct product of λ copies of A. Then

$$|A^\lambda| = \begin{cases} |A|^\lambda & \text{if both } |A| \text{ and } \lambda \text{ are finite} \\ \max\{|A|,\lambda\} & \text{otherwise} \end{cases}$$

In particular, for any field F,

(i) $|F[X]| = \max\{|F|,\aleph_0\}$ (since $F[X] \cong F^{\aleph_0}$ as additive groups)

(ii) If $F^* = t(F^*) \times A^{\lambda}$, where A is countable and λ is infinite, then

$$\lambda = |F|$$

(since $t(F^*)$ is countable and $|F| = |F^*|$)

Proof. Let us first assume that λ is finite. If $|A|$ is finite, then the result is trivial; otherwise (1) implies

$$|A^{\lambda}| = |A|^{\lambda} = |A|$$

as required. Now assume that λ is infinite. Let S be a set with $|S| = \lambda$ and let $\{S_i | i \in I\}$ be the set of all finite subsets of S. Given $s \in S$, put $A_s = A$ so that $A^{\lambda} = \prod_{s\in S} A_s$. Setting $A_i = \prod_{s\in S_i} A_s$, we then have $A^{\lambda} = \bigcup_{i\in I} A_i$. Because $|S| = \lambda$ is infinite, property (4) guarantees that $|I| = \lambda$. Obviously $|A^{\lambda}| \leqslant \sum_{i\in I} |A_i|$. But the finite case proved above shows that $|A_i| = |A|$ if $|A|$ is infinite and $|A_i| < \aleph_0$ if $|A|$ is finite. Therefore, we may apply properties (2) and (3) to deduce that

$$|A^{\lambda}| \leqslant \max\ \{|A|,\lambda\}$$

The opposite inequality being trivial, the result follows. ■

We close by proving the following result

1.4. THEOREM. Let $F(X)$ be the field of rational functions in an indeterminate X over a field F. Then $F(X)^*$ is the direct product of F^* and a free abelian group of rank $\max(|F|,\aleph_0)$

Proof. Note that $F(X)$ is the quotient field of $R = F[X]$. If S denotes the set of irreducible monic polynomials in R, then since R is a PID (and hence UFD) it follows from Lemma 1.1 that

$$F(X)^* \cong U(R) \times G$$

where G is free of rank $|S|$. Because $U(R) \cong F^*$, we are left to verify that $|S| = \max(|F|,\aleph_0)$ or, by Lemma 1.3, that $|S| = |F(X)|$. If F is infinite, then $|S| \geqslant |F| = |F(X)|$ and so $|S| = |F(X)|$. If F is finite, then

$$|F(X)| = \aleph_0 = |S|.$$

as required. ■

2. INFINITE ABELIAN GROUPS

In this section we record some frequently used facts pertaining to infinite abelian groups. Let p denote a prime. The p^n-th complex roots of unity, with n running over all integers ≥ 0, form an infinite multiplicative group. This group is called *quasicyclic* or a *group of type* p^∞ and is denoted by $Z(p^\infty)$. If ε_{p^n} is a primitive p^n-th root of unity, $n = 1,2,\ldots,$ then $Z(p^\infty)$ is a union of the chain

$$1 \subset \langle \varepsilon_p \rangle \subset \langle \varepsilon_{p^2} \rangle \subset \ldots \subset \langle \varepsilon_{p^n} \rangle \subset \ldots$$

and every proper subgroup of $Z(p^\infty)$ is of the form $\langle \varepsilon_{p^n} \rangle$ for some $n \geq 0$.

Under addition the rational numbers form a group called the *full rational group*, denoted by $\mathbb{Q}$. Note that $Z(p^\infty)$ and $\mathbb{Q}$ are obviously divisible groups and it will be shown that there are no divisible groups other than direct products of $Z(p^\infty)$ and $\mathbb{Q}$.

In what follows, all direct products are assumed to be *restricted direct products*. The cardinality of $\mathbb{N}$ is denoted by $\aleph_0$ (aleph-null). A set is *countable* if it is either finite or of cardinality $\aleph_0$.

Let A denote an arbitrary abelian group. Then the *socle* $S(A)$ of A is defined to be the set of all $a \in t(A)$ for which the order of a is a square-free integer. Thus $S(A) = 1$ if and only if A is torsion-free and $S(A) = A$ if and only if A is an *elementary group* in the sense that every element has a square-free order. For a p-group A, we obviously have

$$S(A) = A[p] \quad \text{where} \quad A[p] = \{a \in A \mid a^p = 1\}$$

2.1. LEMMA. If A is a divisible subgroup of an abelian group G, then G splits over A.

Proof. By Zorn's lemma, we may choose a maximal subgroup B of G having trivial intersection with A. We are therefore left to verify that $G = A \cdot B$. We argue by contradiction by assuming that $g \notin A \cdot B$ for some $g \in G$. If

$$\langle g \rangle \cap A \cdot B = 1,$$

then $(B \cdot \langle g \rangle) \cap A = 1$, which is impossible since $B\langle g \rangle \supset B$. Thus $g \notin A \cdot B$ but $g^n \in A \cdot B$.

Choose m to be the least positive integer for which $g^m \in A \cdot B$. Setting $g_1 = g^{m/p}$, where p is a prime divisor of m, it follows that $g_1^p \in A \cdot B$ but $g_1 \notin A \cdot B$. Because A is divisible, there exist $a \in A$, $b \in B$ such that

$$g_1^p = a^p b;$$

hence $g_2^p = b$ where $g_2 = g_1 a^{-1} \notin A \cdot B$. In view of our choice of B, we may choose a nonidentity element $a' \in A \cap \langle g_2, B \rangle$. Because $A \cap B = 1$ and

$$g_2^p \in B,$$

a' can be written in the form

$$a' = g_2^k b' \qquad 0 < k < p,\ b' \in B$$

Since $(k,p) = 1$, there exist $\ell, s \in \mathbb{Z}$ such that $\ell k + sp = 1$. But $g_2^{sp} \in B$ and $g_2^k = a'(b')^{-1} \in A \cdot B$, hence $g_2 = g_2^{\ell k} g_2^{sp} \in A \cdot B$. This contradiction completes the proof of the lemma. ∎

We now provide a description of divisible groups.

2.2. THEOREM. Let G be a divisible abelian group. Then

(i) G is a direct product of quasicyclic and full rational groups.

(ii) The cardinal numbers of the sets of components $Z(p^\infty)$ (for every p) and $\mathbb{Q}$ form a complete and independent system of invariants for G.

Proof. Because $t(G)$ is obviously divisible, Lemma 2.1 tells us that

$$G = t(G) \times A$$

where A is torsion-free and, clearly, again divisible. Put $T = t(G)$ and let T_p be the p-component of T. Then $G = (\prod_p T_p) \times A$ and so we need only show that T_p is a direct product of groups $Z(p^\infty)$, and A is a direct product of groups $\mathbb{Q}$.

Let $\{g_i\}_{i \in I}$ be a maximal independent system in the socle of T_p. Owing to divisibility, for each $i \in I$, there is an infinite sequence $g_{i_1}, \ldots, g_{in}, \ldots$ in T_p such that

$$g_{i1} = g_i, \quad g_{i,i+1}^p = g_{in} \qquad (n = 1,2,\dots)$$

Thus every g_i can be embedded in a subgroup $B_i \cong Z(p^\infty)$ of T_p, namely,

$$B_i = \langle g_{i1},\dots,g_{in},\dots\rangle$$

Since $\langle g_i\rangle$ is the socle of B_i and the $g_i (i \in I)$ are independent, the B_i generate their direct product $B = \prod_{i\in I} B_i$ in T_p. Now B is divisible, hence a direct factor of T_p (by Lemma 2.1). Because B contains the socle of T_p, we have $B = T_p$ and T_p is a direct product of groups $Z(p^\infty)$.

Now choose a maximal independent system $\{a_j\}$, $j \in J$ in A. Because A is torsion-free and divisible, for every positive $n \in \mathbb{Z}$ there is exactly one $x \in A$ satisfying $x^n = a_j$, which shows that every a_j can be embedded in a subgroup $A_j \cong \mathbb{Q}$ of A. Since $\{a_j\}$ is an independent system, the A_j generate their direct product $X = \prod_{j\in J} A_j$ in A. This is a direct factor of A containing a maximal independent system of A, hence $X = A$, and A is a direct product of groups $\mathbb{Q}$.

Let $r(G)$ and $r_p(G)$ be the torsion-free rank and the p-rank of G, respectively. Then the numbers of direct factors isomorphic to $Z(p^\infty)$ or to $\mathbb{Q}$ in a decomposition of G into groups $Z(p^\infty)$, $\mathbb{Q}$ are obviously equal to $r_p(G)$ and $r(G)$. Note further that the ranks $r(G)$ and $r_p(G)$ are invariants of G. Furthermore, they form a complete system of invariants, since if given $r_p(G)$ and $r(G)$, we can uniquely reconstruct G as the direct product of $r_p(G)$ copies of $Z(p^\infty)$ for every p and $r(G)$ copies of $\mathbb{Q}$. Their independence being obvious, the result follows. ■

Let G be an abelian p-group. The *height* of an element $g \in G$ is defined to be the largest integer r such that $g \in G^{p^r}$, if a largest r exists; otherwise the height of g is infinity. We are now ready to prove the following important result.

2.3. THEOREM (Kulikov(1945)). Let G be an abelian p-group. Then G is a direct product of cyclic groups if and only if G is the union of an ascending chain

$$G_1 \subseteq G_2 \subseteq \cdots \subseteq G_n \subseteq \cdots$$

of subgroups such that the heights of elements $\neq 1$ in G_n are less than a finite bound k_n.

Proof. If G is a direct product of cyclic groups, then as G_n we may take the product of all direct factors of order $\leqslant p^n$. To prove the converse, denote by x_1 an element of order p in G_1 which has maximal height in G. Assume that for all ordinals $\alpha < \beta$ we have chosen in G the elements x_α satisfying the following properties:

(1) All elements x_α are of order p.

(2) If $x_\alpha \in G_n - G_{n-1}$, and if $A_\alpha = \langle x_{\alpha'} | \alpha' < \alpha \rangle$, then

(a) $A_\alpha \supseteq G_{n-1}[p]$

(b) $x_\alpha \notin A_\alpha$, and among all elements in $G_n - A_\alpha$, x_α is of maximal height in G.

If $A_\beta \neq G[p]$, then x_β can be chosen as follows. By 2(a), there exists n such that all $x_\alpha \in G_n$ but not all $x_\alpha \in G_{n-1}$. Assume that $A_\beta \neq G_n[p]$. Then, among the elements of order p in $G_n - A_\beta$, pick x_β with maximal height in G. If $A_\beta = G_n[p]$, then pick a similar element in the smallest subgroup G_k which satisfies $G_k[p] \supseteq G_n[p]$. Thus there exists an ordinal γ such that

$$G[p] = \langle x_\alpha | \alpha < \gamma \rangle$$

Invoking (1) and (2b), we conclude that

(3) $G[p] = \prod_{\alpha<\gamma} \langle x_\alpha \rangle$

Denote by h_α the height of x_α in G and let $y_\alpha \in G$ be such that $y_\alpha^{p^{h_\alpha}} = x_\alpha$. Owing to (3), the cyclic groups $\langle y_\alpha \rangle$ generate a subgroup F such that

(4) $$F = \prod_{\alpha<\gamma} \langle y_\alpha \rangle \quad \text{and} \quad G[p] \subseteq F$$

We now claim that any element $z \in G[p] \subseteq F$ has height in F equal to that in G. Indeed, by (3), $z = x'_{\alpha_1} x'_{\alpha_2} \cdots x'_{\alpha_n}$ where $x'_{\alpha_i} \neq 1 (1 \leqslant i \leqslant n)$ and x'_{α_i} is a power of x_{α_i}. Hence x'_{α_i} has height h_{α_i} both in G and in F. Now the height h of z in F is equal to $\min\{h_{\alpha_i} | 1 \leqslant i \leqslant n\}$ by (4). The height of z in G cannot be less than h, and we shall show that it cannot be greater. To this end, let k be such that $h = h_{\alpha_k}$ but $h_{\alpha_i} > h$ for $i > k$. Then in the product

$$z = (x'_{\alpha_1} \cdots x'_{\alpha_k})(x'_{\alpha_{k+1}} \cdots x'_{\alpha_n})$$

the second factor is either 1 (when $k=n$) or its height in G is strictly greater than h; the first factor is not contained in A_{α_k} and its height in G cannot be greater than $h_{\alpha_k} = h$, by (2b). Hence the height of z in G cannot be greater than h, so z has the same height both in G and in F.

To complete the proof, we are left to verify that $G = F$. Assume by way of contradiction that $G \neq F$. Then we may choose an element g of minimal order, say p^s, in $G - F$ (plainly, $s \geq 2$). The element $g^{p^{s-1}}$ is of order p, hence belongs to F. Since $g^{p^{s-1}}$ has the same height in F and in G, there exists $f \in F$ such that $f^{p^{s-1}} = g^{p^{s-1}}$. It follows that the order of gf^{-1} is at most p^{s-1}, so $gf^{-1} \in F$. The latter, however, is contrary to the assumption that $g \notin F$. Thus $G = F$ and the result follows. ∎

An abelian group G is said to be *bounded* if $G^n = 1$ for some $n \geq 1$.

2.4. COROLLARY (Baer (1934), Prüfer (1923)). Any bounded abelian group is a direct product of cyclic groups.

Proof. Because G is bounded, G is torsion and the p-components of G are again bounded. Thus we may assume that $G^{p^n} = 1$ for some $n \geq 1$. It follows that the heights of all nonidentity elements in G are $\leq n$. Thus the sequence $\{G_n\}_{n=1}^{\infty}$, where $G_n = G$ for all n, satisfies the hypothesis of Theorem 2.3. ∎

A subgroup H of an abelian group G is said to be *pure* in G if

$$G^n \cap H = H^n$$

for all $n \geq 1$. The following properties are direct consequence of the definition.

(a) Any direct factor of G is pure in G

(b) If G/H is torsion-free, then H is pure in G

(c) If G is torsion-free, then a subgroup H of G is pure if and only if G/H is torsion-free

(d) Purity is transitive, i.e. if K is pure in H and H is pure in G, then K is pure in G.

2.5. LEMMA. (i) If A is a pure subgroup of G such that G/A is a direct product of cyclic groups, then G splits over A.

(ii) If A is a bounded pure subgroup of G, then G splits over A

Proof. (i) We first demonstrate that every coset of G modulo A contains an element of the same order as this coset. Fix $\bar{g} \in G/A$. If $\bar{g}$ is of infinite order, then so is every representative of the coset $\bar{g}$. If $\bar{g}$ is of finite order n, then for any representative $t \in \bar{g}$, we have $t^n \in A$. Purity ensures that $a^n = t^n$ for some $a \in A$. Then $h = ta^{-1} \in \bar{g}$ is of order $\leqslant n$, hence of order n.

Write $G/A = \prod \langle g_\alpha \rangle$ where $g_\alpha \in \bar{g}_\alpha$ is of the same order as $\bar{g}_\alpha$. Then $G = A \cdot B$ where B is the subgroup of G generated by the g_α. If

$$g_{\alpha_1}^{k_1} \cdots g_{\alpha_n}^{k_n} \in A$$

then

$$(\bar{g}_{\alpha_1})^{k_1} \cdots (\bar{g}_{\alpha_n})^{k_n} = 1$$

and therefore $(\bar{g}_{\alpha_i})^{k_i} = 1 (1 \leqslant i \leqslant n)$. Thus $g_{\alpha_i}^{k_i} = 1$ and so $A \cap B = 1$, as required.

(ii) By hypothesis, there exists $n \geqslant 1$ such that $A^n = 1$. Put

$$H = A \cdot G^n, \ \bar{H} = H/G^n \ \text{and} \ \bar{G} = G/G^n$$

It is routine to verify that $\bar{H}$ is a pure subgroup of $\bar{G}$. Now $\bar{G}/\bar{H}$ is a bounded group, hence a direct product of cyclic groups (Corollary 2.4). We may therefore apply (i) to infer that $\bar{G} = \bar{H} \cdot \bar{F}$, $\bar{H} \cap \bar{F} = 1$, for a suitable subgroup $\bar{F}$ of $\bar{G}$. Let F be the inverse image of $\bar{F}$ in G. Then $G = H \cdot F = A \cdot F$ and

$$H \cap F = G^n$$

Bearing in mind that A is a pure subgroup, we have $A \cap G^n = A^n = 1$. Therefore

$$A \cap F \subseteq H \cap F \subseteq G^n \cap A = 1$$

as we wished to show. ■

As a final result of this section, we prove

2.6. THEOREM (Baer (1936), Fomin (1937)). Let $t(G)$ be the torsion subgroup of an abelian group G. If $t(G)$ is a direct product of a divisible and bounded group, then G splits over $t(G)$.

Proof. By hypothesis, there exist subgroups A and B of G, where A is divisible and B bounded, such that $t(G) = A \times B$. By Lemma 2.1, there is a subgroup C of G such that $G = A \times C$. This implies at once that

$$t(G) = A \times t(C),$$

so $t(C) \cong t(G)/A \cong B$. Because $t(C)$ is a pure subgroup of C, it follows from Lemma 2.5 that $C = t(C) \times D$ for some subgroup D of C. Thus

$$G = A \times t(C) \times D = t(G) \times D$$

as required. ■

3. THE DIRICHLET-CHEVALLEY-HASSE UNIT THEOREM

Throughout this section, R denotes the ring of integers of an algebraic number field F of degree r, $\sigma_1,\ldots,\sigma_r$ all distinct $\mathbb{Q}$-homomorphisms $F \longrightarrow \mathbb{C}$, $d(F)$ the discriminant of F, $[r_1,r_2]$ the signature of F, $t(F^*)$ the torsion subgroup of F^* and S a finite set of primes of F which contain all archimedean primes Our goal is to establish the celebrated Dirichlet's Unit Theorem which asserts that $U(R)$ is a finitely generated group of torsion-free rank $r_1 + r_2 - 1$.

Let I be a nonzero ideal of R. Because the additive group R/I is finite and R is a free $\mathbb{Z}$-module of rank r, I must also be $\mathbb{Z}$-free of rank r. Hence there exists a $\mathbb{Z}$-basis $x_1,\ldots,x_r$ of R and integers ℓ_i such that

$$\ell_1 x_1,\ldots,\ell_r x_r$$

is a $\mathbb{Z}$-basis of I. Moreover, we obviously have

$$N(I) = \ell_1 \ell_2 \ldots \ell_r$$

3.1. LEMMA. Let I be a nonzero ideal of R and let $y_1,\ldots,y_r$ be a $\mathbb{Z}$-basis of I. Then

$$D_{F/\mathbb{Q}}(y_1,\ldots,y_r) = N(I)^2 \cdot d(F)$$

Proof. Let $x_1,\ldots,x_r$ be a $\mathbb{Z}$-basis of R such that $\ell_1 x_1,\ldots,\ell_r x_r$ is a

$\mathbb{Z}$-basis of I for some integers $\ell_i > 0$. Then

$$D_{F/\mathbb{Q}}(\ell_1 x_1,\ldots,\ell_r x_r) = (\prod_{i=1}^{r} \ell_i)^2 D_{F/\mathbb{Q}}(x_1,\ldots,x_r) = N(I)^2 \cdot d(F)$$

and, on the other hand,

$$D_{F/\mathbb{Q}}(\ell_1 x_1,\ldots,\ell_r x_r) = D_{F/\mathbb{Q}}(y_1,\ldots,y_r),$$

as required. ∎

We need a simple observation connecting the norm of an element of F with the norm of the fractional ideal generated by it.

3.2. LEMMA. Let x be a nonzero element of F and $I = Rx$ the fractional ideal generated by x. Then $N(I) = |N_{F/\mathbb{Q}}(x)|$.

Proof. Since both sides of the asserted equality are multiplicative in x (Proposition 2.6.6 and Corollary 3.3.17) we may assume that $x \in R$. If

$$x_1,\ldots,x_r$$

is a $\mathbb{Z}$-basis of R, then $x_1 x,\ldots,x_r x$ is a $\mathbb{Z}$-basis of I. Therefore, by Lemma 3.1,

$$N(I)^2 = D_{F/\mathbb{Q}}(x_1 x,\ldots,x_r x)\, d(F)^{-1}$$

On the other hand, by Proposition 2.7.7,

$$\begin{aligned} D_{F/\mathbb{Q}}(x_1 x,\ldots,x_r x) &= \det(\sigma_i(x_j x))^2 \\ &= \det(\sigma_i(\sigma_{ij} x))^2 \cdot \det(\sigma_i(x_j))^2 \\ &= (N_{F/\mathbb{Q}}(x))^2 \cdot d(F) \end{aligned}$$

Thus $N(I)^2 = (N_{F/\mathbb{Q}}(x))^2$, and since $N(I) > 0$ then $N(I) = |N_{F/\mathbb{Q}}(x)|$. ∎

To take advantage of the result above, we need the following

3.3. LEMMA. (i) If I is a nonzero ideal of R and $m = N(I)$, then $mR \subseteq I$

(ii) For any integer $m > 0$, there are only finitely many ideals of R with norm equal to m

(iii) There exists an integer $\mu > 0$ such that for any nonzero ideal I of R there exists an element $a \in I$, $a \neq 0$, such that

$$|N_{F/\mathbb{Q}}(a)| \leqslant N(I)\mu$$

Proof. (i) Because the additive group R/I is of order m, we have

$$m(R/I) = 0$$

and thus $mR \subseteq I$.

(ii) Let I be an ideal of R with $N(I) = m$. By (i), $mR \subseteq I$ and so, by Lemma 3.3.8, I divides mR. Since mR has only finitely many divisors, the required assertion follows.

(iii) Given a $\mathbb{Z}$-basis $x_1, \ldots, x_r$ of R, choose $\mu \in \mathbb{Z}$ such that

$$\mu \geq \prod_{j=}^{r} \left(\sum_{i=1}^{r} |\sigma_j(x_i)| \right)$$

To prove that μ has the required property, fix a nonzero ideal I of R and choose $k \geq 1$ such that $k^r \leq N(I) < (k+1)^r$. Let S be the set of elements of the form

$$\sum_{i=1}^{r} d_i x_i \qquad (d_i \in \mathbb{Z},\ 0 \leq d_i \leq k)$$

Since $|S| = (k+1)^r > |R/I|$, there exist $b, c \in S$, $b \neq c$, and $a_i \in \mathbb{Z}$ with $|a_i| \leq k$ such that

$$b - c = \sum_{i=1}^{r} a_i x_i \in I$$

Setting $a = b - c$, it follows that

$$|N_{F/\mathbb{Q}}(a)| = \prod_{j=1}^{r} \left| \sum_{i=1}^{r} a_i \sigma_j(x_i) \right| \leq \prod_{j=1}^{r} k \left(\sum_{i=1}^{r} |\sigma_j(x_i)| \right)$$

$$= k^r \prod_{j=1}^{r} \left(\sum_{i=1}^{r} |\sigma_j(x_i)| \right) \leq N(I)\mu ,$$

as required. ∎

We are now ready to prove the following important result.

3.4. THEOREM (Dedekind (1871), Kronecker (1882)). Let R be the ring of integers of an algebraic number field. Then the class group $C(R)$ of R is finite.

Proof. Let an integer μ be as in Lemma 3.3(iii). Then, by Lemma 3.3(ii), there are only finitely many nonzero ideals of R, say $I_1, I_2, \ldots, I_k$, such that $N(I_i) \leq \mu$. Let I be any fractional ideal of R. Then there exists $c \in R$, $c \neq 0$, such that cI^{-1} is an ideal of R. By Lemmas 3.2 and 3.3(iii), there exists an element $b \in cI^{-1}$, $b \neq 0$ such that

$$N(Rb) \leqslant N(cI^{-1})\mu$$

We conclude therefore that

$$N(Ibc^{-1})\cdot N(Rc) = N(Ibc^{-1}\cdot Rc) = N(Rb)N(I)$$
$$\leqslant N(cI^{-1})N(I)\mu$$
$$N(Rc)\mu$$

whence $N(Ibc^{-1}) \leqslant \mu$. However, $Ibc^{-1} \subseteq R$, so $Ibc^{-1} = I_i$ for some

$$i \in \{1,2,\ldots,k\}$$

Thus the images of I and I_i in $C(R)$ are the same, proving that $|C(R)| \leqslant k$. ∎

We now return to our investigation of units of R. An element $x \neq 0$ in F is said to be an *S-unit* if $p(x) = 1$ for all primes p of F not contained in S. One can easily verify that the S-units form a subgroup of F^*. We shall refer to this subgroup as the S-unit group of F and denote it by F^*_S. Thus, if S consists of all archimedean primes, then

$$|S| = r_1 + r_2 \quad \text{and} \quad U(R) = F^*_S$$

by Theorem 2.6.1 and Corollary 2.6.2.

3.5. LEMMA (The Product Formula). Let V be the set of all normalized valuations of F and let x be a nonzero element of F. Then $v(x) = 1$ for all but a finite number of $v \in V$ and

$$\prod_{v\in V} v(x) = 1$$

Proof. Let $Rx = P_1^{n_1}P_2^{n_2}\ldots P_k^{n_k}$ be the factorization into product of nonzero powers of distinct prime ideals of R. For any nonzero prime ideal P of R, let v_P be the normalized P-adic valuation of F. Then $v_P(x) = 1$ for all $P \notin \{P_1,\ldots,P_k\}$. This proves the first statement since V has only finitely many, namely $r_1 + r_2$, nondiscrete valuations.

To prove the required formula, we first observe that

$$\prod_{i=1}^{k} v_{P_i}(x) = \prod_{i=1}^{k} N(P_i)^{-n_i} = N(Rx)^{-1}$$

On the other hand, the product $v(x)$ over all normalized archimedean valuations

v equals $|N_{F/\mathbb{Q}}(x)|$. The required assertion is therefore a consequence of Lemma 3.2. ■

3.6. LEMMA. Let c be a positive real number. Then there exists only finitely many $x \in R$ such that $|x_i| \leqslant c$ for all conjugates x_i of x.

Proof. Put $x_i = \sigma_i(x)$, $1 \leqslant i \leqslant r$, and $A = \{x \in R \mid |\sigma_i(x)| \leqslant c$ for all $i \in \{1,\dots,r\}\}$. To show that A is finite, we need only verify that the set of polynomials

$$\{\prod_{i=1}^{r}(X-x_i) \mid x \in A\}$$

is finite. If $s_1, s_2, \dots, s_r$ are the elementary symmetric functions in r variables, then for all $x \in A$,

$$|s_k(x_1,\dots,x_r)| \leqslant \ell \qquad (1 \leqslant k \leqslant r)$$

where $\ell = \max\{rc, \binom{r}{2}c, \dots, \binom{r}{k}c^k, \dots, c^r\}$. Because $s_k(x_1,\dots,x_r) \in \mathbb{Z}, 1 \leqslant k \leqslant r$, the result follows. ■

Turning our attention to the torsion subgroup $t(F^*)$ of F^*, we now prove

3.7. LEMMA. (i) $t(F^*)$ is a finite cyclic group of even order.

(ii) $x \in t(F^*)$ if and only if $x \in R$ and all conjugates of x have absolute value 1.

Proof. (i) Because $(F:\mathbb{Q}) = r < \infty$, the orders of elements of $t(F^*)$ are bounded. Let $\varepsilon \in t(F^*)$ be an element of maximal order m. If $x \in t(F^*)$, then $x^m = 1$ by our choice of m. Hence $x \in \langle\varepsilon\rangle$ and so $t(F^*) = \langle\varepsilon\rangle$. Moreover, $-1 \in t(F^*)$ and therefore $t(F^*)$ is of even order.

(ii) If $x \in t(F^*)$, then $x \in R$ and all conjugates of x are in $t(F^*)$, hence have absolute value 1. Conversely, suppose that $x \in R$ is such that all conjugates of x have absolute value 1. Because $x, x^2, \dots, x^n, \dots$ all share this property, it follows from Lemma 3.6 that $x^t = x^s$ for some integers t, s with $t < s$. But then $x^{s-t} = 1$ and $x \in t(F^*)$, as required. ■

3.8. LEMMA. (i) If c is a positive real number, then the number of ideals of R with their norm bounded by c is finite.

(ii) There exists a constant $c \in \mathbb{R}$ with the following property: if $a \in F$ and

$$1/2 < |N_{F/\mathbb{Q}}(a)| < 1$$

then there exists $\varepsilon \in U(R)$ such that

$$v(\varepsilon a) \leqslant c \quad \text{for all normalized archimedean valuations} \quad v$$

Proof. (i) Because the norm is multiplicative and its values are natural numbers, it suffices to prove that only finitely many prime ideals of R have their norm bounded by c. Assume by way of contradiction that there exists an infinite family $\{P_i | i \in I\}$ of prime ideals of R such that $|N(P_i)| \leqslant c$ for all $i \in I$. Choose a positive integer m with $c \leqslant m$ and consider any set of $m + 1$ distinct elements $x_1, \dots, x_{m+1}$ of R. Then, for any $P \in \{P_i | i \in I\}$, there exist x_i, x_j with $0 \neq x_i - x_j \in P$. However, the set of differences $x_k - x_s$ is finite, hence there are infinitely many such P, a contradiction.
(ii) Let $w_1, w_2, \dots, w_r$ be a $\mathbb{Z}$-basis of R, let V be the set of all normalized archimedean valuations of F and let

$$X = \{\sum_{i=1}^{r} c_i w_i \,|\, |c_i| \leqslant 1, \ c_i \in \mathbb{Q}\}$$

Let d denote an upper bound of $\{v(x) | v \in V, x \in X\}$. We first claim that if $m \in \mathbb{N}$, $a \in F$ and $|N_{F/\mathbb{Q}}(a)| \leqslant m^r$, then for a suitable nonzero $b \in R$, we have either $v(ab) \leqslant dm$ or $v(ab) \leqslant dm^2$, according as to whether v corresponds to a real or complex embedding of F. Indeed, if $a \in R$ then the sums

$$c_1 w_1 + \dots + c_r w_r \quad (0 \leqslant c_i \leqslant m, \ c_i \in \mathbb{Q})$$

cannot all be distinct (mod Rc) for, by Lemma 3.2, there are

$$(1+m)^r > |N_{F/\mathbb{Q}}(a)| = N(Ra)$$

of them. Thus, by taking an appropriate difference of two such sums, we derive

$$ab = c_1 w_1 + \dots + c_r w_r$$

for some $0 \neq b \in R$ and some $c_i \in \mathbb{Q}$ such that $|c_i| \leqslant m$. Now $v(m^{-1}ab) \leqslant d$ since $m^{-1}ab \in X$, so $v(ab) \leqslant dm$ or $v(ab) \leqslant dm^2$ according as to whether v corresponds to a real or complex embedding of F. If $a \notin R$, then write

$$a = a_0/t$$

with $a_0 \in R$ and $t \in \mathbb{N}$. Clearly

$$|N_{F/\mathbb{Q}}(a_0)| = |N_{F/\mathbb{Q}}(a)||N_{F/\mathbb{Q}}(t)| \leqslant (mt)^r$$

and so, by the preceding argument, $v(a_0 b) \leqslant dmt$ or $v(a_0 b) \leqslant d(mt)^2$ for some nonzero b in R. This substantiates our claim by substituting $a_0 = at$.

By the foregoing, given $a \in F$ with $1/2 < |N_{F/\mathbb{Q}}(a)| < 1$, there is a nonzero $b = b(a)$ in R such that $v(ab) \leqslant d$ for all $v \in V$. This ensures that

$$|N_{F/\mathbb{Q}}(ab)| \leqslant d^{r_1+r_2},$$

whence $|N_{F/\mathbb{Q}}(b)| \leqslant 2d^{r_1+r_2}$ since $|N_{F/\mathbb{Q}}(a)| > 1/2$. Because $|N_{F/\mathbb{Q}}(b)| = N(Rb)$, it follows from (i) that the numbers $b = b(a)$ generate only a finite number of principal ideals, say $Rb_1, Rb_2, \ldots, Rb_k$. Therefore, for each $a \in F$ satisfying our condition, we have $b(a) = \varepsilon b_i$ for some $i \in \{1,2,\ldots,k\}$ and some unit ε of R. Thus we have

$$v(\varepsilon a) = v(ab)v(\varepsilon b^{-1}) = v(ab)v(b_i^{-1}) \leqslant dv(b_i^{-1}) \qquad \text{for all } v \in V$$

Setting $c = \max\{dv(b_i^{-1}) | 1 \leqslant i \leqslant k\}$, the result follows. ■

The next observation will be used in order to establish independence of certain units.

3.9. LEMMA. Let $(a_{ij}) \in M_n(\mathbb{R})$ be such that $a_{ij} < 0$ for $i \neq j$ and

$$a_{i1} + \ldots + a_{in} > 0, \; 1 \leqslant i \leqslant n$$

Then $\det(a_{ij})$ is nonzero.

Proof. Assume by way of contradiction that $\det(a_{ij}) = 0$. Then we can find a system $x_1, \ldots, x_n$ of numbers at least one of which is positive and such that

$$\sum_{j=1}^{n} a_{ij}x_j = 0 \qquad (i = 1,2,\ldots,n)$$

If $x_t = \max\{x_1, \ldots, x_n\}$ then $x_t > 0$ and

$$0 = \sum_{i=1}^{n} a_{tj}x_j = a_{tt}x_t + \sum_{j\neq t} a_{tj}x_j \geqslant x_t \sum_{j=1}^{n} a_{tj} > 0,$$

a contradiction. ■

Let $\varepsilon_1, \varepsilon_2, \ldots, \varepsilon_n$ be a system of S-units of F. We say that this system is *independent* if

$$\varepsilon_1^{\ell_1}\varepsilon_2^{\ell_2}\cdots\varepsilon_n^{\ell_n} = 1 \quad \text{implies} \quad \ell_1 = \ell_2 = \cdots = \ell_n = 0 \qquad (\ell_i \in \mathbb{Z})$$

Expressed otherwise, $\varepsilon_1,\ldots,\varepsilon_n$ is an independent system if the group

$$\langle\varepsilon_1,\ldots,\varepsilon_n\rangle$$

generated by the ε_i is free abelian of rank n. The following easy consequence of Lemma 3.9 is very useful.

3.10. COROLLARY. Let $v_1,\ldots,v_s$, $s = r_1 + r_2$, be all normalized archimedean valuations of F and let $\varepsilon_1,\varepsilon_2,\ldots,\varepsilon_{s-1}$ be units of R such that

$$v_j(\varepsilon_i) < 1 \qquad \text{for } j \neq i,\ 1 \leqslant i \leqslant s-1,\ 1 \leqslant j \leqslant s$$

Then the units $\varepsilon_1,\varepsilon_2,\ldots,\varepsilon_{s-1}$ are independent.

Proof. Consider the homomorphism $f : U(R) \longrightarrow \mathbb{R}^s$ given by

$$f(r) = (\log v_1(r),\ \log v_2(r),\ldots,\log v_s(r))$$

It suffices to verify that $f(\varepsilon_1), f(\varepsilon_2),\ldots,f(\varepsilon_{s-1})$ are $\mathbb{R}$-linearly independent, or equivalently that $\det(\log v_j(\varepsilon_i) \neq 0$, $i,j \in \{1,2,\ldots,s-1\}$. Because

$$v_j(\varepsilon_i) < 1$$

for $i \neq j$, we have $\log v_j(\varepsilon_i) < 0$ for $i \neq j$. On the other hand, by Lemma 3.5, we have

$$\sum_{j=1}^{s-1} \log v_j(\varepsilon_i) = -\log v_s(\varepsilon_i) > 0 \qquad (1 \leqslant i \leqslant s-1)$$

The desired conclusion is therefore a consequence of Lemma 3.9. ■

As a final preparatory step, we examine certain topological properties of $\mathbb{R}^n$. A subset X of $\mathbb{R}^n$ is said to be *bounded* if there is a real number c such that for each $(x_1,\ldots,x_n) \in X$,

$$|x_i| \leqslant c \qquad \text{for all } i \in \{1,2,\ldots,n\}$$

The following two properties of $\mathbb{R}^n$ are standard (see Bourbaki (1966)):

(a) A subset X of $\mathbb{R}^n$ is compact if and only if X is closed and bounded.

(b) A subgroup G of $\mathbb{R}^n$ is discrete if and only if, for any compact subset X of $\mathbb{R}$, the intersection $G \cap X$ is finite.

A typical example of a discrete subgroup of $\mathbb{R}^n$ is $\mathbb{Z}^n$. The following lemma provides a survey of all discrete subgroups of $\mathbb{R}^n$.

3.11. LEMMA. Let G be a discrete subgroup of $\mathbb{R}^n$. Then G is generated as a $\mathbb{Z}$-module by m vectors which are $\mathbb{R}$-linearly independent (so $m \leq n$).

Proof. Let $\{x_1, x_2, \ldots, x_m\}$ be a maximal set of $\mathbb{R}$-linearly independent elements of G and let

$$X = \{x \in \mathbb{R}^n | x = \sum_{i=1}^{m} \lambda_i x_i,\ 0 \leq \lambda_i \leq 1\}$$

Obviously X is closed and bounded, hence compact and so $G \cap X$ is finite. Given $g \in G$, it follows from the maximality of $\{x_1, x_2, \ldots, x_m\}$ that $g = \sum_{i=1}^{m} \mu_i x_i$ for some $\mu_i \in \mathbb{R}$. For each $j \in \mathbb{Z}$, put

$$g_j = jg - \sum_{i=1}^{m} [j\mu_i]x_i$$

where $[\lambda]$ denotes the largest integer less than or equal to $\lambda \in \mathbb{R}$. Then

$$g_j \in G \cap X$$

and $g = g_1 + \sum_{i=1}^{m} [\mu_i]x_i$. Because each x_i lies in $G \cap X$, we infer that the $\mathbb{Z}$-module G is generated by the finite set $G \cap X$. Moreover, since $\mathbb{Z}$ is infinite, there exist distinct integers j and k such that $g_j = g_k$. Therefore, by the definition of g_j,

$$(j-k)\mu_i = [j\mu_i] - [k\mu_i]$$

which ensures that the μ_i's are rational. Hence the $\mathbb{Z}$-module G is generated by a finite number of elements which are $\mathbb{Q}$-linear combinations of the x_i's. Thus $dG \subseteq \bigoplus_{i=1}^{m} \mathbb{Z}x_i \subseteq G$ for some nonzero $d \in \mathbb{Z}$, proving that G is a free abelian group of rank m.

Choose a $\mathbb{Z}$-basis $\{y_1, y_2, \ldots, y_m\}$ of G such that $\{d_1 y_1, \ldots, d_m y_m\}$ is a $\mathbb{Z}$-basis of $\bigoplus_{i=1}^{m} \mathbb{Z}x_i$ for some $d_i \in \mathbb{Z}$. Then the elements $d_1 y_1, \ldots, d_m y_m$ are $\mathbb{R}$-linearly independent and hence so are $y_1, \ldots, y_m$. ∎

We are at last in a position to attain our main objective, which is to prove the following result.

3.12. THEOREM (Chevalley-Hasse, Chevalley (1940)). Let S be a finite set of primes of F which contains all archimedean primes. Then F_S^* is a direct product of $t(F^*)$ and a free abelian group of rank $|S| - 1$.

Proof. For any $p \in S$, choose the normalized valuation v in p and denote by V the set of all such v. Then $|S| = |V| = s$, say. Consider the homomorphism

$$\begin{cases} F_S^* \longrightarrow \mathbb{R}^s \\ x \longmapsto [\log v(x)]_{v \in V} \end{cases}$$

We wish to show that $\operatorname{Ker} f = t(F^*)$ and that Imf is free of rank $s-1$. Because $t(F^*)$ is finite by Lemma 3.7(i), it will follow that $F_S^* \cong t(F^*) \times f(F_S^*)$, as required. Note that $t(F^*) \subseteq \operatorname{Ker} f$ because for any valuation w of F and any $x \in t(F^*)$, $w(x) = 1$. The opposite containment is a consequence of Lemma 3.7(ii) and Corollary 2.6.2. We are therefore left to verify that $f(F_S^*)$ is free of rank $s - 1$. For the sake of clarity we divide the proof into three steps.

Step 1. Here we prove that $f(F_S^*)$ *is free of rank* $\leqslant s - 1$.

Owing to Lemma 3.5, $f(F_S^*)$ is contained in the hyperplane defined by the equation $X_1 + \dots + X_s = 0$. Therefore $f(F_S^*)$ contains at most $(s-1)$ $\mathbb{R}$-linearly independent vectors. Let A be a bounded subset of $\mathbb{R}^n$ and let $B = \{x \in F_S^* | f(x) \in A\}$ By Lemma 3.11, it suffices to show that B is a finite set.

Because A is bounded, there exists a real number $\alpha > 1$ such that for all $x \in B$

$$\alpha^{-1} < v(x) < \alpha \qquad \text{for all } v \in V \qquad (1)$$

Let $P_1, P_2, \dots, P_k$ be prime ideals of R corresponding to the P_i-adic valuations in V. Then, for any $x \in F_S^*$, $Rx = P_1^{n_1} \dots P_k^{n_k}$ for some $n_i \in \mathbb{Z}$. It follows from (1) that if $x \in B$, then $-\alpha < N(P_i)^{-n_i} < \alpha$, which shows that each n_i takes only finitely many values. Consequently, there are only finitely many principal fractional ideals of the form Rx, $x \in B$, say $Rx_1, \dots, Rx_n$. Therefore, if $x \in B$, then $x = rx_i$ for some $r \in U(R)$ and some $i \in \{1, \dots, n\}$. But then (1) implies that $\alpha^{-1} v(r) < v(r)v(x_i) = v(x) < \alpha$ and hence that $v(r) < \alpha^2$ for all $v \in V$. Because V includes all archimedean valuations, the required assertion follows by virtue of Lemma 3.6.

Step 2. Here we verify that there exists $s-1$ *units in* $U(R), s = r_1 + r_2 > 1$, *whose images in* $\mathbb{R}^s$ *are* $\mathbb{R}$*-linearly independent.*

Denote by $v_1, v_2, \ldots, v_s$ all normalized archimedean valuation of F ordered so that $v_1, \ldots, v_{r_1}$ correspond to the real embeddings of F into $\mathbb{C}$. By the proof of Corollary 3.10, it suffices to exhibit units $\varepsilon_1, \varepsilon_2, \ldots, \varepsilon_{s-1}$ of R such that

$$v_j(\varepsilon_i) < 1 \qquad \text{for } j \neq i,\ 1 \leqslant i \leqslant s-1,\ 1 \leqslant j \leqslant s \tag{2}$$

To this end, let $0 < a_i < b_i$ $(1 \leqslant i \leqslant s)$ be given real numbers . We claim that there exists a in F such that

$$a_i < v_i(a) < b_i \qquad (1 \leqslant i \leqslant s) \tag{3}$$

Indeed, define

$$A_i = \begin{cases} a_i\ , & 1 \leqslant i \leqslant r_1 \\ a_i^{1/2}, & r_1 + 1 \leqslant i \leqslant s \end{cases}$$

and

$$B_i = \begin{cases} b_i\ , & 1 \leqslant i \leqslant r_1 \\ b_i^{1/2}, & r_1 + 1 \leqslant i \leqslant s \end{cases}$$

Then (3) can be rewritten in the form

$$A_i < |\sigma_i(a)| < B_i \qquad 1 \leqslant i \leqslant s \tag{4}$$

where σ_i is the embedding corresponding to v_i. We now take an arbitrary $\mathbb{Z}$-basis of R, say $w_1, w_2, \ldots, w_r$ and consider the system of linear equations

$$x_1\sigma_i(w_1) + \ldots + x_r\sigma_i(w_r) = h_i \qquad (1 \leqslant i \leqslant r)$$

where for $i = s+1, s+2, \ldots, r$ the embeddings σ_i are defined by

$$\sigma_i(x) = \overline{\sigma_{i-r_2}(x)},$$

and h_i are real numbers with $A_i < h_i < B_i$, $1 \leqslant i \leqslant s$ and $h_i = h_{i-r_2}$ for $i > s$.

By Cramer's rule this system has a real solution. If we now choose rational numbers $y_1, \ldots, y_r$ with the differences $|x_i - y_i|$ small enough to satisfy the inequalities

$$A_i < |y_i\sigma_i(w_1) + \dots + y_r\sigma_i(w_r)| < B_i \qquad (i = 1,2,\dots,r)$$

(where for $i = s+1,\dots,r$ we define $A_i = A_{i-r_2}$, $B_i = B_{i-r_2}$), then the number $a = y_1w_1 + \dots + y_rw_r$ satisfies (4) and hence (3).

Finally, denote by c the constant of Lemma 3.8.(ii). Owing to (3), we may find $a_1,\dots,a_s$ in F satisfying the inequalities

$$c < v_i(a_j) < dc \qquad (1 \leqslant i,j \leqslant s, i \neq j) \qquad (5)$$

$$1/(2c^{s-1}) < v_j(a_j) < 1/(dc)^{s-1} \qquad (1 \leqslant j \leqslant s)$$

where the constant $d > 1$ is chosen to satisfy $d^{s-1} < 2$. Then

$$1/2 < |N_{F/\mathbb{Q}}(a_i)| < 1$$

and so, by Lemma 3.8(ii), for a certain $\varepsilon_i \in U(R)$ we obtain

$$v_j(\varepsilon_i a_i) < c \qquad (1 \leqslant j \leqslant s)$$

This proves (2) by applying (5).

Step 3. Completion of the proof.

We first note that $U(R)$ is a subgroup of F_S^* and, by Steps 1 and 2, $F_S^*/U(R)$ has torsion-free rank at most $(s-1)-(r_1+r_2-1) = s - r_1 - r_2 = m$, say. Therefore, it suffices to find elements $a_1,a_2,\dots,a_m$ in F_S^* which are independent modulo $U(R)$. Let $v_1,\dots,v_m$ be the discrete valuations in V and let $P_1,P_2,\dots,P_m$ be the correpsonding prime ideals of R. Owing to Theorem 3.4, the class number $h(R)$ of R is finite. Setting $k = h(R)$ it follows that $P_i^k = a_iR$, $1 \leqslant i \leqslant m$, where the a_i are some S-units. Assume that

$$a = a_1^{n_1}\dots a_m^{n_m} \in U(R) \qquad \text{for some } n_i \in \mathbb{Z}$$

Then $Ra = (P_1^{n_1}\dots P_m^{n_m})^k = R$ and so, by the uniqueness of factorization, we obtain $n_1 = n_2 = \dots = n_m = 0$. This completes the proof of the theorem. ■

3.13. COROLLARY (Dirichlet's Unit Theorem, Dirichlet (1840)). Let R be the ring of integers of an algebraic number field F and let $[r_1,r_2]$ be the signature of F. Then $U(R)$ is a direct product of $t(F^*)$ and a free abelian group of rank $r_1 + r_2 - 1$.

Proof. Apply Theorem 3.12 to the special case where S consists of all archimedean primes of F. ■

4. THE TORSION SUBGROUP

Our aim in this section is to investigate the structure of the torsion subgroup $t(F^*)$ of the multiplicative group F^* of a field F. We begin with the case where F is algebraically closed. In what follows, all direct products are assumed to be restricted direct products.

4.1. PROPOSITION. Let F be an algebraically closed field.

(i) If $\mathrm{char}F = 0$, then $t(F^*) \cong \prod_p Z(p^\infty) \cong \mathbb{Q}/\mathbb{Z}$

(ii) If $\mathrm{char}F = q > 0$, then $t(F^*) \cong \prod_{p \neq q} Z(p^\infty)$

Proof. (i) Suppose that $\mathrm{char}F = 0$. Then we obviously have

$$t(F^*) \cong t(\mathbb{C}^*) = \prod_p Z(p^\infty)$$

Furthermore, the map $\mathbb{Q} \longrightarrow t(\mathbb{C}^*)$, $r \longmapsto e^{2ir\pi}$ is clearly a surjective homomorphism. Since the kernel of this map is $\mathbb{Z}$, the assertion follows.

(ii) Suppose that $\mathrm{char}F = q > 0$ and let p be an arbitrary prime. If $p = q$, then

$$X^{p^n} - 1 = (X-1)^{p^n}$$

for all $n \geqslant 1$ and so the p-component of $t(F^*)$ is 1. On the other hand, if $p \neq q$, then $t(F^*)$ and $t(\mathbb{C}^*)$ have isomorphic p-components. Now apply (i). ■

4.2. PROPOSITION. A torsion group is isomorphic to $t(F^*)$ for some field F of characteristic 0 if and only if it is isomorphic to a subgroup of $\mathbb{Q}/\mathbb{Z}$ with nontrivial 2-component.

Proof. Let F be a field of characteristic 0. Because $-1 \in t(F^*)$, the group $t(F^*)$ contains an element of order 2. If $\bar{F}$ is the algebraic closure of F, then by Proposition 4.1(i),

$$t(F^*) \subseteq t(\bar{F}^*) \cong \mathbb{Q}/\mathbb{Z}$$

Hence $t(F^*)$ is isomorphic to a subgroup of $\mathbb{Q}/\mathbb{Z}$ with nontrivial 2-component.

Conversely, assume that G is a subgroup of $\mathbb{Q}/\mathbb{Z}$ with nontrivial 2-component.

We realize G as a multiplicative group of complex roots of unity and select a finite or infinite ascending chain of cyclic groups $<\varepsilon_k>$ of even order n_k such that

$$G = \bigcup_k <\varepsilon_k>$$

Define the field F to be the union of the tower

$$\mathbb{Q}(\varepsilon_1) \subseteq \dots \subseteq \mathbb{Q}(\varepsilon_k) \subseteq \dots$$

Then $t(F^*)$ is the union of $t(\mathbb{Q}(\varepsilon_k))^*$, $k = 1,2,\dots$. Because n_k is even, we have $t(\mathbb{Q}(\varepsilon_k)^*) = <\varepsilon_k>$ and hence $G = t(F^*)$, as required. ∎

Turning our attention to finite groups, we next prove

4.3. PROPOSITION. A finite group is isomorphic to $t(F^*)$ for some field F if and only if it is cyclic and its order is either even or of the form $2^n - 1$ for some integer $n \geqslant 1$.

Proof. If G is a cyclic group of even order n and ε_n a primitive complex n-th root of unity, then G is isomorphic to the torsion subgroup of $\mathbb{Q}(\varepsilon_n)^*$. On the other hand, if G is cyclic of order $2^n - 1$, then by Proposition 2.5.1, $G \cong F^* = t(F^*)$ where F is the field of 2^n elements.

Conversely, let F be a field such that $t(F^*)$ is finite. By Proposition 2.5.5, $t(F^*)$ must be cyclic. If $\text{char}F \neq 2$, then $-1 \in t(F^*)$ is of order two, hence $t(F^*)$ is of even order. Finally, let $\text{char}F = 2$ and let ε be a generator of $t(F^*)$. Denote by E the smallest subfield of F containing ε. Then we have $t(E^*) = t(F^*)$ and E is a finite field, say of 2^n elements. Hence $t(F^*)$ is of order $2^n - 1$, as required. ∎

We close by proving the following result.

4.4. PROPOSITION. (i) If a subgroup of $\mathbb{Q}/\mathbb{Z}$ with trivial 2-component is isomorphic to $t(F^*)$ for some field F, then it is isomorphic to K^* for some subfield K of F.

(ii) A subgroup G of $\mathbb{Q}/\mathbb{Z}$ with trivial 2-component is isomorphic to the multiplicative group of some field if and only if G is the union of an ascending chain of cyclic groups whose orders are of the form $2^n - 1$.

Proof. (i) Let G be a subgroup of $\mathbb{Q}/\mathbb{Z}$ with trivial 2-component and with $G \cong t(F^*)$ for some field F. Because G has trivial 2-component, we must have char$F = 2$. Let K be the smallest subfield of F containing $t(F^*)$. Then clearly $K^* = t(F^*) \cong G$.

(ii) Let G be a subgroup of $\mathbb{Q}/\mathbb{Z}$ with trivial 2-component. Assume that $G \cong F^*$ for some field F. Then char$F = 2$ and F^* is torsion. Therefore F is a union of an ascending chain of its finite subfields. This shows that G satisfies the required property, by applying Proposition 2.5.1.

Conversely, assume that G is the union of an ascending chain of cyclic groups G_k of order $2^k - 1$, $k = 1,2,\ldots$. Then the field F which is the union of an ascending chain of Galois fields F_k of degree 2^k, $k = 1,2,\ldots,$ is such that $G \cong F^*$. ■

5. GLOBAL FIELDS

For any field F, let $F(X)$ be the field of rational functions in an indeterminate X over F. A *global field* is defined to be either an algebraic number field or else a finite extension of $\mathbb{F}_q(X)$ where $\mathbb{F}_q$ is the Galois field of q elements. In what follows, all direct products are assumed to be restricted direct products. As usual, for any abelian group A and any cardinal λ, we write A^λ for the direct product of λ copies of A. Our aim is to determine the isomorphism class of any given global field.

5.1. LEMMA. $\mathbb{Q}^* \cong \mathbb{Z}_2 \times \mathbb{Z}^{\aleph_0}$

Proof. The field $\mathbb{Q}$ is the quotient field of a unique factorization domain $\mathbb{Z}$. We may choose as a complete set of nonassociated primes of $\mathbb{Z}$ all prime natural numbers. Because their cardinality is $\aleph_0$ and $U(\mathbb{Z}) \cong \mathbb{Z}_2$, the result follows by virtue of Lemma 1.1. ■

We are now ready to prove

5.2. THEOREM. (Skolem (1947)). Let F be an algebraic number field. Then

$$F^* \cong \mathbb{Z}_m \times \mathbb{Z}^{\aleph_0} \qquad \text{(for some even } m)$$

and, conversely, for any even m, $\mathbb{Z}_m \times \mathbb{Z}^{\aleph_0}$ is isomorphic to F^* for some algebraic number field F.

Proof. Denote by R the ring of integers of F. Then F is the quotient field of R and R is a Dedekind domain. Invoking Lemma 1.2, we conclude that $F^* \cong U(R) \times P(R)$. Since $P(R)$ is free, it follows from the Dirichlet's Unit Theorem that $F^* \cong \mathbb{Z}_m \times \mathbb{Z}^{\lambda}$ for some even m and some cardinal λ. Since

$$\mathbb{Q}^* \subseteq F^*, \ \lambda \geqslant \aleph_0$$

by Lemma 5.1. On the other hand, $|F| = \aleph_0$ and so $\lambda \leqslant \aleph_0$. Hence $\lambda = \aleph_0$ and $F^* \cong \mathbb{Z}_m \times \mathbb{Z}^{\aleph_0}$. Conversely, let $G = \mathbb{Z}_m \times \mathbb{Z}^{\aleph_0}$ for some even m. Then $G \cong F^*$ where $F = \mathbb{Q}(\varepsilon_m)$ and ε_m a primitive m-th root of unity. ∎

5.3. LEMMA. Let E/F be an arbitrary field extension and let v be a valuation of E which is trivial on F. Then v is trivial on the algebraic closure of F in E.

Proof. Let $a \neq 0$ be an element of E such that

$$a^n = \lambda_0 + \lambda_1 a + \dots \lambda_{n-1} a^{n-1} \qquad (\lambda_i \in F, \ n \geqslant 1)$$

Because v is trivial on F, v is nonarchimedean by Proposition 2.1.7. Thus

$$\begin{aligned} v(a^n) &\leqslant \max\{v(\lambda_0), v(\lambda_1)v(a), \dots, v(\lambda_{n-1})v(a)^{n-1}\} \\ &= \max\{1, v(a), \dots, v(a)^{n-1}\} \end{aligned}$$

This shows that $v(a) \leqslant 1$, for if $v(a) > 1$, then $v(a)^n > v(a)^i > 1, 1 \leqslant i \leqslant n-1$, contrary to the above inequality. Applying the same argument to a^{-1}, we have $v(a^{-1}) = 1/v(a) \leqslant 1$, whence $v(a) \geqslant 1$. Hence $v(a) = 1$ and the result follows. ∎

5.4. LEMMA. Let F be an arbitrary field. Given $c > 1$ in $\mathbb{R}$ and

$$0 \neq a = f(X)/g(X) \in F(X),$$

define

$$\deg a = \deg f(X) - \deg g(X)$$

and put

$$v_c(a) = c^{\deg a}, \ v_c(0) = 0$$

(i) The map $v_c : F(X) \longrightarrow \mathbb{R}$ is a valuation which is trivial on F and whose equivalence class is independent of the choice of $c > 1$ in $\mathbb{R}$.

(ii) If v is a nontrivial valuation of $F(X)$ which is trivial on F, then either $v = v_c$ for some $c > 1$ in $\mathbb{R}$ or $v = v_P$ for some P-adic valuation v_P, where P is a prime ideal of $F[X]$.

Proof. (i) Straightforward.

(ii) First assume that $v(X) > 1$ and put $c = v(X)$. Given

$$f(X) = c_0 + c_1 X + \dots + c_n X^n \in F[X] \qquad (c_n \neq 0)$$

we have

$$v(f(X)) = v(X)^n = c^{\deg f(X)}$$

because v is nonarchimedean and $v(c_n X^n) > v(c_i X^i)$, $0 \leqslant i \leqslant n-1$. Hence $v = v_c$.

Next assume that $v(X) \leqslant 1$. Then, for any $f(X) \in F[X]$, $v(f(X)) \leqslant 1$. Because v is nontrivial, we must have $v(f(X)) < 1$ for some $f(X) \in F[X]$. The set of all such polynomials is an ideal P generated by an irreducible polynomial $p(X)$. If $v(p(X)) = c$ and $f(X) = p(X)^{\nu} g(X)$, $(p(X), g(X)) = 1$, then

$$v(f(X)) = c^{\nu}$$

Thus $v = v_P$ as required. ■

5.5. LEMMA. Let K be a finite field extension of $F(X)$. Then there exists a full set $\{v_i | i \in I\}$ of nonequivalent nontrivial valuations of K which are trivial on F and such that

(i) each v_i is discrete

(ii) for every $a \neq 0$ in K, $v_i(a) \neq 1$ for only finitely many $i \in I$

(iii) for every $a \neq 0$ in K, $\prod_{i \in I} v_i(a) = 1$

(iv) the set $K_0 = \{0\} \cup \{x \in K | v_i(x) = 1$ for all $i \in I\}$ is a subfield of K.

Proof. Denote by $\{v_i | i \in I\}$ any full set of nonequivalent nontrivial valuations of K which are trivial on F. By Lemma 5.4, the restriction w_i of v_i to $F(X)$ is discrete. Because the field extension $K/F(X)$ is finite, it follows from Corollary 2.5.8 that each v_i is discrete. Applying Lemma 5.4, we deduce that, given $a \in F(X)$, $w_i(a) = 1$ for all but a finite number of $i \in I$. Let $a \neq 0$ be an element of K. Then

$$a^n = \lambda_0 + \lambda_1 a + \ldots + \lambda_{n-1} a^{n-1} \qquad (\lambda_i \in F(X),\ n \geq 1)$$

and thus

$$v_i(a^n) \leq \max\{w_i(\lambda_0),\ w_i(\lambda_1)v_i(a), \ldots,\ w_i(\lambda_{n-1})v_i(a)^{n-1}\}$$
$$= \max\{1, v_i(a), \ldots,\ v_i(a)^{n-1}\}$$

where the latter equality holds for all but a finite number of $i \in I$. Therefore, $v_i(a) \leq 1$ for all but a finite number of $i \in I$. Because the same argument applies to a^{-1}, we deduce that $v_i(a) \neq 1$ for only finitely many $i \in I$.

We shall prove (iii) in the special case where $K = F(X)$ (for the general case, we refer to a classical book of Artin (1968,pp.235-236). Select a real number d, $0 < d < 1$. Let v_c be as in Lemma 5.4 and let $\{v_i | i \in I\}$ consist of v_c for $c = d^{-1}$ and all the P-adic valuations v_P, where for a fixed prime ideal $P = (p(X))$ of $F[X]$, v_P is determined by

$$v_P(x)(p(X)) = d^{\deg p(X)}$$

To prove the required formula, we may assume that $a = p(X)$ is an irreducible polynomial. If v_i is distinct from v_c and v_P, then $v_i(a) = 1$. On the other hand,

$$v_c(a)v_P(a) = c^{\deg p(X)} d^{\deg p(X)} = 1,$$

proving (iii).

Let K_0 be the set of all $a \in K$ for which $v_i(a) \leq 1$ for all $i \in I$, and let x,y be two elements of K_0. It follows at once that $-x, xy$ and $x + y$ are also in the set K_0. If $x \in K_0$ and $x \neq 0$, then (iii) implies that $v_i(x) = 1$ for all $i \in I$. It now follows that x^{-1} is in K_0. Hence K_0 forms a subfield of K and it consists of 0 and those elements x of K which satisfy $v_i(x) = 1$ for all $i \in I$. So the lemma is true. ∎

We are now ready to prove the following important result .

5.6. THEOREM. Let F be an arbitrary field, let K be a finite field extension of $F(X)$ and let L be the algebraic closure of F in K. Then the extension L/F is finite and

$$K^* \cong L^* \times A$$

where A is a free abelian group.

Proof. Assume by way of contradiction that L/F is an infinite extension. Then there exists an infinite chain

$$F(\lambda_1) \subset F(\lambda_1,\lambda_2) \subset \ldots \subset F(\lambda_1,\ldots,\lambda_n) \subset \ldots \subset L$$

Since $F(X)(\lambda_1,\ldots,\lambda_n) = F(\lambda_1,\ldots,\lambda_n)(X)$, we see that $F(\lambda_1,\ldots,\lambda_n)$ is the algebraic closure of F in $F(X)(\lambda_1,\ldots,\lambda_n)$. Hence

$$F(X)(\lambda_1) \subset F(X)(\lambda_1,\lambda_2) \subset \ldots \subset F(X)(\lambda_1,\ldots,\lambda_n) \subset \ldots \subset K$$

is a strictly increasing chain of subfields of K containing $F(X)$, contrary to the assumption that the extension $K/F(X)$ is finite.

Let $\{v_i | i \in I\}$ be the set of valuations of K exhibited in Lemma 5.5. Consider the homomorphism

$$\begin{cases} K^* \longrightarrow \prod_{i\in I} v_i(K) \\ k \longmapsto (v_i(k)) \end{cases}$$

Because each v_i is discrete and $v_i(k) \neq 1$ for only finitely many $i \in I$, the image of this homomorphism is a free abelian group. It therefore suffices to verify that $K_0 = L$.

Owing to Lemma 5.3, we have $L \subseteq K_0$. To prove the opposite containment, assume by way of contradiction that $k \in K_0$ is transcendental over F. Then, from the equation satisfied by k with respect to $F(X)$, it follows that X must be algebraic over $F(k)$. Because $F(k) \subseteq K_0$, we conclude, from Lemma 5.3, that $X \in K_0$. But, by Lemma 5.4, $v_c(X) > 1$ and hence $v_i(X) > 1$ for some $i \in I$, a contradiction. Hence $K_0 = L$ and the result follows. ■

5.7. COROLLARY. Let F be an algebraic number field and let K be a finite field extension of $F(X)$. Then

$$K^* \cong \mathbb{Z}_m \times \mathbb{Z}^{\aleph_0} \qquad \text{for some even } m$$

Proof. Applying Theorems 5.2 and 5.6, we have

$$K^* \cong \mathbb{Z}_m \times \mathbb{Z}^{\lambda}$$

for some even m and some infinite cardinal λ. Because $|F| = \aleph_0$ and

$|F(X)| = |K|$, it follows from Lemma 1.3 that $\lambda = |K| = \aleph_0$, as required. ■

5.8. COROLLARY. Let K be a finite extension of $\mathbb{F}_q(X)$, $q = p^n$, p prime. Then

$$K^* \cong \mathbb{Z}_{p^m-1} \times \mathbb{Z}^{\aleph_0}$$

for some $m \geqslant n$

Proof. By Theorem 5.6, there exists $m \geqslant n$ such that for $\ell = p^m$

$$K^* \cong F_\ell^* \times \mathbb{Z}^\lambda$$

for some cardinal λ. Now $F_\ell^* \cong \mathbb{Z}_{p^m-1}$ and, by Theorem 1.4, λ is infinite. Therefore, by Lemma 1.3, $\lambda = |K| = \aleph_0$ and the result follows. ■

6. ALGEBRAICALLY CLOSED, REAL CLOSED AND THE RATIONAL p-ADIC FIELDS

We use the same notation and conventions adopted in the previous sections. In particular, all direct products are assumed to be restricted direct products, and for any cardinal λ and a group A, A^λ is the direct product of λ copies of A.

6.1. THEOREM. Let F be an algebraically closed field.

(i) If char$F = 0$, then

$$F^* \cong (\mathbb{Q}/\mathbb{Z}) \times \mathbb{Q}^{|F|}$$

(ii) If char$F = q > 0$ and F is algebraic over its prime subfield, then

$$F^* \cong \prod_{p \neq q} Z(p^\infty)$$

(iii) If char$F = q > 0$ and F is not algebraic over its prime subfield, then

$$F^* \cong \left(\prod_{p \neq q} Z(p^\infty)\right) \times \mathbb{Q}^{|F|}$$

Conversely, for any infinite cardinal λ and any prime q, the groups

$$(\mathbb{Q}/\mathbb{Z}) \times \mathbb{Q}^\lambda \quad \text{and} \quad \left(\prod_{p \neq q} Z(p^\infty)\right) \times \mathbb{Q}^\lambda$$

are isomorphic to multiplicative groups of suitable algebraically closed fields.

Proof. Since F is algebraically closed, each polynomial $X^n - a$ $(a \in F)$ has a root in F and so F^* is divisible. Therefore, by Theorem 2.2,

$$F^* \cong t(F^*) \times \mathbb{Q}^\lambda$$

for some cardinal λ. Because $\mathbb{Q}$ is of torsion-free rank 1, λ is the torsion-free rank of F^*. If $\mathrm{char}F = 0$, then $\mathbb{Q}^* \subseteq F^*$ and therefore λ is infinite, by virtue of Lemma 5.1. Thus, by Lemma 1.3(ii), $\lambda = |F|$. This establishes (i), by appealing to Proposition 4.1(i).

Suppose that $\mathrm{char}F = q > 0$. If the field F is algebraic over its prime subfield, then F^* is a torsion group and so $F^* = t(F^*)$. This proves (ii), by applying Proposition 4.1(i).

If F is not algebraic over its prime subfield, then λ is infinite by Theorem 1.4. Therefore $\lambda = |F|$, by Lemma 1.3(ii). Thus (iii) is verified, by virtue of Proposition 4.1(ii).

Conversely, let λ be any infinite cardinal and q a prime. Choose any set S with $|S| = \lambda$, and define K to be $\mathbb{Q}$ or $\mathbb{F}_q$, the field of q elements. Denote by F the algebraic closure of the purely transcendental extension $K(S)$ of K generated by S. Because the algebraic closure of any field of infinite cardinality is of the same cardinality, we have

$$|F| = |K(S)| = |S| = \lambda$$

The required assertion is therefore a consequence of (i) and (iii). ■

Turning our attention to real closed fields, we first recall the following facts

(a) If F is real closed, then any element of F is either a square or the negative of the square (Theorem 2.23.5)

(b) If F is real closed, then every polynomial of odd degree with coefficients in F has a root belonging to F (Theorem 2.23.5)

(c) If F is real closed, then $\sqrt{-1} \notin F$ and $F(\sqrt{-1})$ is algebraically closed. Conversely, every algebraically closed field E of characteristic 0 contains a real closed subfield F such that $E = F(\sqrt{-1})$. (Theorems 2.23.10 and 2.23.16).

We are now ready to prove the following result which is a slightly sharpened version of a theorem of Fuchs (1958).

6.2. THEOREM. If F is a real closed field, then

$$F^* \cong \mathbb{Z}_2 \times \mathbb{Q}^{|F|}$$

Conversely, for any infinite cardinal λ, the group $\mathbb{Z}_2 \times \mathbb{Q}^\lambda$ is isomorphic to the multiplicative group of some real closed field.

Proof. Let E be the algebraic closure of a real closed field F. If ε is a primitive n-th root of unity in E, then for $n \neq 2$,

$$1^2 + \varepsilon^2 + \ldots + \varepsilon^{2(n-1)} = (\varepsilon^{2n} - 1)/(\varepsilon^2 - 1) = 0$$

and hence $\varepsilon \notin F$. Thus, ± 1 are the only roots of unity in F, i.e.

$$t(F^*) \cong \mathbb{Z}_2$$

Because $t(F^*)$ is finite, $F^* \cong \mathbb{Z}_2 \times A$ for some torsion-free group A (Theorem 2.6). By the properties (a) and (b), the group A must be divisible. Hence, by the argument of Theorem 6.1, $A \cong \mathbb{Q}^{|F|}$ and so $F^* \cong \mathbb{Z}_2 \times \mathbb{Q}^{|F|}$.

Conversely, given an infinite cardinal λ, we may find an algebraically closed field E with $|E| = \lambda$. By the property (c), $E = F(\sqrt{-1})$ for some real closed subfield F of E. Because $|E| = |F| = \lambda$, we have $F^* \cong \mathbb{Z}_2 \times \mathbb{Q}^\lambda$ and the result follows. ■

For the rest of the section, p denotes a fixed rational prime, $\mathbb{Z}_p$ the ring of p-adic integers and $\mathbb{Q}_p$ the rational p-adic field. To avoid confusion, we temporarily abandon our convention that $\mathbb{Z}_n$ denotes the cyclic group of order and write $\mathbb{Z}/n\mathbb{Z}$ instead of $\mathbb{Z}_n$. For convenience let us recall the following piece of information.

Let I be an ideal of an arbitrary ring S and let $\bigcap_{n=1}^{\infty} I^n = 0$. From the descending chain

$$I \supseteq I^2 \supseteq \ldots$$

of ideals of S we obtain a chain of natural ring epimorphisms:

$$\longrightarrow S/I^n \xrightarrow{\alpha_n} S/I^{n+1} \longrightarrow \ldots \longrightarrow S/I^2 \longrightarrow S/I \longrightarrow 0$$

The corresponding projective limit ring $\varprojlim S/I^n$ is the set

$$\{(\ldots, a_n, a_{n-1}, \ldots, a_2, a_1) \mid a_n \in S/I^n,\ \alpha_n(a_n) = a_{n-1} \quad \text{for } n = 2,3,\ldots\}$$

with componentwise addition and multiplication. There is a natural embedding of S into $\varprojlim S/I^n$, namely

$$s \longmapsto (\ldots, s+I^n, \ldots, s+I^2, s+I)$$

and we identify S with its image in $\varprojlim S/I^n$. The ring $\mathbb{Z}_p$ of p-adic integers is given by

$$\mathbb{Z}_p = \varprojlim \mathbb{Z}/p^n\mathbb{Z}$$

Thus $\mathbb{Z}_p$ is a subring of $\prod_{n\geq 1}(\mathbb{Z}/p^n\mathbb{Z})$ and we denote by π_n the restriction to $\mathbb{Z}_p$ of the natural projection

$$\prod_{n\geq 1} \mathbb{Z}/p^n\mathbb{Z} \longrightarrow \mathbb{Z}/p^n\mathbb{Z}$$

6.3. LEMMA. The sequence

$$0 \longrightarrow p^n\mathbb{Z}_p \longrightarrow \mathbb{Z}_p \xrightarrow{\pi_n} \mathbb{Z}/p^n\mathbb{Z} \longrightarrow 0$$

is exact (in particular, $\mathbb{Z}_p/p^n\mathbb{Z}_p \cong \mathbb{Z}/p^n\mathbb{Z}$)

Proof. Let $\phi_n : \mathbb{Z}/p^n\mathbb{Z} \longrightarrow \mathbb{Z}/p^{n-1}\mathbb{Z}$ be the natural epimorphism. Suppose that $x = (x_n) \in \mathbb{Z}_p$ is such that $px = 0$. Then $px_{n+1} = 0$ for all n, whence $x_{n+1} = p^n y_{n+1}$ for some $y_{n+1} \in \mathbb{Z}/p^{n+1}\mathbb{Z}$. Because $x_n = \phi_{n+1}(x_{n+1})$, x_n is multiple of p^n and hence is 0.

It is clear that $p^n\mathbb{Z}_p \subseteq \operatorname{Ker}\pi_n$. Conversely, if $x = (x_m) \in \operatorname{Ker}\pi_n$, then $x_m \equiv 0 \pmod{p^n}$ for all $m \geq n$ and so $x_m = p^n y_{m-n}$ for some $y_{m-n} \in \mathbb{Z}/p^{m-n}\mathbb{Z}$. The elements y_i determine an element y of $\mathbb{Z}_p = \varprojlim \mathbb{Z}/p^i\mathbb{Z}$ which satisfies $p^n y = x$. So the lemma is proved. ■

6.4. LEMMA (i) $U(\mathbb{Z}_p) = \mathbb{Z}_p - p\mathbb{Z}_p$

(ii) Each nonzero element of $\mathbb{Z}_p$ can be uniquely written in the form

$$p^n u \qquad (u \in U(\mathbb{Z}_p),\ n \geq 0)$$

Proof. (i) It suffices to verify that $U(A_n) = A_n - pA_n$ where $A_n = \mathbb{Z}/p^n\mathbb{Z}$. If $x \in A_n$ and $x \notin pA_n$, then its image in $A_1 = \mathbb{F}_p$ is distinct from 0 and hence a unit. Then there exist $y, z \in A_n$ such that $xy = 1 - pz$ whence

$$xy(1 + pz + \ldots + p^{n-1}z^{n-1}) = 1$$

proving that $x \in U(A_n)$. The opposite containment being obvious, (i) is established.

(ii) Let x be a nonzero element of $\mathbb{Z}_p$. Then there exists a greatest integer n such that $x_n = \pi_n(x)$ is 0. Then, by Lemma 6.3, we have $x = p^n u$ where u is not divisible by p. But then, by (i), $u \in U(\mathbb{Z}_p)$ and the uniqueness of the given decomposition being obvious, the result follows. ■

We now turn our attention to $\mathbb{Q}_p$. It is clear that

$$\mathbb{Q}_p = \mathbb{Z}_p[p^{-1}]$$

and that each element of $\mathbb{Q}_p^*$ can be uniquely written in the form

$$p^n u \qquad (n \in \mathbb{Z},\ u \in U(\mathbb{Z}_p))$$

This shows that

$$\mathbb{Q}_p^* = \langle p \rangle \times U(\mathbb{Z}_p) \cong \mathbb{Z} \times U(\mathbb{Z}_p)$$

and so the multiplicative structure of $\mathbb{Q}_p$ is entirely reflected by that of $U(\mathbb{Z}_p)$.

6.5. THEOREM. The following holds:

$$\mathbb{Q}_p^* \cong \begin{cases} \mathbb{Z} \times \mathbb{Z}_p \times \mathbb{Z}/(p-1)\mathbb{Z}, & \text{if } p \neq 2 \\ \mathbb{Z} \times \mathbb{Z}_2 \times \mathbb{Z}/2\mathbb{Z}, & \text{if } p = 2 \end{cases}$$

Proof. We need only verify that

$$U(\mathbb{Z}_p) \cong \begin{cases} \mathbb{Z}_p \times \mathbb{Z}/(p-1)\mathbb{Z}, & \text{if } p \neq 2 \\ \mathbb{Z}_2 \times \mathbb{Z}/2\mathbb{Z}, & \text{if } p = 2 \end{cases} \qquad (1)$$

To this end, put $U = U(\mathbb{Z}_p)$ and $U_n = 1 + p^n\mathbb{Z}_p$, $n \geqslant 1$. Then U_n is the kernel of the homomorphism $U \longrightarrow U(\mathbb{Z}/p^n\mathbb{Z})$ induced by π_n. In particular, U/U_1 is isomorphic to $\mathbb{F}_p^*$ and hence is cyclic of order $p - 1$. The groups U_n form a decreasing chain of open subgroups of U and we have $U = \varprojlim U/U_n$. If $n \geqslant 1$, then the map $(1+p^n x) \longmapsto x \pmod{p}$ determines an isomorphism

$$U_n/U_{n+1} \cong \mathbb{Z}/p\mathbb{Z}$$

since

$$(1+p^n x)(1+p^n y) \equiv 1 + p^n(x+y) \pmod{p^{n+1}}$$

By induction, it follows that $|U_1/U_n| = p^{n-1}$. Because the order of U/U_1 is coprime to p, we obtain exact splitting sequence

$$1 \longrightarrow U_1/U_n \longrightarrow U/U_n \longrightarrow \mathbb{F}_p^* \longrightarrow 1$$

Therefore U/U_n contains a unique subgroup $V_n \cong \mathbb{F}_p^*$ and the projection

$$U/U_n \longrightarrow U/U_{n-1}$$

isomorphically maps V_n onto V_{n-1}. Since $U = \varprojlim U/U_n$, passing to limit, we obtain a subgroup V of U such that $U = V \times U_1$ and $V \cong \mathbb{F}_p^*$. The conclusion is that

$$U \cong \mathbb{Z}/(p-1)\mathbb{Z} \times U_1 \qquad (2)$$

Let $x \in U_n - U_{n+1}$, where $n \geqslant 1$ if $p \neq 2$ and $n \geqslant 2$ if $p = 2$. We now claim that

$$x^p \in U_{n+1} - U_{n+2} \qquad (3)$$

Indeed, by hypothesis, $x = 1 + kp^n$, where $k \not\equiv 0 \pmod p$. Applying binomial theorem, we derive

$$x^p = 1 + kp^{n+1} + \dots + k^p p^{np}$$

and the powers of p which are not indicated are $\geqslant 2n + 1$ and hence $\geqslant n + 2$. Furthermore, since $n \geqslant 2$ if $p = 2$, we have $np \geqslant n + 2$. Therefore

$$x^p \equiv 1 + kp^{n+1} \pmod{p^{n+2}},$$

proving (3).

It will now be shown that

$$U_1 \cong \begin{cases} \mathbb{Z}_p, & \text{if } p \neq 2 \\ (\mathbb{Z}/2\mathbb{Z}) \times \mathbb{Z}_2, & \text{if } p = 2 \end{cases} \qquad (4)$$

which will complete the proof, by invoking (1) and (2). Suppose first that $p \neq 2$. Choose any element $x \in U_1 - U_2$, say $x = 1 + p$. Then, by (3),

$$x^{p^i} \in U_{i+1} - U_{i+2}$$

Let x_n be the image of x in U_1/U_n. Then $x_n^{p^{n-2}} \neq 1$ and $x_n^{p^{n-1}} = 1$. But U_1/U_n is of order p^{n-1}, hence U_1/U_n is a cyclic group generated by x_n. Let $Q_{n,x} : \mathbb{Z}/p^{n-1}\mathbb{Z} \longrightarrow U_1/U_n$, $z \longmapsto x_n^z$ be an isomorphism and let

$$f_1 : \mathbb{Z}/p^n\mathbb{Z} \longrightarrow \mathbb{Z}/p^{n-1}\mathbb{Z},\ f_2 : U_1/U_{n+1} \longrightarrow U_1/U_n$$

be natural homomorphisms. Then $\theta_{n,x} f_1 = f_2 \theta_{n+1,x}$ and hence the isomorphisms $\theta_{n,x}$ determine an isomorphism of $\mathbb{Z}_p = \varprojlim \mathbb{Z}/p^{n-1}\mathbb{Z}$ onto

$$U_1 = \varprojlim U_1/U_n$$

which proves (4) for $p \neq 2$.

Finally, assume that $p = 2$. Choose $x \in U_2 - U_3$, or equivalently, put $x \equiv 5 (\mathrm{mod}\ 8)$. We may determine, in the above manner, isomorphisms

$$\theta_{n,x} : \mathbb{Z}/2^{n-2}\mathbb{Z} \longrightarrow U_2/U_n$$

and hence an isomorphism $\theta_x : \mathbb{Z}_2 \longrightarrow U_2$. The homomorphism

$$U_1 \longrightarrow U_1/U_2 \cong \mathbb{Z}/2\mathbb{Z}$$

induces an isomorphism of $\{\pm 1\}$ and $\mathbb{Z}/2\mathbb{Z}$. Therefore

$$U_1 = \{\pm 1\} \times U_2 \cong \mathbb{Z}/2\mathbb{Z} \times \mathbb{Z}_2,$$

proving (4) and hence the result. ■

7. LOCAL FIELDS

Recall that a field F is said to be local if there is a complete discrete valuation v of F such that the residue class field of v is finite. We know, from Proposition 3.8.16 that F is local if and only if it is one of the following types:

(i) A finite extension of the rational p-adic field $\mathbb{Q}_p$

(ii) A formal power series field in one variable over a finite field.

The cases (i) and (ii) occur according to whether F and the residue class field of v have unequal or equal characteristic. For this reason, we shall refer to (i) and (ii) as "unequal characteristic case" and "equal characteristic case," respectively. Since all our valuations will be nonarchimedean, we will write these valuations additively. Thus, in accordance with our previous conventions, we use the term "principal valuation" instead of "discrete valuation". For convenience, we divide this section into three subsections.

A. Preparatory results

Throughout, F denotes a field and $v : F \longrightarrow \mathbb{R}$ a fixed principal valuation of F. In case v can be chosen to be such that F is complete with respect to v, then we say that F is a complete discrete valued field. In what follows, R,P, and $\bar{F}$ denote the valuation ring, the unique maximal ideal and the residue class field of v, respectively. Thus, by definition,

$$R = \{x \in F | v(x) \geqslant 0\}, \ P = \{x \in F | v(x) > 0\}, \ \text{and} \ \bar{F} = R/P$$

Recall that the unit group $U(R)$ of R coincides with $R - P$, i.e.

$$U(R) = \{x \in F | v(x) = 0\}$$

The limit of any sequence of elements of F with respect to v will be also called a *P-adic limit.*

In what follows, we write $\bar{F}^{+}$ for the additive group of $\bar{F}$ and π for a prime element of F (i.e. an element of F of least positive value).

7.1. LEMMA. (i) $P = (\pi)$ and all ideals of R are of the form (π^n), $n \geqslant 0$
(ii) $F^* = <\pi> \times U(R) \cong \mathbb{Z} \times U(R)$.

Proof. Replacing v be an equivalent valuation, if necessary, we may assume that $v(\pi) = 1$. Hence, for any $x \in F^*$, if $v(x) = n$, then

$$v(x\pi^{-n}) = v(x) - nv(\pi) = 0,$$

so $x\pi^{-n} = u \in U(R)$. Thus we have $x = \pi^n u$, $n = v(x)$, $u \in U(R)$ and it is clear that such representation is unique once π have been chosen. This obviously implies (i) and (ii). ∎

The natural ring epimorphism $R \longrightarrow R/P = \bar{F}$ induces a group epimorphism

$$U(R) \longrightarrow \bar{F}^*$$

whose kernel is

$$1 + P = \{1+x | x \in P\}$$

Therefore, we have an exact sequence

$$1 \longrightarrow 1 + P \longrightarrow U(R) \longrightarrow \bar{F}^* \longrightarrow 1$$

The elements of $1 + P$ are called *1-units* and, for convenience, in future we write $U_1(R)$ instead of $1 + P$. More generally, for any $n \geqslant 1$, we put

$$U_n(R) = 1 + P^n$$

7.2. LEMMA. (i) For any $n \geqslant 1$, $U_n(R)/U_{n+1}(R) \cong \bar{F}^+$

(ii) For any $u_n \in U_n(R)$ there exists a unique $u_{n+1} \in U_{n+1}(R)$ and $a_n \in R$, determined uniquely mod P, such that

$$u_n = (1+a_n\pi^n)u_{n+1}$$

(iii) If F is complete, $u = u_1 \in U_1(R)$ and, for each $n \geqslant 1$, S_n is a residue system of $\bar{F}$ in R, then u has a unique representation

$$u = \prod_{n=1}^{\infty}(1+a_n\pi^n) \qquad (a_n \in S_n)$$

Proof. (i) Owing to Lemma 7.1(i), $P^n = (\pi^n)$ for all $n \geqslant 1$. Accordingly, a typical element of $U_n(R)$ is $1 + \pi^n x$, $x \in R$. Consider the map

$$\begin{cases} U_n(R) \xrightarrow{\psi} \bar{F}^+ \\ 1 + \pi^n x \longmapsto xP \end{cases}$$

Because of the identity

$$(1+\pi^n x)(1+\pi^n y) = 1 + \pi^n(x+y+\pi^n xy)$$

the map ψ is a homomorphism. By definition of ψ, ψ is surjective and $\text{Ker}\psi$ is obviously equal to $U_{n+1}(R)$.

(ii) By (i), any residue system of $\bar{F}$ in R yields a transversal

$$1 + \pi^n s \qquad (s \in S)$$

for $U_{n+1}(R)$ in $U_n(R)$. Therefore, given $u_n \in U_n(R)$, we may write uniquely $u_n = (1+\pi^n s)u_{n+1}$ with $s \in S$ and $u_{n+1} \in U_{n+1}(R)$, proving (ii).

(iii) Put $\lambda_n = \prod_{i=1}^{n}(1+a_i\pi^i)$. Then, by (ii), $u = \lambda_n u_{n+1}$ and, because

$$\lim_{n\to\infty} u_n = 1$$

we have $u = \lim_{n\to\infty} \lambda_n$. Since each $a_n \in S_n$ is uniquely determined by u, the assertion follows. ■

We next provide circumstances under which $U(R)$ splits over $U_1(R)$.

7.3. LEMMA. Assume that F is complete and $\bar{F}$ is perfect. Then

(i) There is a subgroup L of $U(R)$ such that

$$U(R) = L \times U_1(R) \cong \bar{F}^* \times U_1(R)$$

(ii) $F^* = \langle\pi\rangle \times L \times U_1(R) \cong \mathbb{Z} \times \bar{F}^* \times U_1(R)$

Proof. Apply Lemma 7.1(ii), the fact that $U(R)/U_1(R) \cong \bar{F}^*$ and Theorem 3.8.3. ■

Let $\mathbb{Z}_p$ be the ring of p-adic integers. Then any $g \in \mathbb{Z}_p$ can be represented as the p-adic limit of a convergent sequence of rational integers g_n:

$$g = \lim_{n\to\infty} g_n$$

For example, by Lemma 3.8.1, we can take for the g_n the partial sums in the p-adic series representation

$$g = \sum_{n=0}^{\infty} a_n p^n$$

with coefficients a_n taken from $\{0,1,\ldots,p-1\}$. The p-adic convergence of the sequence g_n is described by a system of congruences

$$g_{n+1} \equiv g_n \pmod{p^{\lambda_n}} \qquad \text{for all } n \geq 0$$

where $\lambda_n \longrightarrow \infty$ as $n \longrightarrow \infty$. Furthermore, if

$$g = \lim_{n\to\infty} g_n'$$

then

$$g_n' \equiv g_n \pmod{p^{\mu_n}} \qquad \text{for all } n \geq 0$$

where $\mu_n \longrightarrow \infty$ as $n \longrightarrow \infty$. Applying Lemma 3.8.2, we now provide the following simple observation which will play a crucial role in our subsequent investigations.

7.4. LEMMA. Assume that F is complete and $\operatorname{char}\bar{F} = p > 0$. Given $g \in \mathbb{Z}_p$, choose any sequence g_n of rational integers such that $g = \lim_{n\to\infty} g_n$.

(i) For any $\lambda \in U_1(R)$, $\lim_{n\to\infty} \lambda^{g_n}$ exists and is independent of the choice of the sequence g_n.

(ii) For all $n \geq 1$, $U_n(R)$ is a $\mathbb{Z}_p$-module, where for $\lambda \in U_n(R)$ and $g \in \mathbb{Z}_p$,

$$\lambda^g = \lim_{n\to\infty} \lambda^{g_n}$$

(iii) For all rational integers $m \not\equiv 0 \pmod{p}$ and all $n \geq 1$, the map

$$\begin{cases} U_n(R) \longrightarrow U_n(R) \\ x \longmapsto x^m \end{cases}$$

is an automorphism.

Proof. (i) Given $\lambda \in U_1(R)$, it follows from Lemma 3.8.2 that

$$\lambda^{p^n} \equiv 1 \pmod{P^{n+1}} \qquad \text{for all } n \geq 0$$

Hence, if $h \in \mathbb{Z}$ is such that $h \equiv 0 \pmod{p^n}$, then

$$\lambda^h \equiv 1 \pmod{P^{n+1}} \qquad \text{for all } n \geq 0$$

Thus, in the notation preceding Lemma 7.4, we derive

$$\lambda^{g_{n+1}} \equiv \lambda^{g_n} \pmod{P^{\lambda_n+1}}$$

$$\lambda^{g'_n} \equiv \lambda^{g_n} \pmod{P^{\mu_n+1}}$$

for all $n \geq 0$. This shows that the sequence λ^{g_n} converges P-adically, and hence has a limit in F depending only on g, and not on the choice of the particular sequence g_n converging to g.

(ii) If $\lambda \equiv 1 \pmod{P^n}$, then all the $\lambda^{g_n} \equiv 1 \pmod{P^n}$ as well and hence

$$\lambda^g \equiv 1 \pmod{P^n},$$

proving that $\lambda^g \in U_n(R)$. The rest follows from the laws for operating with limits and from the power rules for rational integral exponents.

(iii) Let $m \in \mathbb{Z}$ be such that $m \not\equiv 0 \pmod{p}$. Then m is a unit of $\mathbb{Z}_p$ and therefore the action of m on $U_n(R)$ induces an automorphism of $U_n(R)$. But, by definition of the action of $\mathbb{Z}_p$ on $U_n(R)$, for all $\lambda \in U_n(R)$, λ^m is the ordinary m-th power of λ. So the lemma is true. ∎

Before proving our next result, let us recall the following piece of information. Let $\mathbb{Q}_p$ be the rational p-adic field. A finite extension F of $\mathbb{Q}_p$ is said to be a *P-adic field* if P is the maximal ideal of the valuation ring R attached to F. In case $F/\mathbb{Q}_p$ is unramified (equivalently, P is generated by p) we say that F is a p-adic field. Hence a finite extension F of $\mathbb{Q}_p$ is a p-adic field if and only if p is a prime element of F. In what follows we write ε_{p^n} for a primitive p^n-th root of unity.

7.5. LEMMA. Let F be a p-adic field. Then $F(\varepsilon_p) = F(\lambda)$ where λ is a prime element of $F(\varepsilon_p)$ such that

(i) $\lambda^{p-1} = -p$

(ii) $\lambda = (1-\varepsilon_p)u$ for some 1-unit u of $F(\varepsilon_p)$

Proof. Put $\pi_1 = 1 - \varepsilon_p$ and $P_1 = (\pi_1)$. Then, by Proposition 3.8.20, π_1 is a prime element of $F(\varepsilon_p)$ and so $F(\varepsilon_p)$ is a P_1-adic field. If $p = 2$, then the result is obvious by taking $\lambda = -2$. Therefore we may assume that p is odd in which case $(-1)^{p-1} = 1$. Applying equation

$$\frac{1-(1-\pi_1)^p}{\pi_1} = p - \binom{p}{2}\pi_1 + \dots + (-1)^{p-1}\pi_1^{p-1} = 0$$

established in the course of the proof of Proposition 3.8.20, we deduce that

$$-p \equiv (-1)^{p-1}\pi_1^{p-1} = \pi_1^{p-1} \pmod{P_1^p}$$

which in turn implies

$$-p = \pi_1^{p-1}u_0 \qquad \text{for some 1-unit } u_0 \text{ in } F(\varepsilon_p)$$

By Lemma 7.4(iii), there exists a 1-unit u in $F(\varepsilon_p)$ such that $u^{p-1} = u_0$. Setting $\lambda = \pi_1 u$, we therefore derive $\lambda^{p-1} = -p$. This shows that λ is a prime element of $F(\varepsilon_p)$ which satisfies (i) and (ii). Because $F(\varepsilon_p)/F$ is totally ramified (Proposition 3.8.20), we also have $F(\varepsilon_p) = F(\lambda)$, by appealing to Lemma 3.8.9(i). ■

7.6. LEMMA. Let F be a P-adic field with prime element π and write the prime decomposition of p in F in the form

$$-p = \pi^e u \qquad (u \text{ is a unit of } R)$$

Then $\varepsilon_p \in F$ if and only if the following two conditions hold:

(i) $p-1$ divides e

(ii) $x^p - ux \in P$ for some unit x of R

Proof. Let F_0 denotes the inertia field of the extension $F/\mathbb{Q}_p$. Assume that conditions (i) and (ii) hold. By (ii), $x^{p-1} \equiv u \pmod{p}$ so that

$$-p \equiv \pi^e x^{p-1} \pmod{p}$$

and thus

$$-p = \pi^e x^{p-1} u_0 \qquad \text{for some 1-unit } u_0 \text{ in } F$$

Because $p - 1$ divides e, $\pi^e = (\pi^k)^{p-1}$ for some $k \geq 1$ and, by Lemma 7.4(iii), $y^{p-1} = u_0$ for some 1-unit y in F. Setting $\lambda = \pi^k xy$, we therefore derive $-p = \lambda^{p-1}$. Therefore, by Lemma 7.5, $F_0(\varepsilon_p) = F_0(\lambda) \subseteq F$ and hence $\varepsilon_p \in F$.

Conversely, assume that $\varepsilon_p \in F$ and hence that $F_0(\varepsilon_p) \subseteq F$. Since F_0 is unramified and, by Proposition 3.8.20, $F_0(\varepsilon_p)$ is totally ramified of degree $p - 1$, it follows that $p - 1$ divides e. Moreover P^k, $k = e/(p-1)$, is the prime ideal attached to $F_0(\varepsilon_p)$. Owing to Lemma 7.5, we now have

$$F_0(\varepsilon_p) = F(\lambda) \quad \text{with} \quad \lambda^{p-1} = -p$$

where $\lambda \in P^k$. From this it follows that $\lambda = \pi^k u_0$ for some $u_0 \in U(R)$ and hence $-p = \pi^e u_0^{p-1}$. Hence $u = u_0^{p-1}$ and therefore $x = u_0$ satisfies (ii), as required. ∎

As a final preparatory result, we prove

7.7. LEMMA. Let F be a P-adic field with the valuation ring R and residue class field $\bar{F} = R/P$, let S be a unique multiplicative residue system of $\bar{F}$ in R and let $U_k(R) = 1 + P^k$, $k \geq 1$. Denote by e the absolute ramification index of F.

(i) If $\varepsilon_{p^n} \in F$, then $\phi(p^n)$ divides e and

$$\varepsilon_{p^n} \in U_k(R) - U_{k+1}(R) \qquad \text{for } k = e/\phi(p^n)$$

(ii) $S^* = S - \{0\}$ consists precisely of all p'-roots of unity of F.

Proof. Property (ii) is a consequence of the fact that $S^* \cong \bar{F}^*$ (hence S^* is of p'-order) and of the characterization of S given by Theorem 3.8.3(ii).

Let $\varepsilon_{p^n} \in F$ and let F_0 be the inertia group of F. Then, by the argument of Lemma 7.6, $\phi(p^n)$ divides e and P^k, $k = e/\phi(p^n)$, is the prime ideal attached to $F_0(\varepsilon_{p^n})$. But, by Proposition 3.8.20, $1 - \varepsilon_{p^n}$ is a prime element of $F_0(\varepsilon_{p^n})$. Thus

$$1 - \varepsilon_{p^n} \in P^k - P^{k+1}$$

and the result follows. ■

B. The equal characteristic case.

Throughout this section, F denotes a local field and R, P and $\bar{F}$ are the valuation ring, the unique maximal ideal and residue class field attached to it. By definition, $\bar{F}$ is a finite field, say of p^f elements. Our aim is to examine the multiplicative structure of F in the equal characteristic case, i.e. when $\text{char}F = p$. Owing to Theorem 3.8.3, there is a unique residue system, say S, which is a field. Hence $S \cong \bar{F}^*$ and, by Lemma 7.3, we have

$$F^* \cong \mathbb{Z} \times \mathbb{Z}/(p^f-1)\mathbb{Z} \times U_1(R) \tag{1}$$

Here again we abandon the convention that $\mathbb{Z}_n$ denotes a cyclic group of order in order to avoid a possible confusion with the ring $\mathbb{Z}_p$ of p-adic integers. Owing to (1), the multiplicative structure of F is entirely reflected by that of $U_1(R)$. By Lemma 7.4, $U_1(R)$ is a $\mathbb{Z}_p$-module and the precise structure of this module is given by the following classical result.

7.8. THEOREM. (Hensel (1916)). Let $\lambda_1,\ldots,\lambda_f$ be an $\mathbb{F}_p$-basis of S and let

$$e_{ik} = 1 + \lambda_i \pi^k \qquad (1 \leqslant i \leqslant f,\ k \not\equiv 0 \pmod p)$$

where π is a prime element of F. Then each $u \in U_1(R)$ has a unique representation in the form

$$u = \prod_{k \not\equiv 0 \pmod p} \prod_{i=1}^{f} e_{ik}^{\alpha_{ik}} \qquad (\alpha_{ik} \in \mathbb{Z}_p) \tag{2}$$

where the infinite products appearing in (2) are to be understood as limits of the sequences of their finite partial products with respect to the valuation of F. Thus $U_1(R)$ is isomorphic to the (unrestricted) direct product of $\aleph_0$ copies of the additive group of $\mathbb{Z}_p$.

Proof. We first claim that if T_m is a transversal for $U_{m+1}(R)$ in $U_m(R)$, then $T_m^p = \{x^p \mid x \in T_m\}$ is a transversal for $U_{pm+1}(R)$ in $U_{pm}(R)$. Indeed, if

$$t = 1 + r\pi^m \qquad (r \in R)$$

then because $\text{char}F = p$, we have

$$t^p = 1 + r^p\pi^{mp} \in U_{mp}(R)$$

Since the map $rP \longmapsto r^pP$ is an automorphism of $\bar{F}$, our claim follows by virtue of Lemma 7.2(i).

Starting with the transversal T_k for all $k \not\equiv 0(\text{mod } p)$, we thus obtain specific transversals T_m,

$$T_k,\ T_{pk} = T_k^p, \ldots, T_{p^nk} = T_k^{p^n}, \ldots \tag{3}$$

for all possible m, each exactly once.

By assumption, each element s of S can be uniquely written in the form

$$s = \sum_{i=1}^{f} \alpha_i\lambda_i \qquad (\alpha_i \in \mathbb{Z},\ 0 \leqslant \alpha_i < p)$$

Let us fix $k \not\equiv 0(\text{mod } p)$. Then, in view of the isomorphism

$$\begin{cases} \bar{F}^+ \longrightarrow U_k(R)/U_{k+1}(R) \\ sP \longmapsto (1+\pi^k s)U_{k+1}(R) \end{cases}$$

established in Lemma 7.2(i), the power products

$$\prod_{i=1}^{f} e_{ik}^{\alpha_i} \equiv 1 + s\pi^k (\text{mod } U_{k+1}(R)) \qquad (\alpha_i \in \mathbb{Z},\ 0 \leqslant \alpha_i < p)$$

constitute a transversal, say T_k, for $U_k(R)$ in $U_{k+1}(R)$. Applying (3), we now obtain specific transversals T_m for $U_{m+1}(R)$ in $U_m(R)$ for all possible m, each exactly once.

We now invoke Lemma 7.2(iii) to express each $u \in U_1(R)$ uniquely in the form

$$u = \prod_{\nu=0}^{\infty} \prod_{k\not\equiv 0(\text{mod } p)} \prod_{i=1}^{f} e_{ik}^{\alpha_{ik}^{(\nu)}p^\nu} \qquad (\alpha_{ik}^{(\nu)} \in \mathbb{Z},\ 0 \leqslant \alpha_{ik}^{(\nu)} < p)$$

By rearrangement, we obtain, from the nature of action of $\mathbb{Z}_p$ on $U_1(R)$, the unique representation

$$u = \prod_{k\not\equiv 0(\text{mod } p)} \prod_{i=1}^{f} e_{ik}^{\alpha_{ik}} \qquad (\alpha_{ik} = \sum_{\nu=0}^{\infty} \alpha_{ik}^{(\nu)}p^\nu \text{ in } \mathbb{Z}_p)$$

thus completing the proof. ■

C. The unequal characteristic case.

Throughout, F denotes a local field and R, P and $\bar{F}$ the valuation ring, the unique maximal ideal and residue class field attached to it. Our aim is to examine the multiplicative structure of F in the unequal characteristic case, i.e.

when $\mathrm{char}F = 0$ and $\mathrm{char}\bar{F} = p > 0$. Owing to Proposition 3.8.16, F is a finite extension of the rational p-adic field $\mathbb{Q}_p$. Let us fix the following notations:

$n = (F:\mathbb{Q}_p)$

$e = e_{F/\mathbb{Q}_p}$, $f = f_{F/\mathbb{Q}_p}$

π is a prime element of F

$\mathbb{Z}_p$ is the ring of p-adic integers

$\mathbb{Z}/n\mathbb{Z}$ is the cyclic group of order n

By definition of f, $\bar{F}$ is a finite field of p^f elements and, by Proposition 3.8.15, we have

$$n = f \quad \text{and} \quad Rp = P^e$$

Owing to Theorem 3.8.3, there is a unique multiplicative residue system S of $\bar{F}$ in R. We denote by S^* the set of nonzero elements of S. Thus

$$S^* \cong \bar{F}^* \cong \mathbb{Z}/(p^f-1)\mathbb{Z}$$

and, by Theorem 3.8.3, S^* contains all p'-roots of unity in F. Note also that, by Lemma 7.3(ii),

$$F^* \cong \mathbb{Z} \times \mathbb{Z}/(p^f-1)\mathbb{Z} \times U_1(R) \tag{1}$$

and so we need only to investigate the structure of the $\mathbb{Z}_p$-module $U_1(R)$. This will be achieved with the aid of the following preliminary result.

7.9. LEMMA. Let $x, r \in R$ be such that

$$x \equiv 1 + r\pi^k (\mathrm{mod}\ P^{k+1})$$

and write $-p = \pi^e u$ for some $u \in U(R)$.

(i) If $k < e/(p-1)$, then

$$x^p \equiv 1 + r^p\pi^{pk} (\mathrm{mod}\ P^{pk+1})$$

(ii) If $k = e/(p-1)$, then

$$x^p \equiv 1 + rp\pi^k + r^p\pi^{pk} = 1 + (r^p - ur)\pi^{pk} (\mathrm{mod}\ P^{pk+1})$$

(iii) If $k > e/(p-1)$, then

$$x^p \equiv 1 + rp\pi^k = 1 - ur\pi^{k+e} (\text{mod } P^{k+e+1})$$

Proof. We may harmlessly assume that $x = 1 + r\pi^k$. Indeed, because

$$x \equiv 1 + r\pi^k (\text{mod } P^{k+1}),$$

we have $x = 1 + r'\pi^k$ with $r' \equiv r (\text{mod } P)$ and the expressions appearing in the assertions (i), (ii), and (iii) on the right depend only on the residue class mod P.

Invoking the binomial theorem, we have

$$x^p = 1 + pr\pi^k + \binom{p}{2} r^2\pi^{2k} + \ldots + \binom{p}{p-1} r^{p-1}\pi^{(p-1)k} + r^p\pi^{pk}$$

Now $p = -\pi^e u$ and the middle binomial coefficients are divisible by p. Thus the terms following the 1 are contained respectively in

$$P^{k+e}, P^{2k+e}, \ldots, P^{(p-1)k+e}, P^{pk}$$

Accordingly, for $pk < k + e$, that is, for $k < e/(p-1)$, all these terms, except for the final $r^p\pi^{pk}$, are contained in P^{pk+1}, proving (i). In case

$$pk = k + e,$$

that is, for $k = e/(p-1)$, all the terms except for the initial $pr\pi^k$ and the final $r^p\pi^{pk}$, are contained in P^{pk+1}, proving (ii). Finally, for $pk > k + e$, that is, for $k > e/(p-1)$, all terms, except for the initial $pr\pi^k$, are contained in P^{k+e+1}, proving (iii). ■

We are now ready to establish the following classical result.

7.10. THEOREM. (Hensel (1916)).

(i) For any $t > e/(p-1)$, $(U_1(R):U_t(R)) = p^{(t-1)f}$ and $U_t(R)$ is a free $\mathbb{Z}_p$-module of rank n.

(ii) $U_1(R)$ is a direct product of a cyclic group of order p^μ for some $\mu \geq 0$ and a free $\mathbb{Z}_p$-module of rank n.

Proof. We first demonstrate that (ii) is a consequence of (i). Indeed, $\mathbb{Z}_p$ is a principal ideal domain and so by (i), $U_1(R)$ is a finitely generated $\mathbb{Z}_p$-module whose free factor is of rank n. Let x be an element of the torsion submodule of $U_1(R)$ and let $\alpha = p^n u$, $n \geq 0$, $u \in U(\mathbb{Z}_p)$, be an arbitrary element of $\mathbb{Z}_p$. Then $x^a = 1$ implies $(x^{p^n})^u = 1$ and hence $x^{p^n} = 1$, since u acts as an automorphism of $U_1(R)$. Consequently, the torsion submodule of $U_1(R)$ consists of roots of unity in F whose order is a power of p. Owing to Proposition 3.8.20, there are only finitely many of them and hence they constitute a cyclic group of order p^μ for some $\mu \geqslant 0$. This proves that (ii) is a consequence of (i), by invoking the fact that $U_1(R)$ is a direct product of the torsion submodule and a free module.

To prove (i), we first observe that the equality

$$(U_1(R):U_t(R)) = p^{(t-1)f}$$

follows from the isomorphism

$$U_i(R)/U_{i+1}(R) \cong \bar{F}^+$$

(Lemma 7.2(i)) and the fact that $\bar{F}$ is a finite field of p^f elements.

Now choose an $\mathbb{F}_p$-basis $\omega_1, \omega_2, \ldots, \omega_f$ of $\bar{F}$ and let $\lambda_1, \lambda_2, \ldots, \lambda_f$ be elements of S such that ω_i is the image of λ_i, $1 \leqslant i \leqslant f$. For any $u \in U(R)$, the map $r + P \longmapsto -ur + P$ is an automorphism of the group $\bar{F}^+$. Therefore, by Lemma 7.9(iii), for any $m \geqslant t > e/(p-1)$, if T_m is a transversal for $U_{m+1}(R)$ in $U_m(R)$, then T_m^p is a transversal for $U_{m+e+1}(R)$ in $U_{m+e}(R)$. Hence, if T_k is a transversal for $U_{k+1}(R)$ in $U_k(R)$ for $t \leqslant k < t + e$, then

$$T_k,\ T_k^p,\ T_k^{p^2},\ \ldots \qquad (t \leqslant k < t + e)$$

exhaust all transversals for $U_{m+1}(R)$ in $U_m(R)$ with $m \geqslant t$. Invoking the argument employed in the proof of Theorem 7.8, we infer that each element of $U_t(R)$ can be uniquely written in the form

$$\prod_{t \leqslant k \leqslant t+e} \prod_{i=1}^{f} e_{ik}^{\alpha_{ik}} \qquad (\alpha_{ik} \in \mathbb{Z}_p)$$

with

$$e_{ik} = 1 + \lambda_i \pi^k \qquad (1 \leqslant i \leqslant f,\ t \leqslant k < t + e)$$

Thus the e_{ik} form a free basis for $U_t(R)$ and since their number is $n = ef$, the result follows. ■

7.11. COROLLARY. Let F be a finite extension of $\mathbb{Q}_p$. Then, with the above notation, there exists $\mu \geq 0$ such that

(i) $F^* \cong [\mathbb{Z}/p^{\mu}(p^f-1)\mathbb{Z}] \times \mathbb{Z} \times \underbrace{(\mathbb{Z}_p \times \ldots \times \mathbb{Z}_p)}_{n \text{ factors}}$

(ii) $t(F^*) \cong \mathbb{Z}/p^{\mu}(p^{f-1})\mathbb{Z}$

(iii) $\mu = 0$ if and only if $(p-1) \nmid e$ or $x^{p-1} + u \in U(R)$ for all $x \in U(R)$, where $p = \pi^e u$ with $u \in U(R)$

(iv) $\phi(p^{\mu})$ divides e and the p-component of $t(F^*)$ is contained in $U_k(R)$, where $k = e/\phi(p^{\mu})$, but not contained in $U_{k+1}(R)$.

Proof. Property (i) is a consequence of (1) and Theorem 7.10(ii), while (ii) is a consequence of (i). Because $\mu = 0$ if and only if F does not contain a primitive p-th root of unity, the assertion (iii) follows from Lemma 7.6. Finally, (iv) is a consequence of Lemma 7.7(i). ∎

7.12. COROLLARY. With the preceding notation, for any $m = p^k m_0$, $(p,m_0) = 1$,

$$F^*/(F^*)^m \cong [\mathbb{Z}/(m,p^{\mu}(p^f-1)\mathbb{Z}] \times (\mathbb{Z}/m\mathbb{Z}) \times \underbrace{(\mathbb{Z}/p^k\mathbb{Z} \times \ldots \times \mathbb{Z}/p^k\mathbb{Z})}_{n \text{ factors}}$$

In particular,

$$(F^*:(F^*)^m) = (m,p^{\mu}(p^f-1))mp^{nk}$$

Proof. Let $G = \langle g \rangle$ be a cyclic group of order s. Then $G^m = \langle g^m \rangle$ is of order $s/(m,s)$ and hence $(G:G^m) = (m,s)$. If G is an infinite cyclic group, then clearly $(G:G^m) = m$. On the other hand, we have $m\mathbb{Z}_p = p^k\mathbb{Z}_p$, because m_0 is a unit of $\mathbb{Z}_p$. Thus $\mathbb{Z}_p/m\mathbb{Z}_p \cong \mathbb{Z}/p^k\mathbb{Z}$ is cyclic of order p^k. The desired conclusion is therefore a consequence of Corollary 7.11(i). ∎

7.13. COROLLARY. Let ε be a primitive m-th root of unity over $\mathbb{Q}_p$ with $m = p^s k$, $(k,p) = 1$, and let t be the order of p modulo k. Then

$$\mathbb{Q}_p(\varepsilon)^* \cong [\mathbb{Z}/p^s(p^t-1)\mathbb{Z}] \times \mathbb{Z} \times \underbrace{(\mathbb{Z}_p \times \ldots \times \mathbb{Z}_p)}_{\phi(p^s)t \text{ factors}}$$

if p is odd or $s \geq 1$ and

$$\mathbb{Q}_2(\varepsilon)^* \cong [\mathbb{Z}/2(2^t-1)\mathbb{Z}] \times \mathbb{Z} \times \underbrace{(\mathbb{Z}_2 \times \ldots \times \mathbb{Z}_2)}_{t \text{ factors}}$$

if $p = 2$ and $s = 0$.

Proof. Let ε_k and ε_{p^s} be primitive k-th and p^s-th roots of unity, respectively. Owing to Corollary 3.8.21, $\mathbb{Q}_p(\varepsilon_k)$ is the inertia field of the extension $\mathbb{Q}_p(\varepsilon)/\mathbb{Q}_p$ and

$$e_{\mathbb{Q}_p(\varepsilon)/\mathbb{Q}_p} = \phi(p^s) \quad \text{and} \quad f_{\mathbb{Q}_p(\varepsilon)/\mathbb{Q}_p} = t \tag{2}$$

Now $\mathbb{Q}_p(\varepsilon) = \mathbb{Q}_p(\varepsilon_k)(\varepsilon_{p^s})$ and $\mathbb{Q}_p(\varepsilon_k)$ is a p-adic field. Hence

$$(\mathbb{Q}_p(\varepsilon):\mathbb{Q}_p(\varepsilon_k)) = \phi(p^s) \tag{3}$$

by Proposition 3.8.20. If $\mathbb{Q}_p(\varepsilon)$ contains $\varepsilon_{p^{s+1}}$, then

$$\mathbb{Q}_p(\varepsilon) = \mathbb{Q}_p(\varepsilon_k)(\varepsilon_{p^{s+1}})$$

and, by (3), $\phi(p^s) = \phi(p^{s+1})$ which is possible only in the case $p = 2$ and $s = 0$. Hence, if p is odd or $s \geqslant 1$, then in the notation of Corollary 7.11, $\mu = s$. This establishes the case where p is odd or $s \geqslant 1$, by applying (2).

In case $p = 2$ and $s = 0$, $\mathbb{Q}_2(\varepsilon)$ contains -1, but does not contain a primitive 4-th root of unity, by virtue of (3). Thus, in the notation of Corollary 7.11, $\mu = 1$ and the result follows again by applying (2). ■

8. EXTENSIONS OF ALGEBRAIC NUMBER FIELDS

Let $F \subset E$ be algebraic number fields. Our aim is to discover the isomorphism class of the factor group E^*/F^*. As a preparation for the proof of the main result, we shall establish certain properties which will also be used in our subsequent investigations.

In what follows, unless explicitely stated otherwise, $F \subset E$ are algebraic number fields and $R \subset S$ are their rings of integers. We say that a prime P of R *splits completely* in S if PS is a product of distinct prime ideals of residue degree 1. Thus, if P splits completely in S, then P does not ramify in S and PS is not a prime ideal. We assume that the reader is familiar with a standard fact (see Serre (1979)) that all almost all prime ideals of R do not ramify in S. Our first aim is to prove that there are infinitely many prime ideals of R which split completely in S.

8.1. LEMMA. There exists $s \in S$ such that $E = F(s)$.

Proof. Because E/F is a separable extension, $E = F(\lambda)$ for some $\lambda \in E$. Let

$$r_nX^n + r_{n-1}X^{n-1} + \dots + r_0 = 0 \qquad (r_i \in R,\ r_n \neq 0)$$

be an equation satisfied by λ. Then the element $s = r_n\lambda$ satisfies the equation

$$X^n + r_{n-1}X^{n-1} + r_{n-2}r_nX^{n-2} + \dots + r_0r_n^{n-1} = 0,$$

which shows that $s \in S$ and $E = F(s)$. ■

Let M be any nonzero R-module contained in R. The set

$$M^* = \{x \in E \mid Tr_{E/F}(xM) \subseteq R\}$$

is called the *codifferent* of M over F. It is obvious that M^* is an R-module which may be equal to 0 (e.g. take $M = E$).

8.2. LEMMA. Let s be an element of S with $E = F(s)$ and let $f(X)$ be the minimal polynomial over R for s. Then $R[s]^*$ is generated as an R-module by the set

$$s^j/f'(s) \qquad 0 \leq j \leq n-1,\ n = (E:F)$$

Proof. Let A be the R-module generated by the above elements. Applying the Lagrange interpolation formula successively to the polynomials $X, X^2, \dots, X^{n-1}$ and $X^n - f(X)$, we obtain the identities:

$$\sum_{j=1}^{n} \frac{s_d^k}{f'(s_j)} \cdot \frac{f(X)}{X-s_j} = \begin{cases} X^{1+k}, & k = 0,1,\dots,n-2 \\ \\ X^n - f(X), & k = n-1 \end{cases}$$

in which s_j, $1 \leq j \leq n$, denote the conjugates of s and $s = s_1$. Putting $X = 0$, we immediately derive $Tr_{E/F}(s^j/f'(s)) \in R$. Therefore $s^j/f'(s) \in R[s]^*$ which implies that $A \subseteq R[s]^*$.

To prove the opposite containment, fix $b \in R[s]^*$. If we put

$$g(X) = \sum_{i=1}^{n} b_if(X)/(X-s_i),$$

where the b_i's are all conjugates of b, then we have

$$g(X) = \sum_{j=1}^{n} c_j \sum_{k=0}^{j-1} X^kTr_{E/F}(bs^{j-k-1})$$

where the c_i are defined by $c_n = 1$ and $f(X) = X^n + c_{n-1}X^{n-1} + \ldots + c_0$. Because, by hypothesis, $Tr_{E/F}(bR[s]) \subseteq R$ we see that the coefficients of $g(X)$ lie in R. But then

$$bf'(s) = g(s) \in R[s]$$

i.e. $b \in A$ and thus $R[s]^* \subseteq A$. ■

8.3. COROLLARY. With the assumptions and notation of Lemma 8.2,

$$f'(s)S \subseteq R[s]$$

Proof. Since $S \subseteq R[s]^*$, we have $f'(s)S \subseteq f'(s)R[s]^* \subseteq R[s]$. ■

Let $I(R)$ and $I(S)$ be the groups of fractional ideals of R and S, respectively. Then the *norm homomorphism*

$$N_{E/F} : I(S) \longrightarrow I(R)$$

is defined as follows. Let Q be a prime ideal of S with residue degree f over R. Then we put

$$N_{E/F}(Q) = P^f$$

where $P = Q \cap R$. This defines a unique homomorphism from $I(S)$ to $I(R)$ since $I(S)$ is freely generated by all such Q.

Let A be a subring of S containing R. Then the greatest common divisor of ideals of S contained in A is called the *conductor* of A.

8.4. LEMMA. Let P be a prime ideal in R, s an element of S with $E = F(s)$, $A = R[s]$ and C the conductor of A. Then the following conditions are equivalent:

(i) $PS \cap A = P[s]$

(ii) The embedding of A in S induces an isomorphism

$$S/PS \cong A/(A \cap PS)$$

(iii) The embedding of A in S induces isomorphisms

$$S/P^mS \cong A/(A \cap P^mS) \qquad (m = 1,2,\ldots,)$$

(iv) For all $m \geqslant 1$, $P^mS \cap A = P^m[s]$

(v) The prime ideal P does not divide the norm $N_{E/F}(C)$.

Proof. (i) ⇒ (ii): The induced homomorphism $A/(A \cap PS) \longrightarrow S/PS$ is obviously injective. Because $|S/PS| = N(PS) = N(P)^n$, where $n = (E:F)$, and by hypothesis,

$$|A/(A \cap PS)| = |A/P[s]| = N(P)^n$$

the assertion follows.

(ii) ⇒ (iii): The case $m = 1$ being true by hypothesis, we argue by induction on m. So assume that (iii) holds for m and choose $x \in P-P^2$. We may write $xR = PI$ where $P \nmid I$. Let $y \in I - P$, let y' be defined by $yy' \equiv 1 (\text{mod } P^m)$ and let $b \in S$. By our assumption, there is a polynomial $f(X)$ over R such that $b \equiv f(s)(\text{mod } P^mS)$. Therefore the element $b_1 = (b-f(s))y^m x^{-m}$ lies in S. If we now choose a polynomial $f_1(X) \in R[X]$ with $b_1 \equiv f_1(s)(\text{mod } PS)$, then

$$b = f(s) + b_1 x^m y^{-m} \equiv f(s) + (xy')^m f_1(s)(\text{mod } P^{m+1}S),$$

proving (iii).

(iii) ⇒ (iv): By hypothesis, the index of $A \cap P^mS$ in A is $N(P)^{mn}$. On the other hand, the index of $P^m[s]$ in A is again $N(P)^{mn}$. The desired conclusion is therefore a consequence of $P^m[s] \subseteq A \cap P^mS$

(iv) ⇒ (i): Trivial

(iii) ⇒ (v): Let $f(X)$ be the minimal polynomial for s over R, write

$$N_{E/F}(f'(s))R = P^mI \qquad (P \nmid I)$$

and fix $b \in I-P$. By hypothesis, for every $x \in S$ there is an element $y \in A$ such that $b(x-y)/N_{E/F}(f'(s))$ lies in S. Therefore $b(x-y)/f'(s)$ is in S, hence

$$b(x-y) \in f'(s)S \subseteq A$$

by Corollary 8.3. Accordingly $by \in A$ and we obtain $bS \subseteq A$, proving that $b \in C$. Since $b \notin P$, we also have $N_{E/F}(b) = b^n \notin P$, but $N_{E/F}(b) \subseteq N_{E/F}(C)$ and so P does not divide $N_{E/F}(C)$.

(v) ⇒ (iii): If P does not divide $N_{E/F}(C)$ and $PS = \prod_{i=1}^{r} P_i^{e_i}$, then no P_i can divide C. Owing to Chinese remainder theorem, there exists an element b such that

$$b \equiv 0 (\text{mod } C) \ , \ b \equiv 1 (\text{mod } P_i) \qquad (1 \leqslant i \leqslant r)$$

Having such a $b = f(s)$, $f(X) \in R[X]$, we may write each $x \in S$ in the form $x = g(s)/b = g(s)/f(s)$ for a suitable $g(X) \in R[X]$. Since the ideal bS is relatively prime to PS, there exists $r \in \mathbb{N}$ with $f(s)^r \equiv 1 (\text{mod } PS)$. Thus

$$x \equiv g(s)/f(s)^{r-1} \ (\text{mod } PS)$$

as required. ■

We are now ready to prove the following result

8.5. THEOREM. Let $F \subset E$ be algebraic number fields, $R \subset S$ their rings of integers, let $s \in S$ be such that $E = F(s)$ and let $f(X) \in R[X]$ be the minimal polynomial of s over R. Let C be the conductor of $R[s]$, let P be a prime ideal of R not dividing $N_{E/F}(C)$ and let $\pi : R[X] \longrightarrow R[X]/P[X] = K[X]$ be the homomorphism induced by the natural map $R \longrightarrow R/P = K$. Write

$$\pi(f) = f_1^{a_1} f_2^{a_2} \cdots f_m^{a_m}$$

where the f_i's are distinct irreducible monic polynomials in $K[X]$. Then

$$PS = P_1^{a_1} P_2^{a_2} \cdots P_m^{a_m}$$

where the P_i's are distinct prime ideals of S, the residue degree $f_{E/F}(P_i)$ of P_i is equal to the degree n_i of f_i and

$$P_i = PS + F_i(s)S \qquad (1 \leqslant i \leqslant m)$$

where $F_i[X] \in R[X]$ is such that $\pi(F_i) = f_i$.

Proof. Put $K_i = K[X]/f_iK[X]$ and consider the natural homomorphism

$$\pi_i : K[X] \longrightarrow K_i \qquad (1 \leqslant i \leqslant m)$$

Then the kernel of $\psi_i = \pi_i \circ \pi$ is the ideal $I_i = P[X] + F_i(X)R[X]$ of $R[X]$. The principal ideal generated by f in $R[X]$ is divisible by I_i and so ψ_i induces a surjective homomorphism

$$R[X]/fR[X] \longrightarrow K_i$$

Since $R[X]/fR[X] \cong R[s]$, we obtain a surjective homomorphism

$$\lambda_i : R[s] \longrightarrow K_i$$

with $\mathrm{Ker}\lambda_i = P[s] + F_i(s)R[s]$.

Let $b \in S$ and choose (using Lemma 8.4(ii) a polynomial $v \in R[X]$ such that $b \equiv v(s)(\mathrm{mod}\ PS)$. Define $\mu_i : S \longrightarrow K_i$ by

$$\mu_i(b) = \lambda_i(v(s)) \qquad (1 \leqslant i \leqslant m)$$

This is well-defined because from $v(s) \equiv 0(\mathrm{mod}\ PS)$ we obtain, by applying Lemma 8.4(i), $v(s) \in PS \cap R[s] = P[s]$ and so $\lambda_i(v(s)) = 0$.

The homomorphism μ_i is surjective, since so is λ_i. Setting $P_i = \mathrm{Ker}\mu_i$, we see that P_i is a nonzero prime ideal in S. Moreover, we have

$$P_i = PS + F_i(s)S \qquad (1 \leqslant i \leqslant m)$$

Indeed,

$$\begin{aligned} P_i &= \{x \in S \mid x \equiv v(s)(\mathrm{mod}\ PS),\ v \in R[X],\ v(s) \in P[s] + F_i(s)R[s]\} \\ &= PS + P[s] + F_i(s)R[s] \\ &= PS + F_i(s)R[s] \\ &= PS + F_i(s)S, \end{aligned}$$

since Lemma 8.4(ii) implies $S = R[s] + PS$ which in turn gives

$$F_i(s)S \subseteq F_i(s)R[s] + PF_i(s)S \subseteq F_i(s)R[s] + PS$$

Assume by way of contradiction that $P_i = P_j$ for $i \neq j$. Because $(f_i, f_j) = 1$, there exist polynomials $g_1, g_2 \in R[X]$ and $g_3 \in P[X]$ such that

$$g_1(X)f_i(X) + g_2(X)f_j(X) = 1 + g_3(X) \qquad (1)$$

But $F_i(s) \in P_i$, $F_j(s) \in P_j$ and $g_3(s) \in P_i$, so putting $X = s$ in (1), we obtain $1 \in P_i$, a contradiction. Hence all the P_i's, $1 \leqslant i \leqslant m$, are distinct.

Since the element

$$F_1(s)^{a_1} \ldots F_m(s)^{a_m} \qquad (2)$$

belongs to PS, the prime ideals $P_1, \ldots, P_m$ are the only possible prime ideals dividing PS. Therefore there exist positive integers $e_1, \ldots, e_m$ such that

$$PS = P_1^{e_1} \ldots P_m^{e_m}$$

The residue degree of P_i equals $(K_i:K)$ and this is equal to n_i. We are therefore left to verify that $a_i = e_i$ for all $i \in \{1,\dots,m\}$.

The ideal $P_i^{e_i}$ divides the product (2) but, as we have seen P_i can divide only the factor $F_i(s)^{a_i}$ of that product. Therefore $P_i^{e_i}$ divides $F_i(s)^{a_i}S$. It follows that the ideal $PS + F_i(s)^{a_i}S$ is divisible $P_i^{e_i}$ and since it divides $P_i^{a_i} = P^{a_i}S + F_i(s)^{a_i}S$, we have $a_i \geq e_i$, $1 \leq i \leq m$. On the other hand, if

$$n = (E:F),$$

then $n = e_1n_1 + \dots + e_mn_m$ (Corollary 3.7.5) and obviously $a_1n_1 + \dots + a_mn_m = n$. Thus we must have $a_i = e_i$ and the result follows. ■

As an application of the preceding result, we prove

8.6. THEOREM. Let $F \subset E$ be algebraic number fields and let $R \subset S$ be their rings of integers. Then there exist infinitely many prime ideals of R which split completely in S.

Proof. Let $M/\mathbb{Q}$ be a finite normal extension with $E \subseteq M$ and let L be the ring of integers of M. If $p\mathbb{Z}$ is a prime ideal in $\mathbb{Z}$ which splits completely in L, then every prime ideal of R lying above $p\mathbb{Z}$ splits completely in S, by the multiplicative property of the residue degree. We may therefore assume that $F = \mathbb{Q}$ (hence $R = \mathbb{Z}$) and that the extension E/F is normal: Because almost all prime ideals P of $\mathbb{Z}$ do not ramify in S, almost all prime ideals P of $\mathbb{Z}$ do not ramify in S, and satisfy the hypothesis of Theorem 8.5. Moreover, since E/F is normal, then in the notation of Theorem 8.5, the polynomials f_i, $1 \leq i \leq m$, have the same degree (see Corollary 3.7.5). Invoking Theorem 8.5, we are therefore left to verify the following assertion: If $f(X)$ is a nonconstant polynomial over $\mathbb{Z}$, then its values are divisible by infinitely many rational primes.

Assume by way of contradiction that the values of some $f(X) \in \mathbb{Z}[X]$ are divisible only by a finite number of primes, say $p_1,p_2,\dots,p_r$. Select x_0 so that $f(x_0)$ is different from 0, -1 and 1. Then we have

$$f(x_0) = \pm p_1^{a_1}\cdots p_r^{a_r}$$

for suitable $a_i \geqslant 0$ not all of which are zero. If now t is an arbitrary rational integer and $x = x_0 + tp_1 \cdots p_r f(x_0)$, then for large t we have

$$f(x) = \pm p_1^{b_1} \cdots p_r^{b_r} \equiv f(x_0) (\text{mod } p_1 \cdots p_r f(x_0))$$

which implies $a_i = b_i$ for $i = 1,2,\ldots,r$. Hence $f(X) = \pm f(x_0)$ for all large t's, which is a contradiction. ■

8.7. COROLLARY. Let $F \subset E$ be algebraic number fields and let $R \subset S$ be their rings of integers. Then there exist infinitely many prime ideals P of R for which SP is a product of some $n \geqslant 2$ distinct prime ideals of S.

Proof. Since every prime ideal P of R which splits completely in S satisfies the above property, the result follows by virtue of Theorem 8.6. ■

Let R be an arbitrary Dedekind domain, let K be the quotient field of R and let L/K be a finite separable field extension. Then the integral closure S of R in L is also a Dedekind domain. Let $I(R)$ and $I(S)$ be the groups of fractional ideals of R and S, respectively.

8.8. LEMMA. The map $i_{S/R} : I(R) \longrightarrow I(S)$ defined by

$$i_{S/R}(A) = SA$$

is an injective homomorphism, which preserves inclusion and maps principal ideals into principal ideals.

Proof. Every fractional ideal A of R is of the form

$$A = Rx_1 + \ldots + Rx_n \qquad (x_i \in K)$$

Hence $SA = Sx_1 + \ldots + Sx_n$ is a fractional ideal of S. It is obvious that $i_{S/R}$ preserves inclusion and maps principal ideals into principal ideals. Furthermore, the equality $(AB)S = (AS)(BS)$ shows that $i_{S/R}$ is a homomorphism.

We now show that

$$S \cap K = R \tag{3}$$

Indeed, the inclusion $R \subseteq S \cap K$ is trivial. On the other hand, every element of $S \cap K$ is integral over R and lies in K. Since R is integrally closed, we conclude that $S \cap K \subseteq R$, proving (3).

Now assume that $i_{S/R}(A) = S$, the identity element of $I(S)$. Then $AS = S$ and, by (3),

$$R = S \cap K = AS \cap K \supseteq A$$

proving that A is an ideal of R. But

$$i_{S/R}(A^{-1}) = S^{-1} = S$$

and we may repeat this argument to conclude that A^{-1} is also an ideal of R. Hence $A = R$ and the result follows. ■

8.9. COROLLARY. Let A be a fractional ideal of R such that $SA = Sx$ for some $x \in K$. Then $A = Rx$.

Proof. By hypothesis, $i_{S/R}(A) = Sx = i_{S/R}(Rx)$. Hence, by Lemma 8.8, $A = Rx$. ■

We have now come to the demonstration for which this section has been developed. The result below is essentially due to May(1979).

8.10. THEOREM. Let $F \subset E$ be algebraic number fields. Then

$$E^*/F^* \cong A \times B$$

where A is free abelian of rank $\aleph_0$ and B is finite.

Proof. Let R and S be rings of integers of F and E, respectively, and let $I(R)$ and $I(S)$ be their groups of fractional ideals. Denote by $P(R)$ and $P(S)$ the subgroups of principal fractional ideals of $I(R)$ and $I(S)$, respectively. Owing to Lemma 8.8, the map $i_{S/R} : I(R) \longrightarrow I(S)$ defined by $i_{S/R}(X) = SX$ is an injective homomorphism which preserves inclusion and maps $P(R)$ into $P(S)$. We may therefore regard $I(R)$ as embedded in $I(S)$ and $P(R)$ in $P(S)$. Let $C \subseteq I(R)$ be the free subgroup generated by the prime ideals of R that do not ramify in S. For each such prime ideal of R choose a prime ideal of S lying above it. The remaining prime ideals of S generate a complement D of C in $I(S)$. Furthermore, by Corollary 8.7, D is free of infinite rank. Because only finitely many prime ideals of R ramify in S, $I(R)/C$ is finitely generated. Therefore $I(S)/I(R)$ is the product of a finite group and a free group of infinite rank. Now $P(S) \cap I(R) = P(R)$ (Corollary 8.9), so

$$P(S)/P(R) \cong P(S)I(R)/I(R) \subseteq I(S)/I(R)$$

Moreover, $I(S)/P(S)\cdot I(R)$ is finite since so is $I(S)/P(S)$ (Theorem 3.4). Thus $P(S)/P(R)$ is the product of a finite group and a free group of infinite rank.

Owing to Lemma 1.2, there are natural isomorphisms

$$F^*/U(R) \cong P(R) \quad \text{and} \quad E^*/U(S) \cong P(S)$$

Moreover, $F^* \cap U(S) = U(R)$, and the induced inclusion $F^*/U(R) \subseteq E^*/U(S)$ agrees with $P(R) \subseteq P(S)$. Because $P(S)/P(R)$ is the product of a finite group and a free group of infinite rank, the same applies to the isomorphic group $E^*/U(S)F^*$. But

$$U(S)F^*/F^* \cong U(S)/U(R)$$

and $U(S)$ is finitely generated, by the Dirichlet unit theorem. It follows that E^*/F^* is the product of a finite group and a free group of infinite rank. Since E^*/F^* is countable, this rank must be equal to $\aleph_0$ and the result follows. ∎

9. BRANDIS'S THEOREM

Let E/F be a field extension with $E \neq F$. Our aim is to prove a theorem due to Brandis which asserts that if F is infinite, then E^*/F^* is not finitely generated. As a preparation for the proof of the main result, we shall establish certain properties of simple transcendental extensions.

9.1. LEMMA. Let $F(\lambda)/F$ be a simple transcendental extension where F is an arbitrary field and let μ be an element of $F(\lambda)$ not in F. Write

$$\mu = f(\lambda)g(\lambda)^{-1}$$

where $f(\lambda)$ and $g(\lambda)$ are polynomials in λ with no common factor of positive degree in λ. If $n = \max(\deg f, \deg g)$, then λ is algebraic over $F(\mu)$ and $(F(\lambda):F(\mu)) = n$.

Proof. We may write

$$f(\lambda) = \alpha_0 + \alpha_1\lambda + \dots + \alpha_n\lambda^n, \quad g(\lambda) = \beta_0 + \beta_1\lambda + \dots + \beta_n\lambda^n$$

where either $\alpha_n \neq 0$ or $\beta_n \neq 0$. The relation $\mu = f(\lambda)g(\lambda)^{-1}$ gives

$$f(\lambda) - \mu g(\lambda) = 0 \qquad .$$

and

$$0 = (\alpha_n - \mu\beta_n)\lambda^n + (\alpha_{n-1} - \mu\beta_{n-1})\lambda^{n-1} + \dots + (\alpha_0 - \mu\beta_0)$$

Furthermore, $\alpha_n - \mu\beta_n = 0$ since α_n or $\beta_n \neq 0$ and $\mu \notin F$. Therefore λ is a root of the polynomial of degree n : $\sum_{i=0}^{n} (\alpha_i - \mu\beta_i)X^i$ with coefficients in $F(\mu)$. We are therefore left to verify that the given polynomial is irreducible.

First, it is obvious that μ is transcendental over F, because λ is algebraic over $F(\mu)$; hence μ algebraic over F implies λ algebraic over F, contrary to assumption. Then ring $F[\mu,X] = F(\mu)[X]$ is the polynomial ring in two indeterminates μ,X and hence the theorem on unique factorization into irreducible elements holds in $F[\mu,X]$. Observe also that a polynomial in $F[\mu,X]$ of positive degree in X is irreducible in $F(\mu)[X]$ if it is irreducible in $F(\mu)[X]$. Now

$$f(\mu,X) = \Sigma(\alpha_i - \mu\beta_i)X^i = f(X) - \mu g(X)$$

is of degree 1 in μ. Hence, if $f(\mu,X)$ is reducible in $F(\mu)[X]$, then it has a factor $h(X)$ of positive degree in X. But this implies that $f(X)$ and $g(X)$ are divisible by $h(X)$, contrary to assumption. ∎

9.2. THEOREM (Lüroth). Let F be an arbitrary field and let $F(\lambda)/F$ be a simple transcendental extension. Then any field K with $F \subset K \subseteq F(\lambda)$ is also a simple transcendental extension : $K = F(\gamma)$, γ transcendental.

Proof. By hypothesis, K contains an element β not in F, so by Lemma 9.1, $P = F(\lambda)$ is algebraic over $F(\beta)$ and hence is algebraic over $K \supseteq F(\beta)$. Let the minimal polynomial of λ over K be

$$f(X) = X^n + \gamma_1 X^{n-1} + \dots + \gamma_n$$

The γ_i have the form $\mu_i(\lambda)\nu_i(\lambda)^{-1}$ where μ_i,ν_i are polynomials in the transcendental element λ. Multiplication of $f(X)$ by a suitable polynomial in λ will give a polynomial

$$f(\lambda,X) = c_0(\lambda)X^n + c_1(\lambda)X^{n-1} + \dots + c_n(\lambda) \qquad (1)$$

in $F[\lambda,X]$, which is a primitive polynomial in X in the sense that the highest common factor of the $c_i(\lambda)$ is 1. Observe also that $\gamma_i = c_i(\lambda)c_0(\lambda)^{-1} \in K$ and not all of these are in F because λ is transcendental over F. Thus one of the γ's has the form $\gamma = g(\lambda)h(\lambda)^{-1}$ where $g(\lambda), h(\lambda)$ have no common factor of positive degree in λ and $\max(\deg g, \deg h) = m > 0$. Owing to Lemma 9.1,

$$g(X) - \gamma h(X)$$

is irreducible in $F(\gamma)[X]$ and $[P : F(\gamma)] = m$. Since $K \supseteq F(\gamma)$ and $(P:K) = n$, clearly $m \geqslant n$. We shall show that $m = n$ and this will prove that $K = F(\gamma)$.

Since λ is a root of $g(X) - \gamma h(X)$ and the coefficient of this polynomial are contained in K, we have $g(X) - \gamma h(X) = f(X)q(X)$ in $K[X]$. We have $\gamma = g(\lambda)h(\lambda)^{-1}$ and we can replace the coefficients of f and q by their rational expressions in λ and then multiply by a suitable polynomial in λ to obtain a relation in $F[\lambda,X]$ of the form

$$K(\lambda)[g(X)h(\lambda) - g(\lambda)h(X)] = f(\lambda,X)q(\lambda,X) \tag{2}$$

where $f(\lambda,X)$ is the primitive polynomial given in (1). We may therefore deduce that $k(\lambda)$ is a factor of $q(\lambda,X)$ and so cancelling this we may assume the relation is

$$g(X)h(\lambda) - g(\lambda)h(X) = f(\lambda,X)q(\lambda,X) \tag{3}$$

Now the degree in λ of the left-hand side is at most m. Because

$$\gamma = g(\lambda)h(\lambda)^{-1}$$

with $(g(\lambda),h(\lambda)) = 1$ and $\max(\deg g, \deg h) = m$, the λ-degree of $f(\lambda,X)$ is at least m. It therefore follows that it is exactly m and $q(\lambda,X) = q(X) \in F[X]$. Hence the right-hand side of (3) is primitive as a polynomial in X. This holds also for the left-hand side. By symmetry, the left-hand side is primitive as a polynomial in λ also, and this ensures that $q(X) = q$ is a nonzero element of F. Then (3) implies that the X-degree and λ-degree of $f(\lambda,X)$ are the same. Thus $m = n$ and the result follows. ■

We are now ready to prove our main result.

9.3. THEOREM. (Brandis (1965)). Let E/F be a field extension, where F is an aribtrary infinite field and $E \neq F$. Then E^*/F^* is not finitely generated.

Proof. Because $E \neq F$, we may choose $\lambda \in E$ such that $1, \lambda$ are F-linearly independent. Hence the elements $1 + \alpha\lambda$, $\alpha \in F$, of E^* and mutually incongruent modulo F^*. Because F is infinite, we conclude that E^*/F^* is also infinite.

Assume by way of contradiction that E^*/F^* is finitely generated, say

$$E^*/F^* = \langle u_1F^*, \ldots, u_kF^* \rangle$$

Because E^*/F^* is infinite, we may assume that u_1F^* is of infinite order. Let F_0 be the prime subfield of F, and put $E_1 = F_0(u_1)$, $F_1 = E_1 \cap F$. Then

$$F_1 \subset E_1$$

and

$$E_1^*/F_1^* \cong E_1^*F^*/F^* \subseteq E^*/F^*$$

which shows that E_1^*/F_1^* is also finitely generated. Because no power of u_1 is in F_0, E_1 is an infinite field. Thus, if $\mathrm{char}F_0 = p > 0$, then u_1 is transcendental over F_0. We are therefore left to verify the following assertion : E^*/F^* is not finitely generated, provided $E = F_0(u)$, $F \subset E$, F_0 is a prime subfield of E and u is transcendental over F_0 if $\mathrm{char}F_0 = p > 0$.

If $F_0 = \mathbb{Q}$ and u is algebraic over $\mathbb{Q}$, then the result follows by virtue of Theorem 8.10. If u is transcendental over F_0 and $F = F_0$, then the result follows by applying Lemma 1.4. Thus, by Theorem 9.2, we may assume that $E = F_0(u)$ and $F = F_0(v)$ where both u and v are transcendental over the prime field F_0.

Write $v = v(u) = p(u)/q(u)$, where $p(u), q(u) \in F_0[u]$, $(p(u), q(u)) = 1$, and where the highest common factor of coefficients of $p(u)$ and $q(u)$ is 1. The embedding of F into E is given by the map

$$f(v) \longmapsto f^*(u) = f(v(u))$$

Let $g(v) = a_0 + a_1v + \ldots + a_nv^n \in F_0[v]$ with $a_n \neq 0$. Then

$$g^*(u) = \{a_0q(u)^n + a_1q(u)^{n-1}p(u) + \ldots + a_np(u)^n\}/q(u)^n$$

Setting

$$z_g(u) = a_0 q(u)^n + a_1 q(u)^{n-1} p(u) + \dots + a_n p(u)^n \tag{4}$$

it suffices, in view of Lemma 1.1, to show that there exist infinitely many irreducible polynomials $g(v) \in F_0[v]$ for which $z_g(u)$ is not irreducible.

To this end, put $r = \deg p$, $s = \deg q$ and $h = \max(r,s)$. We claim that, if $g(v)$ is irreducible, then

$$h \geqslant 2 \tag{5}$$

$$\deg z_g(u) = hn \quad \text{if} \quad g(v) \neq v,\ v - 1 \tag{6}$$

Indeed, by Lemma 9.1, $(E:F) = h$ and hence $h \geqslant 2$, since $E \neq F$. To prove (6), it suffices to show that $\deg z_g(u) \geqslant hn$ since the inequality $\deg z_g(u) \leqslant hn$ is obvious. Suppose firstly that $r \neq s$. Then the term u^{hn} in (4) comes exactly from one of the summands $a_0 q(u)^n$ or $a_n p(u)^n$. By hypothesis, $a_n \neq 0$ and since $g(v) \neq v$ and $g(v)$ is irreducible, we also have $a_0 \neq 0$. Thus

$$\deg z_g(u) \geqslant hn,$$

proving the case $r \neq s$. In case $r = s$, the coefficient of u^{hn} is equal to

$$a_0 + a_1 + \dots + a_n = g(1) \neq 0$$

since $g(v)$ is irreducible and $g(v) \neq v - 1$. This proves (5) and (6).

Let $w(u)$ be an irreducible polynomial in $F_0[u]$ with $w(u) \nmid q(u)$. We claim that there is an irreducible polynomial $g(v) \in F_0[v]$ with $w(u) | z_g(u)$. Indeed, let λ be a root of $w(u)$ in the algebraic closure $\bar{F}_0$ of F_0. Since $w(u) \nmid q(u)$, $v(\lambda) = \mu \in \bar{F}_0$. We choose $g(v)$ to be irreducible and such that $g(\mu) = 0$. Then we have

$$g(\mu) = g(v(\lambda)) = g^*(\lambda) = 0$$

and therefore $z_g(\lambda) = 0$. It follows that $w(u) | z_g(u)$, as claimed.

For any natural number m, there exists an irreducible polynomial of degree m over F_0 (e.g. if $F_0 = \mathbb{Q}$, then take $X^m - 2$ and if $F_0 = \mathbb{F}_p$, then take the minimal polynomial corresponding to the extension $\mathbb{F}_{p^m}/\mathbb{F}_p$). We choose a sequence of infinitely many distinct irreducible polynomials $w_1(u), w_2(u), \dots$, whose degrees are not divisible by h, for example, of degrees

$$1, h+1, 2h+1, \ldots$$

By the foregoing, for each polynomial $w_i(u)$ we obtain a polynomial $g_i(v)$ such that $w_i(u) | z_{g_i}(u)$. When $g_i(v) \neq v, v-1$ it follows from (6) that

$$\deg w_i(u) \neq \deg z_{g_i}(u)$$

and thus $z_{g_i}(u)$ is reducible. Each $z_{g_i}(u)$ contains only a finite number of prime factors and therefore there are infinitely many distinct $z_{g_i}(u)$. Since all the $z_{g_i}(u)$ are reducible, the result follows. ■

10. FIELDS WITH FREE MULTIPLICATIVE GROUPS MODULO TORSION

Let G be an abelian group and let $t(G)$ be the torsion subgroup of G. For convenience, the identity group will be regarded as a free group (of rank 0). We say that G is *free modulo torsion* if $G/t(G)$ is free. Thus G is free modulo torsion if and only if $G = t(G) \times A$ where A is a free subgroup of G. The problem that motivates this section is to discover circumstances under which the multiplicative group F^* of a field F is free modulo torsion. A typical example of such a field F is an algebraic number field (see Theorem 5.2). More generally, if F is finitely generated (i.e. finitely generated over its prime subfield), then F^* is free modulo torsion, by virtue of Theorem 10.1 below. What can be said about infinitely generated fields? The main results of this section assert the following:

(i) Assume that F is a field such that for every finite field extension E, E^* is free modulo torsion. If K is any field generated over F by algebraic elements whose degrees over F are bounded, then K^* is free modulo torsion.

(ii) Assume that F is a field such that for every finite field extension E, E^* is free modulo torsion and E contains only finitely many roots of unity. If K is any abelian extension of F, then K^* is free modulo torsion.

We present below an improved version of the original proof of (i) and (ii). This version was furnished by May in a private communication to the author.

10.1. THEOREM (May(1972)). Let E/F be a finitely generated field extension and let F_1 be the algebraic closure of F in E. Then $(F_1:F) < \infty$ and there exists a free abelian group A such that $E^* \cong F_1^* \times A$. In particular, suppose that E is a finitely generated field. Then

$$E^* \cong W \times A$$

where W is a finite cyclic group and A is a free abelian group.

Proof. That the extension F_1/F is finite is a consequence of Theorem 2.24.7. If E/F is algebraic, then $E = F_1$ and there is nothing to prove. We may therefore assume that the transcendency degree n of E over F is $\geqslant 1$. If $n = 1$, then E is a finite extension of $F(X)$ and the result follows by virtue of Theorem 5.6. We may choose a subfield L of E with $L \supseteq F$ and with L being a purely transcendental extension of degree 1 over F. Then E has transcendency degree $n-1$ over L and hence, by induction, $E^* \cong L_1^* \times A$, where L is the algebraic closure of L in E and A is a free abelian group.

Now consider the chain $F \subseteq L \subseteq L_1$. Because $L \cong F(X)$, L_1 is a finite extension of $F(X)$, hence by Theorem 5.6, $L_1^* \cong L_2^* \times A_1$ where L_2 is the algebraic closure of F in L_1 and A_1 is a free abelian group. But L_2 obviously equals F_1, hence the first part of the theorem.

Suppose that E is a finitely generated field. Then we may apply the foregoing to the special case where $F = \mathbb{Q}$ or $F = \mathbb{F}_p$, the field of p elements. If $F = \mathbb{Q}$, then F_1 is an algebraic number field and the required assertion follows by virtue of Theorem 5.2. If $F = \mathbb{F}_p$, then F_1 is a finite field and hence F_1^* is cyclic. This completes the proof. ■

The rest of the section is devoted to providing circumstances under which an infinitely generated field F has the property that F^* is free modulo torsion. The following observation will clear our path.

10.2. LEMMA. Let S be an arbitrary set and let $F(S)$ be a purely transcendental extension of F generated by S. Then

$$F(S)^* \cong F^* \times A$$

for some free abelian group A.

Proof. Let $R = F[S]$ be the polynomial ring generated by S. Then R is a *UFD* (see Theorem 1.2.20). Since $F(S)$ is the quotient field of R, it follows from Lemma 1.1 that

$$F(S)^* \cong U(R) \times A$$

where A is a free abelian group. Since $U(R) \cong F^*$, the result follows. ∎

10.3. COROLLARY. Assume that F is a field such that F^* is free modulo torsion. If E/F is an arbitrary purely transcendental extension of F, then E^* is free modulo torsion.

Proof. Direct consequence of Lemma 10.2. ∎

10.4. LEMMA. Let E/F be a finite field extension and let T/F^* be the torsion subgroup of E^*/F^*. If E^* is free modulo torsion, then

$$E^* = A \times T$$

for some free abelian subgroup A of E^*.

Proof. Let N denote the norm mapping from E^* to F^*, let H be the kernel of N and let $n = (E:F)$. If $\lambda \in E^*$, then $N(\lambda^n N(\lambda)^{-1}) = 1$ and thus $(E^*)^n \subseteq HF^*$. If $\lambda \in H \cap T$, then $\lambda^m \in F^*$ for some positive integer m, hence

$$1 = N(\lambda^m) = \lambda^{mn}$$

Thus we conclude that $H \cap T = t(H)$.

Let $f : E^* \longrightarrow E^*/T$ be the natural homomorphism. Because E^* is free modulo torsion, it follows that $H/t(H)$ is free. But

$$f(H) = HT/T \cong H/(H \cap T) = H/t(H)$$

and what we have shown above implies that $(E^*/T)^n \subseteq f(H)$. Because E^*/T is torsion-free, we deduce that E^*/T is free and the result follows. ∎

The following example illustrates that the above result fails for infinite algebraic extensions.

10.5. EXAMPLE (May(1980)). Let $\{x_p | p \text{ prime or } p = 0\}$ be algebraically independent transcendentals over $\mathbb{Q}$, let F be the field generated over $\mathbb{Q}$ by $\{x_p\}$, and let E be the field generated over $\mathbb{Q}$ by $\{y_p\}$, where

$$y_0^2 = x_0 \quad \text{and} \quad y_p^p = x_p(y_0 + 1) \qquad \text{for } p \text{ prime}$$

Then $y_0 + 1 \notin T$, but

$$y_p^p \equiv y_0 + 1 (\text{mod } T)$$

for every prime p. Hence E^*/T is not free. But, by Corollary 10.3, E^* is free modulo torsion since $\{y_p\}$ are algebraically independent transcendentals over $\mathbb{Q}$. Thus T cannot have a complement in E^*, since such a complement would necessarily be free. ■

If F is any field, let $\bar{F}$ denote $F^*/t(F^*)$ and let $\bar{\lambda}$ be the image of $\lambda \in F^*$ in $\bar{F}$. Observe that if E/F is a field extension, then by identifying $F^*t(E^*)/t(E^*)$ with $\bar{F}$ we may regard $\bar{F}$ as a subgroup of $\bar{E}$.

10.6. LEMMA. Let E/F be a field extension. Suppose that $\lambda^m = \mu$ with $\lambda \in E^*$, $\mu \in F^*$ and $m \geqslant 1$. Assume that m is the order of $\bar{\lambda}$ in $\bar{E}$ modulo $\bar{F}$. Then the polynomial $X^m - \mu$ is irreducible over F.

Proof. Assume that $f(X) \in F[X]$ is a divisor of $X^m - \mu$ of degree

$$j \in \{1,2,\ldots,m-1\}$$

Then all the roots of $f(X)$ (in a splitting field of f) are of the form $\lambda\delta^i$ with $i \in \{0,1,\ldots,m-1\}$ and with $\delta^m = 1$. Hence the product of these roots is of the form $\lambda^j\varepsilon$, where ε is a root of unity. But then $\lambda^j\varepsilon \in F$ and this would imply that $\bar{\lambda}^j \in \bar{F}$, contrary to the way in which m was chosen. ■

The discussion has now reached a point where, in order to make further progress, we need to bring in certain group-theoretic facts.

Let B,H be subgroups of an abelian group A. Then H is said to be a *B-high subgroup* if

$$H \cap B = 1, \quad \text{and if} \quad H \subset H_1 \subseteq A \quad \text{implies} \quad H_1 \cap B \neq 1$$

That is, H is maximal with respect to the property of having trivial intersection with B. Then, in particular, $HB = H \times B$. The existence of B-high subgroups, for every B, is guaranteed by Zorn's lemma. Moreover, H may be chosen so as to contain a prescribed subgroup G of A with $G \cap B = 1$. Indeed, the set of all subgroups A that contain G and have trivial intersection with B is nonempty and is inductive; thus it contains a maximal member H.

10.7. LEMMA. Let B,C be subgroups of an abelian group A where C is a B-high subgroup of A. Then

(i) $a \in A$, $a^p \in C$ (p prime) implies $a \in B \times C \subseteq A$

(ii) $A = B \times C$ if and only if $a^p = bc$ ($a \in A$, $b \in B$, $c \in C$) implies $(b')^p = b$ for some $b' \in B$.

Proof. (i) If $a \in C$, there is nothing to prove. If $a \notin C$, then $\langle C,a\rangle$ contains, by the choice of C, an element $1 \neq b \in B$, i.e. $b = ca^k$ for some $c \in C$ and integer k. Therefore $(k,p) = 1$ because of $a^p \in C$ and $B \cap C = 1$. It follows that $rk + sp = 1$ for some $r,s \in \mathbb{Z}$, and so

$$a = (a^k)^r(a^p)^s = (bc^{-1})^r(a^p)^s \in B \times C$$

as required.

(ii) Assume that $A = B \times C$. Then, given $a \in A$, we have $a = b'c'(b' \in B$, $c' \in C)$ and hence $a^p = (b')^p(c')^p = bc$ implies $(b')^p = b$. Conversely, if $a^p = bc$ implies $(b')^p = b$ for some $b' \in B$, then $a(b')^{-1}$ satisfies the hypothesis of (i), and so $a(b')^{-1} \in B \times C$, $a \in B \times C$. This shows that the quotient group $A/(B \times C)$ contains no elements of prime order, and hence is torsion-free. But if $x \in A$ is arbitrary, not in $B \times C$, then $\langle C,x\rangle$ intersects B in a nonidentity element b'', $c''x^\ell = b''$ for some $c'' \in C$ and integer ℓ. Moreover, $\ell \neq 0$ because of $B \cap C = 1$, thus $x^\ell = b''(c'')^{-1} \in B \times C$, and $A/(B \times C)$ is a torsion group. The conclusion is that $A = B \times C$, as asserted.■

10.8. LEMMA. Let A be an abelian group and let n be a positive integer. If A^n splits over its torsion subgroup, then A splits over its torsion subgroup.

Proof. We may harmlessly assume that $n = p$ is a prime. By hypothesis, $A^p = t(A^p) \times B$ where B is torsion-free. Obviously we have $t(A^p) = t(A)^p$. Let G be a $t(A)$-high subgroup of A containing B. If $a \in A$ and

$$a^p = bc(b \in t(A),\ c \in G),$$

then $b \in t(A^p) = t(A)^p$. Hence, by Lemma 10.7(ii), $A = t(A) \times G$ and the result follows. ■

We now apply Lemma 10.8 to prove

10.9. COROLLARY. Let B be a subgroup of an abelian group A. If A/B is bounded and B is free modulo torsion, then A is free modulo torsion.

Proof. By hypothesis, $A^n \subseteq B$ for some positive integer n. Because B is free modulo torsion, so is A^n. Hence A^n splits over $t(A^n)$ and, by Lemma 10.8, A splits over $t(A)$, say $A = t(A) \times F$, where F is torsion-free. Since $A^n = t(A)^n \times F^n$, we see that F^n is free. But then the map $F \longrightarrow F^n$, $x \longmapsto x^n$ is an isomorphism, since F is torsion-free. Hence $F \cong F^n$ is free, as asserted. ■

Let K be an extension field of F, and let τ be an ordinal. Following May (1980), we shall say that a family of subfields of K

$$\{E_\alpha | \alpha < \tau\}$$

forms a *continuous chain of finite extensions from F to K* if the following conditions hold:

$$E_\beta \subseteq E_\alpha \qquad \text{for } \beta \leqslant \alpha < \tau$$

$$(E_{\alpha+1}:E_\alpha) < \aleph_0 \qquad \text{for } \alpha + 1 < \tau$$

$$E_\alpha = \bigcup_{\beta<\alpha} E_\beta \qquad \text{for } \alpha \text{ a limit ordinal}$$

and

$$K = \bigcup_{\alpha<\tau} E_\alpha$$

10.10. LEMMA. Let $\{E_\alpha | \alpha < \tau\}$ be a continuous chain of finite extensions from F to K. For each $\alpha < \tau$, put $W_\alpha = E^*_\alpha t(K^*)$ and let T_α/E^*_α be the torsion subgroup of K^*/E^*_α. Assume that E^*_α is free modulo torsion and that T_α/W_α is bounded for every α. Then K^* is free modulo torsion.

Proof. Using the obvious analogue of the definition above, we note that $\{T_\alpha | \alpha < \tau\}$ forms a continuous chain of subgroups from T_0 to K^* (We do not require $T_{\alpha+1}/T_\alpha$ to be finite). Because E^*_0 is free modulo torsion, so is W_0. Since T_0/W_0 is bounded, it follows from Corollary 10.9 that T_0 is free modulo torsion.

Let B_0 be a free complement in T_0 for the torsion subgroup of T_0 (which in fact equals the torsion subgroup of K^*). It suffices to show that $T_{\alpha+1}/T_\alpha$ is free for every α such that $\alpha + 1 < \tau$, for then we can choose a free abelian group $B_{\alpha+1}$ such that $T_{\alpha+1} = B_{\alpha+1} \times T_\alpha$. The group $\prod_{\alpha<\tau} B_\alpha$ is then easily seen to be a free complement in K^* for the torsion subgroup of K^*.

To prove that $T_{\alpha+1}/T_\alpha$ is free, let T'_α/E^*_α be the torsion subgroup of $E^*_{\alpha+1}/E^*_\alpha$. Then

$$T_\alpha \cap E^*_{\alpha+1} = T'_\alpha$$

and Lemma 10.4 implies that $E^*_{\alpha+1}/T'_\alpha$ is free, hence $T_\alpha E^*_{\alpha+1}/T_\alpha$ is free. But $T_{\alpha+1}/T_\alpha$ is torsion-free, and $T_{\alpha+1}/T_\alpha E^*_{\alpha+1}$ is bounded since $T_{\alpha+1}/W_{\alpha+1}$ is bounded by hypothesis. Hence $T_{\alpha+1}/T_\alpha$ is free, by Corollary 10.9. The proof is thus complete. ■

As a final preparatory result, we prove

10.11. LEMMA. Let $F \subseteq E \subseteq K$ be a chain of fields, let K be generated over F by a family of elements that are algebraic over F of degree bounded by some positive integer n, and assume that $i \in F (i^2 = -1)$ if char$F \neq 2$. If T/E^* is the torsion subgroup of K^*/E^*, then $T/E^*t(K^*)$ is bounded.

Proof. Enlarge K by adjoining all conjugates of the given generators. Then K is still generated by elements of degree $\leqslant n$ and it suffices to prove the lemma for this new K, which is then normal over F.

If σ is an automorphism of some field extension between K and F, then it extends to K by normality. Because the extension permutes conjugates of the generators, the order of σ must divide $n!$. To complete the proof, it suffices to show that every element of $T/E^*t(K^*)$ has order dividing $(n!)^2$.

Let $v \in T$ be of order p^r modulo $E^*t(K^*)$, where p is a prime. Then $v^{p^r} = u\varepsilon$ for some $u \in E^*$ and some root of unity ε. We may assume that ε has p-power order $< p^r$ by absorbing an appropriate root of unity factor into v.

First consider the case where $p = $ char$K$. Because K is generated over E by elements of degree $\leqslant n$, there is a uniform power $p^s \leqslant n$ such that the p^s-power of each generator is separable over E. Then every element of K^{p^s} is separable over E. If $s < r$, then v^{p^s} is separable over E and is a root of $X^{p^{r-s}} - u$. Thus $v^{p^s} \in E$ contrary to the order of v modulo $E^*t(K^*)$. Therefore $r \leqslant s$ and hence $p^r | n!$

Now we consider the case where $p \neq \mathrm{char}K$. Let ε_{p^s} be a primitive p^s-th root of unity and let s be the largest integer $\leqslant r$ such that $\varepsilon_{p^s} \in K$, and put $\eta = \varepsilon_{p^s}$. Put $d = (E(\eta,v):E)$ and let N be the norm mapping for this extension. Then $N(v)^{p^r} = u^d N(\eta)$, where $N(\eta)$ is a root of unity. We claim that $p^r | d$. Assume the contrary. Then

$$p^{r-1} = xp^r + yd$$

for some integers x and y. Therefore

$$u^{p^{r-1}} = (u^x)^{p^r}(N(v)^y)^{p^r} N(\eta)^{-y},$$

hence $u = w^p\eta_1$ for some $w \in E$ and some root of unity η_1. But then

$$v^{p^{r-1}} = w\eta_2$$

for some root of unity η_2, contrary to the order of v modulo $E^*t(K^*)$. Thus $p^r | d$. Let t be the smallest integer $\geqslant (1/2)r$. Since

$$d = (E(\eta,v):E(\eta))(E(\eta):E)$$

it follows that p^t divides at least one of the two degrees. In the second case, the extension $E(\eta)/E$ is cyclic (recall that $\varepsilon_4 \in F$ if $p = 2$), hence there is an automorphism of order p^t. The first case will require more work.

Let p^k be the order of v modulo $E(\eta)^*$. Then $v^{p^k} \notin E(\eta)^p$ since otherwise $\varepsilon_p \in K - E(\eta)$, contrary to the choice of η. Thus $X^{p^k} - v^{p^k}$ is irreducible over $E(\eta)$. But normality of K over $E(\eta)$ implies that $\varepsilon_{p^k} \in K$ which in turn implies that $\varepsilon_{p^k} \in E(\eta)$ by choice of η. Hence the Galois group of $E(v,\eta)/E(\eta)$ is cyclic of order p^k. But p^t divides the degree in this case, hence there is an automorphism of order p^t in this case also. But then $p^t | n!$, hence $p^r | (n!)^2$ and the result follows. ■

We have done most of the work to demonstrate the following remarkable result.

10.12. THEOREM (May(1980)). Assume that F is a field such that for every finite extension E, E^* is free modulo torsion. If K is any field generated over F by algebraic elements whose degrees over F are bounded, then K^* is free modulo torsion.

Proof. Let K and F satisfy the hypotheses of the theorem. We may add $i(i^2 = -1)$ to K if the characteristic of K is $\neq 2$, since the original K will be free modulo torsion if the enlarged one is . Now enlarge F by adding i to it. The hypothesis on F will carry over to a finite extension. We shall use induction on $(K:F)$. Clearly we may assume that $(K:F) \geqslant \aleph_0$. Let τ be the first ordinal with $|\tau| = (K:F)$. If K is generated over F by elements whose degrees are bounded by n, then we may choose a family $\{x_\alpha | \alpha < \tau\}$ of such generators. For each $\alpha < \tau$, let E_α be generated over F by $\{x_\beta|_\beta < \alpha\}$. Then it is clear that $\{E_\alpha | \alpha < \tau\}$ is a continuous chain of finite extensions from F to K. Because $(E_\alpha:F) < |\tau|$, it follows by induction that E_α^* is free modulo torsion. Lemma 10.11 now applies to show that T_α/W_α is bounded, hence Lemma 10.10 finishes the proof. ■

10.13. COROLLARY. Let F be a finitely generated field, and assume that K is a field extension generated over F by algebraic elements of bounded degrees. Then K^* is free modulo torsion.

Proof. Direct consequence of Theorems 10.1 and 10.12. ■

10.14. COROLLARY. Assume that F is a field such that for every finite field extension E, E^* is the (restricted) direct product of cyclic groups. If K is any field generated over F by algebraic elements whose degrees over F are bounded, then K^* is the direct product of cyclic groups.

Proof. By Theorem 10.12, K^* is free modulo torsion. Hence it suffices to show that there is no prime number $p \neq \mathrm{char}K$ such that $\varepsilon_{p^n} \in K$ for every n, where ε_{p^n} is a primitive p^n-th root of unity. This is true for every finite extension of F by hypothesis, hence its failure for K would imply the existence of automorphisms of K/F of arbitrary large order. The latter being impossible, by virtue of the first paragraph in proof of Lemma 10.11, the result follows. ■

10.15. COROLLARY. (Schenkman(1964)). Let K be any field generated over $\mathbb{Q}$ by algebraic elements of bounded degree. Then K^* is the direct product of cyclic groups.

Proof. Put $F = \mathbb{Q}$ and observe that, by Theorem 5.2, F satisfies the hypothesis of Corollary 10.14. ■

10.16. COROLLARY (Schenkman(1964)). A multiplicative group generated by algebraic numbers of bounded degree is the direct product of cyclic groups.

Proof. Direct consequence of Corollary 10.15 and the fact that a subgroup of the direct product of cyclic groups is the direct product of cyclic groups. ■

10.17. COROLLARY (Čarin(1954)). Suppose that K is an algebraic extension of $\mathbb{Q}$ and that the degrees of elements of K over $\mathbb{Q}$ are bounded. Then K^* is the direct product of cyclic groups.

Proof. This is a special case of Corollary 10.16. ■

Returning to Theorem 10.12, we now demonstrate that it is not sufficient to assume that just F^* is free modulo torsion.

10.18. THEOREM (May(1972)). There exists a field F and a quadratic field extension K such that the following holds:

(i) F^* is free modulo torsion, but K^* is not free modulo torsion

(ii) F is algebraic over $\mathbb{Q}$ and every finite field extension of F contains only finitely many roots of unity.

Proof. Let i denote the usual complex number. Construct a sequence $\{\alpha_n\}$ of complex numbers by letting

$$\alpha_0 = (2+i)(2-i)^{-1},$$

and by choosing α_n, $n \geqslant 1$, such that

$$\alpha_n^4 = \alpha_{n-1}$$

Put $K_0 = \mathbb{Q}(i)$ and define K_n, $n \geqslant 1$, by $K_n = K_0(\alpha_n)$. Then we have an ascending chain of fields

$$K_0 \subseteq K_1 \subseteq \dots \subseteq K_n \subseteq \dots$$

and we define K by

$$K = \bigcup_{n=0}^{\infty} K_n$$

Clearly K^* is not free modulo torsion. Denote by σ complex conjugation. Let us show that σ induces an automorphism of each K_n, and hence an automorphism of K. Bearing in mind that $\sigma(\alpha_0) = \alpha_0^{-1}$ and $\alpha_n^{4n} = \alpha_0$, we have

$$(\alpha_n \sigma(\alpha_b))^{4n} = 1$$

Because $\alpha_n \sigma(\alpha_n)$ is a real number, $\sigma(\alpha_n) = \pm\alpha_n^{-1}$ and therefore $\sigma(\alpha_n) \in K_n$ for every n. Thus σ induces an automorphism of K_n of order 2.

Let F (respectively, F_n) be the fixed field of σ on K (respectively, K_n). Then

$$(K:F) = (K_n:F_n) = 2$$

and $F \cap K_n = F_n$. Furthermore, it is obvious that $F_{n-1} \subseteq F_n (n \geqslant 1)$ and that

$$F = \bigcup_{n=0}^{\infty} F_n$$

Since α_0 is not a square in K_0, it follows that $X^{4^n} - \alpha_0$ is irreducible over K_0. Thus $(K_n:K_0) = 4^n$ and therefore $(K_n:K_{n-1}) = 4$ $(n \geqslant 1)$. Moreover, by looking at the extension K_n/E_{n-1}, it is clear that $(F_n:F_{n-1}) = 4$.

Let the field C be obtained from $\mathbb{Q}$ by adjoining all roots of unity. It will be shown that

$$K \cap C = K_0 \quad \text{or} \quad K \cap C = K_0(\sqrt{5}) \tag{1}$$

Since $(K:\mathbb{Q}) = 2^{\infty}$ (in the sense that every finite subextension of $K/\mathbb{Q}$ has degree a power of 2 and these degrees are unbounded) the Galois theory of $C/\mathbb{Q}$ tells us that $K \cap C$ is generated over K_0 by 2^r-th roots of unity for appropriate integers r, and by the elements of form $\sqrt{p}$ for appropriate odd primes p.

We next examine how primes of K_0 ramify in K. To this end, we first observe that $5\alpha_n$ is an integral generator for K_n over K_0 whose minimal polynomial f is $X^{4^n} - 5^{4^n-1}(2+i)$. Bearing in mind that

$$f'(5\alpha_n) = 4^n(5\alpha_n)^{4^n-1}$$

we conclude that the only primes of K_0 that can ramify in K are

$$1 + i,\ 2 + i, \text{ and } 2 - i$$

However, the presence of $\sqrt{p}$ in K would imply the ramification of the Gaussian prime(s) lying above p, thus only $\sqrt{5}$ can lie in K. To examine the 2^r-th roots of unity, it suffices to show that $\varepsilon_8 \notin K$, where ε_8 is a primitive 8-th root of unity. Assume the contrary. Then there is an n such that

$$K_0 \subset \mathbb{Q}(\varepsilon_8) \subseteq K_n$$

But it is clear that the primes $2 + i$ and $2 - i$ ramify totally from K_0 to K_n. Therefore they would also ramify from K_0 to $\mathbb{Q}(\varepsilon_8)$. This is contrary to the standard fact that every odd prime is unramified from $\mathbb{Q}$ to $\mathbb{Q}(\varepsilon_8)$. Thus (1) holds and it follows directly from (1) that every finite extension of K contains only finitely many roots of unity.

We now show that F^* is free module torsion. Owing to Lemma 10.10 and Theorem 10.1, we need only verify that the torsion subgroup of $\bar{F}/\bar{F}_n$ is bounded. Assume that $v \in F^*$ and $\bar{v}$ has order m modulo $\bar{F}_n$. Then $v^m = u\varepsilon$ for some root of unity ε and some $u \in F_n^*$. Since F is contained in $\mathbb{R}$, we have $\varepsilon = \pm 1$. Therefore we may assume that $v^m = u$, in which case Lemma 10.6 guarantees that $X^m - u$ is irreducible over F_n. Since $(F:F_n) = 2^\infty$, we deduce that m is a power of 2. It will be shown that it is impossible for $m = 4$, which will imply that 2 is the exponent of the torsion subgroup of $\bar{F}/\bar{F}_n$.

Assume by way of contradiction that $m = 4$. Then $(F_n(v):F_n) = 4$ and so $(K_n(v):K_n) = 4$ since $F \cap K_n = F_n$. We argue that

$$K_n(v) = K_{n+1} \tag{2}$$

Choose t such that K_t contains both $K_n(v)$ and K_{n+1}. To establish (2), it suffices to verify that the intermediate fields of the extension $K_t(\varepsilon)/K_n(\varepsilon)$ are nested. But $K_t(\varepsilon) = K_n(\varepsilon, \alpha_t)$, where $\alpha_t^{4^t} \in K_n(\varepsilon)$. Therefore the Galois group of this extension is cyclic of order a power of two. Thus the subgroups are nested and hence the same applies to the intermediate fields, proving (2).

We may apply Theorem 2.15.4. to the extension K_{n+1}/K_n since it is cyclic of degree 4. Then $u = \alpha_n^k\beta^4$, where $\beta \in K_n$, and where k is relatively prime to 4. Let N denote the norm mapping from K_n to K_0. Since u is a real number, one easily sees that $N(u)$ must also be real, hence $N(u) \in \mathbb{Q}$. We may write $N(u) = 5^s b$, where s is some integer and b is a nonzero rational number whose prime factorization does not involve 5. Now the minimal polynomial of α_n over K_0 is $X^{4^n} - \alpha_0$, hence $N(\alpha_n) = -\alpha_0$. It follows that

$$5^s b = N(u) = N(\alpha_n)^k N(\beta) = -\alpha_0^k N(\beta)^4$$

and thus

$$N(\beta)^4 = -(2+i)^{s-k}(2-i)^{s+k}b$$

Applying the unique factorization in $\mathbb{Q}(i)$, we infer that 4 divides $s-k$ and also $s+k$. But then 2 divides k, contrary to the fact that $(k,4) = 1$. This contradiction completes the proof. ■

We now head towards our second major result which is Theorem 10.21 below. The following two preliminary results will clear our path.

10.19. LEMMA. Let $F \subseteq E \subseteq K$ be a chain of fields such that K/F is abelian, $i \in F(i^2 = -1)$ if $\mathrm{char}F \neq 2$ and such that E contains only finitely many roots of unity. If T/E^* is the torsion subgroup of K^*/E^*, then $T/E^*t(K^*)$ is bounded.

Proof. Enlarge K by adding all roots of unity. This new K is an abelian extension of F, and it suffices to prove the lemma for this new K.

Let p be a prime and let $v \in T$ be of order p^r modulo $E^*t(K^*)$. Denote by ε_{p^r} a primitive p^r-th root of unity. It suffices to show that $\varepsilon_{p^r} \in E$ because of the assumption on E. We have $v^{p^r} = u$, $u \in E$, where any root of unity factor on the right could be absorbed in v since K contains all roots of unity. If $u = u_1^p$ for some $u_1 \in E$, then $v^{p^{r-1}} = u_1\varepsilon_p^j$, contrary to the order of v modulo $E^*t(K^*)$. Thus $u \notin E^p$ and hence by Lemma 2.16.5, $X^{p^r} - u$ is irreducible over E.

Because K/E is separable, we have $p \neq \mathrm{char}K$. Since K/E is abelian, $E(v)/E$ is normal, hence $X^{p^r} - u$ splits in $E(v)$. Consequently, $\varepsilon_{p^r} \in E(v)$.

Note that $\varepsilon_p \in E$ since the degree of ε_p over E is relatively prime to p. To complete the proof, we assume that $\varepsilon_{p^r} \notin E$ and derive a contradiction.

The polynomial $X^{p^r} - u$ must be reducible then over $E(\varepsilon_{p^r})$, hence $u = u_1^p$ with $u_1 \in E(\varepsilon_{p^r})$. Accordingly,

$$v^{p^{r-1}} = u_1 \varepsilon_p^j$$

which implies that $E(v^{p^{r-1}}) \subseteq E(\varepsilon_{p^r})$. Since $(E(v^{p^{r-1}}) : E) = p$, the Galois theory of $E(\varepsilon_{p^r})/E$ implies that

$$E(v^{p^{r-1}}) = E(\varepsilon_{p^s})$$

for some s such that $\varepsilon_{p^{s-1}} \in E$. (The Galois group of $E(\varepsilon_{p^r})/E$ is cyclic of p-power order). Owing to Corollary 2.15.6,

$$v^{p^{r-1}} = \varepsilon_{p^s}^k u_2$$

for some k and some $u_2 \in E$. This contradicts the order of v modulo $E^*t(K^*)$ and the proof is therefore complete. ■

10.20. LEMMA. Let $F_1 \subseteq F_2 \subseteq \ldots$ be subfeilds of a field K, and put

$$F_\infty = \bigcup_{n \geqslant 1} F_n$$

Then there exist subfields $K_1 \subseteq K_2 \subseteq \ldots$ such that

$$K_n \cap F_\infty = F_n \quad \text{for } n \geqslant 1 \quad \text{and} \quad K = \bigcup_{n \geqslant 1} K_n$$

Proof. Invoking Zorn's lemma, we may choose a maximal subfield K_1 of K such that

$$K_1 \cap F_\infty = F_1 \quad \text{and} \quad K_1 F_2 \cap F_\infty = F_2$$

Again, applying Zorn's lemma, we may choose a maximal subfield K_2 if K such that

$$K_2 \supseteq K_1, \; K_2 \cap F_\infty = F_2 \quad \text{and} \quad K_2 K_3 \cap F_\infty = F_3$$

We may therefore continue by induction to construct $K_1 \subseteq K_2 \subseteq \ldots$ such that

$$K_n \cap F_\infty = F_n \;, \; K_n F_{n+1} \cap F_\infty = F_{n+1}$$

and K_n is maximal with respect to these two properties. Given $\alpha \in K$, we must show that $\alpha \in \bigcup_{n\geq 1} K_n$. Observe that α cannot be transcendental over $\bigcup_{n\geq 1} K_n$, since otherwise $K_1(\alpha)$ is easily seen to contradict the maximality of K_1. Hence, there is some $m \geq 1$ such that α is algebraic over K_m. Let $n \geq m$. We may assume that $K_n(\alpha) \neq K_n$ since otherwise we are done. Owing to the maximality of K_n, we must have either $K_n(\alpha) \cap F_\infty \supset F_n$ or $K_n(\alpha)F_{n+1} \cap F_\infty \supset F_{n+1}$. First assume that

$$K_n(\alpha) \cap F_\infty \supset F_n$$

Choose $\beta \in K_n(\alpha) \cap F_\infty$, $\beta \notin F_n$. Then $\beta \in K_j$ for some $j > n$. Note that $\beta \notin K_n$, hence

$$(K_n(\alpha):K_n) > (K_n(\alpha):K_n(\beta))$$

and therefore

$$(K_n(\alpha):K_n) > (K_j(\alpha):K_j)$$

Now assume that $K_n(\alpha)F_{n+1} \cap F_\infty \supset F_{n+1}$. Then

$$K_{n+1}(\alpha) \cap F_\infty \supset F_{n+1}$$

since $K_{n+1}(\alpha) \supset K_n(\alpha)F_{n+1}$. Applying the previous case, we conclude that

$$(K_n(\alpha):K_n) \geq (K_{n+1}(\alpha):K_{n+1}) > (K_j(\alpha):K_j)$$

for some $j > n$. Therefore, in either case, this reduction of the degree of α implies that $\alpha \in \bigcup_{n\geq 1} K_n$ So the lemma is true. ■

10.21. THEOREM (May(1980)). Assume that F is a field such that for every finite field extension E, E^* is free modulo torsion and E contains only finitely many roots of unity. If K is any abelian field extension of F, then K^* is free modulo torsion.

Proof. Let K and F staisfy the hypotheses of the theorem. We may enlarge both K and F by adding $i(i^2 = -1)$ (if $\operatorname{char} F \neq 2$) since the hypothesis on F carries over to a finite extension and it suffices to prove that the enlarged K^* is free modulo torsion.

We shall use induction on $(K:F)$, which we may assume is infinite. First we shall assume that K contains only finitely many roots of unity. Let τ be the first ordinal with $|\tau| = (K:F)$, and choose a family of generators $\{x_\alpha | \alpha < \tau\}$ of K over F. For each $\alpha < \tau$, let E_α be generated over F by $\{x_\beta | \beta < \alpha\}$. Then $\{E_\alpha | \alpha < \tau\}$ is a continuous chain of finite extensions from F to K. Since E_α is an abelian extension of F, and because $(E_\alpha:F) < |\tau|$, it follows by induction that E_α^* is free modulo torsion. Hence Lemmas 10.19 and 10.10 finish the proof in this case.

Now assume that K may contain infinitely many roots of unity. Denote by F_∞ the field generated over F by the roots of unity that lie in K. The hypothesis on F imply that we may choose subfields

$$F \subseteq F_1 \subseteq F_2 \subseteq \dots$$

such that

$$\bigcup_{n \geqslant 1} F_n = F_\infty$$

and F_n contains only finitely many roots of unity for each n. Applying Lemma 10.20, we obtain subfields $K_1 \subseteq K_2 \subseteq \dots$ such that

$$\bigcup_{n \geqslant 1} K_n = K \quad \text{and} \quad K_n \cap F_\infty = F_n \qquad \text{for all } n \geqslant 1$$

Put $K_0 = F$. Then it is clear that we may choose a continuous chain of finite extensions from K_n to K_{n+1} for every $n \geqslant 0$. Hence, we obtain a continuous chain $\{E_\alpha | \alpha < \tau\}$ (for some ordinal τ) of finite extensions from F to K such that given $\alpha < \tau$, there exists $n \geqslant 1$ such that $E_\alpha \subseteq K_n$. Therefore each E_α contains only finitely many roots of unity. Since E_α is an abelian extension of F with $(E_\alpha:F) \leqslant (K:F)$, what we have proved in the previous paragraph quarantees that E_α^* is free modulo torsion. Applying Lemmas 10.19 and 10.10, the result follows. ■

10.22. COROLLARY. (May(1972)). Let F be a finitely generated field and let K be any abelian field extension of F. Then K^* is free modulo torsion.

Proof. Any finite field extension of F is again finitely generated. The desired conclusion is therefore a consequence of Theorems 10.1 and 10.21. ■

10.23. COROLLARY (May(1976)). Let R be a finitely generated integral domain, and let $\bar{R}$ be the domain obtained by adjoining all roots of unity to R. Then

$U(\bar{R})$ is free modulo torsion.

Proof. Let F be the quotient field of R and let $\bar{F}$ be the field obtained by adjoining all roots of unity to F. Then F is a finitely generated field and $\bar{F}$ is an (infinite) abelian extension of F. Hence, by Corollary 10.22, $\bar{F}^*$ is free modulo torsion. Since $U(\bar{R}) \subseteq \bar{F}^*$, the result follows. ■

We close by giving an example which illustrates that in the statement of Theorem 10.21 neither the hypothesis on the commutativity of the Galois group, nor the hypothesis concerning roots of unity can be dropped.

10.24. EXAMPLE (May(1972)). Let the field K be generated over $\mathbb{Q}$ by

$$\{\varepsilon_p, \alpha_p \mid p \text{ a prime}\}$$

where ε_p is a primitive p-th root of unity and $\alpha_p^p = 2$. Let M be the subfield of K generated over $\mathbb{Q}$ by ε_p, where p ranges over all prime numbers. Then $K/\mathbb{Q}$ is a Galois extension whose Galois group is solvable, since K/M is abelian and $M/\mathbb{Q}$ is abelian. Moreover, any finite extension of $\mathbb{Q}$ contains only finitely many roots of unity. However, K^* is clearly not free modulo torsion. This shows that we cannot replace "abelian" by "solvable" in the statement of Theorem 10.21.

We have K/M abelian, and further, if M_1 is any finite extension of M, then M_1^* is free modulo torsion. This follows from Corollary 10.22 since M_1 is an abelian extension of a finite extension of $\mathbb{Q}$. But K^* is not free modulo torsion. Thus we cannot drop the assumption on roots of unity. ■

11. A NONSPLITTING EXAMPLE

Fuchs(1960) asked whether the multpilicative group of a field must split over its torsion subgroup. A negative answer was provided by Cohn (1962), who utilized his embedding theorem to give an example of a field F for which F^* does not split over $t(F^*)$. The existence of such an example for algebraic extensions of $\mathbb{Q}$ was indicated by May (1972). All the details below were furnished by May in a private communication to the author.

11.1. THEOREM (May). There exists an algebraic field extension $F/\mathbb{Q}$ such that

F^* does not split over its torsion subgroup.

Proof. Let F be generated over $\mathbb{Q}$ by $\{\varepsilon_p, \alpha_p \mid p \text{ odd prime}\}$, where ε_p is a primitive p-th root of unity and $\alpha_p^p = 2\varepsilon_p$ for every odd prime p. Then F is obviously an algebraic extension of $\mathbb{Q}$. Denote by M the subfield generated over $\mathbb{Q}$ by all ε_p, p odd prime.

Assume by way of contradiction that F^* splits over $t(F^*)$. The residue class of 2 modulo torsion has a p-th root for all odd primes p. Thus some coset representative in F^* must have this property. Therefore there exists a root of unity ε such that 2ε has a p-th root in F^* for all odd primes p. Now fix p not dividing the order of ε. Then 2 has a p-th root in F^*. Since $2\varepsilon_p$ has a p-th root in F^* (namely, α_p), we see that ε_p has a p-th root, that is $\varepsilon_{p^2} \in F$.

We claim that $\alpha_p \in M(\varepsilon_{p^2})$. We may suppose $\alpha_p \notin M$. Since $\varepsilon_p \in M$, it follows that $\alpha_p^p \notin M^p$, hence $(M(\alpha_p):M) = p$, by virtue of Lemma 2.16.3. But by the same argument for all odd $q \neq p$, $M(\alpha_q) = M$ or else $(M(\alpha_q):M) = q$. Since in the latter case, $M(\alpha_q)/M$ is an abelian extension of degree q, the Galois group of F/M is $\prod_{\alpha_q \notin M} \mathbb{Z}_q$. Now the only roots of unity in $\mathbb{Q}(\varepsilon_n)$ are those of orders dividing n (if n is even) or $2n$ (if n is odd). Hence

$$\varepsilon_{p^2}^p = \varepsilon_p \notin M^p$$

and so

$$(M(\varepsilon_{p^2}):M) = p$$

Therefore the Galois theory of F/M implies $M(\varepsilon_{p^2}) = M(\alpha_p)$. Thus $\alpha_p \in M(\varepsilon_{p^2})$. But $M(\varepsilon_{p^2})/\mathbb{Q}$ is an abelian extension (it is generated by roots of unity). Thus any subextension is normal. In particular,

$$\mathbb{Q}(\alpha_p \varepsilon_{p^2}^{-1})/\mathbb{Q}$$

is normal. However,

$$(\alpha_p \varepsilon_{p^2}^{-1})^p = \alpha_p^p \varepsilon_p^{-1} = 2\varepsilon_p \varepsilon_p^{-1} = 2$$

and thus

$$\alpha_p \varepsilon_{p^2}^{-1} = \sqrt[p]{2}$$

On the other hand, $\mathbb{Q}(\sqrt[p]{2})/\mathbb{Q}$ is not a normal extension. Indeed, it has degree p, but normality would imply $\varepsilon_p \in \mathbb{Q}(\sqrt[p]{2})$ since $\sqrt[p]{2}\,\varepsilon_p$ is a conjugate of $\sqrt[p]{2}$. This cannot occur since ε_p has degree $p-1$ over $\mathbb{Q}$, which provides the desired contradiction. ■

12. EMBEDDING GROUPS

Throughout this section, all groups are assumed to be abelian and, for any given group G, $t(G)$ denotes the torsion subgroup of G. We say that G is *locally cyclic* if every finitely generated subgroup of G is cyclic. By Proposition 4.1, if G is contained in the multiplicative group of a field, then $t(G)$ is locally cyclic. Owing to Cohn(1962), the converse is also true. In fact, Cohn proved that if $t(G)$ is locally cyclic, then G can be embedded in a special manner as a subgroup of the multiplicative group of a field. Subsequently May (1972) provided a refinement of Cohn's theorem which we present below.

We begin by proving some group-theoretic results.

12.1. LEMMA. Every group can be embedded as a subgroup of a divisible group.

Proof. The infinite cyclic group $\mathbb{Z}$ can clearly be embedded in a divisible group, namely $\mathbb{Q}$. Hence every free group is embeddable in a divisible group. Given an arbitrary group G, we may write $G \cong F/N$ for a suitable free group F and a subgroup N of F. If we embed F in a divisible group D, then G will be isomorphic to the subgroup F/N of the divisible group D/N. ■

12.2. LEMMA. Let G_1 be a group. Then there exists a group G such that $G_1 \subseteq G$, $t(G_1) = t(G)$, $G/t(G)$ is divisible and $G_1/t(G)$ contains a basis for $G/t(G)$.

Proof. Suppose A is torsion-free with $A \subseteq D$, where D is divisible. We claim that there exists a torsion-free divisible group D_1 with $A \subseteq D_1 \subseteq D$ and D_1/A is torsion. Indeed, $t(D)$ is divisible and $A \cap t(D) = 1$. The projection $A \times t(D) \longrightarrow t(D)$ extends to a homomorphism $D \longrightarrow t(D)$ since $t(D)$ is divisible. Let X be the kernel of $D \longrightarrow t(D)$. Then $A \subseteq X$. Let D_1 be the group with

$A \subseteq D_1 \subseteq X$ such that D_1/A is the torsion subgroup of X/A. Then D_1 is divisible and D_1 is torsion-free since $D_1 \cap t(D) = 1$, as claimed.

By Lemma 12.1, we have $G_1 \subseteq D$ where D is divisible. Then

$$G_1/t(G_1) \subseteq D/t(G_1),$$

where $D/t(G_1)$ is divisible. Since $G_1/t(G_1)$ is torsion-free, the preceding paragraph implies the existence of D_1 such that $G_1 \subseteq D_1 \subseteq D$ and such that $D_1/t(G_1)$ is torsion-free divisible and D_1/G_1 is torsion. Put $G = D_1$. Then $t(G) = t(G_1)$ and $G/t(G)$ is divisible. If $\{x_i\}$ is a basis for $G/t(G)$, then since G/G_1 is torsion, for any $i \in I$ there exists $m_i > 0$ such that $\{x_i^{m_i}\}$ is a basis for $G_1/t(G)$. ∎

We are now ready to prove

12.3. THEOREM (May (1972)). Let G be an abelian group such that $t(G)$ is locally cyclic. Then there is a field L and a group H such that $L^* \cong G \times H$, where H is a free abelian group if $t(G)$ has nontrivial 2-component and H is the direct product of a free abelian group with a cyclic group of order 2 if $t(G)$ has trivial 2-component.

Proof. If $t(G)$ has trivial 2-component, we may replace G by the direct product of G with a cyclic group of order 2. Therefore we may harmlessly assume that $t(G)$ has nontrivial 2-component.

For the sake of clarity, we divide the rest of the proof into three steps. The first two steps treat the special case where $G/t(G)$ is divisible. The third step utilizes the construction given in the previous steps to complete the proof in the general case.

Step 1. Assume that $G/t(G)$ is divisible. Then, by Theorem 2.2, there exists a well-ordered set I and an inner direct product decomposition

$$G/t(G) = \prod_{\alpha \in I} Q_\alpha$$

where each Q_α is isomorphic to the additive group of $\mathbb{Q}$. Given $\alpha \in I$, let G_α denote the inverse image modulo $t(G)$ of $\prod_{\beta \leqslant \alpha} Q_\beta$. Then

$$G = \bigcup_{\alpha \in I} G_\alpha$$

Now fix a sequence of positive integers $n_1, n_2, \ldots$ such that $n_1 = 1$, $n_i | n_{i+1}$ for every i and such that any given positive integer divides some n_i. For every $\alpha \in I$, choose an identification of Q_α with the additive group of $\mathbb{Q}$. Then Q_α is generated by $\{n_i^{-1} | i \geqslant 1\}$. Let $g_{\alpha,i} \in G_\alpha$ be such that $g_{\alpha,i}$ reduces modulo $t(G)$ to n_i^{-1} in Q_α. Our choice of $g_{\alpha,i}$ then ensures that G is generated by $t(G)$ and

$$\{g_{\alpha,i} | \alpha \in I,\ i \geqslant 1\}$$

subject to relations of the form

$$g_{\alpha,i+1}^{m_{i+1}} = g_{\alpha,i}\, u_{\alpha,i}$$

where $u_{\alpha,i} \in t(G)$ and $m_{i+1} = n_{i+1}/n_i$. Owing to Proposition 4.2, we may choose a cyclotomic extension of $\mathbb{Q}$, say K, such that $t(K^*) \cong t(G)$. Fix an isomorphism of $t(K^*)$ with $t(G)$, and let $\varepsilon_{\alpha,i}$ be the root of unity in K that corresponds to $u_{\alpha,i}$. Next choose elements $t_{\alpha,1}$ $(\alpha \in I)$ that are algebraically independent over K. For each $\alpha \in I$, inductively choose elements $t_{\alpha,i}$ $(i \geqslant 2)$ such that

$$t_{\alpha,i}^{m_i} = t_{\alpha,i-1}\, \varepsilon_{\alpha,i-1}$$

If $j \geqslant 1$, then $t_{\alpha,j}^{n_j} = t_{\alpha,}\, \eta_{\alpha,j}$ for some root of unity $\eta_{\alpha,j}$. Therefore, if we choose a positive integer i_α for each α, then $\{t_{\alpha,i_\alpha} | \alpha \in I\}$ is obviously algebraically independent over K. Put

$$L = K(\{t_{\alpha,i} | \alpha \in I,\ i \geqslant 1\})$$

and note that the isomorphism of $t(G)$ with $t(K^*)$ may be extended to a homomorphism $\phi : G \longrightarrow L^*$. Because ϕ is injective on $t(G)$ and on a basis

$$\{g_{\alpha,1} | \alpha \in I\}$$

of G modulo $t(G)$, we see that ϕ is injective. We may therefore regard G as a subgroup of L^*, where G is generated by

$$\{\varepsilon_{\alpha,i}\, t_{\alpha,i} | \alpha \in I,\ i \geqslant 1\}$$

Note that

$$L = K(G) = \bigcup_{\alpha \in I} K(G_\alpha)$$

Step 2. Keeping the assumption that $G/t(G)$ is divisible, we now prove that L^*/G is free. To this end, note that $G_\alpha \subseteq K(G_\alpha)^* \cap G$. To prove the opposite containment, fix $x \in K(G_\alpha)^* \cap G$. Then $x = \omega\varepsilon$, where ε is a root of unity, and where ω is a word in $\{t_{\beta,i_\beta} | \beta \in I\}$ for a suitable choice i_β for every β. Put

$$S = \{t_{\beta,i_\beta} | \beta \leqslant \alpha\} \quad \text{and} \quad S' = \{t_{\beta,i_\beta} | \beta > \alpha\}$$

Then S' is algebraically independent over $K(S)$. Because each $t_{\beta,j} (\beta \leqslant \alpha, j \geqslant 1)$ is algebraic over $K(S)$, $K(G_\alpha)$ is algebraic over $K(S)$ and thus S' is algebraically independent over $K(G_\alpha)$. The conclusion is that ω is a word in S since $\omega \in K(G_\alpha)$ and so $x \in G_\alpha$, proving that

$$K(G_\alpha)^* \cap G = G_\alpha$$

We may therefore regard $\{K(G_\alpha)^*/G_\alpha | \alpha \in I\}$ as a linearly ordered family of subgroups of L^*/G.

Let us construct a free basis for L^*/G by induction over I. At the initial step, we have $K^*/t(G)$. Because K is an abelian field extension of $\mathbb{Q}$, Theorem 10.21 guarantees that $K^*/t(G)$ is free. Choose a free basis for

$$K^*/t(G),$$

fix $\alpha \in I$ and put $H = \bigcup_{\beta<\alpha} G_\beta$. Then

$$(K(H_\alpha)^*/H_\alpha) = \bigcup_{\beta<\alpha} (K(G_\beta)^*/G_\beta)$$

At the inductive step, we may assume that we have constructed bases for

$$K(G_\beta)^*/G_\beta \qquad (\beta < \alpha)$$

such that their union gives a basis for $K(H_\alpha)^*/H_\alpha$. We are left to verify that this basis can be extended to a basis for $K(G_\alpha)^*/G_\alpha$.

To this end, put

$$H_{\alpha,i} = \langle H_\alpha, t_{\alpha,i} \rangle, \quad i \geqslant 1$$

Then

$$H_{\alpha,1} \subseteq H_{\alpha,2} \subseteq \cdots \quad , \quad G_\alpha = \bigcup_{i=1}^{\infty} H_{\alpha,i}$$

and

$$K(G_\alpha) = \bigcup_{i=1}^{\infty} K(H_{\alpha,i})$$

Setting

$$R_i = K(H_\alpha)[t_{\alpha,i}]$$

we see that R_i is a principal ideal domain with quotient field $K(H_{\alpha,i})$, since $t_{\alpha,i}$ is transcendental over $K(H_\alpha)$. Because $t_{\alpha,i+1}$ is integral over R_i and R_{i+1} is integrally closed, it follows that R_{i+1} is the integral closure of R_i in $K(H_{\alpha,i+1})$. Let D_i be the group of fractional ideals of R_i. Then the inclusion $R_i \subseteq R_{i+1}$ induces an injection $D_i \longrightarrow D_{i+1}$ for every i. We denote by D the direct limit of $\{D_i | i \geqslant 1\}$ under these mappings. Let

$$\phi : K(G_\alpha)^* \longrightarrow D$$

be a homomorphism induced by the natural mappings $K(H_{\alpha,i})^* \longrightarrow D_i$, $i \geqslant 1$. Then ϕ induces a homomorphism

$$\psi : K(G_\alpha)^*/G_\alpha \longrightarrow D/\phi(G_\alpha)$$

The kernel of ϕ is precisely $K(H_\alpha)^*$, hence $\mathrm{Ker}\psi = (K(H_\alpha)^* G_\alpha)/G_\alpha$. But

$$K(H_\alpha)^* \cap G_\alpha = H_\alpha$$

hence $\mathrm{Ker}\psi$ may be identified with $K(H_\alpha)^*/H_\alpha$.

We now claim that $D/\phi(G_\alpha)$ is free; if sustained, it will follow that the free basis of $K(H_\alpha)^*/H_\alpha$ could be extended to a free basis of $K(G_\alpha)^*/G_\alpha$. To substantiate our claim, note that the image of $H_{\alpha,i}$ in D_i is the subgroup T_i generated by the prime ideal $< t_{\alpha,i} >$. In D_i we have

$$< t_{\alpha,i} > = < t_{\alpha,i+1} >^{m_{i+1}} ,$$

so we obtain an injective mapping

$$\psi_i : D_i/T_i \longrightarrow D_{i+1}/T_{i+1} \qquad \text{for every } i$$

Because $G_\alpha = \bigcup_{i=1}^{\infty} H_{\alpha,i}$, we have

$$D/\phi(G_\alpha) = \bigcup_{i=1}^{\infty} (D_i/T_i)$$

Each D_i/T_i is free on the prime ideals distinct from $<t_{\alpha,i}>$; thus to prove that $D/\phi(G_\alpha)$ is free, we need only show that the mapping ψ_i splits. Choose a prime ideal P of R distinct from $<t_{\alpha,i}>$, and let $P_1, P_2, \ldots$ be the prime ideals of R_{i+1} extending P. Note that $R_{i+1} = R_i[t_{\alpha,i+1}]$ and that $f(t_{\alpha,i+1}) = 0$, where

$$f = X^{m_{i+1}} - t_{\alpha,i}\varepsilon_{\alpha,i}$$

Because $f'(t_{\alpha,i+1}) \in <t_{\alpha,i+1}>$, it follows that $P_1, P_2, \ldots$ are all unramified over P. Consequently, $P = P_1 P_2 \ldots$ is in D_{i+1}. Define

$$\theta_i : D_{i+1}/T_{i+1} \longrightarrow D_i/T_i$$

by $\theta_i(P_1) = P$, $\theta_i(P_j) = 1$ $(j \geqslant 2)$. Then $\theta_i \circ \psi_i$ is the identity map on D_i/T_i, hence ψ_i splits as desired.

Step 3. We now consider the case of a group G_1 such that $t(G_1)$ is locally cyclic. Owing to Lemma 12.2, there exists an abelian group G such that

$$G_1 \subseteq G,\ t(G_1) = t(G),\ G/t(G) \text{ is divisible}$$

and such that

$$G_1/t(G) \text{ contains a basis for } G/t(G)$$

Note that the last condition ensures that G/G_1 is a torsion group. Let L be constructed from G as before, and let $L_1 = K(G_1)$. To complete the proof, it suffices to show that L_1^*/G_1 is free. The latter will follow provided we prove that $L_1 \cap G = G_1$ since L^*/G is free.

The inclusion $G_1 \subseteq L_1 \cap G$ is trivial, hence assume that $x \in L_1 \cap G$. There exist distinct indices $\alpha_1, \ldots, \alpha_k$, and positive integers $\ell_1, \ldots, \ell_k$ such that, if H is the subgroup generated by $t(G)$ and by the elements

$$t_{\alpha_i,\ell_1}, \ldots, t_{\alpha_k,\ell_k}$$

then $x \in H$, and also $x \in K(H \cap G_1)$. Observe that the transcendency degree of $K(H)$ over K is k. Because G/G_1 is a torsion group, there exists a positive integer r such that $H^r \subseteq H \cap G_1$. However H is free modulo $t(G)$ of rank k, and the same holds for H^r, thus $H \cap G_1$ is free modulo $t(G)$ of rank k.

Hence

$$H \cap G_1 = \langle t(G), h_1, \ldots, h_k \rangle \qquad (h_i \in H)$$

and therefore

$$K(H \cap G_1) = K(h_1, \ldots, h_k)$$

But $K(H)$ is an algebraic field extension of $K(H \cap G_1)$ since $H^r \subseteq H \cap G_1$. Thus, k is also the transcendency degree of $K(H \cap G_1)$ over K, from which it follows that $h_1, \ldots, h_k$ are algebraically independent over K. Because

$$x^r \in H \cap G_1,$$

we have $x^r = \omega\varepsilon$, where ε is a root of unity and ω is a word in $h_1, \ldots, h_k$. Bearing in mind that $x \in K(h_1, \ldots, h_k)$, it follows by unique factorization that x is a product of a root of unity and word in $h_1, \ldots, h_k$. This shows that $x \in G_1$ and completes the proof. ■

We close by remarking that the following problem is still wide open.

Problem. Let G be a group whose torsion subgroup is locally cyclic. What are necessary and sufficient conditions for G to be isomorphic to the multiplicative group of a field?

13. MULTIPLICATIVE GROUPS UNDER FIELD EXTENSIONS

Let E/F be a field extension. We wish to examine the change in multiplicative groups in going from F^* to E^*. The problem of relating the group-theoretic structure of E^* to that of F^* is quite complicated even when F^* has particular simple structure and the extension is quadratic. The simple structure that we shall consider for F^* is the direct product of a finite cyclic group and a free abelian group. What can be said about the structure of E^*? First let us observe a trivial fact. If $(E:F) = n < \infty$ and A is a free direct factor of F^*, then E^* has a free direct factor isomorphic to A. Indeed, let ψ be the composite

$$E^* \longrightarrow F^* \longrightarrow A$$

of the norm followed by the projection map. Then E^* has a free direct factor isomorphic to $\psi(E^*)$. But the image of the norm map contains $(F^*)^n$, hence $\psi(E^*) \cong A$. Therefore, to relate the structure of F^* to that of E^*, we must compare the sizes of complementary factors. It will be demonstrated that even in the case where E is a quadratic extension of F, the complementary factor for E^* may be essentially as arbitrary as possible for a subgroup of the multiplicative group of a field.

13.1. THEOREM. (May (1979)). Let G be any abelian group whose torsion subgroup is locally cyclic with nontrivial 2-component. Then there exists a field K and a quadratic field extension L such that $K^* \cong \mathbb{Z}_2 \times A$ for some free abelian group A, while $L^* \cong G \times B$ for some free abelian group B.

Proof. Let L be the field constructed in the proof of Theorem 12.3. Then G is a subgroup of L^*, $L = \mathbb{Q}(G)$, L^*/G is a free abelian group, and there exists a torsion-free basis for G consisting of elements that are algebraically independent over $\mathbb{Q}(t(G))$ (called K in the proof of Theorem 12.3).

Let H be any finitely generated subgroup of G and write

$$H = \langle \varepsilon \rangle \times \langle h_1 \rangle \times \dots \times \langle h_n \rangle,$$

where ε is a root of unity and $h_1,\dots,h_n$ are torsion-free elements of G. We assert that $h_1,\dots,h_n$ are algebraically independent over $\mathbb{Q}(\varepsilon)$. Indeed, by the remark on the torsion-free basis for G, there exist elements $g_1,\dots,g_m \in G$ that are algebraically independent over $\mathbb{Q}(\varepsilon)$, and a positive integer k such that

$$\langle h_1^k,\dots,h_n^k \rangle \subseteq \langle g_1,\dots,g_m \rangle$$

But any basis for $\langle g_1,\dots,g_m \rangle$ is clearly algebraically independent. Applying the stacked basis theorem (see Lang (1965, Theorem 5, p. 393)), it therefore follows that $\langle h_1^k,\dots,h_n^k \rangle$ has a basis that is algebraically independent. The conclusion is that $h_1,\dots,h_n$ are algebraically independent over $\mathbb{Q}(\varepsilon)$.

It follows from the above that the automorphism of $\mathbb{Q}(\varepsilon)$ sending ε to ε^{-1} can be extended to a unique automorphism σ_H of $\mathbb{Q}(H)$ such that $\sigma_H(h) = h^{-1}$ for all $h \in H$. If $H_1 \subseteq H_2$ are two finitely generated subgroups of G, then σ_{H_1}

is the restriction of σ_{H_2} to $\mathbb{Q}(H_1)$. Let $\sigma = \cup\sigma_H$, where the union is taken over all finitely generated subgroups H of G. Then σ is an automorphism of L such that $\sigma(g) = g^{-1}$ for any $g \in G$. Denote by K the fixed field of σ. Since σ has order 2, we have $(L:K) = 2$. Moreover, $K^* \cap G = \langle -1 \rangle$ and so $K^*/\langle -1 \rangle$ is isomorphic to a subgroup of the free abelian group A and

$$L^* \cong G \times B$$

for a suitable free abelian group B. ■

Let p be a prime and n a nonzero integer. The symbol $(\frac{n}{p})$ defined by

$$(\frac{n}{p}) = \begin{cases} 1 & \text{if } n \equiv x^2 \pmod p \\ -1 & \text{if } n \not\equiv x^2 \pmod p \end{cases}$$

is known as the *quadratic symbol*, and depends only on the residue class of n modulo p.

13.2. LEMMA. Let p be a prime, let ε be a primitive p-th root of unity and let

$$a = \sum_n (\frac{n}{p})\varepsilon^n$$

the sum being taken over nonzero residue classes modulo p. Then

$$a^2 = (\frac{-1}{p})p$$

and, in particular,

$$\mathbb{Q}(\sqrt{p}) \subseteq \mathbb{Q}(\varepsilon, i)$$

Proof. We have

$$a^2 = \sum_{n,m} (\frac{n}{p})(\frac{m}{p})\varepsilon^{n+m} = \sum_{n,m} (\frac{nm}{p})\varepsilon^{n+m}$$

If n ranges over nonzero residue classes, so does nm for any fixed m, and thus replacing n by nm gives

$$\begin{aligned} a^2 &= \sum_{n,m} (\frac{nm^2}{p})\varepsilon^{m(n+1)} = \sum_{n,m} (\frac{n}{p})\varepsilon^{m(n+1)} \\ &= \sum_m (\frac{-1}{p})\varepsilon^0 + \sum_{n\neq -1} (\frac{n}{p}) \sum_m \varepsilon^{m(n+1)} \end{aligned}$$

However, $1 + \varepsilon + \dots + \varepsilon^{p-1} = 0$, and the sum on the right over m therefore yields -1. Accordingly,

$$a^2 = \left(\frac{-1}{p}\right)(p-1) + (-1)\sum_{n\neq -1}\left(\frac{n}{p}\right)$$
$$= p\left(\frac{-1}{p}\right) - \sum_n \left(\frac{n}{p}\right)$$
$$= p\left(\frac{-1}{p}\right),$$

as asserted. ■

As a preliminary to our final result, let us recall the following piece of information. Let G be an abelian group and let p be a prime. Given $g \in G$, the greatest nonnegative integer r for which $x^{p^r} = g$ is solvable for some $x \in G$, is called the *p-height* (or simply the *height*) of g. If $x^{p^r} = g$ is solvable for all r, we say that g is of *infinite p-height*. By the *circle group* we mean the unit circle in the complex numbers

$$\{\alpha \in \mathbb{C} | \ |\alpha| = 1\}$$

13.3. THEOREM. (May (1979)). Let G be any countable torsion-free abelian group. Then there exists a field K, algebraic over $\mathbb{Q}$, and a quadratic field extension L such that $K^* \cong \mathbb{Z}_2 \times A$, for some free abelian group A, while $L^* \cong \mathbb{Z}_4 \times G \times B$ for some free abelian group B.

Proof. We shall construct L as a subfield of $\mathbb{C}$. By Dirichlet's theorem on primes in arithmetic progressions, there are infinitely many primes p with $p \equiv 1 (\text{mod } 12)$. Because such primes also satisfy $p \equiv 1 (\text{mod } 4)$, there is a factorization

$$p = (a + bi)(a - bi),$$

where $a + bi$ are nonassociated primes in the Gaussian integers. Note further that

$$(a+bi)(a-bi)^{-1} \qquad (1)$$

lies on the unit circle, and that these elements for various p are independent free generators. Let G_0 be the subgroup of G generated by some basis for G. Since G_0 is of countable rank, we may assume that G_0 is generated by a subset of the elements given above. Bearing in mind that the circle group is divisible, we may in fact assume that G is realized as a subgroup of the unit circle. Let us choose a chain of subgroups such that

$$(G_{n+1}:G_n) = p_n$$

for some prime p_n and such that

$$\bigcup_n G_n = G$$

Given $n \geqslant 0$, choose $\alpha_n \in G_{n+1}$ with $G_{n+1} = \langle G_n, \alpha_n \rangle$. Then $\alpha_n \notin G_n$ and $\alpha_n^{p_n} \in G_n$. Now define

$$L_0 = \mathbb{Q}(i),\ L_n = L_0(G_n),\ n \geqslant 1, \text{ and } L = L_0(G)$$

Then we must have

$$L_{n+1} = L_n(\alpha_n) \text{ and } \bigcup_n L_n = L$$

Let the field C be obtained from $\mathbb{Q}$ by adjoining all roots of unity. It will be shown, by induction on n, that there exist free groups $B_0, B_1, \ldots$ such that

$$L_n^* = \langle i \rangle \times G_n \times \prod_{j \leqslant n} B_j \tag{2}$$

and such that

$$((L_n \cap C):L_0) \text{ is a power of } 2 \tag{3}$$

By the unique factorization in the Gaussian integers, the elements in (1) used earlier to give a basis for G_0 are part of a free basis for $L_0^*/\langle i \rangle$. In fact, it is immediate that there exists a free group B_0 such that $L_0^* = \langle i \rangle \times G_0 \times B_0$, and such that all odd primes $p \not\equiv 1 (\text{mod } 12)$ form part of a set of free generators for B_0. Since $L_0 \cap C = L_0$, the initial step of the induction is completed.

Now assume that (2) and (3) hold for n. We shall write $\alpha = \alpha_n$, $p = p_n$ and $\alpha^p = \beta$. Since G is torsion-free, β has p-height 0 in G_n, hence $\beta \notin L_n^p$, and therefore $(L_{n+1}:L_n) = p$. We claim that $((L_{n+1} \cap C):L_0)$ is a power of 2, and that $t(L_{n+1}^*) = \langle i \rangle$. Assume that

$$((L_{n+1} \cap C):(L_n \cap C)) = m > 1$$

This is a normal extension (being contained in C), so $L_n(L_{n+1} \cap C) \supseteq L_n$ is a normal extension of degree m. Because the extension is contained in L_{n+1}, we deduce that $m = p$, and that L_{n+1} is a normal extension of L_n. Now the polynomial $X^p - \beta$ splits in L_{n+1}, so L_{n+1} contains a primitive p-th root of unity. This root of unity must therefore lie in L_n since its degree over L_n divides $p-1$. Invoking the decomposition of L_n^*, we infer that $p = 2$, which shows that $((L_{n+1} \cap C):L_0)$ is a power of 2.

To prove that $t(L_{n+1}^*) = \langle i \rangle$, first assume that $\varepsilon \in t(L_{n+1}^*)$ is a primitive q-th root of unity for q an odd prime. Then $\mathbb{Q}(\varepsilon) \subseteq L_{n+1} \cap C$ and hence, by what we have shown above, $q-1$ is a power of 2. Therefore $q \not\equiv 1 \pmod{12}$. Furthermore, by Lemma 13.2, $\sqrt{q} \in \mathbb{Q}(i,\varepsilon)$ and thus $\sqrt{q} \in L_{n+1}$. By the way B_0 was chosen, q has height 0 in B_0, hence height 0 in L_n^*. This shows that

$$(L_n(\sqrt{q}):L_n) = 2,$$

so $p = 2$ and

$$L_n(\sqrt{q}) = L_{n+1} = L_n(\alpha)$$

Hence, by Theorem 2.15.4, $\sqrt{q} = \alpha\gamma$ for some $\gamma \in L_n^*$ which in turn implies that $q = \beta\gamma^2$, where $\beta \in G_n$. However, this is impossible since q has height 0 in L_n^*/G_n. The conclusion is that $t(L_{n+1}^*)$ has trivial q-component for odd primes q. Now assume that a primitive 8-th root of unity ε_8 lies in $t(L_{n+1}^*)$. Because $\varepsilon_8 \notin L_n$, we must have $p = 2$ and $L_{n+1} = L_n(\varepsilon_8)$. Again, by Theorem 2.15.4, we have $\varepsilon_8 = \alpha\gamma$, hence $1 = \beta^4\gamma^8$. Taking into account that G is torsion-free and the 2-height of β in G_n is 0, we conclude that the 2-height of β^4 in G_n is 2. Hence the 2-height of β^4 in L_n^* is 2. However, the 2-height of γ^8 in L_n^* is at least 3. This contradiction shows that

$$\varepsilon_8 \notin t(L_{n+1}^*)$$

Therefore $t(L_{n+1}^*) = \langle i \rangle$, thus substantiating our claim.

It is now an easy matter to prove the remaining part of the induction step. Indeed, by Theorem 2.16.26 and by what we have just shown, we have

$$t(L_{n+1}^*/L_n^*) = G_{n+1}L_n^*/L_n^*$$

Invoking Theorem 8.10, we therefore deduce that $L^*_{n+1}/G_{n+1}L^*_n$ is free. Hence

$$L^*_{n+1} = (G_{n+1}L^*_n) \times B_{n+1}$$

for some free group B_{n+1}. Consider the group G_{n+1}/L^*_n. Because G_{n+1} is torsion-free and G_{n+1}/G_n is torsion, we have

$$G_{n+1} \cap (\langle i \rangle \times \prod_{j \leqslant n} B_j) = 1$$

and thus

$$G_{n+1}L^*_n = \langle i \rangle \times G_{n+1} \times \prod_{j \leqslant n} B_j$$

It follows that

$$L^*_{n+1} = \langle i \rangle \times G_{n+1} \times \prod_{j \leqslant n+1} B_j$$

and the induction is finished.

Finally, define $B = \bigoplus_{j<\omega} B_j$. Then B is free and it is clear that

$$L^* = \langle i \rangle \times G \times B$$

As in the proof of Theorem 13.1, L is closed under complex conjugation because G is contained in the unit circle and $L = L_0(G)$. Put $K = L \cap \mathbb{R}$ so that $(L:K) = 2$. Taking into account that $K^* \cap G = 1$, it follows that the projection

$$\langle i \rangle \times G \times B \longrightarrow \langle i \rangle \times B$$

is injective on K^*. Hence $K^* \cong \mathbb{Z}_2 \times A$ for some free group A, thus completing the proof. ■

14. NOTES

The subject matter of this chapter is at the crossroads of two areas: field theory and abelian group theory. As is often the case, this provides a vivid interplay and is a source of enrichment for both disciplines. From the viewpoint of the field theory, we start with a class of fields and then give the structure of their multiplicative groups. From the viewpoint of abelian group theory, we start with an abelian group satisfying certain hypotheses, and ask whether it is the multiplicative group of some field, or at least a subgroup of such a multiplicative group. A detailed discussion of some initial progress may be found in Fuchs(1973).

The following is a more up-to-date historical account.

Suppose that K is an algebraic extension of the countable field F. If F is the field of rational numbers and if every element of K is algebraic of bounded degree over F, then a theorem of Čarin (Corollary 10.17) states that K^* is the (restricted) direct product of cyclic groups. This theorem was generalized by Schenkman (Corollary 10.16) who showed that it is sufficient to assume that K is generated over the rational numbers by elements of bounded degree. Schenkman further conjectured that his theorem remained true if F was allowed to be any countable field such that F^* was a direct product of cyclic groups. May (1972) showed that this conjecture is incorrect as stated, but that it is true if F has the property that, whenever E is a finite field extension of F, E^* is the direct product of cyclic groups.

Fuchs (1960) asked whether the multiplicative group of a field must split over its torsion subgroup. A negative answer was provided by Cohn (1962), who utilized his embedding theorem to give an example of a field F for which F^* does not split over $t(F^*)$. The existence of such an example for algebraic extensions of $\mathbb{Q}$ was indicated by May (1972) (a detailed proof is contained in Theorem 11.1).

Let G be an abelian group. Then a necessary condition for G to be the multiplicative group of some field is that the torsion subgroup of G be locally cyclic. Cohn (1962) proved that if the torsion subgroup of G is locally cyclic, then G can be embedded in a special manner as a subgroup of the multiplicative group of a field. Subsequently May (1972) established a refinement of Cohn's theorem, by proving that if the torsion subgroup of G is locally cyclic, then there exists a field L and a group H such that $L^* \cong G \times H$. Moreover, H is either a free abelian group or the direct product of a free abelian group with a cyclic group of order 2 (see Theorem 12.3).

Let K be a field and L an extension field. Fuchs (1973) has suggested studying the change in multiplicative groups in going from K^* to L^*. May (1979) made a significant contribution by proving Theorems 13.1 and 13.3.

Let F be a field. When is F^* free modulo torsion? Of course, if F is an algebraic number field, then F^* is the direct product of cyclic groups and hence free modulo torsion. On the other hand, there exists an infinite algebraic extension $F/\mathbb{Q}$ such that F^* does not split over $t(F^*)$, and *a posteriori* F^* cannot be free modulo torsion. This shows that the above problem is very delicate even for infinite algebraic extensions of $\mathbb{Q}$. In the light of this, the results of May (Theorems 10.12 and 10.21) are outstanding by any account.

Let K/F be a field extension and let $t(K^*/F^*)$ be the torsion subgroup of K^*/F^*. Gay and Vélez (1981) determined the structure of the group $t(K^*/F^*)$ in case $K = F(\alpha)$, where α is a root of irreducible polynomial $X^m - a \in F[X]$ and char$F \nmid m$. Subsequently de Orozco and Vélez (1984) generalized this result by treating abelian extensions. The statements of these theorems are too technical to quote here and for details we refer to their original papers.

Bibliography

Akizuki, Y
(1936) Eine homomorphe Zuordnung der Elemente der galoischen Gruppe zu den Elementen einer Untergruppe der Normklassengruppe, Math. Ann. 112, 566-571

Artin, E
(1946) Galois theory, Notre Dame Mathematical Lectures No 2, Notre Dame, Indiana
(1967) Algebraic numbers and algebraic functions, Gordon and Breach, New York.

Artin, E and Whaples, G
(1945) Axiomatic characterization of fields by the product formula for valuations, Bull. Amer. Math. Soc. 51, 469-492.

Baer, R
(1934) Der Kern, eine charakteristische Untergruppe, Compositio Math, 1, 254-283
(1936) The subgroup of the elements of finite order of an abelian group, Ann. Math. 37, 766-781.

Bernstein, L
(1971) Infinite non-solubility classes of the Delaunay-Nagell diophantine equation $x^3 + my^3 = 1$, J. London Math. Soc. (2), 118-120.
(1975) Units and their norm equation in real algebraic number fields of any degree, Sympos. Math. Vol. XV, 307-340.

Bourbaki, N
(1958) Algébre, Ch.8, Modules et anneaux semi-simples, Hermann, Paris
(1959) Algébre, Hermann, Paris, Chaps. 4 and 5
(1961) Algébre commutative, Hermann, Paris, Chaps 1 - 4
(1964) Algébre commutative, Hermann, Paris, Chaps 5 and 6
(1965) Algébre commutative, Hermann, Paris, Chap. 7
(1966) Éléments de mathematique, Topologie Generale, Hermann, Paris
(1972) Commutative algebra, Addison-Wesley, Reading, Mass.

Brandis, R
(1965) Über die multiplikative struktur von korpererweiterungen, Math. Z. 87, 71-73.

Carin, V S
(1954) On automorphism groups of nilpotent groups, Ukrain. Math. J. 6. 295-304.

Chase, S U and Rosenberg, A
(1966) A theorem of Harrison, Kummer theory, and Galois algebras, Nagoya Math. J. Vol. 27, 663-685.

Chevalley, C
(1940) La theorie du corps de classes, Ann. Math. 41, 394-418.

Cohn, J H E
(1967) The diophantine equation $x^3 = dy^3 + 1$, J. London Math. Soc. 42, 750-752.

Cohn, P M
(1962) Eine bemerkung über die multiplikative gruppe eines körpers, Arch. Math. 13, 344-348.

Darbi, G
(1926) Sulla Riducibilita delle Equazioni Algebriche, Annali di Mat. pura e appl., Ser. 4, 4, 185-208.

Dedekind, R
(1871) Über die Theorie der ganzen algebraischen Zahlen, Supplement XI to Dirichlet's "Vorlesungen über Zahlentheorie", 2nd. ed.
(1900) Über die Anzahl der Idealklassen in reinen kubischen Zahlkörpern, J. Reine Angew Math., 121, 40-123.

Degert, G
(1958) Über die Bestimmung der Grundeinheit gewisser reellquadratischer Zahlkörper, Abh. Math. Sem. Univ. Hamburg 22, 92-97.

Delaunay, B N
(1928) Vollständige Lösung der unbestimmten Gleichung $X^3q + Y^3 = 1$ in ganzen Zahlen, Math. Z., 28, 1-9.

Delaunay, B N and Faddeev, D K
(1964) The theory of irrationals of the third degree, Amer. Math. Soc. Transl. Math. Monographs, Vol. 10.

de Orozco, M A and Velez, W Y
(1982) The lattice of subfields of a radical extension, J. Number Theory 15, 388-405.
(1984) The torsion group of a field defined by radicals, J. Number Theory 19, 283-294.

Dicker, R M
(1968) A set of independent axioms for a field and a condition for a group to be the multiplicative group of a field, Proc. London Math. Soc. 3, 18, 114-124.

Dirichlet, P G Lejeune
(1840) Sur la theorie des nombres, C R Acad. Sci. Paris 10, 258-288.

Eilenberg, S and MacLane, S
(1947a) Cohomology theory in abstract groups, I, Ann. of Math. 48, 51-78.
(1947b) Cohomology in abstract groups II, Ann. of Math. 48, 326-341.
(1948) Cohomology and Galois theory I, Trans. Amer. Math. Soc. 64, 1-20.

Fomin, C V
(1937) Über periodische Untergruppen der unendlichen abelschen Gruppen, Mat. Sb. 2, 1007-1009.

Fuchs, L
(1958) Abelian groups, Publ. House of the Hungar. Acad. Sci., Budapest.
(1960) Abelian groups, Pergamon Press, London.
(1970) Infinite abelian groups, Vol I, Academic Press, New York .

(1973) Infinite abelian groups, Vol. II, Academic Press, New York

Gay, D and Vélez, W Y
(1978) On the degree of the splitting field of an irreducible binomial, Pacif. J. Math. Vol. 78, No 1, 117-120.
(1981) The torsion group of a radical extension, Pacif. J. Math. Vol.92, No 2, 317-327.

Gilbarg, D
(1942) The structure of the group of p-adic 1-units, Duke Math. J. 9, 262-271

Greither, C and Harrison, D K
(1986) A Galois correspondence for radical extensions of fields, J. Pure Appl. Algebra 43, 257-270.

Guan Chi-Wen
(1963) On the structure of multiplicative groups of discretely valued complete fields, Scientia Sinica, Vol. XII, No 8, 1079-1103.

Harrison, D K
(1963) Abelian extensions of arbitrary fields, Trans. Amer, Math. Soc. Vol.106, 230-235.

Hensel, K
(1916) Die multiplikative Darstellung der algebraischen Zahlen für den bereich eines Beliebigen Primteilers, J. Reine Angew Math. 146, 189-215.

Hochschild, G
(1950) Local class field theory, Ann. of Math, Vol. 51, No 2, 331-347.

Isaacs, I M
(1970) Degrees of sums in a separable field extension, Proc. Amer. Math. Soc.25, 638-641.

Jacobson, N
(1968) The theory of rings, Amer. Math. Soc.

Janusz, G J
(1973) Algebraic number fields, Academic Press, New York and London

Kaplansky, I
(1972) Fields and rings, Chicago Lectures in Mathematics. The University of Chicago Press, Chicago and London.

Kneser, M
(1975) Lineare Abhängigkeit von Wurzeln, Acta Arith. 26, 307-308.

Kronecker, L
(1982) Grundzuge einer arithmetischer Theorie der algebraischen Grossen, J. Reine Angew. Math. 92, 1-122.

Kulikov, Ya.
(1945) On the theory of abelian groups of arbitrary power, Matem. Sb. 16, 129-162.

Lang, S
(1965) Algebra, Addison-Wesley, Reading, Mass.
(1970) Algebraic number theory, Addison-Wesley, Reading, Mass.

MacLane, S
(1963) Homology, Grundlechren der Mathematik, 114, Springer, Berlin.

May, W
(1970) Unit groups of infinite abelian extensions, Proc. Amer. Math. Soc. Vol. 25, No 3, 680-683.
(1972) Multiplicative groups of fields, Proc. London Math. Soc. (3), 24, 295-306.
(1976) Group algebras over finitely generated rings, J. Algebra, Vol. 39, No 2, 483-511.
(1979) Multiplicative groups under field extension, Canad. J. Math. Vol. XXXI, No 2, 436-440.
(1980) Fields with free multiplicative groups modulo torsion, Rocky Mountain J. Math. Vol. 10, No 3, 599-604.

Nagell, T
(1925) Solution complete de quelques équations cubiques á deux indeterminées, J. Math. Pures Appl. (9), 4, 204-270.

Nakayama, T
(1936) Über die Beziehungen zwischen den Faktoren-system und der Normklassengruppe eines galoisschen Erweiterungskorpers, Math. Ann. 112, 85-91.

Ostrowski, A
(1913) Über eininge Fragen der allgemeinen Körpertheori, J. Reine Angew. Math. 143, 255-284.
(1917) Über sogenannte perfekte Körper, J. Reine Angew. Math. 147, 191-204.
(1918) Über eininge Lösungen der Funktionalgleichung $\phi(x)\phi(y) = \phi(xy)$, Acta Math. 41, 271-274.
(1935) Untersuchungen zur arithmetischer Theorie der Körper, Math. Z. 39, 269-320, 321-361, 361-404.

Pontryagin, L S
(1977) Topological groups, Gordon and Breach, New York, London.

Prüfer, H
(1923) Untersuchungen über die Zerlegbarkeit der abzählbaren primaren abelschen Gruppen, Math. Z., 17, 35-61.

Richaud, C
(1866) Sur la resolution des equations $x^2 - Ay^2 = \pm 1$, Atti Acad. Pontif. Nuovi Lincei, 177-182.

Rudman, R J
(1973) On the fundamental unit of a purely cubic field, Pacif. J. Math. 46, 253-256.

Schenkman, E
(1964) On the multiplicative group of a field, Arch. Math. 15, 282-285.

Schinzel, A
(1977) Abelian binomials, power residues and exponential congruences, Acta Arithmetica, XXXII, 245-274.

Serre, J-P
(1979) Local fields, GTM 67, Springer-Verlag, New York, Heidelberg, Berlin.

Skolem, T
(1947) On the existence of a multiplicative basis, Norske Vid. Selsk. Forh. 2, 4-7.

Stender, H J
(1969) Über die Grundeinheit für spezielle unendliche Klassen reiner Kübischer Zahlkorper, Abh. Math. Sem. Univ. Hamburg, 33, 203-215.

Teichmuller, O
(1940) Über die sogenannte nichtkommutative Galoissche Theorie und die Relation $\xi_{\lambda,\mu,\nu}\,\xi_{\lambda,\mu\,\nu,\pi}\,\xi^{\lambda}_{\mu,\nu,\pi} = \xi_{\lambda,\mu,\nu\pi}\,\xi_{\lambda\mu,\nu,\pi}$, Deutsche Mathematik vol. 5, 138-149.

Vélez, W Y
(1980) On normal binomials, Acta Arithmetica, XXXVI, 113-124.

Wada, H
(1970) A table of fundamental units of purely cubic fields, Proc. Japan Acad. 46, 1135-1140

Waterhouse, W C
(1974) Profinite groups are Galois groups, Proc. Amer. Math. Soc. Vol. 42, No 2, 639-640.

Weiss, E
(1969) Cohomology of groups, Academic Press, New York.

Witt, E
(1935) Konstruktion von galoisschen Korpern der Charakteristik p zu vorgegebener Gruppe der Ordnung p^f, J. fur Math., 174(4), 237-245.
(1936) Zyklische Körper and Algebren der Charakteristik p vom Grad p^n. Struktur diskret bewerteter perfekter Körper mit vollkommenem Restklassenkörper der Charakteristik p, J. fur Math. 116 (3), 126-140.

Wójcik, J
(1982) Contributions to the theory of Kummer extensions, Acta Arithmetica, XL, 155-174.

Yokoi, H
(1968) On real quadratic fields containing units with norm -1, Nagoya Math. J. 33, 139-152.
(1974) The diophantine equation $x^3 + dy^3 = 1$ and the fundamental unit of a pure cubic field $\mathbb{Q}(\sqrt[3]{d})$, J. Reine Angew, Math., 268/269, 174-179.

Notation

Number systems

$\mathbb{N}$	the natural numbers
$\mathbb{Z}$	the integers
$\mathbb{Q}$	the rational numbers
$\mathbb{Q}_p$	the rational p-adic numbers
$\mathbb{R}$	the real numbers
$\mathbb{C}$	the complex numbers
$\mathbb{F}_q$	the field of q elements

Set theory

$\|X\|$	the cardinality of the set X
$Y-X$	the complement of X in Y
$\aleph_0$	the cardinal of $\mathbb{N}$
$\subset$	proper inclusion
$\subseteq$	inclusion
$g \circ f$	the composite of g and f

Number theory

$a\|b$	a divides b
$a \nmid b$	a does not divide b
(a,b)	the greatest common divisor of a and b
$(\frac{n}{p})$	the quadratic symbol

Group theory

A^n	$= \{a^n \mid a \in A\}$
$A[n]$	$= \{a \in A \mid a^n = 1\}$
$t(A)$	the torsion subgroup of A
$Z(p^\infty)$	the group of the p^n-th complex roots of unity with n running over $\mathbb{N}$
A_p	the p-component of A
$G = \varprojlim G_i$	projective limit

Rings and modules

$A \otimes B$	the tensor product
char$R$	the characteristic of R
$\bigoplus_{i \in I} R_i$	the direct sum
$\prod_{i \in I} R_i$	the direct product of rings
$Z(R)$	the centre of R
$U(R)$	the unit group of R
$R[X]$	the polynomial ring on X over R
$J(R)$	the Jacobson radical of R
RG	the group ring of G over R
$R[A]$	the smallest subring containing A and R
R_S	the quotient ring associated with S
$U(m)$	the unit group of $\mathbb{Z}/m\mathbb{Z}$
$\mathbb{Z}_p$	the ring of p-adic integers
PID	principal ideal domain
UFD	unique factorization domain
(r)	the principal ideal generated by r
$M_n(R)$	the ring of $n \times n$ matrices over R
$C_A(B)$	the centralizer of B in A
$\dim_F A$	the F-dimension of A

Notation

Fields and Dedekind domains

$GF(p)$ or $\mathbb{F}_p$	the Galois field of p elements
E/F	the field extension
$(E:F)$	the degree of E over F
$\mathrm{Gal}(E/F)$	the Galois group of E over F
$N_{E/F}(\alpha)$	the norm of α
ε_n	a primitive n-th root of unity
$K(S)$	the smallest subfield containing S and K
$K(\alpha_1,\ldots,\alpha_n)$	the smallest subfield containing K and $\alpha_1,\ldots,\alpha_n$
$Tr_{S/R}(\lambda)$	the trace of λ
$D_{S/R}$	the discriminant of S over R
$d(K)$	the discriminant of the field K
$I(R)$	the monoid of fractional ideals of R
$P(R)$	the group of principal fractional ideals of R
$C(R)$	the class group of R
$N(I)$	the norm of the ideal I
$[r_1,r_2]$	the signature of F
F^*	the multiplicative group of F
$t(F^*)$	the torsion subgroup of F^*
F_S^*	the S-unit group of F
$\mathbb{Q}(\sqrt{d})$	the quadratic field
$(a\|p)$	the Legendre symbol
ε_d	the canonical fundamental unit of $\mathbb{Q}(\sqrt{d})$
$\mathbb{Q}(\sqrt[3]{d})$	the pure cubic field
$i_{S/R} : C(R) \longrightarrow C(S)$	the natural map
$U_n(R)$	$= 1 + P^n$
M^*	the codifferent of M
$N_{E/F} : I(S) \longrightarrow I(R)$	the norm homomorphism
$W(R)$	the Witt ring of R
$W_n(R)$	the n-th Witt ring of R

$s : W(R) \longrightarrow W(R)$	the shift map
$s_n : W_n(R) \longrightarrow W_n(R)$	the shift map
$(E:F)_s$	the separable degree
$(E:F)_i$	the inseparable degree
$F^{p^{-\infty}}$	$= \bigcup_{n=1}^{\infty} F^{p^{-\infty}}$
F^G	the fixed field of G
$Tr(f)$	the trace of f
$\det(f)$	the determinant of f
$\sqrt[n]{F^*}$	the subgroup of E^* consisting of all n-th roots of elements of F^*
deg α	the degree of α
$\mathrm{Cog}(E/F)$	the torsion subgroup of E^*/F^*
$e_{E/F}$	the remaification index
$f_{E/F}$	the residue degree
R_v	the valuation ring of v
P_v	the maximal ideal of v
R_v/P_v or $\bar{F}$	the residue class field of v

Cohomology theory

A^G	the fixed module of a G-module A
$\mathrm{Hom}(A,B)$	the group of all abelian group homomorphisms $A \longrightarrow B$
$C^n(G,A)$	the group of all n-cochains of G in A
$Z^n(G,A)$	the group of all n-cocycles of G in A
$B^n(G,A)$	the group of all n-coboundaries of G in A
$H^n(G,A)$	the n-th cohomology group of G in A
Inf	the inflation map
Res	the restriction map
δf	coboundary corresponding to f
$[A]$	the Brauer class of A
$Br(F)$	the Brauer group of F

$E^{\alpha}G$	the crossed product of G over E
$Br(E/F)$	the subgroup of $Br(F)$ consisting of those $[A]$ which are split by E.
$Br_G(E)$	the Brauer classes of G-normal algebras
$Br_F(E)$	the image of the canonical homomorphism $Br(F) \longrightarrow Br(E)$
$H_0^3(G\ E^*)$	$= Br_G(E)/Br_F(E)$

Index